THE OBSERVATION AND ANALYSIS OF STELLAR PHOTOSPHERES

Fourth Edition

This textbook describes the equipment, observational techniques, and analysis used in the investigation of stellar photospheres. Now in its fourth edition, the text has been thoroughly updated and revised to be more accessible to students. New figures have been added to illustrate key concepts, while diagrams have been redrawn and refreshed throughout. The first half of the book develops the tools of analysis, and the second half demonstrates how they can be applied. Topics covered include radiation transfer, models of stellar photospheres, spectroscopic equipment, how to observe stellar spectra, and techniques for measuring stellar temperatures, radii, surface gravities, chemical composition, velocity fields, and rotation rates. Up-to-date results for real stars are included. This textbook is for starting graduate students or advanced undergraduates; in addition to its use in university courses, it includes a wealth of reference material useful to researchers.

DAVID F. GRAY is a Professor Emeritus of Astronomy at Western University, London, Ontario, Canada, where he has held positions since 1966. He was president of IAU Commission 36, on the Theory of Stellar Atmospheres 1988–1991, and served on the observing-time allocation panels of the Hubble Space Telescope and the Canada–France–Hawaii telescope. He is a member of the Canadian Astronomical Society, the International Astronomical Union, Sigma Xi Honorary Society, and the American Astronomical Society, and has served on numerous organizing committees for astronomical symposia. He was an invited lecturer at the Canary Islands Winter School of Astrophysics, the Catania Observatory, and the Beijing Observatory.

THE OBSERVATION AND ANALYSIS OF STELLAR PHOTOSPHERES

Fourth Edition

DAVID F. GRAY
Western University, Ontario, Canada

CAMBRIDGE
UNIVERSITY PRESS

CAMBRIDGE
UNIVERSITY PRESS

University Printing House, Cambridge CB2 8BS, United Kingdom

One Liberty Plaza, 20th Floor, New York, NY 10006, USA

477 Williamstown Road, Port Melbourne, VIC 3207, Australia

314–321, 3rd Floor, Plot 3, Splendor Forum, Jasola District Centre, New Delhi – 110025, India

103 Penang Road, #05–06/07, Visioncrest Commercial, Singapore 238467

Cambridge University Press is part of the University of Cambridge.

It furthers the University's mission by disseminating knowledge in the pursuit of education, learning, and research at the highest international levels of excellence.

www.cambridge.org
Information on this title: www.cambridge.org/highereducation/isbn/9781009078016
DOI: 10.1017/9781009082136

First edition © John Wiley & Sons, Inc. 1976
Second edition © Cambridge University Press 1992
Third edition © D. F. Gray 2005
Fourth edition © David F. Gray 2022

First published by John Wiley and Sons, Inc. 1976
Second edition published by Cambridge University Press, 1992
Third edition 2005
Fourth edition 2022

A catalogue record for this publication is available from the British Library.

Library of Congress Cataloging-in-Publication Data
Names: Gray, David F., 1938– author.
Title: The observation and analysis of stellar photospheres / David F. Gray, Western University, Ontario.
Description: Fouth edition. | Cambridge, United Kingdom ; New York, NY : Cambridge University Press, 2022. | Includes bibliographical references and index.
Identifiers: LCCN 2021029884 (print) | LCCN 2021029885 (ebook) | ISBN 9781009078016 (paperback) | ISBN 9781009082136 (epub)
Subjects: LCSH: Stellar photospheres. | Stellar photospheres–Observations.
Classification: LCC QB809 .G67 2022 (print) | LCC QB809 (ebook) | DDC 523.8/2–dc23
LC record available at https://lccn.loc.gov/2021029884
LC ebook record available at https://lccn.loc.gov/2021029885

ISBN 978-1-009-07801-6 Paperback

Contents

Preface to the First Edition

The remarkable nature of stars is transmitted to us by the light they send. The light escapes from the outer layers of the star – called, by definition, the atmosphere. The complete atmosphere of a star can be viewed comprehensively as a transition from the stellar interior to the interstellar medium. And yet almost the whole visible stellar spectrum comes from a relatively thin part called the photosphere. Obviously we cannot disconnect the photosphere from the adjacent portions of the atmosphere, but in actual fact it is the only region we can study extensively for most stars. It is for this reason that the photosphere has taken its place as the central theme of this book.

Several books have appeared during the last decade dealing with the *theory* of stellar atmospheres. These works are for the most part excellent. It is to the material largely omitted by these books that the present treatise is directed. My students and I have felt for some time the need of a book that presents the basics of the field through the eyes of an observer and analyzer of stellar atmospheres. An introduction to a subject, in my opinion, should be presented in a way that can be understood by a reader who has not studied the topic before. It follows that the material should be presented in as simple and straightforward a manner as possible. The Fourier transform (as covered in Chapter 2) is a unifying theme helping to accomplish this aim. Transforms lead naturally into the material on data collection, optical instruments, the instrumental profile, line absorption coefficients, velocity fields, and spectral line analysis. In addition, I have selected and developed topics that I consider to be important to those of us who look at stars and attempt to understand what we see. At the same time, I have tried to present the material in the least complicated manner. The word "complicated" is affixed to things that are difficult to understand. Complicated things consequently are often unsuitable topics for the novice. We should seek not the most general case conceivable, but the least complicated case that is serviceable (a version of the principle of minimum assumption).

The development of each of the main topics starts at an elementary level, proceeds with a discussion of the topic, and ends by pointing the direction to the more advanced literature. It should be easy to expand from this book into the areas holding an attraction for you. Realizing that astrophysics is a very dynamic field, I have documented (or otherwise made clear) the source of the material used in examples. When no source is indicated, the material is from observations or calculations of my own. The references in general have been selected because they are good illustrations of the material being discussed or because they have a basic lasting approach to the subject. I have also biased the referencing toward good starting points in the literature and toward review articles and journals to which the student is likely to have access. The references are listed at the end of each chapter and ordered according the author's name and date of publication.

The first two chapters contain preparatory material. The main theme starts in Chapter 3 with a discussion of spectroscopic tools. Generally the continuous spectrum topics are developed first, followed by the somewhat more involved subject of the line spectrum. From Chapter 14 (Chemical Analysis) through to Chapter 18, the material is oriented completely toward analysis and deduction. These later chapters are closely interlinked with the preceding chapters.

The book is suitable as a text for a one-year course and as a reference to the more advanced reader.

Preface to the Second Edition

Wonderful growth has occurred in our understanding of stellar photospheres during the 15 years since the appearance of the first edition of "Photospheres." I have managed to retain the same chapter names and the general plan of the first edition, and many of the equation numbers are also the same. But a significant portion of the material is new or revised. A revolution in light detectors has given us hundreds of times greater efficiency in measuring stellar spectra; Chapter 4 on detectors has been re-done. The astronomical literature is burgeoning with new results on the structure of photospheres, chemical abundances, radius measurements, stellar rotation, and photospheric velocity fields. Many of these results have been incorporated in this second edition, of course. At the same time, I stayed with my original purpose of making this volume an introduction to the subject. Unhappily, this means leaving out numerous exciting topics. My book *Lectures* (Gray 1988) takes up some of these, and it is recommended as a second installment, after the material in "Photospheres" has been mastered.

More than ever, the reader should keep in mind the fundamental nature of the stellar photosphere: of interest in its own right, with marvelous and intriguing physics, yet the link between the interior and chromospheres, coronae, and interstellar surroundings, and the source of most of our basic information about stars and stellar systems.

Once again, I thank my students and colleagues for their help and patience.

Reference

Gray, D.F. 1988. *Lectures on Spectral-Line Analysis: F, G, and K Stars* (Arva, Ontario: The Publisher).

Preface to the Third Edition

Studies of stars continue to flourish. More detailed and sophisticated observations continue to be made, analysis tools are honed, understanding grows for more complex situations. The beauty of the stars is integrated more fully into our lives. I hope you enjoy the many revisions incorporated in this third edition of *Photospheres*: new figures, more complete data sets, reorganization of some of the material, a cleaner presentation of several topics, and some exercises and questions to help you probe the material.

I extend my thanks to J. Power for proofreading and to the many others who have guided my thoughts, given me corrections and suggestions, and contributed their efforts to knowing the stars.

Preface to the Fourth Edition

Welcome again in this fourth edition to the wonderful world of stars and stellar spectroscopy. As you work your way through this material, you will gain insight into how stars behave and how we deduce their physical characteristics. If you are not careful, you will soon find yourself becoming emotionally attached to stars. And why not? There are many interesting subdisciplines in modern astronomy, but when we gaze into the dark night sky, it is the panorama of *stars* that makes our hearts beat faster.

> Twinkle, twinkle on, oh bright little star,
> we see your light travelling oh so far.
> The light you send us we work to decode,
> analyzed and studied in spectroscopic mode.
>
> Your light spread out in a glorious rainbow,
> reveals your nature in a wonderful light show.
> From continuum on through each spectral line,
> send us your secrets about how you do shine.
>
> As we learn more about just what you are,
> you're ever more awesome, oh bright little star.
> Twinkle, twinkle on in the dark night;
> inspire and teach us with your little light.

My heartfelt thanks to the friends and colleagues who have helped, guided, and encouraged me over the years.

You will notice the quaint watermark, "from David F. Gray Stellar Photospheres," on many of the figures. Should you choose to use figures from the book in talks, lectures, or websites, this watermark will help you remember where the figures came from.

May your journey to understand stars be pleasant and fulfilling.

1

Background

Starlight tells us about the nature of stars. Spectroscopy is the key by which the starlight is decoded. From stellar spectra, both lines and continua, we extract the fundamental stellar properties of surface gravity, temperature, and chemical composition. From there we can often deduce the mass, the most fundamental property of all. But there is much more. In the spectral lines, we see signatures of the velocity fields pervading stellar atmospheres, which in turn uncover details of surface convection and the mottled appearance of stellar surfaces. Chemical markers tell us about deep convection. The broadening of spectral lines tells us the rotation rates of stars, and sometimes time variations of lines allow us to map spots on the surface. Stars couple convection and rotation to generate magnetic fields. Magnetic fields, together with winds of escaping material, dissipate angular momentum, and this can be studied using the measured rotation rates. Knowledge of chemical composition informs us about nuclear processes, energy generation, and element building. Differences in chemical composition help us map out the dynamical structure of the Milky Way and teach us about star formation processes. And the list goes on. Stellar spectroscopy is the connecting link between the observations and the rest of stellar astrophysics.

The topics brought together in this chapter are background material to set the stage.

What Is a Stellar Photosphere?

The photosphere is the region of a star's atmosphere from which the major portion of the light comes. The whole atmosphere is a transition region from the stellar interior to the interstellar medium. One way to quantify this description is to look at the change in average kinetic temperature with height, the Sun being a close-at-hand example. Figure 1.1 shows the solar temperature profile with the four basic sections labeled, sub-photosphere, photosphere, chromosphere, and corona. The visible spectrum comes from the photosphere. When we speak of the size of a star, for example its radius, we mean the size of the apparent stellar disk as defined by the photosphere. When we talk about a star's temperature, we fundamentally mean the characteristic temperature of the photosphere. And if we think about the surface of a star, it is the photosphere we have in mind. Light from the sub-photospheric layers does not penetrate to the surface, so we cannot see those layers. The outer layers, the chromosphere and the corona, emit most of their light in the ultraviolet (UV) and x-ray regions of the spectrum.

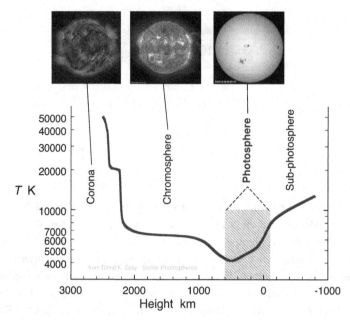

Figure 1.1 The temperature in the outer layers of the Sun is shown as a function of geometrical height in kilometers. The center of the Sun is off to the right. Visible light comes from the photosphere. Based on Vernazza *et al.* (1973). Images, left: inner corona Fe IX/X 171 Å line; center: chromosphere at 304 Å; right: photosphere in white light with large sunspots. Images courtesy of SOHO consortium (ESA & NASA).

The solar photosphere is ~700 km thick, as illustrated in Figure 1.1. It is only by virtue of its small extent compared to the much larger solar radius, $R_\odot = 696\,100$ km, that the edge of the Sun appears sharp. The circumpunct, $\odot$, is used to denote solar values. The geometrical extent of the photosphere in other stars differs inversely with the *surface gravity* defined by

$$g = g_\odot \frac{\mathfrak{M}}{R^2}, \tag{1.1}$$

where $g_\odot = 2.740 \times 10^4$ cm/s^2; $\log g_\odot = 4.438$ is the surface gravity of the Sun, and the mass, $\mathfrak{M}$, and radius, R, of the star are in solar units (see Appendix A). In astronomical culture, centimeter–gram–second (cgs) units are frequently used as the default unit system. This is especially true for surface gravity numbers, where $\log g$ is assumed to be in cm/s^2 units.

The overall thickness of the photosphere depends upon the opacity of the gases forming it. In cool stars, the opacity is larger and the photosphere thinner when the gas is richer in elements other than hydrogen and helium. In the somewhat arcane astronomical terminology, these elements are called "metals."

The third physical variable strongly affecting the nature of the atmosphere is its characteristic temperature. Typically the temperature drops by somewhat more than a factor of two from the bottom to the top of the photosphere, as in Figure 1.1. Instead of choosing a temperature at some depth to characterize the temperature of a star, it is customary to use the *effective temperature*, T_{eff}. The effective temperature is defined in terms of the total power per unit area radiated by the star,

$$\int_0^\infty \mathfrak{F}_\nu \, d\nu \equiv \sigma T_{\text{eff}}^4, \tag{1.2}$$

where the total radiant power per unit area is given by the integral, and $\sigma = 5.6704 \times 10^{-5} \, \text{erg}/(\text{s cm}^2 \, \text{K}^4)$ is the Stefan–Boltzmann constant. Here $\mathfrak{F}_\nu$, or alternatively $\mathfrak{F}_\lambda$, is the flux leaving the stellar surface and is formally defined in Chapter 5. For now, think of it as "the amount of light" in units of $\text{erg}/(\text{s cm}^2 \, \text{Hz})$ or $\text{erg}/(\text{s cm}^2 \, \text{Å})$. Equation 1.2 has the form of the Stefan–Boltzmann law, which we will meet in Chapters 5 and 6, making T_{eff} the temperature of a black body having the same power output per unit area as the star. However, the distribution of power across the spectrum may differ dramatically from a black body at the same effective temperature.

The total power emitted by the star is termed the *luminosity*. Luminosity is related to the effective temperature through the surface area, namely,

$$L = 4\pi R^2 \int_0^\infty \mathfrak{F}_\nu \, d\nu = 4\pi R^2 \sigma T_{\text{eff}}^4 \tag{1.3}$$

for a spherical star of radius R.

Traditionally and historically, the more empirical parameters of spectral type, magnitude, and color index have been used to describe stellar temperature and surface gravity.

Spectral Types

The most powerful way to analyze starlight is through spectroscopy. The first step in many cases is to classify a star by spectral type. Thanks to many studies done over past decades, knowing the spectral type tells us what kind of star we are dealing with, hot or cool, big or small, and so on. The concept of a spectral type is to classify stars into groups according to the strengths of their spectral lines. The lines in stellar spectra are overwhelmingly seen as absorption lines, i.e., regions where light is missing. The strongest of these were first seen in the solar spectrum by Joseph Fraunhofer (1787–1826) and are sometimes referred to as "Fraunhofer lines." Some of these are listed in Appendix D with their Fraunhofer letter designations, many of which are still in common use, such as the sodium D lines.

The format of a spectral type consists of a capital letter followed by an Arabic number representing the fraction toward the next letter class, followed by a Roman numeral. The letter is the temperature classification, the Roman numeral the luminosity classification. Spectrograph dispersions of ~100 Å/mm, giving ~2 Å resolution, are best for classification. Higher resolution resolves details of line profiles giving unwanted complications; lower resolution does not adequately separate individual lines. The traditional approach uses photographic records, called spectrograms, that are centered in the blue region of the spectrum where photographic emulsions are most sensitive, but modern electrical detectors work equally well or better. Spectrograms are in black and white, rather than in color, because color photography is much less sensitive to the faint starlight. But the position or wavelength of a line essentially tells us the color, so color photography would be redundant.

In the traditional approach, line strengths and ratios are estimated visually by comparing an untyped spectrogram with standards. The more modern approach uses graphs of the spectra. Establishing a grid of standard stars is a major undertaking with deep historical roots (see Huggins & Huggins 1899, Draper 1935, Garrison 1984, Hearnshaw 1986, Keenan 1985, 1987). Figure 1.2 shows a set of standard spectra in order of decreasing effective temperature. The height of the spectrum has no meaning; it just helps us see the spectral lines. Those spectra toward the cool end of the sequence are referred to as "late-type" spectra, and those on the hot end as "early-type" spectra. In principle the spectral classification separates stars into about 60 temperature steps. These range (hottest to coolest) from O2 to M8 through the standard (temperature) sequence: O, B, A, F, G, K, M with 10 subdivisions per letter group. But some subclasses are rarely used. For example, the main subtypes for the cooler stars are G0, G2, G5, G8, K0, K1, K2, K3, K4, K5, M0, M1, M2, M3, and M5. In a few cases, half intervals are given full weight as a subclass, for example, O9.5 and B2.5. Statistically smooth relations exist between these main subtypes

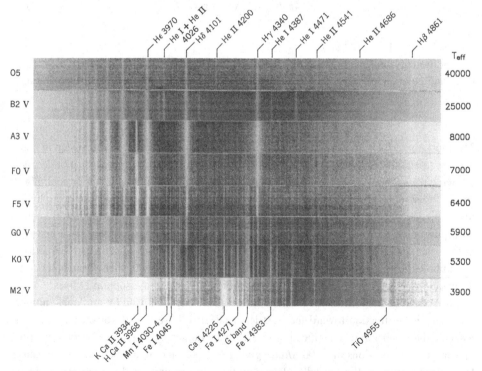

Figure 1.2 The temperature classification sequence is illustrated by these standard spectral types from Abt *et al.* (1968). Notice how the hydrogen lines reach maximum absorption in the A type and how most lines increase in strength toward cooler stars. Specific lines used for classification are labeled. Direct photographs of spectra are "negatives," meaning light is dark and vice versa.

and other temperature indicators such as color indices (to be discussed shortly). The Sun's spectral type is G2 V by definition. The effective temperature associated with each spectral type requires a separate calibration, and is discussed in Chapter 14.

Spectral classification of very cool stars includes extension to classes L (2200–1300 K) and T (1300–800 K), and on down to planet temperatures (see Geballe *et al.* 2002, Leggett *et al.* 2002). There are also side branches: R stars parallel K stars, but show carbon molecular bands; N stars parallel M stars, but show very strong carbon molecular bands; S stars parallel M stars, but have the usual titanium oxide bands overshadowed by those of zirconium oxide (see Hearnshaw 1986, Gray & Corbally 2009).

Luminosity-sensitive features in the spectrum show markedly less variation, and only a few classes are recognized: 0, I, Ia, Ib, II, III, IV, and V in order of decreasing luminosity. The names associated with these numerals are 0 hypergiants, I supergiants, II bright giants, III giants, IV subgiants, and V dwarfs. Sometimes subdwarfs are denoted as class VI. Luminosity effects are illustrated in Figure 1.3 and their physical causes are discussed in later chapters. Sometimes the suffixes listed in Table 1.1 are used to supply additional information.

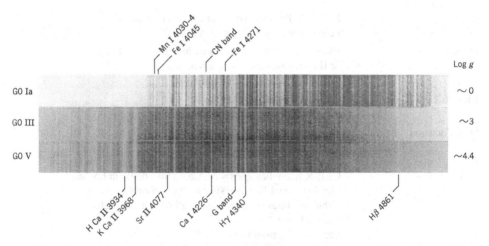

Figure 1.3 Luminosity classification is illustrated by these standard spectra from Abt *et al.* (1968). Specific lines used for classification are labeled.

A summary of criteria, based on Morgan *et al.* (1943), Keenan and McNeil (1976), and Morgan *et al.* (1978) is given in Table 1.1. Before classification can begin, the observer must acquire spectrograms of the standard stars using the same equipment and technique used with the program stars. The first step is to identify the lines whose strengths serve as the criteria for classification. We easily recognize the hydrogen lines from class O to F5, and the less conspicuous lines can be referred to them. As seen in Figures 1.2 and 1.3, H$_\beta$ remains distinguishable as the most intense line at the long-wavelength end of the spectrogram as late as K0. From F5 to M, the G band near the center of the spectrogram and the H and K lines on the short-wavelength end can be used as guideposts.

The spectrum is first classed approximately, following the flow chart in Figure 1.4. Does it have strong hydrogen lines, many lines, or does it show molecular bands? This rough

Table 1.1 *Spectral classification criteria*

Temperature criteria	
O4–B0	He I λ4471/He II λ4541 increase with type
	He I + He II λ4026/He II λ4200 increase with type
B0–A0	H lines increase with type
	He I lines reach maximum at B2
	Ca II K line becomes visible at B8
A0–F5	Ca II K/H$_\delta$ increase with type
	Neutral metals become stronger
	G band visible starting at F2
F5–K2	H lines decrease with type
	Neutral metals increase with type
	G band strengthens with type
K2–M5	G band changes appearance
	Ca I λ4226 increases rapidly with type
	TiO starts near M0 in dwarfs, K5 in giants
Luminosity criteria	
O9–A5	H and He lines weaken with increasing luminosity
	Fe II becomes prominent in A0–A5
F0–K0	Blend $\lambda\lambda$4172–9 increases with luminosity (early F)
	Sr II λ4077 increases with luminosity
	CN λ4200 increases with luminosity
K0–M6	CN λ4200 increases with luminosity
	Sr II increases with luminosity
Suffix notation	
e	Emission lines are present
f	He II λ4686 and/or CIII λ4650 in emission; mostly for O stars
k	Ca II K line when unexpected, e.g., interstellar in hot stars
m	Metallic; metal lines are stronger than normal
n	Nebulous; lines are broad and shallow; usually high rotation
nn	Very nebulous!
p	Peculiar; spectrum is abnormal
q	Queer; unusual emission; evolved from Q novae designation (archaic)
s	Sharp; lines are sharp, usually for early-type stars with low rotation
v	Variable; spectrum changes with time
w	Wolf–Rayet bands present (archaic)
Old prefix notation	
c for supergiants; *g* for giants; *d* for dwarfs	

classification is refined in successive steps. It is generally considered unwise to attempt an immediate accurate classification. Although key features are singled out, it is the appearance of the whole spectrum that is ultimately compared to the standards. Some care must be exercised, especially with early-type stars, because high rotation (see Chapter 18) can wash out weaker classification lines. Most modern spectral types are on the Morgan–Keenan system (MK) or a close derivative of it. See Johnson and Morgan (1953), Keenan (1963), and Gray and Corbally

(2009). A review of spectral classification is given by Morgan and Keenan (1973) and there is also a symposium dealing with this topic (Garrison 1984). Additional classification information can be found in the comprehensive objective-prism survey by Houk and Cowley (1975). Early O star classification is discussed by Walborn *et al.* (2002).

SPECTRAL CLASSIFICATION

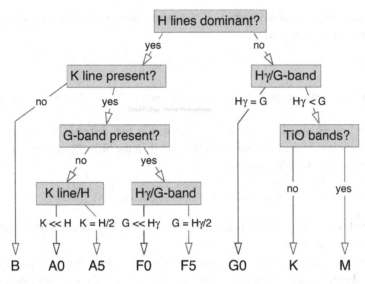

Figure 1.4 A flow chart organizes the classification basics. Start at the top.

Most catalogues of stars list spectral types, e.g., *The Bright Star Catalogue* (Hoffleit & Warren 1995), *Geneva Photometric Catalogue 1988* (Rufener 1988), or the SIMBAD database (simbad.u-strasbg.fr/simbad/).

Magnitudes and Color Indices

Stellar brightness is often expressed in magnitude units, defined by

$$m = -2.5 \log \int_0^\infty F_\lambda \, W(\lambda) \, d\lambda + \text{constant} \tag{1.4}$$

in which the amount of light or flux from the star, F_λ, is recorded in the spectral window specified by the response function $W(\lambda)$. The visual magnitude, often denoted by m_V or V, has its spectral window, $W_V(\lambda)$, centered on the yellow–green wavelengths near 5480 Å. Should we wish to include radiation at all wavelengths, i.e., $W(\lambda) = 1$ everywhere, the corresponding magnitude is called the *bolometric* magnitude. Even for this case, there remains an arbitrary zero-point constant. The constant is adjusted to suit an adopted set of standard stars (e.g., Landolt 2009, Smith *et al.* 2002), or, more recently, a well-calibrated standard star, usually Vega (α Lyr, HR 7001, A0 V), discussed in Chapter 10. In the latter case, all magnitudes in all color bands are set to zero for an A0 V star having $V = 0$. Vega itself has $V = 0.03$, so any magnitude defined for it is 0.03. Notice that the minus sign implies smaller magnitudes for brighter stars. The bright star Arcturus, for example, has $V = -0.05$, while the fainter star, σ Dra, has $V = 4.68$.

In many situations, there is no need to use the constant in Equation 1.4 because we deal with magnitude differences. Simplify the notation in Equation 1.4 and write

$$m = -2.5 \log F + \text{constant}. \tag{1.5}$$

That is, replace the integral by the flux, F, that is measured in the $W(\lambda)$ band. Now if two stars have magnitudes m_1 and m_2,

$$m_2 - m_1 = -2.5 \log \frac{F_2}{F_1}, \tag{1.6}$$

and the constant is gone. Equation 1.6 tells us that magnitude *differences* correspond to flux *ratios*.

Equation 1.6 can also be used on the same star in two different wavelength regions. A classic example is to measure the flux in a window in the blue, F_B, and compare it to the flux measured in the visible window, F_V. Then

$$m_B - m_V = B - V = -2.5 \log \frac{F_B}{F_V}. \tag{1.7}$$

The magnitude difference, $B - V$, is termed a color index. As you might have guessed, it is an empirical measure of a star's temperature.

Many standard magnitude systems have been set up. These differ in the wavelength positions and number of windows. One of the earliest is the red, green, ultraviolet or RGU system of Becker (1948). See also Becker and Stock (1954) and Becker 1963). The most commonly encountered systems are the UBV system (Johnson & Morgan 1953, Nicolet 1978) and its extension to longer wavelengths in the RIJKLMNH magnitudes (Johnson 1966, Johnson *et al.* 1966, Morel & Magnenat 1978, Landolt 1992), the uvby system (Strömgren 1966, Hauck & Mermilliod 1980, Perry *et al.* 1987), the Geneva $UBVB_1B_2V_1G$ system (Rufener 1988, Rufener & Nicolet 1988), and the DDO system (McClure & van den Bergh 1968). The UBVRI response functions are shown in Figure 1.5.

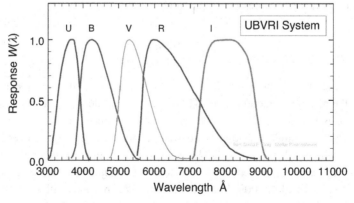

Figure 1.5 The UBVRI windows or response functions are shown normalized to unity. Based on Johnson (1966) and Bessell (2005).

More recently developed is the Sloan system (Fukugita *et al.* 1996, Ahumada *et al.* 2020). Its response functions are shown in Figure 1.6. There is less overlap between adjacent bandpasses compared to the UBVRI system. The Sloan observations extend to very faint magnitudes and cover hundreds of thousands of stars. Transformation equations between the UVBRI and Sloan systems are given by Krisciunas *et al.* (1998) and by Smith *et al.* (2002). A comprehensive review of photometric systems is given by Bessell (2005).

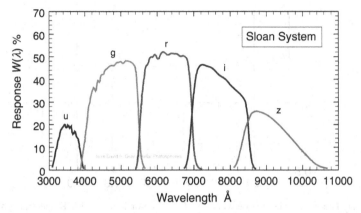

Figure 1.6 Response curves for the Sloan system are shown. These curves are the combined response of specially designed filters and the silicon response of a CCD light detector. Based on Fukugita *et al.* (1996).

Spectral types and color indices both give an empirical handle on stellar temperatures. Figure 1.7 shows the relationship between spectral type and the $B - V$ color index. Similar plots could be made for other color indices. A color index has the advantage of resolving finer temperature differences. The span from $B0$ to $M0$, for example has ~40 spectral-type classes, while the same span in $B - V$ can be divided into ~180 steps. Color indices can also be measured for considerably fainter stars, primarily a result of the ~1000 Å bands compared to the ~1–2 Å resolution elements in classification spectra. The primary disadvantage of color indices is interstellar extinction. Dust in space dims stars by progressively larger amount for shorter wavelengths. This makes stars seen through dust look redder or cooler. The extinction also makes stars look fainter. The spectral lines used for classification are not affected by extinction. The stars used for Figure 1.7 are nearby and have no interstellar reddening.

If both the color index and a spectral type are measured, we can go through Figure 1.7 and use the spectral type to predict the color index. Then we can compare the prediction with the measured value. The difference, observed minus predicted, is called the color excess, E_{B-V}, and it turns out to be related to the total extinction. In the *UBV* system, the *V* magnitude extinction, A_V, is

$$A_V = 3.1\, E_{B-V} = 3.1 \left[(B - V)_{\text{observed}} - (B - V)_{\text{predicted}} \right]. \tag{1.8}$$

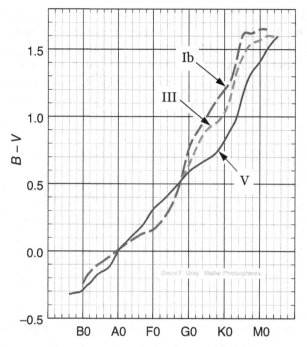

Figure 1.7 The mean relations between the $B - V$ color index and MK spectral types are shown based on the tabulation of Fitzgerald (1970).

The 3.1 coefficient in this relation is a nominal value (Hiltner & Johnson 1956, Whitford 1958, Schultz & Wiemer 1975, Krełowski & Papaj 1993). Larger values are found in some directions. For stars nearer than ~100 parsecs, there is little or no extinction.

There is one more important application for Equation 1.6 to which we now turn.

Stellar Distances and Absolute Magnitudes

How bright a star appears in the sky depends on how bright it really is and how far away it is from us. We can get the real relative brightness by mathematically allowing for the distance. The basic way to measure stellar distance is through the parallax introduced by the Earth orbiting the Sun. The effect is seen as an apparent displacement on the sky of nearby stars relative to stars much farther away. During the course of a year, nearby stars in the orbital plane of the Earth, the ecliptic plane, trace out a small straight line in the plane. Stars perpendicular to the plane at the north and south ecliptic poles trace out small circular motions (neglecting the small 0.017 eccentricity of the Earth's orbit). At positions in between, the apparent motion is an ellipse with its major axis parallel to the ecliptic, the so-called parallactic ellipse. Half the major axis of the ellipse is the parallax. From the star's perspective, this is the angular size of the Earth's orbit, or to be more precise, the angle subtended by one astronomical unit (a.u.) as seen from the star. The farther away a star is, the smaller its parallax. Call the parallax angle p arcseconds. The distance in units of parsecs (pc) is $r = 1/p$. The parsec is the standard unit of distance in astronomy. Since there are (to six figures) 206 265 seconds in a radian, one parsec equals 206 265 astronomical units. Related numbers are given in Appendix A.

The largest stellar parallax is only 0.742 arcsec or $r = 1.35$ pc. The solar "neighborhood" includes distances to ~100 pc. The distance between spiral arms in our galaxy is ~1 kpc, and we are ~8 kpc from the galactic center.

The standard way to accommodate the distance factor is to mathematically move all stars to the adopted standard distance of 10 pc. The magnitude at 10 pc is called the *absolute magnitude*, denoted by M. So Equation 1.6 for this case is

$$M - m = -2.5 \log\left(\frac{r}{10}\right)^2,$$

where we have used the fact that the flux is inversely related to the square of the distance. Or

$$M = m + 5 - 5 \log r = m + 5 + \log p. \tag{1.9}$$

If m is dimmed by interstellar extinction, M has to made brighter by the amount of the dimming using a relation like Equation 1.8.

Now, by using M, we can compare stars as if they really were all at the same distance. This leads to important results such as the HR diagram.

The Hertzsprung–Russell Diagram

When Rosenberg (1910), Hertzsprung (1911, but hard to find), and Russell (1914) first plotted the absolute magnitudes of stars as a function of their spectral type, they discovered a curious thing. The stars did not populate all portions of the diagram, but instead it showed most stars lying along a strip, now called the "main sequence," implying that the brightness of stars is directly connected to their temperature. Since, according to Equation 1.3, the radius factors in, there must also be a constraint on the radius. A few stars above the main sequence, i.e., brighter than main-sequence stars, must have larger radii. These revelations led to major steps in understanding stellar energy production, stellar lifetimes, and stellar evolution, and formed the backbone for physical analysis of stars. Figure 1.8 shows a modern summary of the Hertzsprung–Russell or HR diagram. We now know that the main sequence is a mass sequence, with larger mass stars lying higher up the main sequence. Most stars are found on the main sequence because this is the stable hydrogen-burning stage that lasts most of a star's lifetime. They start on the zero-age main sequence (ZAMS in the figure) and slowly move to the average main sequence shown by the solid line. Stars above the main sequence are in the late stages of their lives, which have short duration compared to their main-sequence lifetimes.

Figure 1.8a is an "observational" version of the HR diagram, with absolute magnitude shown as a function of spectral type. A more "theoretically oriented" version of the diagram plots the logarithm of the luminosity against the logarithm of the effective temperature. The lines in the figure show mean relations. Figure 1.8b shows some individual star positions in the same diagram. While most dwarfs lie very close to the class V line, evolved stars have a continuous distribution in luminosity and do not cluster around the mean lines. As we have seen, a color index can also be a measure of temperature and can be used as the abscissa instead of spectral type.

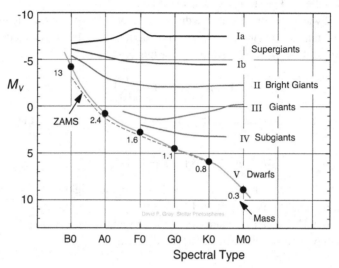

Figure 1.8a This HR diagram shows absolute visual magnitude as a function of spectral type. The zero-age main sequence is labeled ZAMS. The main sequence is a mass sequence with values of mass shown at several positions. The luminosity classes are labeled with their Roman numerals. Evolved stars populate the luminosity classes above the main sequence.

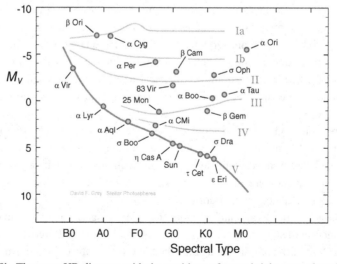

Figure 1.8b The same HR diagram with the positions of some bright stars plotted.

The HR diagram is so fundamental in astronomy that other graphs showing stellar parameters as a function of temperature are usually plotted with the temperature increasing to the left.

We now move on to some of the physical laws involved with stellar photospheres.

The Gas Laws

The photospheres of stars are hot gas. The gas pressure in most stellar photospheres is $\lesssim 10^5$ dyne/cm^3 (1 N/m^2), or ~0.1 the pressure of the terrestrial atmosphere. The perfect gas law is exactly what we need since it holds in the limit of low pressures. The form of the perfect gas law we use most frequently is

$$P = NkT. \tag{1.10}$$

Here N is the number of particles per cm^3, Boltzmann's constant is $k = 1.3806 \times 10^{-16}$ erg/K, and T is the absolute temperature in degrees kelvin (K). If there are different kinds of particles in the gas, we can assign a partial pressure to each such that the total gas pressure, P_g, comprises the sum of all the partial pressures over all the j elements,

$$P_g = \sum_j P_j = kT \sum_j N_j.$$

This is referred to as Dalton's law. In a stellar photosphere, one of these partial pressures is for free electrons from the ionized gases,

$$P_e = N_e kT, \tag{1.11}$$

where N_e is the number of electrons per cm^3. We refer to P_e as the electron pressure, dropping the adjective "partial."

The physical density, ρ g/cm^3, is given by

$$\rho = \sum_j N_j m_j, \tag{1.12}$$

with m_j being the weight of the jth constituent of the gas, given by the atomic weight in atomic mass units (amu) multiplied by 1.6605×10^{-24} g/amu.

Photospheric gases have temperatures of thousands of degrees. The gas particles are moving fast. What these velocities look like is next on our list.

Velocity and Speed Distributions

Gas pressure is produced by the motions of the gas particles. The partial pressure for the jth constituent is $P_j = N_j m_j v_{rms}/3$, where $N_j m_j$ is the density of the jth component of the gas and v_{rms} is the root-mean-square speed of this same component (Equation 1.17 below). The velocities of particles in a hot gas are distributed in a Gaussian or Maxwellian distribution. This is nature's signature of randomness, stemming from the huge number of collisions taking place in the gas. The orientation of the coordinate system is arbitrary since this thermal velocity distribution is isotropic. Each of the j components making up the gas has its own velocity distribution (omit the subscript j for simplicity) given by

$$N(v)\, dv = \left(\frac{m}{2\pi kT}\right)^{3/2} e^{-mv^2/2kT}\, dv$$

$$= \frac{1}{\sqrt{\pi}\, v_0} e^{-(v/v_0)^2} dv, \tag{1.13}$$

where $N(v)$ is normalized to unity and

$$v_0 = \sqrt{\frac{2kT}{m}} \tag{1.14}$$

is the velocity dispersion, k is Boltzmann's constant, m is the mass of the gas particle, and T is the absolute temperature of the gas. Notice that this velocity distribution is symmetric around zero velocity, where it has its maximum, and that v_0 is the $1/e$ width. Examples are shown in Figure 1.9.

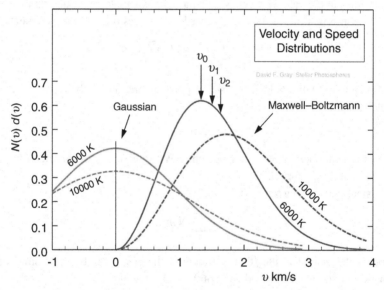

Figure 1.9 Gaussian velocity distributions are shown centered on zero velocity for $T = 6000$ K and for $T = 10\,000$ K, using the mass of iron atoms. The corresponding Maxwell–Boltzmann speed distributions are also shown. The positions of v_0, v_1, and v_2 are shown for the 6000 K temperature. All distributions are normalized to unit area.

Since the distribution is isotropic and we are interested in spectroscopic Doppler shifts, we can take v to be along the line of sight to the star, i.e., the radial velocity, $v = v_R$. Each velocity produces a Doppler shift, $\Delta\lambda = \lambda\, v_R/c$, for light of wavelength λ, where c is the velocity of light, 2.9979×10^{10} cm/s. In other words, the thermal *Doppler-shift distribution* is also a Gaussian. As we shall see in Chapter 11, this Doppler-shift Gaussian is the foundation of intrinsic spectral line shapes.

Although Equation 1.13 is our main interest, in situations where energy, momentum, or pressure aspects of the gas are being considered, the speed distribution, also called the Maxwell–Boltzmann distribution, is useful. It is

$$N(v)\, dv = \left(\frac{2}{\pi}\right)^{1/2} \left(\frac{m}{kT}\right)^{3/2} v^2\, e^{-mv^2/2kT} dv$$

$$= \frac{4}{\sqrt{\pi}} \left(\frac{1}{v_0}\right)^3 v^2 e^{-(v/v_0)^2} dv. \tag{1.15}$$

The maximum of this speed distribution occurs at v_0, as already given in Equation 1.14. The average speed is

$$v_1 = \left(\frac{8}{\pi}\frac{kT}{m}\right)^{1/2} = 1.128\, v_0, \tag{1.16}$$

and the root-mean-square velocity (speed) is

$$v_2 = \left(\frac{3kT}{m}\right)^{1/2} = 1.225\, v_0. \tag{1.17}$$

The positions of v_0, v_1, and v_2 are shown in Figure 1.9 for the 6000 K case.

Among other things, high particle velocities mean energetic collisions that lead to excitation and ionization of the photospheric gas.

Atomic Excitation

The atomic process of forming spectral lines depends completely on the energy-level structure (orbits mainly) of the chemical. An absorption line corresponds to using a photon to lift an electron from some energy level to a higher one. Since the energy levels are fairly narrow, only photons in a narrow wavelength range can do this. How narrow is discussed in Chapter 11. If the average energy for the lower level is W_l and in the upper level W_u, the photon energy needed must be very close to $h\nu$ or $hc/\lambda = W_u - W_l$, where $h = 6.6261 \times 10^{-27}$ erg s is Planck's constant and c is the velocity of light. In most stellar-photosphere situations, there are plenty of photons available to do this, so the strength of the line pivots on the number of atoms in the lower level. The lower level for most lines is an excited state. Excitation is the process of lifting electrons from their lowest energy level, the ground state, to some higher state. Excitation is accomplished mainly by thermal collisions, and it is this process that we discuss here.

In Figure 1.10, we have a schematic energy-level diagram of an atom. The energy levels are labeled with the principal quantum number, and we are interested in excitation to the nth level, where the energy above the ground level is called the excitation potential, χ. If the energy imparted to the electron is $\geq I$, the ionization potential, the electron is lost to the atom, which is then ionized. Although all of the bound levels have negative energy, the ground level is most negative, so that χ and I are positive numbers.

The number per cubic centimeter of a species excited to the nth level is proportional to the statistical weight g_n and the so-called Boltzmann factor,

$$N_n = \text{constant}\ g_n\ e^{-\chi_n/kT}.$$

The statistical weight is $2J + 1$, where J is the inner quantum number (obtainable from Moore 1959, for example). For hydrogen, $g_n = 2n^2$. The absolute temperature, T, is in kelvin as usual.

We may want the ratio of populations in two levels, say m and n. Then we write

$$\frac{N_n}{N_m} = \frac{g_n}{g_m} e^{-(\chi_n - \chi_m)/kT}.$$

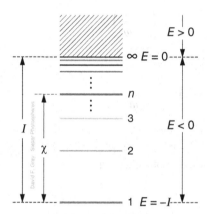

Figure 1.10 A schematic one-dimensional energy-level diagram shows the excitation potential, χ, for an arbitrary level of quantum number n and the level corresponding to the ionization energy I. The ground state has $n = 1$. The energy of the electron, as indicated on the right, is negative for all bound states and positive in the continuum (hatched area). The energy is zero at the ionization limit, where $n = \infty$.

Similarly, the number of atoms in a level n expressed as a fraction of all atoms of the same species is

$$\frac{N_n}{N} = \frac{g_n\, e^{-\chi_n/kT}}{g_1 + g_2 e^{-\chi_2/kT} + g_3 e^{-\chi_3/kT} + \cdots}$$

$$= \frac{g_n}{u(T)} e^{-\chi_n/kT}. \tag{1.18}$$

The denominator is lumped together in $u(T) = \sum g_i\, e^{-\chi_i/kT}$, called the *partition function*. In practice, the partition function is not an infinite series, but is truncated because the highest energy levels are essentially removed by collisions. Partition functions have been calculated for most elements of interest, and a tabulation can be found in Appendix C.

Equation 1.18 is often expressed as a power of 10,

$$\frac{N_n}{N} = \frac{g_n}{u(T)} 10^{-\log e\, \chi_n/kT} = \frac{g_n}{u(T)} 10^{-\theta\chi_n}, \tag{1.19}$$

with $\theta = \log e/kT = 5040/T$ per eV. Here $k = 8.6173 \times 10^{-5}$ eV/K, i.e., k must be in units of electronvolts per degree to be compatible with χ_n in electronvolts. Equation 1.18 or 1.19 is called the *excitation* or Boltzmann equation.

Figure 1.11 illustrates an important fact about excitation. The fraction excited to some level can vary by many orders of magnitude while the number in the ground state remains nearly constant.

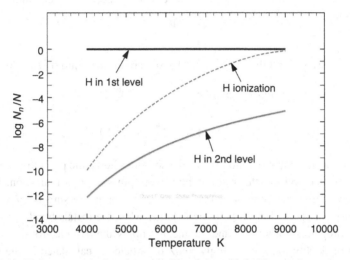

Figure 1.11 Hydrogen excited to the second level ($\chi = 10.2$ eV) can change by orders of magnitude, but is so small in absolute terms that the fraction in the ground state (1st level) remains sensibly at unity. For temperatures higher than these, hydrogen is ionized. The dashed curve is for $P_e = 100$ dyne/cm^2.

Next we consider the case where the energy given to the electron, either by a collision or a photon, equals or exceeds I, i.e., the process of ionization.

Ionization

The ionization process involves removing one or more electrons from the atom. How much of an element is ionized affects the number of absorbers and consequently the strengths of spectral lines of that species. Ionization generates free electrons that make up the electron pressure. More ionizations, larger P_e, as in Equation 1.11. The ratio of first ions, N_1, to neutral atoms, N_0, is given by

$$\frac{N_1}{N_0} P_e = \frac{(2\pi m_e)^{3/2} (kT)^{5/2}}{h^3} \frac{2u_1(T)}{u_0(T)} e^{-I/kT}. \tag{1.20a}$$

Here m_e is the electron mass and h is Planck's constant. The ratio of partition functions for the ion and the neutral atom is $u_1(T)/u_0(T)$, and I is the ionization potential, as in Figure 1.10. Ionization potentials are given in Appendix C. This equation was first derived by M.N. Saha (1920, 1921) and sometimes bears his name, but we will refer to it as the *ionization equation*. Derivation details can be found in Aller (1963) or in statistical mechanics expositions such as that by Tolman (1938). See Struve and Zebergs (1962) and Hearnshaw (1986) for historical details.

The ionization equation can also be written as

$$\log \frac{N_1}{N_0} = -\log P_e - \theta I + 2.5 \log T + \log \frac{u_1}{u_0} - 0.1762, \tag{1.20b}$$

where, as before, $\theta = 5040/T$. It is often convenient to simplify the equation to

$$\frac{N_1}{N_0} = \frac{\Phi(T)}{P_e}, \tag{1.21}$$

where all the constants and the temperature factors are lumped into $\Phi(T)$, i.e.,

$$\Phi(T) = 0.6668 \frac{u_1}{u_0} T^{5/2} \, 10^{-5040I/T}$$

$$= 1.2024 \times 10^9 \frac{u_1}{u_0} \theta^{-5/2} \, 10^{-\theta I}. \tag{1.22}$$

Casting the ionization equation into this form also brings home the point that the ionization is directly influenced by the electron pressure, more pressure, less ionization. In effect, a higher P_e forces free electrons back into the ions. Electron pressure is larger when the surface gravity is larger. Electron pressure is larger when the metal abundances are larger, i.e., when hydrogen is not ionized and the metals supply the free electrons.

The ratio of doubly ionized to singly ionized particles is calculated in the same way, $N_2/N_1 = \Phi(T)/P_e$, but now the ionization potential in Equation 1.22 is for the ion, and the partition function ratio is $u_2(T)/u_1(T)$. Virtually all of an element is accounted for by including three stages of ionization, as can be seen in Figure 1.12. The transition from neutral to first ion, from first to second ion, and so on occurs fairly rapidly along the temperature coordinate.

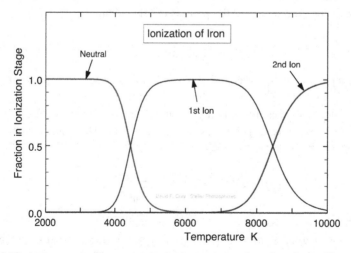

Figure 1.12 As temperature increases, ionization occurs rather abruptly, as illustrated by iron in this example. These curves are for unit P_e.

When calculating the number of absorbers available to produce a spectral line, Equations 1.19 and 1.20 are combined to give the number in the level n as a fraction of the total number of the element,

$$\frac{N_n}{N_{\text{total}}} = \frac{N_n}{N_0 + N_1 + N_2 + \cdots}. \tag{1.23}$$

If the level n is in the neutral atom, write

$$\frac{N_n}{N_{\text{total}}} = \frac{N_n/N_0}{1 + N_1/N_0 + (N_2/N_1)(N_1/N_0) + \cdots}$$

or if the level n is in the first ion, write

$$\frac{N_n}{N_{\text{total}}} = \frac{N_n/N_1}{N_0/N_1 + 1 + N_2/N_1 + \cdots}.$$

The excitation ratio N_n/N_0 or N_n/N_1 is found from Equation 1.19 with $N = N_0$, or for N_n/N_1 with $N = N_1$, etc., and each of the ionization ratios, $N_1/N_0, N_2/N_1, \cdots$, is found with Equation 1.20 or 1.21.

At this point, we now have the basic framework in place, except to say, we should always be cognizant of the resources and publications around us.

Key Links and Bibliography

The amount of information and data accessible on line is extensive and useful. A few key links are given here.

ADS (Astrophysics Data Systems) for literature: ui.adsabs.harvard.edu/
> Find papers, books, abstracts using searchable fields including author, date, title words, abstract words, etc. separately or in combination. View and/or download most papers.

NIST (National Institute of Standards and Technology) physical database: physics.nist.gov/
> An extensive source for physical constants, calibrations, atomic data, and related items.

SIMBAD (Set of Identifications, Measurements, and Bibliography for Astronomical Data) a stellar database: simbad.u-strasbg.fr/simbad/
> A basic source reference for stellar parameters, including positional information, spectral type, magnitudes, parallax, velocities, metallicity, rotation rates, etc. for almost 6 000 000 stars. Biographical links are included. Caution, the data are sometimes incomplete.

VALD (Vienna Atomic Line Database) atomic line data: www.astro.uu.se/valdwiki/
> A good source for line identification, wavelength, excitation potential, log $g_n f$ values, damping constants, etc. for stellar spectra.

A small selection of other useful references follows.

Astrophysical Quantities (Allen 1973, Cox 2000): Reference volume giving astronomical data on all topics ranging from physical constants through basic stellar parameters to clusters of galaxies, and containing many useful numbers for studies in stellar photospheres.

A Multiplet Table of Astrophysical Interest (Moore 1959): Multiplets are listed for about 26 000 lines of most common elements. Gives wavelength, intensity class, excitation potential, inner quantum number *J*, atomic configuration, and multiplet number.

The Bright Star Catalogue (Hoffleit & Warren 1995): Lists 9091 stars by HR = BS number. Gives HD and ADS numbers, equatorial coordinates for 1900 and 2000, precession rates, visual magnitude, color index, spectral type, proper motion, parallax, radial velocity, rotational velocity, and useful notes.

Sloan Digital Sky Survey (SDSS, Fukugita *et al.* 1996): Contains a wide range of photometric data, accurate radial velocities, and gives low-resolution spectra for several thousand stars.

The Hipparcos and Tycho Catalogues (Perryman *et al.* 1997): Primarily lists astrometric data, parallaxes and proper motions, but also summarizes spectral types, magnitudes, and color indices for over one million stars.

The Gaia Catalogue (Gaia Collaboration 2016): The big brother of Hipparcos, mainly astrometric data, but also gives magnitudes, color indices, and modest-resolution photometrically calibrated spectroscopic data for about a hundred million stars, used mainly for radial velocity measurements.

Lines of the Chemical Elements in Astronomical Spectra (Merrill 1958): Summarizes many interesting facts about the appearance of each element in stellar spectra. Extensive referencing to the early literature. Merrill gives useful energy-level diagrams in his Appendix A.

Additional surveys and catalogues are cited throughout the text as appropriate. The appendices at the end of the book have many useful numbers.

Questions and Exercises

1. Along the main sequence, the mass, temperature, color indices, spectral type, rotation rates, the strength of convection, and to a lesser degree the surface gravity all change. What makes us think that the spectral type depends primarily on temperature rather than the other parameters?

2. Look in the SIMBAD database and find the apparent visual magnitude, spectral type, and parallax of Vega. Compute Vega's absolute visual magnitude and place your results on the HR diagram in Figure 1.8. Do the same for ξ Boo A and for α UMa. Do the positions you find agree with the luminosity classification given in SIMBAD? For comparison, read Section 1 in 2018 *ApJ* **869**, 81.

3. It is common to have binary stars so close together that their images are unresolved and they look like a single star. Assume the two components of a binary are the same, both dwarfs, same spectral type, same brightness. Use Equation 1.6 to estimate their position in the HR diagram relative to their single-star position.

4. For a pure isothermal hydrogen gas having a temperature of 4000 K, calculate and graph the speed distribution using Equation 1.15. What is the most probably speed for your distribution? Compare your results to Figure 1.9.

5. Add to your graph in Exercise 4 the velocity distribution given by Equation 1.13. What is the most probable velocity for this distribution?

6. Calculate the velocity of an iron atom that has enough kinetic energy to ionize hydrogen. Ionization potentials can be found in Appendix C. What gas temperature would this correspond to if this were the root-mean-square (rms) velocity? What would the temperature be if this velocity corresponded to three times the rms velocity, i.e., only atoms in the upper tail of the velocity distribution did the ionizing?

7. Compute N_0, N_1, and N_2 as fractions of the total number of particles for iron. Take $P_e = 100$ dyne/cm^2. The ionization potentials for iron can be found in Appendix C. For convenience, ignore the temperature dependence of the partition functions and simply use g_0 from Appendix C. Compare your results with Figure 1.12. Explain any differences in qualitative physical terms.

8. Develop some introductory skills in spectral classification by constructing a conceptual HR diagram for the Pleiades. We will follow the historical path, using photographic images characteristic of the era when the MK system was developed. Figure E.1.1 is a photograph of the Pleiades open star cluster taken with a Schmidt telescope (to get a wide 5° field of view). Since the cluster stars are all at nearly the same distance, they have the same distance dimming (Equation 1.9). So the apparently brightest cluster star is truly the brightest cluster star, and so on for less-bright stars.

   The next exposure, Figure E.1.2, is of the same star cluster but a 10° prism was placed over the front end of the telescope, so each star generates a spectrum. In times of yore these old historical photographic spectra were inspected through a hand-held magnifying glass or a low-power microscope. To make it easier for this exercise, a selection of the spectra have been extracted and are displayed in Figure E.1.3. With the aid of the line map and the flow chart in Figure 1.4, find the spectral types of these selected spectra.

   Then go back to Figure E.1.1 and rank these same stars according to their brightness (image strength). Make a pseudo HR diagram by plotting the brightness rank as ordinate, brightest on top, as a function of spectral type. Can you see the main sequence?

   This kind of objective-prism technique has the advantage of recording many spectra at once, but the disadvantage of having the spectra of the bright stars over-exposed, while the spectra of the faint stars are under-exposed.

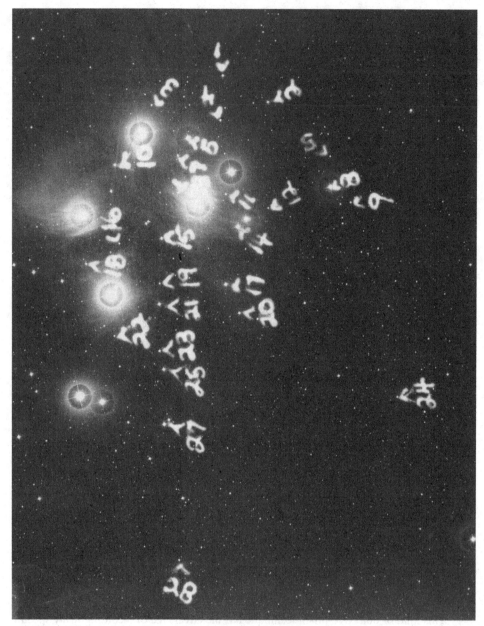

Figure E.1.1 A direct photographic image of the Pleiades.

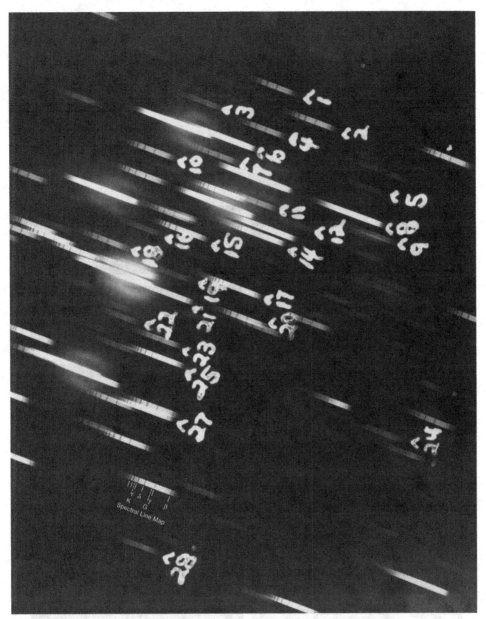

Figure E.1.2 A spectroscopic image of the Pleiades.

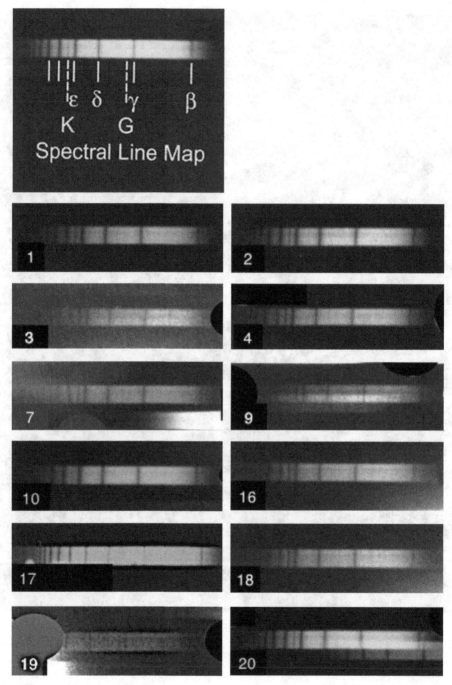

Figure E.1.3 A map of the spectral lines needed for simple classification along with 18 spectra extracted from Figure E.1.2.

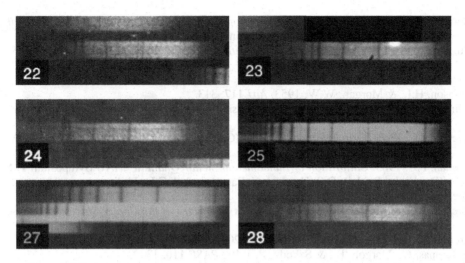

Figure E.1.3 (*cont.*)

References

Abt, H.A., Meinel, A.B., Morgan, W.W., & Tapscott, J.W. 1968. *An Atlas of Low-Dispersion Grating Stellar Spectra* (Tucson: Kitt Peak National Observatory).

Ahumada, R., *et al.* 2020. *ApJS* **249**, 3.

Allen, C.W. 1973. *Astrophysical Quantities* (London: Athlone), 3rd ed.

Aller, L.H. 1963. *Astrophysics: The Atmospheres of the Sun and Stars* (New York: Ronald), 2nd ed.

Becker, W. 1948. *ApJ* **107**, 278.

Becker, W. 1963. *Basic Astronomical Data* (Chicago: University of Chicago Press), K.A. Strand, ed., Chapter 13.

Becker, W. & Stock, J. 1954. *ZfAp* **34**, 1.

Bessell, M.S. 2005. *ARAA* **43**, 293.

Cox, A.N. (ed.) 2000. *Allen's Astrophysical Quantities* (New York: AIP, Springer-Verlag).

Draper, H. 1935. *IAU Trans.* **5**, 180.

FitzGerald, M.P. 1970. *A&A* **4**, 234.

Fukugita, M., Ichikawa, T., Gunn, J.E., *et al.* 1996. *AJ* **111**, 1748.

Gaia Collaboration 2016. *A&A* **595**, 1.

Garrison, R.F. (ed.) 1984. *The MK Process and Stellar Classification* (Toronto: David Dunlap Observatory).

Geballe, T.R., Knapp, G.R., Leggett, S.K., *et al.* 2002. *ApJ* **564**, 466.

Gray, R.O. & Corbally, C.J. 2009. *Stellar Spectral Classification* (Princeton: Princeton University Press).

Hauck, B. & Mermilliod, M. 1980. *A&AS* **40**, 1.

Hearnshaw, J.B. 1986. *The Analysis of Starlight, One Hundred and Fifty Years of Astronomical Spectroscopy* (Cambridge: Cambridge University Press).

Hertzsprung, G. 1911. *POPot* 22, **1**, #63.

Hiltner, W.A. & Johnson, H.L. 1956. *ApJ* **124**, 367.

Hoffleit, D. & Warren, W.H. Jr. 1995. *The Bright Star Catalogue* (VizieR Online Data Catalogue), 5th ed.

Houk, N. & Cowley, A.P. 1975. *University of Michigan Catalogue of Two-Dimensional Spectral Types for the HD Stars* (Washington, DC: Comnet-Publicate).

Huggins, W. & Lady Huggins 1899. *Publications of Sir William Huggins' Observatory*, Vol. I (London: W. Wesley & Son).

Johnson, H.L. 1966. *ARAA* **4**, 193.

Johnson, H.L. & Morgan, W.W. 1953. *ApJ* **117**, 313.

Johnson, H.L., Mitchell, R.I., Iriarte, B., & Wisniewski, W.Z. 1966. *Commun Lunar Planet Lab* **4**, 99.

Keenan, P.C. 1963. *Basic Astronomical Data* (Chicago: University of Chicago Press), K.A. Strand, ed., Chapter 8.

Keenan, P.C. 1985. *Calibration of Fundamental Stellar Quantities,* IAU Symposium **111** (Dordrecht: Reidel), D.S. Hayes, L.E. Pasinetti, & A.G.D. Philip, eds., p. 121.

Keenan, P.C. 1987. *PASP* **99**, 713.

Keenan, P.C. & McNeil, R.C. 1976. *An Atlas of Spectra of the Cooler Stars* (Columbus: Ohio State University Press).

Krełowski, J. & Papaj, J. 1993. *PASP* **105**, 1209.

Krisciunas, K., Margon, B., & Szkody, P. 1998. *PASP* **110**, 1342.

Landolt, A.U. 1992. *AJ* **104**, 340.

Landolt, A.U. 2009. *AJ* **137**, 4186.

Leggett, S.K., Golimowski, D.A., Fan, X., *et al.* 2002. *ApJ* **564**, 452.

McClure, R.D. & van den Berg, S. 1968. *AJ* **73**, 313.

Merrill, P.W. 1958. *Lines of the Chemical Elements in Astronomical Spectra*, Pub. No. 610 (Washington, DC: Carnegie Institution of Washington).

Moore, C.E. 1959. *A Multiplet Table of Astrophysical Interest*, National Bureau of Standards Technical Note 36 (Washington, DC: US Department of Commerce Office of Technical Services).

Morel, M. & Magnenat, P. 1978. *A&AS* **34**, 477.

Morgan, W.W. & Keenan, P.C. 1973. *ARAA* **11**, 29.

Morgan, W.W., Keenan, P.C., & Kellman, E. 1943. *An Atlas of Stellar Spectra* (Chicago: University of Chicago Press).

Morgan, W.W., Abt, H.A., & Tapscott, J.W. 1978. *Revised MK Spectral Atlas for Stars Earlier than the Sun* (Williams Bay: Yerkes Observatory; Chicago: University of Chicago; Tucson: Kitt Peak National Observatory).

Nicolet, B. 1978. *A&AS* **34**, 1.

Perry, C.L., Olsen, E.H., & Crawford, D.L. 1987. *PASP* **99**, 1184.

Perryman, M.A.C., Lindegren, L., Kovalevsky, J., *et al.* 1997. *A&A* **323**, L49.

Rosenberg, H. 1910. *Ast Nach* **186**, 71.

Rufener, F. 1988. *Catalogue of Stars Measured in the Geneva Observatory Photometric System* (Sauverny, Switzerland: Observatoire de Genève), 4th ed.

Rufener, F. & Nicolet, B. 1988. *A&A* **206**, 357.

Russell, H.N. 1914. *Pop Ast* **22**, 331.

Saha, M.N. 1920. *Phil Mag* **40**, 472.

Saha, M.N. 1921. *Proc R Soc London A*, **99**, 135.

Schultz, G.V. & Wiemer, W. 1975. *A&A* **43**, 133.

Smith, J.A., *et al.* 2002. *AJ* **123**, 2121.

Strömgren, B. 1966. *ARAA* **4**, 433.

Struve, O. & Zebergs, V. 1962. *Astronomy of the 20th Century* (New York: Macmillan).

Tolman, R.C. 1938. *The Principles of Statistical Mechanics* (London: Oxford University Press).

Vernazza, J.E., Avrett, E.H., & Loeser, R. 1973. *ApJ* **184**, 605.

Walborn, N.R., Howarth, I.D., Lennon, D.J., *et al.* 2002. *AJ* **123**, 2754.

Whitford, A.E. 1958. *AJ* **63**, 201.

2

Fourier Transforms

There is hardly a phase of modern astrophysics to which Fourier techniques do not lend some insight or practical advantage. We bump into Fourier transforms repeatedly. Diffraction patterns are transforms of the aperture causing them. This is particularly important with radio antennae, telescope arrays, and optical interferometers, where the response pattern is strongly shaped by their configuration. Within the arena of stellar photospheres, Fourier concepts help us understand spectrographs and the stellar spectra they give us, they facilitate our understanding of the intrinsic shapes of spectral lines, and they are powerful helpers when it comes to analyzing line profiles.

This chapter forms an introduction to Fourier transforms for those unfamiliar with them and a useful refresher for those who have studied them in past years. The treatment is highly abbreviated, but covers all the concepts used in the remainder of the book. Those wishing to learn the material in a more rigorous and extensive way are referred to books dealing specifically with Fourier transforms, for instance, the books of Jennison (1961), Bracewell (1965), and Gaskill (1978).

A good place to start is with the basic definition of a Fourier transform.

The Definition

A function's Fourier transform is a specification of the amplitudes and phases of sinusoidals which, when added together, reproduce the function. Only one-dimensional functions are treated here. Expansion to two or higher dimensions can be done by the reader with modest effort. Take a function $F(x)$ and write

$$f(\sigma) = \int_{-\infty}^{\infty} F(x)\, e^{2\pi i x \sigma}\, dx. \tag{2.1}$$

This new function, $f(\sigma)$, is the Fourier transform of $F(x)$. The variables x and σ are called a Fourier pair. The inverse transform is

$$F(x) = \int_{-\infty}^{\infty} f(\sigma)\, e^{-2\pi i x \sigma}\, d\sigma. \tag{2.2}$$

Upper and lower case symbols denote, respectively, the functions and their transforms. Thus $f(\sigma)$ and $F(x)$ are transforms of each other. Notice the change in sign of the exponent between Equations 2.1 and 2.2. A derivation of the relations 2.1 and 2.2 can be found in Kharkevich (1960).

Suppose we expand Equation 2.1 using Euler's formula, that is, $e^{ix} = \cos x + i \sin x$. We obtain four integrals, provided we allow $F(x)$ to be a complex function, $F(x) = F_R(x) + i F_I(x)$, consisting of a real, $F_R(x)$, and imaginary, $F_I(x)$, component. Then

$$f(\sigma) = \int_{-\infty}^{\infty} F_R(x) \cos 2\pi x\sigma\, dx + i \int_{-\infty}^{\infty} F_I(x) \cos 2\pi x\sigma\, dx$$

$$+ i \int_{-\infty}^{\infty} F_R(x) \sin 2\pi x\sigma\, dx - \int_{-\infty}^{\infty} F_I(x) \sin 2\pi x\sigma\, dx. \tag{2.3}$$

This equation is useful for visualizing the Fourier transform process. Consider the first integral as an example. It says: multiply the function $F_R(x)$ by a cosine with periodicity $1/\sigma$, with σ being a constant for this integration. The total area resulting from this product is the value needed. Figure 2.1 illustrates this process. The other three integrals in Equation 2.3 can be visualized in a similar manner. Combining the integrals gives $f(\sigma)$ at one value of σ. Changing the frequency of the sines and cosines and repeating the process gives $f(\sigma)$ at a second point, and so on. This process is the essence of the numerical evaluation of Fourier transforms which is discussed later in the chapter.

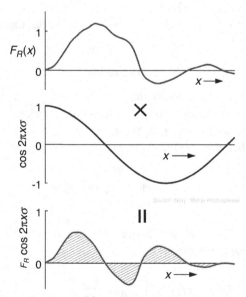

Figure 2.1 The arbitrary function $F_R(x)$ is multiplied by $\cos 2\pi x\sigma$ to give the bottom curve. The area is the transform $f(\sigma)$ for one value of σ.

If we go on to write Equation 2.3 as $f(\sigma) = f_R(\sigma) + i f_I(\sigma)$, we see that $f_R(\sigma)$ is composed of the first and last integrals, while $f_I(\sigma)$ is given by the center two integrals. So in the general case, both $F(x)$ and $f(\sigma)$ are complex functions.

Under some physical circumstances, $F(x)$ may be a real function. One such case is when $F(x)$ is the spectrum of starlight. Since $F_I(x) = 0$, $f(\sigma)$ is given by

$$f(\sigma) = \int_{-\infty}^{\infty} F_R(x) \cos 2\pi x \sigma \, dx + i \int_{-\infty}^{\infty} F_R(x) \sin 2\pi x \sigma \, dx.$$

We see that $f(\sigma)$ is still a complex function. Only if $F(x)$ is an even function, $F_R(x) = F_R(-x)$, as well as real do we find

$$f(\sigma) = \int_{-\infty}^{\infty} F_R(x) \cos 2\pi x \sigma \, dx.$$

Here the evenness of $F_R(x)$ combined with the oddness of $\sin 2\pi x \sigma$ gives exactly as much negative as positive area, causing the imaginary term to be zero.

All complex numbers can be specified in rectangular or polar coordinates. In rectangular form, we have written $f(\sigma) = f_R(\sigma) + i f_I(\sigma)$. We can convert this to polar form in the usual way using an amplitude and phase,

$$f(\sigma) = |f(\sigma)| \, e^{i\psi}, \tag{2.4}$$

where the amplitude is

$$|f(\sigma)| = \left[f_R^2(\sigma) + f(\sigma)_I^2 \right]^{1/2}$$

and

$$\tan \psi = \frac{f_I(\sigma)}{f_R(\sigma)},$$

where ψ is the phase.

We conclude this section by noting that the zero ordinate of the transform is the area under the original function. And because the original function is the transform of the transform, its zero ordinate specifies the area under the transform. This conclusion can be reached trivially from the definitions, Equations 2.1 and 2.2, by setting $\sigma = 0$, $f(0) = \int F(x) \, dx$ and $F(0) = \int f(\sigma) \, d\sigma$. These relations are useful, for example, when the Fourier transform of a spectral line is being investigated. Then the $\sigma = 0$ value is the total absorption caused by the line.

Now let us move on to consider some common transforms.

The Box and the sinc Function

First consider the box shown in Figure 2.2. It might represent the graph of the light distribution in the plane of an opaque screen having a long slit of width W in it. We can denote the function by

$$B(x) = 0 \quad \text{for} \quad -\frac{W}{2} > x > \frac{W}{2}, \text{outside the box,}$$

$$B(x) = 1 \quad \text{for} \quad -\frac{W}{2} \leq x \leq \frac{W}{2}, \text{inside the box.}$$

And immediately the Fourier transform can be written

$$b(\sigma) = \int_{-\infty}^{\infty} B(x)\, e^{2\pi i x\sigma}\, dx = \int_{-W/2}^{W/2} e^{2\pi i x\sigma}\, dx = \left[\frac{e^{2\pi i x\sigma}}{2\pi i\sigma}\right]_{-W/2}^{W/2}$$

$$= W\frac{\sin \pi W\sigma}{\pi W\sigma}$$

$$\equiv W\,\mathrm{sinc}\,\pi W\sigma. \tag{2.5}$$

That is, a general $(\sin x)/x$ form is abbreviated as sinc x.

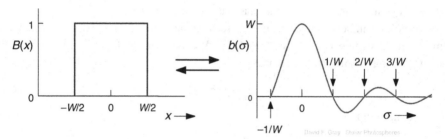

Figure 2.2 The box of width W transforms into $b(\sigma) = W\,\mathrm{sinc}\,\pi W\sigma$ with a characteristic width of $1/W$. Arrows mean "transforms into."

This sinc function is shown on the right in Figure 2.2. The reciprocal behavior in the widths of these functions is particularly important. When W is large the transform is narrow, and vice versa. This illustrates a general behavior between all functions and their transforms. The first zeros of this transform are at $\sigma = \pm 1/W$. The other zeros are equally spaced at multiples of $1/W$ along the σ axis.

The Gaussian Function

As a second function, take the unit-area Gaussian given by

$$G(x) = \frac{1}{\beta\pi^{1/2}}e^{-x^2/\beta^2}. \tag{2.6}$$

(In statistics and some physical situations, the notation is slightly different, with β^2 being replaced by $2\beta^2$, as in Equation 1.13.) Again, we find the transform from the definition:

$$g(\sigma) = \int_{-\infty}^{\infty} G(x)\, e^{2\pi i x\sigma} dx = \frac{1}{\beta\pi^{1/2}}\int_{-\infty}^{\infty} e^{-x^2/\beta^2} e^{2\pi i x\sigma} dx.$$

But this function is real and symmetric, so we can use the expansion as in Equation 2.3 to get

$$g(\sigma) = \frac{1}{\beta \pi^{1/2}} \int\limits_{-\infty}^{\infty} e^{-x^2/\beta^2} \cos 2\pi x \sigma \, dx.$$

Standard integral tables supply the necessary integration with the result that

$$g(\sigma) = e^{-\pi^2 \beta^2 \sigma^2}. \tag{2.7}$$

In short, a Gaussian transforms into a Gaussian. If the dispersion of the original Gaussian is β, the dispersion of the transform is $1/(\pi\beta)$, again an inverse relation.

The Dispersion Function

Another function that comes up, particularly in the study of spectral line profiles, is the dispersion profile defined by

$$F(x) = \frac{1}{\pi} \frac{\beta}{x^2 + \beta^2}. \tag{2.8}$$

Here β is the half-half width of the function, and $F(x)$ has unit area. The transform is

$$f(\sigma) = \frac{1}{\pi} \int\limits_{-\infty}^{\infty} \frac{\beta}{x^2 + \beta^2} e^{2\pi i x \sigma} \, dx = \frac{2}{\pi} \int\limits_{0}^{\infty} \frac{\beta}{x^2 + \beta^2} \cos 2\pi x \sigma \, dx,$$

where we have jumped straight to the cosine integral since this function is real and symmetric. This integral is of the form

$$\int\limits_{0}^{\infty} \frac{\cos mu}{1 + u^2} \, du = \frac{\pi}{2} e^{-m}.$$

We find that $f(\sigma) = e^{-2\pi\beta\sigma}$ for $\sigma \geq 0$ and $f(\sigma) = e^{2\pi\beta\sigma}$ for $\sigma < 0$, or

$$f(\sigma) = e^{-2\pi\beta|\sigma|}, \tag{2.9}$$

an exponential drop off in both directions from $\sigma = 0$.

The Delta Function

Now let us turn to a very useful but slightly more complicated function, the delta function or δ-function. Physically the δ-function represents an impulse. This is equivalent to saying that the physical process takes place so rapidly that it is beyond our power to resolve it. Therefore the integrated property of the impulse is what we observe, which for the δ-function is defined as unity,

$$\int\limits_{-\infty}^{\infty} \delta(x) \, dx = 1. \tag{2.10}$$

Although we cannot resolve the impulse, we can specify its position or time of occurrence. Symbolically we write

$$\delta(x) = 0 \text{ for } x \neq 0,$$

which implies that $\delta(x)$ occurs at $x = 0$. If we write $\delta(x - x_1)$, then we mean the impulse occurs at $x = x_1$.

Mathematically it is convenient to think of some unit area function that can be made infinitesimal in width while maintaining unit area. This process is necessary to provide a rigorous proof of many of the properties of a δ-function. In keeping with the level of treatment in this chapter, we omit these proofs. It should be understood, however, that the very large amplitude implied in making a unit area function go to infinitesimal width is not a practical problem because either the δ-function is used in conjunction with another function or only its integral property is of concern. As an example, consider the product of the function $F(x)$ with the δ-function,

$$\delta(x - x_1)F(x) = \delta(x - x_1)F(x_1), \tag{2.11}$$

since $\delta(x - x_1) = 0$ except at $x = x_1$. In other words, multiplication by the δ-function retains only one point in the original function and that point is at the position of the δ-function.

Further, if we integrate Equation 2.11, we have

$$\int_{-\infty}^{\infty} \delta(x - x_1)F(x)\,dx = F(x_1) \int_{-\infty}^{\infty} \delta(x - x_1)\,dx = F(x_1). \tag{2.12}$$

This interesting property proves useful when the δ-function is applied in analysis. Notice also that Equation 2.12 can be rearranged slightly and written as

$$\int_{-\infty}^{\infty} \delta(x_1 - x)F(x)\,dx = F(x_1).$$

This expression has the form of a convolution which is defined and discussed later in this chapter.

The transform of $\delta(x - x_1)$ is given by

$$\Delta(\sigma) = \int_{-\infty}^{\infty} \delta(x - x_1)\,e^{2\pi i x \sigma}\,dx,$$

but since $\delta(x - x_1) = 0$, except when $x = x_1$, we can write

$$\Delta(\sigma) = e^{2\pi i x_1 \sigma} \int_{-\infty}^{\infty} \delta(x - x_1)\,dx = e^{2\pi i x_1 \sigma}, \tag{2.13}$$

and so the amplitude of $\Delta(\sigma)$ is unity independent of σ, but there is a linear phase term, $2\pi x_1 \sigma$. These properties are shown in Figure 2.3. This function represents the ultimate in the inverse dimension behavior: an infinitesimal width for the function gives an infinite extension in the transform. The linear phase term is shown as a dashed line with a slope of $2\pi x_1$. The relation between the x_1 shift in $\delta(x - x_1)$ and phase term is fundamental, a shift in one domain produces a phase term on the other. Notice the introduction of the vertical arrow in Figure 2.3. It stands for "transforms into."

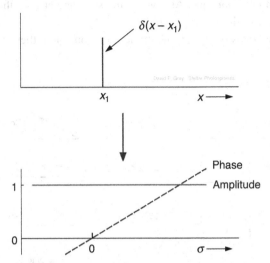

Figure 2.3 A δ-function at $x = x_1$ transforms into a constant amplitude with a linear phase term $2\pi x_1 \sigma$.

The δ-function is a basic building block.

Sines, Cosines, and Delta Functions

We can easily obtain the transforms of sines and cosines using δ-functions and their transforms from Equation 2.13. Suppose, for example, we have two δ-functions symmetrically placed about $x = 0$. That is, $F(x) = \delta(x - x_1) + \delta(x + x_1)$, and because Fourier transforms are linear (Theorem 1 below), we can write the transform as the sum of the individual transforms,

$$f(\sigma) = e^{2\pi i x \sigma} + e^{-2\pi i x \sigma}$$
$$= 2\cos 2\pi x_1 \sigma. \tag{2.14}$$

In a similar way, $F(x) = \delta(x - x_1) - \delta(x + x_1)$ transforms to $f(\sigma) = 2i \sin 2\pi x_1 \sigma$. In short, pairs of δ-functions transform into sinusoidals.

The Shah Function: an Array of Delta Functions

Instead of two δ-functions, imagine an infinite array of them equally spaced at intervals of Δx. Then denoting this array by the Shah function, $III(x)$, we can write

$$III(x) = \sum_{n=-\infty}^{\infty} \delta(x - n\Delta x), \tag{2.15}$$

where n is an integer. The transform of $\text{Ш}(x)$ is another infinite set of δ-functions, that is, $\text{Ш}(\sigma) = \sum \delta(\sigma - n/\Delta x)$, where we have taken the liberty of using the same Ш symbol for the Shah function in both domains. As always, the spacing between the peaks in $\text{Ш}(\sigma)$ is inversely proportional to the spacing in $\text{Ш}(x)$.

Figure 2.4 summarizes some of the basic functions and their transforms now in our arsenal.

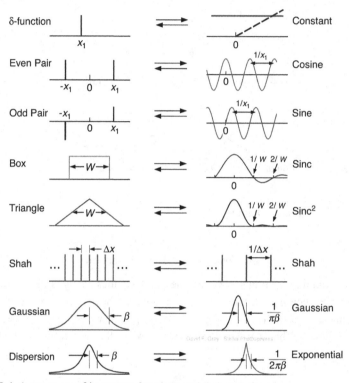

Figure 2.4 A summary of important functions and their transforms.

Data Sampling and Data Windows

An important application of the Shah function is to convert a continuous function to a sampled one. We commonly measure a spectral line profile by taking data points spaced in wavelength by $\Delta\lambda$. Call these data measurements $D(\lambda)$, and write

$$D(\lambda) = \text{Ш}(\lambda)F(\lambda), \tag{2.16}$$

where $F(\lambda)$ is the true spectrum and multiplication by $\text{Ш}(\lambda)$ gives the discrete sampling. We are quite accustomed to thinking of $D(\lambda)$ as the measured line profile. Figure 2.5 shows this process applied to a small section of a stellar spectrum. We can extend this concept and

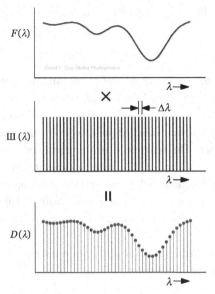

Figure 2.5 The spectrum $F(\lambda)$ is sampled at the spacing $\Delta \lambda$ by the Shah function, $III(\lambda)$, to give the data we record, $D(\lambda)$.

imagine *any* function to be synthesized from δ-functions. These δ-functions can be as closely packed as necessary for the job at hand, and in this context they have amplitudes modulated by the true function as in Equation 2.16.

In addition to sampling the data, we record them over a certain span or window. When we measure a stellar spectrum, our data extend from λ_1 to λ_2, the ends of our data window. Or when we measure a variable star from a start time t_1 to an end time t_2, this time interval forms a window. The window can be expressed in our data string by multiplication with a unit-amplitude box, $B(\lambda)$, or a series of boxes. With this in mind, our spectral data in Equation 2.16 become

$$D(\lambda) = B(\lambda)\ III(\lambda)\ F(\lambda). \tag{2.17}$$

It is $D(\lambda)$, sampled by $III(\lambda)$ and bounded by $B(x)$ that we record and compare to theory. Particularly relevant applications of these concepts arise in Chapters 3, 13, 16, 17, and 18.

Next comes one of the fundamental tools of Fourier transforms.

Convolutions

Multiplication of functions, as in Equation 2.17, clearly needs to be done. If we then ask about the Fourier transform of a product of functions, say $F(x)\ G(x)$, we find that it is a new process called a *convolution*, defined by

$$F(x)G(x) \rightarrow k(\sigma) = \int_{-\infty}^{\infty} f(\sigma_1)\ g(\sigma - \sigma_1)\ d\sigma_1, \tag{2.18}$$

where $f(\sigma)$ and $g(\sigma)$ are the transforms of $F(\lambda)$ and $G(\lambda)$. This integral is abbreviated by writing

$$k(\sigma) = f(\sigma) * g(\sigma),$$

the $*$ denoting convolution. The correspondence of multiplication in one Fourier domain with convolution in the other is called the *convolution theorem*.

It is relatively straightforward to look at Equation 2.18 and understand what convolution means. The function $k(\sigma)$ is to be evaluated point by point, the integration being done over σ_1 with σ being a constant. The function $g(\sigma_1)$ is translated by an amount σ relative to $f(\sigma_1)$, but because of the negative sign of σ_1 in the argument, $g(\sigma - \sigma_1)$ is reversed right for left. The product of these functions is formed, and the resulting net area is $k(\sigma)$ for that σ. Changing σ amounts to placing $g(\sigma - \sigma_1)$ at a new position. Tabulating $k(\sigma)$ then means sliding the reversed $g(\sigma)$ over $f(\sigma)$ and at each step recording the area under the product of the curves.

If we instead move a reversed $f(\sigma)$ over $g(\sigma)$, we obtain exactly the same result for $k(\sigma)$. The order of convolution does not matter,

$$f(\sigma) * g(\sigma) = g(\sigma) * f(\sigma). \tag{2.19}$$

Consider the simple case of the convolution of one unit amplitude box with another, as shown in Figure 2.6. This is a convenient example because the ordinates of each function are constant when they are not zero. When the boxes are separated, the product is zero. When the boxes overlap, the product is unity over the length of overlap. The length of overlap is proportional to the translation (A to B) until one of the boxes is contained within the other (B to C). Then the reduction of common area occurs as the boxes begin to separate (C to D), becoming zero beyond point D. With the exception of the δ-function discussed below, all convolutions span a wider range of the variable than either of the individual functions. The convolution is a blurring process.

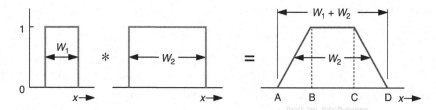

Figure 2.6 The convolution of two boxes is a trapezoid. As the narrower box slides over the wider one, the overlapping area increases linearly until the narrower box is contained within the wider one. This interval corresponds to the width W_1. Then for the dimension $W_2 - W_1$, the overlapping area is constant. Finally the narrower box emerges, mirroring the initial rise.

A special case of interest is when the widths of the boxes are the same, so we have the convolution of two identical boxes or the convolution of the box with itself. Since the unit width box, $B(x)$, transforms into sinc $\pi\sigma$, we deduce from the convolution theorem that the triangle $B(x) * B(x)$ transforms into sinc$^2 \pi\sigma$. Figure 2.7 illustrates this general pattern of analysis for the specific case just cited. There are two paths by which one can get from the function to the transform. One path is usually easier to visualize or to compute than the other.

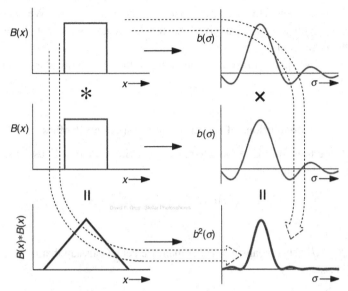

Figure 2.7 In the x domain on the left, the convolution of a box with itself gives the triangle at the bottom. The transform of the triangle is $b^2(\sigma) = \text{sinc}^2 \pi\sigma$. An alternative route is to take the transform of the box, $B(x) \longrightarrow b(\sigma) = \text{sinc } \pi\sigma$ which is then multiplied by itself, according to the convolution theorem, to give the $b^2(\sigma) = \text{sinc}^2 \pi\sigma$ transform at the bottom right

The power of the convolution theorem lies in the fact that visualizing the product of two functions is easier than visualizing the convolution (and it is also easier to compute). We profit from this when we analyze stellar spectra, for example, in Chapters 17 and 18.

Convolution with a Delta Function

If we re-run in our minds the convolution in Figure 2.6, but now with the narrower box being a δ-function, we see that the process simply maps out the second box. In other words, convolution of any function with a δ-function returns the shape of the function without modification. However, if the δ-function is not at zero, the function is shifted by that offset. Write the convolution, while keeping aware that x_1 is the variable here and x is a constant,

$$K(x) = \int_{-\infty}^{\infty} F(x_1)\, \delta(x - x_1)dx_1,$$

but the integrand is zero except when $x_1 = x$, so

$$K(x) = F(x - x_1) \int_{-\infty}^{\infty} \delta(x - x_1)\, dx_1$$

$$= F(x - x_1),$$

since the δ-function has unit area.

Alternatively, go through the transforms. In Equation 2.13, we saw that the transform of a δ-function that is not at the origin results in a phase term $e^{2\pi i x_1 \sigma}$. The product of transforms is then $f(\sigma)e^{2\pi i x_1 \sigma}$, corresponding back in the x domain to a translation of $F(x)$ to the position of the δ-function. This is a special case of Theorem 2 below. We again conclude that convolution with a δ-function gives the function back again, but displaced to the position of the δ-function.

At this point we jump to some convolutions that are particularly important for spectral lines.

Convolutions of Gaussian and Dispersion Profiles

Equation 2.7, the transform of a Gaussian, can be written for two Gaussians, $G_a(x)$ and $G_b(x)$, and multiplied together,

$$g_a(\sigma)g_b(\sigma) = e^{-\pi^2 \beta_a^2} e^{-\pi^2 \beta_b^2}$$

$$= e^{-\pi^2 \beta_c^2},$$

where $\beta_c^2 = \beta_a^2 + \beta_b^2$. That means in the x domain we get a Gaussian that is the convolution of the two Gaussians,

$$G_c(x) = \frac{1}{\beta_c \pi^{1/2}} e^{-x^2/\beta_c^2}$$

$$= G_a(x) * G_b(x). \tag{2.20}$$

The convolution of two Gaussians is a third Gaussian with $\beta_c^2 = \beta_a^2 + \beta_b^2$.

In an identical way, using Equation 2.9, one can show that the convolution of two dispersion profiles is a new dispersion profile with a half-half width given by $\beta_c = \beta_a + \beta_b$.

The convolution between a Gaussian and a dispersion function is called a Hjerting function or a Voigt function. If we let β_G and β_d represent the width parameters for Gaussian and dispersion profiles, then the transform of the Hjerting function is

$$h(\sigma) = e^{-\pi^2 \beta_G^2 \sigma^2} e^{-2\pi \beta_d |\sigma|}. \tag{2.21}$$

The Hjerting function itself is best evaluated numerically, as we do in Chapter 11.

The convolution of two (or more) Hjerting functions leads to a product of transforms of this same form. Since the exponents can be grouped together, it is trivial to show that the result is a new Hjerting function with $\beta_G^2 = \beta_{G1}^2 + \beta_{G2}^2$ and $\beta_d = \beta_{d1} + \beta_{d2}$. These relations are used repeatedly in Chapter 11.

Resolution: Our Blurred Data

All measuring devices imprint their signature on the observations we make with them. This signature is called the instrumental profile or the δ-function response, $I(x)$. To measure $I(x)$

we send through the instrument a δ-function; in practical terms, a spike that is narrow compared to the instrumental profile. For example, this process is applied to spectrographs in Chapter 12, where the spike is a narrow spectral line. We then look at what comes out of the instrument, namely, a blurred spike. We recognize this as a convolution of a δ-function with the instrumental profile, i.e., we have $I(x)$.

Next we consider a more general function that will be processed through the instrument. Think of representing this function with a set of δ-functions as we did in Equation 2.16 and Figure 2.5, but now each δ-function will be blurred by the instrument. Since the convolution process is linear (Theorem 1 below), the result will be the sum of all the blurred δ-functions, each entry in the sum being displaced to the position of the corresponding δ-function. This amounts to simply convolving the function with the instrumental profile.

Therefore, we expand Equation 2.17 to include this convolution and write

$$D(x) = B(x) \, Ш(x) \, F(x) * I(x), \qquad (2.22)$$

where $I(x)$ is the instrumental profile. So not only are our collected data "boxed" and sampled, they are also blurred.

In the Fourier domain, the convolution becomes a product. In most common physical cases, $I(x)$ is roughly like a Gaussian or a dispersion profile. In which case, multiplication implies stronger filtering, i.e., reduction in amplitude, toward higher Fourier frequencies. It is common for signals at high Fourier frequencies to be reduced to the noise level and consequently lost from the observations. In the x or λ domain, the high-frequency features have been "blurred" out by spatial integration across the dimensions of the instrumental profile.

Sampling and Aliasing

Now consider Equation 2.22 specifically with regard to the effect of the $Ш(x)$ sampling. The transform of Equation 2.22 is

$$d(\sigma) = b(\sigma) * Ш(\sigma) * f(\sigma) \times i(\sigma) \qquad (2.23)$$

where $i(\sigma)$ is the transform of $I(x)$. Assume for the present that $B(x)$ is very wide so that $b(\sigma)$ is essentially an impulse. In a like manner, take $i(\sigma)$ to be unity over any σ range we discuss. Then we are essentially back to Equation 2.16. The spacing between δ-functions in $Ш(\sigma)$ is $1/\Delta x$, where Δx is the data spacing in $D(x)$ or $Ш(x)$. The convolution of $f(\sigma)$ with $Ш(\sigma)$ causes $f(\sigma)$ to be reproduced over and over, as shown in Figure 2.8. We see now that if $f(\sigma)$ becomes zero for $\sigma < 0.5/\Delta x$, these replications are separated. But if not, they overlap and there is trouble. The trouble is called *aliasing*. The frequency

$$\sigma_N = \frac{1}{2\Delta x} \qquad (2.24)$$

is termed the cut-off or Nyquist frequency.

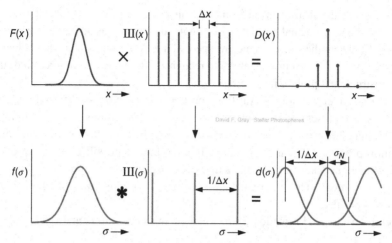

Figure 2.8 Here is a graphical illustration of the sampling of $F(x)$ and the multiplicity of $f(\sigma)$ it causes in $d(\sigma)$. In this example, Δx is too large, causing $d(\sigma)$ to show undesirable overlapping of adjacent peaks. This is aliasing.

Generally we measure $D(x)$ and compute $d(\sigma)$, so we do not know initially if $f(\sigma)$ has significant amplitudes at $\sigma \geq \sigma_N$. On the other hand, from the physics of the situation and the behavior of $f(\sigma)$ for $\sigma < \sigma_N$ we can often confidently expect that $f(\sigma)$ decreases to smaller and smaller values as σ moves away from the position of the peak. In some cases, multiplication with $i(\sigma)$ may force this to happen. For the situation where $f(\sigma)$ does become zero for $\sigma < \sigma_N$ we can manipulate the transform $d(\sigma)$ as needed and unambiguously reconstruct $F(x)$ by taking the inverse transform of $d(\sigma)$ for $-\sigma_N \leq \sigma \leq \sigma_N$.

For the unfortunate case where successive patterns do overlap, there is no longer a unique solution. Many different $F(x)$ functions can be concocted that have the same $d(\sigma)$. These functions differ in the way the overlapped portion is divided, as shown in Figure 2.9.

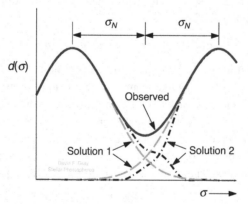

Figure 2.9 In the aliased situation, we obtain in $d(\sigma)$ only the vector sum of the overlapping replications, labeled Observed. Without additional information, it is impossible to separate the overlap in a unique way. Two arbitrary examples of separation are shown, labeled Solution 1 and Solution 2.

This of course is the origin of the term aliasing. The low frequencies of the peak to the right are disguised as high frequencies in the peak to the left, and so on. The only real solution for such a situation is to re-do the measurements with the data spacing, Δx, decreased sufficiently to spread out the replications of $f(\sigma)$.

Additional mixing of the $f(\sigma)$ patterns can occasionally result from the convolution of $b(\sigma)$ in Equation 2.23. In unfavorable circumstances, $b(\sigma)$ may be sufficiently wide to result in aliasing that would otherwise not exist. The extended sidelobes of $b(\sigma) = \text{sinc } \pi\sigma$ can accentuate this problem, and for this reason it may be necessary to modify the ends of $B(x)$ by bringing them smoothly to zero. The smooth wings may still be significant, and so to make certain $b(\sigma) * f(\sigma)$ is no larger than needed, there should be no large signals in $f(\sigma)$ that are unnecessary. Subtraction of the mean value of $F(x)$, for example, will often remove a large spike at $\sigma = 0$. Application of these principles is wide ranging, but is especially important in Chapter 12, where we deal with measuring spectral lines.

Useful Theorems

Physical insight as well as computational saving can be gained by application of theorems. Most of these can be proved by reverting to the defining integrals, Equations 2.1 and 2.2. In all cases, $f(\sigma)$ is the transform of $F(x)$ and the arrow means "transforms to."

***Theorem 1**. Fourier transforms are linear.*
If
$$F(x) = F_1(x) + F_2(x) + \cdots,$$
then
$$f(\sigma) = f_1(\sigma) + f_2(\sigma) + \cdots. \tag{2.25}$$

***Theorem 2**. Translation in one domain gives a phase term in the opposite domain.*
$$F(x - x_1) \longrightarrow e^{2\pi i x_1 \sigma} f(\sigma). \tag{2.26}$$

***Theorem 3**. Scaling the argument in one domain leads to inverse scaling in the other domain of both the argument and the amplitude.*
$$F(ax) \longrightarrow \left|\frac{1}{a}\right| f(\sigma/a). \tag{2.27}$$

***Theorem 4**. Differentiation in one domain is equivalent to multiplication by $2\pi i\sigma$ in the other domain.*
$$\frac{dF(x)}{dx} \longrightarrow 2\pi i\sigma f(\sigma). \tag{2.28}$$
Integration is the opposite of this.

***Theorem 5**. A product of functions in one domain corresponds to a convolution in the other domain.*
$$F(x)G(x) \longrightarrow f(\sigma) * g(\sigma). \tag{2.29}$$

Theorem 6. *Derivatives of convolutions involve differentiation of only one or the other of the convolved functions, but not both.*
If

$$K(x) = F(x) * G(x)$$

then

$$\frac{dK(x)}{dx} = \frac{F(x)}{dx} * G(x) = F(x) * \frac{dG(x)}{dx}. \tag{2.30}$$

Theorem 7. *If $G(x)$ is normalized to unit area, then the convolution $K(x) = F(x) * G(x)$ has the same area as $F(x)$, that is,*

$$\int_{-\infty}^{\infty} K(x)\, dx = \int_{-\infty}^{\infty} F(x)\, dx. \tag{2.31}$$

Theorem 8. *If we call the centroid of a function x', where*

$$x' = \int_{-\infty}^{\infty} x\, F(x)\, dx \Big/ \int_{-\infty}^{\infty} F(x)\, dx,$$

then the centroid of a convolution $K(x) = F(x) * G(x)$ is the sum of the centroids of the two functions being convolved, that is,

$$x'_K = x'_F + x'_G. \tag{2.32}$$

Theorem 9. *Rayleigh's theorem relates the amplitudes of the function and its transform. The integral of the modulus squared or power is the same in both domains,*

$$\int_{-\infty}^{\infty} |F(x)|^2\, dx = \int_{-\infty}^{\infty} |f(\sigma)|^2\, d\sigma. \tag{2.33}$$

Numerical Calculation of Transforms

While transforms of analytic functions are useful, particularly for thinking through a physical situation, numerical calculations are often needed when processing data. So let us turn to the numbers, with an eye toward spectroscopic data. It is possible to use quadrature formulae on Equations 2.1 and 2.2 directly, and such a procedure may be useful in some applications, especially when the data points are spaced non-uniformly along the abscissa, but in most spectroscopic cases we use the fast Fourier transform (FFT). A mild revolution occurred with the invention of the fast Fourier transform, also called the Cooley–Tukey algorithm (Cooley & Tukey 1965). This method of calculating transforms depends on a clever method of nesting sums and avoiding redundancy in calculating sines and cosines that the direct integration does not. The gain in efficiency over direct integration is phenomenal for lengthy calculations. Bell (1972) showed that for N points the number of calculations for the classical method varies as $2N^2$, while the fast Fourier transform equivalent is $3N \log_2 N$. The ratio is $0.46N/\ln N$. Table 2.1 gives examples of this gain.

Table 2.1 *Numbers of Fourier transform computations*

N	Non-FFT	FFT	Gain
128	32 768	2 689	12
256	131 072	6 147	21
512	524 288	13 830	38
1 024	2 097 152	30 734	68
2 048	8 388 608	67 614	124
4 096	33 554 432	147 521	227
8 192	134 217 728	319 629	420
16 384	563 870 912	688 432	780
32 768	2 147 483 648	1 475 212	1456

The values of N for the fast Fourier transform are required to be a power of 2. This requirement is easy to meet for any string of data by simply extending it to the next larger power of 2 using zeros or some other suitable constant. The second requirement of the fast Fourier transform is that all data points be equally spaced along the abscissa. The nesting procedure does not work without even spacing. Many types of data are directly recorded at equal steps. In other cases, treating unequally spaced data as if they were equally spaced results in negligible error.

We can write our data, bearing in mind Equation 2.16, as

$$D(x_j) = D(j\Delta x) = D(j),$$

in which Δx is the step and j is an integer index running from 1 to N for the N data points. The Fourier integral can then be written as the sum

$$d(\sigma) = \sum_{-\infty}^{\infty} D(x_j)\, e^{2\pi i\, x_j\, \sigma}\, \Delta x$$

$$= \sum_{1}^{N} D(j)\, e^{2\pi i\, j\Delta x\, \sigma}\, \Delta x.$$

We can also write $\sigma = k\Delta\sigma$, where $\Delta\sigma$ is the spacing of points in $d(\sigma)$ and k is an integer. Then

$$d(\sigma) = d(k) = \sum_{1}^{N} D(j)\, e^{2\pi i j k \Delta x \Delta \sigma}\Delta x. \tag{2.34}$$

Further, the Nyquist frequency, $\sigma_N = 0.5/\Delta x$, is the highest frequency we can hope to work to. It is usually assumed that if there are N points defining $D(x)$ then N values of $d(\sigma)$ will be computed. Since $d(\sigma)$ will be computed equally far on both sides of $\sigma = 0$, the computations extend to $\pm N/2$ points away from zero, and the step in σ must then be

$$\Delta\sigma = \frac{\sigma_N}{N/2} = \frac{1}{\Delta x N}$$

or

$$\Delta\sigma\, \Delta x = 1/N. \tag{2.35}$$

With this substitution, Equation 2.34 becomes

$$d(k) = \sum_1^N D(j)\, e^{2\pi i j k/N} \Delta x.$$

$$\tag{2.36}$$

It is a summation of this form that is evaluated in the fast Fourier transform techniques, and most versions also assume $\Delta x = 1$ in this expression. The inverse transform corresponding to Equation 2.36 is

$$D(j) = \sum_1^N d(k)\, e^{-2\pi i j k/N} \Delta\sigma.$$

$$\tag{2.37}$$

The same algorithm can be used to evaluate this sum, but $\Delta\sigma$ must be explicitly taken into account using Equation 2.35. This is equivalent to renormalization,

$$D(j) = \frac{1}{\Delta x N} \sum_1^N d(k)\, e^{-2\pi i j k/N}.$$

$$\tag{2.38}$$

Program code for the fast Fourier transform is given in several languages including Python, C, and FORTRAN at https://rosettacode.org/wiki/Fast_Fourier_transform.

Noise Transfer between Domains

Suppose we measure a function, $F(x)$, and take its transform. The measurements will contain noise, and we ask how the noise appears in the transform. Let $E(x)$ be the error or noise in these measurements, and write

$$F(x) = F_0(x) + E(x),$$

where $F_0(x)$ is the error-free function. Because of the linearity (Theorem 1) we can write

$$f(\sigma) = f_0(\sigma) + e(\sigma),$$

where $e(\sigma)$ is the transform of $E(x)$.

Theorem 8 tells us that

$$\int_{-\infty}^{\infty} E^2(x)\, dx = \int_{-\infty}^{\infty} e^2(\sigma)\, d\sigma,$$

but our measurements extend over a finite range in x, call it L, and our transform is unique only over a finite range in σ, say $\pm l/2$, due to the data sampling spacing and the attendant aliases. Therefore we can rewrite Theorem 8 as

$$\int_0^L E^2(x)\, dx = \int_{-l/2}^{l/2} e^2(\sigma)\, d\sigma.$$

$$\tag{2.39}$$

If we denote the average by angle brackets, the relation can also be written

$$\langle E^2 \rangle L = \langle e^2 \rangle l.$$

$$\tag{2.40}$$

This is a fundamental relation that says the mean square error times the length of the data is the same in both domains.

More specifically, the data we transform are usually an array of numbers. Let us also assume the fast Fourier transform is used. Then with N equally spaced data points Δx apart, Equation 2.39 takes the form

$$\sum_1^N E^2(x_j)\, \Delta x = \sum_1^N e^2(\sigma_j)\, \Delta \sigma.$$

In place of the mean square errors, define the standard deviations

$$S_x = \left[\sum_1^N \frac{E^2(x_j)}{N-1} \right]^{1/2},$$

$$S_\sigma = \left[\sum_1^N \frac{e^2(\sigma_j)}{N-1} \right]^{1/2}.$$

The corresponding form for Equation 2.40 is

$$S_x^2 L = S_\sigma^2 l, \tag{2.41}$$

or with $L = \Delta x N$ and $0.5 l = \sigma_N = 0.5/\Delta x$,

$$S_\sigma = S_x (L/l)^{1/2} = S_x (\Delta x L)^{1/2} = S_x \Delta x N^{1/2}. \tag{2.42}$$

As might be expected, the rms noise in the two domains is proportional. If the measured span L is constant but more finely divided by making the sampling interval Δx smaller, S_σ is reduced because the noise is spread over a larger span in σ.

The common practice of padding the true data array to reach an even power of 2 should be taken into account when using Equation 2.42. The added values carry no noise and should not be included in N. It also follows that the noise contribution from a spectral line-profile measurement can be minimized by restricting the number of points to only those defining the profile itself. In some cases, smooth continuum points can be replaced with their mean value.

When the noise $E(x)$ is "white," meaning independent of frequency, the noise is statistically constant with σ. We have seen how it is important to choose Δx small enough so that the whole transform is sensibly within $\sigma_N = 0.5/\Delta x$ (no aliasing). If σ_N is made still larger by oversampling the data, the amplitudes at the high frequencies are pure noise. In other words it is possible to see the noise level in the σ domain (see Figure 12.9, for example), and in some cases it is advantageous to calculate the noise in the original data by working back through Equation 2.42 to get S_x from S_σ. Suppose, for instance, a spectral-line profile is defined by 16 points separated by $\Delta x = \Delta \lambda = 0.04\,\text{Å}$. A transform might be computed using 64 points, 49 of which are continuum values of unity. Assuming the oversampling is sufficient to allow the noise level in the σ domain to be seen at, say, $S_\sigma = 10^{-3}\,\text{Å}$, we then have $S_\lambda = S_\sigma/(\Delta \lambda\, N^{1/2}) = 0.6\%$ error in the original line profile data. We do a similar thing, perhaps only semiconsciously, directly in the λ domain. When $\Delta \lambda$ is small, that is, well below the dimension associated with signal variations, we expect adjacent points should be strongly correlated. Differences in value are then due to noise. We assume white noise and estimate the noise level from the "grass," which is the high-frequency noise.

One other application of transforms should also be mentioned, namely time series.

Time Series Analysis

The existence of periodic variations can often tell us important information about a physical process. For example, spots on the surface of a star carried across the hemisphere facing us by the star's rotation will introduce modulation in the shapes of spectral lines and the brightness of the star. A time-series analysis will give us the period of rotation, and in many cases specific information about the size and location of the spots. The parameter we measure might be an asymmetry in a line profile, a temperature parameter like the ratio of the depths of two spectral lines, a variation in radial velocity, and so on.

In the cleanest case, where the variation can be mapped more-or-less continuously, there will be no ambiguity, and one can establish the pattern of variability by simple inspection of the observations. Choose the first cycle as the reference and compare successive cycles with it by translating each in the time coordinate by Δt_j until a match with the first cycle is obtained. Then from each one we compute a period, $P_j = \Delta t_j/n_j$, where n_j is the number of cycles occurring in the time span Δt_j. Let δt be the error associated with matching each subsequent cycle with the first cycle. Then the error in the period is $\delta P_j = \delta t/n_j$. In other words, we know the period better in proportion to the number of cycles elapsed. This is a very general and powerful result: precise periods come from long time series covering many cycles.

The best estimate of the true period, P, is the weighted average of the N values of P_j. Since each P_j improves in proportion to the number of cycles involved in its calculation, the weights are the squares of the number of cycles, thus

$$P = \sum_1^N n_j^2 P_j \bigg/ \sum_1^N n_j^2 = \sum_1^N n_j \Delta t_j \bigg/ \sum_1^N n_j^2. \tag{2.43}$$

However, in many cases the periodicity of a variation is not immediately obvious. In such cases, a Fourier analysis can be of value. The Fourier transform of a time series is called a *periodogram*. Suppose we have a simple case with one frequency, as shown in Figure 2.10. This oscillation is observed at certain times and over time windows leading to the observations shown in the upper left panel of Figure 2.10. In the time domain we can express these observations, $D(t)$, as

$$D(t) = F(t) \times W(t), \tag{2.44}$$

in which the original signal $F(t)$ is sampled according to the window $W(t)$. Basically $W(t)$ is a statement of the times of observation, which can be irregular owing to weather, availability of equipment, assignment of telescope time, and so on. Naturally, in the Fourier domain, the product becomes a convolution, and the window function, $w(\sigma)$, the transform of $W(t)$, is translated to the positions of the δ-functions representing the sinusoidal oscillation. The two window functions interfere with each other where they overlap; their vector components (real and imaginary) are added. Unless we have extremely thorough data sampling, we should not expect a single peak in the periodogram, but rather we look for the central location of the window pattern (see Gray & Desikachary 1973).

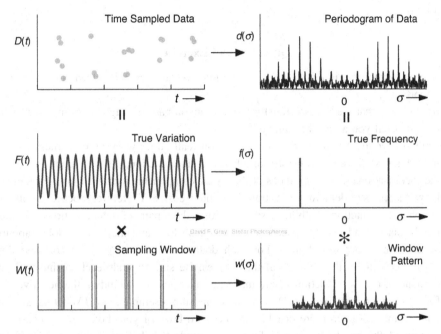

Figure 2.10 Astronomical data are often sparsely sampled and typically give a time series like the one shown in the upper left panel as $D(t)$. This can be thought of as the product of the true variation, the cosine wave $F(t)$, with δ-functions at the times of sampling, $W(t)$, as illustrated in the two lower left panels. In the Fourier domain on the right, the time sampling produces a window pattern (bottom) that is convolved with the two δ-functions corresponding to the cosine wave. This explains the complicated shape of the periodogram.

If the oscillation is undersampled, as in Figure 2.10, the window pattern will have several strong peaks. With poorer sampling, the peaks are closer together and have similar amplitudes. The first peak on either side of the central peak is the first alias, the second peak, the second alias, and so on. Sometimes, if the true frequency is small compared to the frequency spread of the window pattern, the negative-frequency pattern will spill over into positive frequencies and the two patterns will interfere, making analysis a bit more of a challenge. When the physical time signal has a more complex shape than a single sinusoidal, there will be harmonics, each producing its own window pattern. Still more complicated is the case where there are several frequencies actually occurring in the physical measurements, adding more window patterns. These window patterns combine as vectors. In other words, the periodogram is the vector sum of all the window patterns.

Some investigators prefer to use power rather than amplitude in the Fourier domain, which is the square of the amplitude. The reality of a signal can be judged by the visibility of the window pattern compared to the noise. A formal "false-alarm probability," given by e^{-p}, where p is the power at the frequency of interest, has been used in some cases (e.g., Horne & Baliunas 1986).

Now that we have Fourier transforms in hand, we can apply them to spectrographs, our next topic.

Questions and Exercises

1. Fourier transformation is a linear operation. What does this mean and why is it important?
2. If you know that the Fourier transform of a function has no imaginary component, what does that tell you about the function?
3. Convince yourself that Fourier decomposition really does work by constructing a box from cosines. Since the transform of a box is a sinc function, the sinc function must be the specification of the amplitudes of the cosine components forming a box. Further, we know that pairs of delta functions transform into a cosine. Therefore, proceed as follows. Graph a sinc function. Within the sinc envelope, draw pairs of delta functions, symmetrically placed and evenly spaced across the main lobe and the first sidelobe (about a dozen pairs is a good choice). For each delta-function pair, plot the corresponding cosine and finally the sum of all the cosines. Be sure to scale each cosine with the ordinate of the sinc function where the delta functions are, including the negative sign for those in the first sidelobe. Does your cosine sum resemble a box? What needs to be done to make it even more box-like? Is the zero level of your box correct, i.e., at the bottom of the box? Or did you forget to include the delta-function "pair" with no separation at the zero position of the sinc function? Is the area of your box equal to the zero ordinate of your sinc function? Did you remember that your cosine sum is an integration and is to be multiplied by an abscissa increment?
4. It is easy to see that the convolution of two boxes is a triangle. (If you haven't worked this through yet, do so first.) What does the convolution of two identical isosceles triangles look like? What is the transform of this convolution?
5. Write a short computer program using Equation 2.36 so that you can take numerical Fourier transforms. Test your program on boxes and triangles, and Gaussians, for example. Try computing a transform to frequencies two or three time higher than the Nyquist frequency given in Equation 2.24. What do you see?
6. Select two or three line profiles from the Arcturus spectrum in the Hinkle *et al.* (2000) atlas or some other convenient source. Set the continuum to zero by subtracting unity and throwing away the minus sign. Compute the transforms out to the Nyquist frequency. Does the zero ordinate of your transform correctly give the total area of the line profile (called the equivalent width)? If not, check that Δx in Equation 2.36 has been included. Can you see the noise level at the higher frequencies? Is it consistent with Equation 2.42?

References

Bell, R.J. 1972. *Introductory Fourier Transform Spectroscopy* (New York: Academic Press).

Bracewell, R. 1965. *The Fourier Transform and Its Applications* (New York: McGraw-Hill).

Cooley, J.W. & Tukey, J.W. 1965. *Math Computation* **19**, 297.

Gaskill, J.D. 1978. *Linear Systems, Fourier Transforms, and Optics* (New York: Wiley).

Gray, D.F. & Desikachary, K. 1973. *ApJ* **181**, 523.

Hinkle, K., Wallace, L., Valenti, J., & Harmer, D. 2000. *Visible and Near Infrared Atlas of the Arcturus Spectrum 3727–9300 Å* (San Francisco: ASP).

Horne, J.H. & Baliunas, S.L. 1986. *ApJ* **302**, 757.

Jennison, R.C. 1961. *Fourier Transforms and Convolutions for the Experimentalist* (New York: Pergamon).

Kharkevich, A.A. 1960. *Spectra and Analysis* (New York: Consultants Bureau).

3

Spectroscopic Tools

Observations of stellar photospheres require instruments for collecting and analyzing light. Chief among these is the spectrograph, that ingenious device that spreads light out according to color. Low-resolution spectrographs are used to measure the broad continuous spectrum, while high-resolution spectrographs are needed to measure the fine details of spectral lines. Virtually everything we know about stellar photospheres comes from spectroscopic data. So how do these devices work? What in them makes starlight break up into rainbow colors? What distinguishes a low-resolution from a high-resolution spectrograph? What parameters control spectrograph capabilities and performance? Let us find out.

Spectrographs: Some General Relations

The heart of a spectrograph is the dispersing element. There are two devices used to disperse light, i.e., geometrically spread light out by wavelength. The first, dating back to Newton's famous discovery (Newton 1704), is the prism. The second is the diffraction grating, invented in about 1785 (Hopkinson & Rittenhouse 1786). But to make these dispersers useful, the rest of the spectrograph is also needed. The invention of the spectroscope is generally attributed to Fraunhofer (1788–1826). The basic astronomical spectrograph is shown in Figure 3.1. It consists of an entrance slit upon which the stellar image formed by the telescope is placed, a collimator mirror that intercepts and collimates the divergent telescope light beam, a dispersing element (the prism or diffraction grating), and a camera that focuses the dispersed light onto a detector. Since the purpose of the collimator is to make the divergent beam parallel, the distance between the slit and the collimator is the focal length of the collimator, f_{coll}. Similarly, the distance between the camera and the focused spectrum is the focal length of the camera, f_{cam}. The distances between the collimator, disperser, and camera affect the detailed design and optimization of the spectrograph, but do not matter for the basics we are considering at the moment. We concentrate on plane diffraction gratings as the dispersing element since these are almost universally used in astronomical spectrographs.

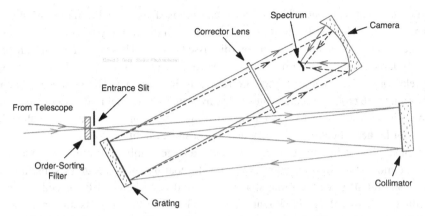

Figure 3.1 This is the optical layout for a typical spectrograph. The stellar seeing disk formed by the telescope is focused on the entrance slit. The beam diverges inside the spectrograph until it reaches the collimator mirror and is collimated. The collimated light is returned to the grating and from it monochromatic collimated beams are sent to the camera. Only two of these beams are indicated by the solid and dashed lines. The spectrum is focused on the curved surface in this type of camera. The camera has a wide field of view, and needs a corrector lens to attain sharp focus across the full field. The filter in front of the entrance slit restricts the light to one of the many spectra produced by the grating.

In *monochromatic* light, a single image of the entrance slit is formed at the focus of the camera. This image is inverted and the magnification along the direction of the entrance slit (perpendicular to the dispersion) is the ratio of camera to collimator focal lengths, f_{cam}/f_{coll}. The magnification of the width of the entrance slit depends not only on the ratio of focal lengths, but also on the orientation of the grating, and this is discussed later in Equation 3.14. When *polychromatic* light is analyzed with the same instrument, the spectrum is composed of a continuum of monochromatic images of the entrance slit spread out across the focal surface of the camera. The smallest detail we can see in the stellar spectrum is set primarily by the width of these monochromatic images. The term "chromatic resolution," or "spectral resolution," or simply "resolution," is used to describe how well the colors are separated. Generally the higher the resolution, the happier the astronomer, and yet, higher resolution implies fewer photons per detector pixel, forcing the fainter stars to slip off our observing list. A quantitative discussion requires a measurement of the monochromatic images, which you probably already recognize as the instrumental profile, $I(\lambda)$, produced by the spectrograph. A detailed discussion of $I(\lambda)$ is given in Chapter 12.

With the general spectrograph layout in mind, let us turn our attention to the plane diffraction grating of the type used in astronomical spectrographs.

Diffraction Gratings

Plane diffraction gratings have proved to be the most popular dispersing element for a number of reasons. First, they have high efficiency, up to ~80%, an important factor when dealing with faint starlight. Plane gratings can be used over a wide range of angles of incidence, making it possible to direct the dispersed light into any of several cameras to conveniently change the linear dispersion (discussed below). These gratings can be made in relatively large formats, ~400 mm, that are needed to make spectrographs efficient with large telescopes. And plane gratings are relatively inexpensive. The commonly encountered gratings are replica gratings constructed by using the original ruled grating as a mold. A replica is far less expensive than an original master grating, and usually has the added bonuses of higher reflection efficiency and less scattered light.

A typical plane reflection grating consists of a large number of parallel grooves ruled with a diamond tool into an aluminum or gold coating on a flat glass substrate. The grooves should all have the same shape, be smooth, and be equally spaced. A greatly magnified view of rulings is shown in Figure 3.2. The best results so far have been obtained on ruling engines that are servo-controlled by interferometer monitors to maintain the strict positional requirement, like the one shown in Figure 3.3. Large gratings are especially important for high-resolution spectrographs, but wear on the diamond ruling tool limits the maximum ruled area. Harrison (1973) and Mi *et al.* (2019) discuss ruling engines and grating quality. A summary from the manufacturing point of view is given by Palmer (2020).

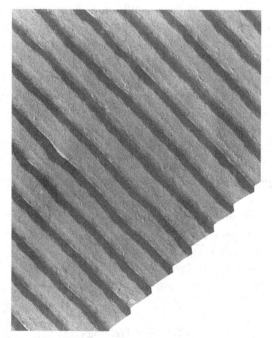

Figure 3.2 This picture through a microscope shows the rulings on a plane reflection grating having 1180 lines per mm. Courtesy of Jarrell-Ash.

Figure 3.3 This ruling engine has its ruling diamond poised above the corner of the grating blank. Part of the interferometer control system can be seen at the lower left. Courtesy of Bausch and Lomb, Analytical Systems Division, Rochester, New York.

How a grating works is most easily understood by first considering the simple *transmission* grating illuminated in collimated monochromatic light as in Figure 3.4. (We will presently convert our results to the reflection gratings actually used in stellar spectrographs.) The slit width is denoted by b, the slit spacing by d. The angle of incidence, α, is measured from the grating normal, as shown in the figure. The diffraction angle, β, we choose to have the same sign as α when it is on the same side of the grating normal, as in Figure 3.4. Because the rulings are so long compared to their width, we can justifiably consider the grating as a one-dimensional object. Further, let us restrict the angles α and β to the plane of the page in Figure 3.4.

The incident plane scalar wave (for a more rigorous vector wave treatment, see Stroke 1967) can be written

$$F(x) = F_0 \, e^{2\pi i \, (x \sin \alpha)/\lambda}, \tag{3.1}$$

in which $x \sin \alpha$ is a linear dimension along the direction of the wave. Time variations in the light wave occur on scales of $\sim 10^{-15}$ s and are completely undetectable during astronomical measurements. Therefore, F_0 is taken to be a constant. Describe the transmission of the grating by

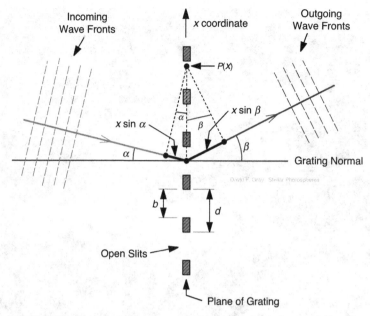

Figure 3.4 The plane transmission grating with slit spacing d and slit width b lies in the plane perpendicular to the page. The linear coordinate x runs across the rulings as shown. Light is incident at an angle α. The angle of diffraction is β. Both of these angles have the same sign as drawn. The dashed-line triangles show the path differences relative to the point labeled $P(x)$.

$$G(x) = 1 \quad \text{for } x \text{ values within the slits}$$
$$= 0 \quad \text{for all other } x \text{ values.}$$

It is the product of $F(x)$ with $G(x)$ that emerges from the back of the grating. The diffracted wave $g(\beta)$, going in any arbitrary direction β, will be the sum of contributions with their proper phase shifts, from all portions of the grating. In other words, we integrate the contributions along the x coordinate,

$$g(\beta) = \int_{-\infty}^{\infty} F(x)\, G(x)\, e^{2\pi i\,(x\sin\beta)/\lambda} dx,$$

where $2\pi(x\sin\beta)/\lambda$ is the phase difference along the grating for the out-going direction β. Putting in $F(x)$ from Equation 3.1, we have

$$g(\beta) = F_0 \int_{-\infty}^{\infty} G(x)\, e^{2\pi i\,(x\sin\alpha + x\sin\beta)/\lambda} dx,$$

and if we choose to measure x in units of λ, while denoting $\sin\alpha + \sin\beta$ by θ, we have

$$g(\theta) = F_0 \int_{-\infty}^{\infty} G(x)\, e^{2\pi i x\theta} dx. \tag{3.2}$$

We now recognize Equation 3.2 to be Equation 2.1, i.e., the light leaving the grating is the Fourier transform of the grating transmission function, $G(x)$.

The description of $G(x)$ in terms of the Shah function and boxes is

$$G(x) = B_1(x) * \text{Ш}(x)\, B_2(x), \tag{3.3}$$

where as usual the star means convolution. The box $B_1(x)$ represents the transmission through a single slit, as in Figure 3.5, and so has a width b. The box $B_2(x)$ matches the total width of the grating, which we call W. The spacing of the δ-functions in $\text{Ш}(x)$ is d. The grating diffraction pattern is shown on the upper right of Figure 3.5. It is immediately clear that the different orders, n, arise from the many rulings (Shah function), the envelope comes from the individual slit width, b, and the full width of the grating, W, sets the width of the individual interference maxima. The amplitude diffraction pattern of the grating follows from Equations 3.2 and 3.3,

$$g(\theta) = \text{Ш}(\theta) * \frac{W \sin \pi\theta W}{\pi\theta W} \times \frac{b \sin \pi\theta b}{\pi\theta b},$$

where $\text{Ш}(\theta) = \sum \delta(\theta - n/d)$. The convolution with a δ-function produces a translation to the position of the δ-function, so

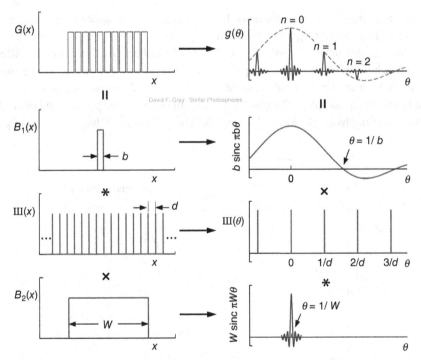

Figure 3.5 The light distribution for a diffraction grating is shown on the upper left as $G(x)$. It is composed of the convolution of the box $B_1(x)$ of width b with $\text{Ш}(x)$ of spacing d multiplied by $B_2(x)$ representing the total width W of the grating. The right side shows the transforms. The final amplitude diffraction pattern, $g(\theta)$, is seen to be the convolution of sinc $\pi\theta W$ with $\text{Ш}(\theta)$, which is multiplied by sinc $\pi\theta b$. The light strength in the diffraction pattern of the grating is proportional to $g^2(\theta)$.

$$g(\theta) = \sum_n \frac{W \sin \pi(\theta - n/d)W}{\pi(\theta - n/d)W} \times \frac{b \sin \pi\theta b}{\pi\theta b}. \qquad (3.4)$$

The square of this wave amplitude is proportional to the light intensity in the usual way for wave optics.

In most applications we need the grating pattern as a function of β, but since $\sin \alpha + \sin \beta = \theta$, we have β, namely $\beta = \sin^{-1}(\theta - \sin \alpha)$. Since α is fixed in any given case, we also know that $d\theta = \cos \beta \, d\beta$. So when β is small, as when gratings are used in low orders, $d\theta$ and $d\beta$ are nearly the same.

Now consider the width of the interference maxima. The zeros are equally spaced in θ from the center of the maximum by $\Delta\theta = 1/W$ or $\cos \beta \Delta\beta = \lambda/W$. The grating (not the spectrograph) resolution is therefore proportional to its overall width, W.

The large maxima in $g(\theta)$ occur at the δ-functions, when $\theta - n/d$ is zero, i.e., $\theta = n/d$. That means each maximum corresponds to a value of the order integer n. If we revert back to not expressing lengths in wavelength units, this can be written

$$\frac{n\lambda}{d} = \theta = \sin \alpha + \sin \beta, \qquad (3.5)$$

where d and λ must be in the same units. Equation 3.5 is so fundamental in working with gratings that it is called the *grating equation*.

For gratings illuminated in *polychromatic* light, the collection of maxima for different wavelengths having the same n make up the nth-order spectrum. The zero-order spectrum is the white light or undispersed order and falls in the $\beta = -\alpha$ direction, i.e., straight through transmission (or the specular reflection for a reflection grating). The position and orientation for two wavelengths is shown in Figure 3.6. The larger the value of n, the larger β

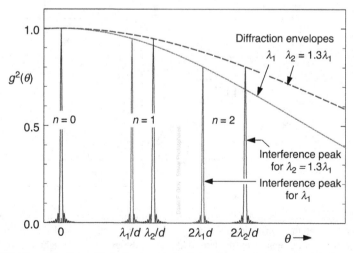

Figure 3.6 The collection of interference peaks for a given order, n, make up the spectrum of that order. Zero order has no dispersion. Dispersion in second order is twice that of first order. The heights of the interference peaks are set by the diffraction envelopes, as shown.

becomes for a given λ, although naturally β is restricted to range between $\pm 90°$. Polychromatic light is spread out in θ proportional to the order n, and each spectrum is seen successively farther from the zero-order position. Unfortunately there is an overlap of wavelengths at a given θ for all combinations of $n\lambda$ having the same value – a result following trivially from the grating equation. Fortunately it is easy to place a filter in the light beam to remove light from unwanted orders. This filter is best placed before the entrance slit of the spectrograph (as in Figure 3.1).

The angular dispersion seen in Figure 3.6 follows from Equation 3.5,

$$\frac{d\theta}{d\lambda} = \frac{n}{d} \quad \text{or} \quad \frac{d\beta}{d\lambda} = \frac{n}{d \cos \beta}. \tag{3.6}$$

When $\beta \sim 0$, the angular dispersion is nearly constant with wavelength for a given order. Equation 3.6 shows that the angular dispersion can be increased by observing in a higher order or by selecting a grating having smaller spacing between rulings. Gratings are made with 1000 rulings/mm or more, and values down to 300 rulings/mm are typical of grating used in low orders, e.g., $n = 1$, 2, or 3. Gratings used in much higher orders, e.g., $n \sim 100$, are called echelles. Their ruling densities are usually lower, the 79 rulings/mm being a popular choice.

If we take the angular dispersion from Equation 3.6 and write $\Delta\lambda = \Delta\theta \, d/n$ and combine it with the angular resolution for the grating, $\Delta\theta = \lambda/W$, we find the chromatic resolution,

$$\Delta\lambda = \frac{\lambda}{W}\frac{d}{n}, \tag{3.7a}$$

and its close relation the resolving power,

$$\frac{\lambda}{\Delta\lambda} = n\frac{W}{d}. \tag{3.7b}$$

The resolving power of the grating is independent of wavelength. Values as large as 10^6 have been attained. As we shall see later in this chapter, the chromatic resolution and the resolving power of the whole spectrograph are much poorer than for the grating itself.

Although the slit diffraction grating is a very good invention, we can improve it with "blazing."

The Blazed Reflection Grating

One of the objectionable features of the simple transmission grating is that the diffraction envelope has its largest maximum at $\theta = 0$, where the chromatic dispersion is zero. A large part of the light is lost to this zero order; bad news when looking at faint sources like stars. Further, as we try to use the higher orders to gain more resolution, we find the light intensity continues to decline, following the decline in the diffraction envelope, reaching uncomfortably low levels. To solve this problem, the diffraction envelope must be shifted relative to the interference pattern to those θ values where we wish to work. The Fourier shift theorem (Chapter 2, Theorem 2) tells us that we can accomplish this by introducing a phase term in $B_1(x)$ of Equation 3.3. When this is done the grating is said to be blazed.

As a thought experiment, we could blaze a transmission grating by placing a small prism in each slit, as shown in Figure 3.7. The beam would be deviated through the angle γ, thus shifting the diffraction envelope associated with $B_1(x)$, but of course the interference pattern from $\text{III}(x)$ is not changed because the slits would still have the same separation and would still be oriented along the same plane.

Our fortunes take an upswing if we turn to reflection gratings. A reflecting surface can be manipulated to give the needed phase shifts by simply replacing the slits and prisms of the transmission grating with tilted "mirrors," i.e., the rulings, the tilt giving the phase shift. Another advantage also occurs at this point: the opaque portion between the slits is no longer needed in order to distinguish one ruling from the next, leading to increased optical efficiency. The grating cross section then makes the conceptual transition shown in Figure 3.8. The individual reflecting surfaces are called "facets." The tilt or facet angle, ϕ, is set by the manufacturer according to the shape and orientation of the diamond in the ruling engine. The front and back sides of the facets are usually at right angles, making $b = d \cos \phi$. This has the effect of making the slits slightly narrower than the limiting case where $b = d$, and the diffraction envelope is correspondingly wider.

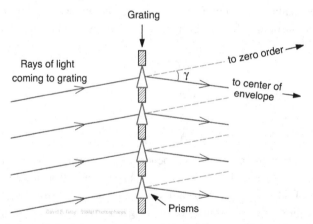

Figure 3.7 A hypothetical transmission grating is shown with a prism in each slit. A monochromatic incident wave experiences an angular deviation of γ. The zero-order image is no longer at the center of the diffraction envelope.

The equations derived for the transmission grating hold for the reflection grating. The angles α and β are both positive when they are on the same side of the grating normal as the facet normal.

The distribution of light in the spectrum produced by a blazed grating for a white-light source can be derived in the following way. Referring back to Equation 3.4 and Figure 3.5, we see that the diffraction *envelope* of $g(\theta)$ is given by

$$b \frac{\sin \pi\theta b}{\pi\theta b} = b \operatorname{sinc} \pi b(\sin \alpha + \sin \beta).$$

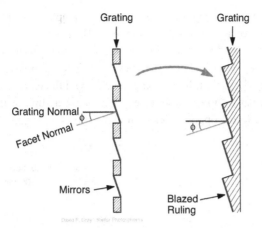

Figure 3.8 On the left is a hypothetical grating with slits replaced by mirrors. The inter-slit spacing is conceptually eliminated in the more realistic grating on the right. The facet normal makes an angle ϕ with the grating normal. The direction of positive angles from the grating normal is defined so that ϕ is positive.

Tilting the reflection facet through the angle ϕ changes the phase of the reflected wave such that α is replaced by $\alpha - \phi$ and β by $\beta - \phi$. Then the envelope light distribution is proportional to

$$g^2(\beta) = \mathrm{sinc}^2 \frac{\pi b}{\lambda} [\sin(\alpha - \phi) + \sin(\beta - \phi)], \qquad (3.8)$$

where we have explicitly converted to b/λ in place of b measured in λ units, and we have squared the amplitude pattern to get the light intensity. Another useful form of this expression is to solve for λ in Equation 3.5 and substitute into Equation 3.8. After expanding the $\alpha - \phi$ and $\beta - \phi$ terms and using the identities $\sin \alpha + \sin \beta = 2 \sin \frac{1}{2}(\alpha + \beta) \cos \frac{1}{2}(\alpha - \beta)$ and $\cos \alpha + \cos \beta = 2 \cos \frac{1}{2}(\alpha + \beta) \cos \frac{1}{2}(\alpha - \beta)$, we obtain

$$g^2(\beta) = \mathrm{sinc}^2 \frac{n \pi b}{d} \left[\cos \phi - \sin \phi / \tan \frac{1}{2}(\alpha + \beta) \right] \qquad (3.9a)$$

or

$$g^2(\beta) = \mathrm{sinc}^2 n \pi \cos \phi \left[\cos \phi - \sin \phi / \tan \frac{1}{2}(\alpha + \beta) \right] \qquad (3.9b)$$

when b has its usual value of $d \cos \phi$. For any given grating, ϕ is specified, so we can plot its "universal blaze curve," giving $g^2(\beta)$ as a function of $\alpha + \beta$ for any order n. Sample curves are shown in Figure 3.9 for the case of $\phi = 20°$. Notice how the width of the blaze distribution is inversely proportional to the order number. The maximum lies at $\alpha + \beta = 2\phi$, which is the configuration for specular reflection off the facet surfaces.

The blaze distribution with wavelength is easily calculated once the configuration of the spectrograph is specified. The value of λ associated with each $\alpha + \beta$ of the curves in

Figure 3.9 is calculated from the grating equation. For example, if we choose $\alpha = \phi = 20°$ and $d = 1/600$ mm, we obtain the results in Figure 3.10. A comparison of this scalar-wave theory to an observed blaze distribution in unpolarized light is made in Figure 3.11. The agreement is reasonable. However, the efficiency does depend on the vector properties of the light, namely, the polarization. Radiation polarized with the electric vector perpendicular to the rulings generally shows a higher efficiency and a shift of the blaze to longer wavelengths compared to the opposite orientation, as shown in Figure 3.12.

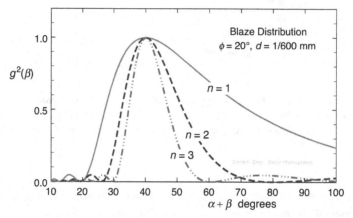

Figure 3.9 Equation 3.9 gives this scalar-wave blaze distribution of light. The first three orders are shown for $\phi = 20°$.

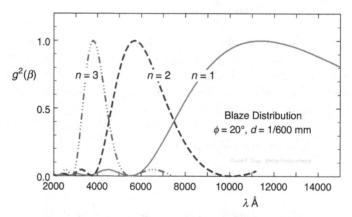

Figure 3.10 Here the blaze distributions of Figure 3.9 are plotted on a wavelength scale using $\alpha = \phi = 20°$ and a ruling density of 600 lines/mm.

The polarizing or analyzing property depends on wavelength and, at about 9600 Å for the grating of Figure 3.12, the sense of the polarization reverses. We conclude that near the peak of the blaze, linear polarization of a few percent can be expected, but several tens of percent can arise somewhat farther from the center of the blaze. Polarization properties of gratings are further discussed by Breckinridge (1971). Polarization introduced by the entrance slit of a spectrograph can also be several percent (Ismail 1986).

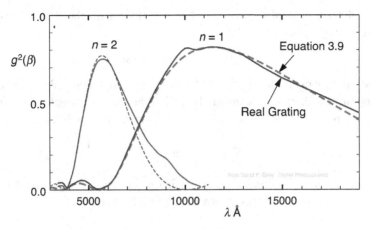

Figure 3.11 The scalar-wave blaze equation, Equation 3.9 (dashed lines) with vertical scaling, is compared to a real grating blaze distribution (solid lines). Jarrell-Ash grating #1300. Data courtesy of Jarrell-Ash.

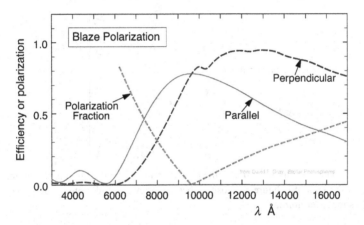

Figure 3.12 Incident light polarized perpendicular to and parallel to the rulings shows rather different blaze distributions. The sum of these curves is the $n = 1$ curve in Figure 3.11. If the incident light is unpolarized, the grating will polarize it. The sense of the polarization reverses as it goes through zero. Adapted from data supplied by Jarrell-Ash.

The Blaze Wavelength

Manufacturers of gratings specify the "blaze wavelength" for the $\alpha = \beta = \phi$ Littrow configuration used in first order. This wavelength is where the blaze distribution peaks. The number helps us choose a grating if we are in the spectrograph-building business or it tells us which grating to use if we are a spectrograph user. The blaze specification states the wavelength region for which the grating is intended to be used. For example, a grating blazed at 12 000 Å would have high efficiency ~12 000 Å in first order, ~6000 Å in second order, and ~4000 Å in third order. This blaze wavelength is related to the facet angle through the grating equation ($n = 1$),

$$\lambda_b = d(\sin \alpha + \sin \beta) = 2d \sin \phi. \tag{3.10}$$

Since most spectrographs deviate from the Littrow configuration, the true blaze will actually appear at a shorter wavelength. Take λ_b' to be the wavelength of the true blaze and β_b the true blaze direction. The grating equation gives the true blaze wavelength,

$$\lambda_b' = \frac{d}{n}(\sin \alpha + \sin \beta_b) = \frac{2d}{n} \sin \frac{1}{2}(\alpha + \beta_b) \cos \frac{1}{2}(\alpha - \beta_b).$$

Since the facet specular reflection gives the blaze, the law of reflection tells us that $\alpha - \phi = \phi - \beta_b$ or $\alpha + \beta_b = 2\phi$, as shown in Figure 3.13. Then

$$\lambda_b' = \frac{2d}{n} \sin \phi \cos (\alpha - \phi). \tag{3.11}$$

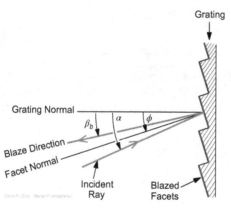

Figure 3.13 The specular reflection at the facet determines the direction of the blaze, β_b. All the angles shown in this figure are positive because they lie on the same side of the grating normal as the facet normal.

Compared to the Littrow case,

$$\frac{\lambda_b'}{\lambda_b} = \cos (\alpha - \phi).$$

For example, if α differs by 15° from the Littrow configuration, a grating that is said to be blazed at 8000 Å will actually have its blaze at 7727 Å. In most spectrographs, the reduction in blaze wavelength is modest.

If a grating is used far from the Littrow configuration, another issue arises, called shadowing.

Shadowing

Reflection gratings are designed to be used with incoming and outgoing beams approximately in the direction of the facet normal. Indeed, when $\alpha \sim \beta \sim \phi$, the material in this section is of little consequence. But in the exceptional application where this is not true, the structure of the rulings can obstruct light. Let us assume that the rulings have 90° between their sides. Figure 3.14, Case A, shows an incoming beam for one orientation of the grating. Here β_b is the direction of the blaze peak. Light does not reach the left-most portion of the facets owing to the shadow cast by the adjacent groove. In effect, the facet has a smaller width given by

$$b = d \cos \phi - d \sin \phi \tan (\alpha - \phi).$$

This is to be compared to the usual unobstructed length of $d \cos \phi$. A smaller b implies a slightly wider blaze distribution, as per Equation 3.9a. In the case of echelle gratings, to be briefly discussed below, ϕ is large, so $\sin \phi$ does not reduce the second term significantly, as it does for low-order gratings. But the tangent factor is normally small, i.e., incoming light is usually not far from the direction of the facet normal. The upshot is that when α is positive, as in Case A, there is a slight reduction in the slit width owing to shadowing.

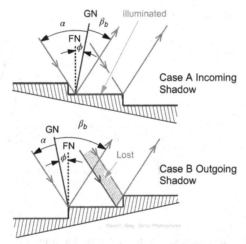

Figure 3.14 The grating can be used in either of two orientations. Notice that the incoming and outgoing beams are in the same direction for both cases, while the grating has been turned over. The effective width of the facet is reduced to the illuminated portion in Case A. In Case B, the light in the shaded strip is lost to reflection off the back of the next facet and scattered into the spectrograph.

If the grating is turned over, i.e., rotated 180°, light will illuminate the full facet, as in Case B of the figure. But now light reflected from the right end of the facet impinges on the backside of the next facet and is lost from the beam. In fact, this kind of shadowing causes more harm than just light loss because the "lost" light becomes scattered light in the spectrograph (see Chapter 12 for a discussion of scattered light). The fraction of the beam lost to outgoing shadowing is given by

$$f = \tan \phi \tan (\beta - \phi).$$

It increases with β, i.e., longer wavelength. In the Littrow configuration, the outgoing beam will be close to the facet normal, making this shadowing tiny.

While shadowing can readily be demonstrated for real gratings, its behavior is more complex than the foregoing geometrical arguments indicate, mainly because of diffraction effects.

Real gratings are also imperfect, sometimes producing ghosts.

Grating Ghosts

Spurious spectral lines can be generated by various imperfections in gratings. One class, the ghosts, arises from periodic errors in the position of the rulings. These errors can usually be traced back to flaws in the ruling engine, such as minute deviations in the pitch of the carriage-advance screw used to space the rulings. The multiple line pattern of the type shown in Figure 3.15 appears in place of each single emission line in the spectrum.

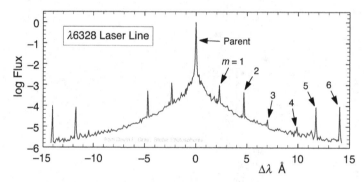

Figure 3.15 Ghosts are weak spurious lines symmetrically placed on each side of the parent line. In this example, some of them are labeled with their order number. Note the logarithmic ordinate. Adapted from Griffin (1968).

Insight into the ghost phenomenon can be gained by looking at the grating in the light of a ghost. We find that the grating appears to be striped, as in Figure 3.16, that is, the light forming the ghost comes from a series of bright stripes. Each of these stripes corresponds to one cycle of the periodic error in the rulings. The set of stripes forms a ghost grating. The grating aperture distribution of Equation 3.3 should then be replaced by

$$G'(x) = G(x)H(x),$$

where $H(x)$ is the periodic modulation by the ghost grating. In analogy with Equation 3.3, write

$$H(x) = B_3(x) * \text{Ш}(x)B_2(x),$$

where $B_3(x)$ describes the illumination across one stripe of the ghost grating, $B_2(x)$ is the box corresponding to the full width of the grating, and $\text{Ш}(x)$ is a new Shah function,

$\sum \delta(x - mD)$, with δ-functions spaced D apart (the periodic error spacing) with m being the order integer.

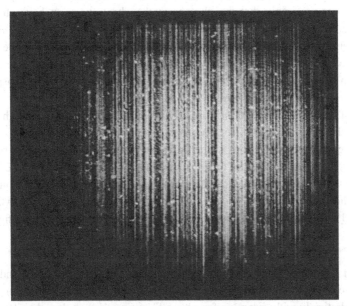

Figure 3.16 A photograph of a large grating in the light of the $m = 1$ ghost shows the ghost grating as a series of stripes superimposed across the face of the main grating. Courtesy of J.B. Breckinridge.

The amplitude spectrum formed is then

$$g'(\theta) = g(\theta) * h(\theta)$$

with $g(\theta)$ given by Equation 3.4 and $h(\theta)$ being the spectrum from the coarse ghost grating. The convolution brings the spectrum of this grating to the position of each line in $g(\theta)$, thus giving the ghosts. The dispersion of the ghost grating is low because of the large spacing D, and in fact we can write

$$\Delta\theta_g = m\lambda_g / D$$

by analogy with Equation 3.5. Since $\Delta\theta_g = 0$ at the center of the parent line according to the convolution above, $\Delta\theta_g$ is a measure of the angular separation of the mth ghost from the parent line. The ghost spacing in wavelength follows from Equation 3.7a,

$$\Delta\lambda_g = \frac{d}{n}\Delta\theta_g = \frac{d\,m}{n\,D}\lambda_g. \tag{3.12}$$

This is the ghost spacing equation. The ghosts associated with each real line are found at the positions given by letting m run through the $\pm$ integers starting at $m = \pm 1$. The spacing can be quite large. If we take $d = 1/600$ mm, $n = 1$, $\lambda = 6000$ Å, and $D = 1$ mm (which is typical of ruling engine screw pitch), $\Delta\lambda_g \pm 10$, ± 20 Å, and so on.

The color of light forming the ghost images is the same as that of the parent line. This follows from the fact that the different order ghosts (m) are really different orders of the same wavelength formed by the ghost grating. Equation 3.12 gives the spacing of the ghosts, but does not imply that the photons in the ghost spectrum have changed their wavelength from that of the parent line. These ghosts are called Rowland ghosts. When more than one periodic error occurs in the rulings, the interference of the two ghost diffraction patterns can cause ghosts very distant from the parent line. These are often called Lyman ghosts. Distant ghosts still have the color of the parent. Visual inspection can be used to identify them on this basis.

The run of ghost intensity with m is usually irregular. Often the $m = \pm 1$ ghosts are the largest, and the general trend is for decreasing strength with increasing m, which must follow from general considerations of the diffraction envelope. The irregular behavior (as seen in Figure 3.15) arises because there is more than one periodic error in the rulings and the interference between them reduces the intensity of some of the ghosts. Non-symmetry in pairs of ghosts results if the ghost grating shows blaze characteristics. Rowland (1893) and Meyer (1934) give a derivation of the ghost strength for the Littrow configuration,

$$\ell_g = \ell_p (\pi n A/d)^2,$$

where ℓ_g and ℓ_p are the brightness of the ghost and parent line, respectively, and A/d is the fractional size of the ruling error. Ghosts are likely to be a bother in high orders since they increase quadratically with order, but for a "good" grating they are less than 10^{-4} as strong as the parent line in first order. Newer gratings have weaker ghosts because of the precise interferometer control of the ruling engine steps.

A second class of grating imperfections gives rise to spurious lines called satellites. Looking in the light of the satellite feature reveals that certain portions of the grating surface are distorted. In effect, one has a little grating superimposed on the main grating but not in the same plane. Satellites are usually much closer to the parent line than the ghost spacing. They are not measurable in the best gratings. Poor mounting of the grating can distort it and produce satellite lines.

The third type of undesirable behavior is called Wood's anomaly (Wood 1935, Breckinridge 1971). It consists of a relatively large discontinuity introduced into an otherwise smooth continuous spectrum. Fortunately they generally lie away from the center of the blaze.

The Rest of the Spectrograph

We now turn our attention from gratings back to spectrographs more generally. Our emphasis is primarily on spectrographs for ground-based telescopes. The stellar "image" formed by the telescope is the seeing disk caused by inhomogeneities and fluctuations in the terrestrial atmosphere, and it is this seeing disk that is placed on the entrance slit of the spectrograph. How we handle the starlight before the grating and after dispersion is important for getting good spectra. Let us start at the back end of the spectrograph by looking into the role of the camera.

The Camera and Linear Dispersion

The fundamental point of the camera is to bring the dispersed light into focus on the light detector, usually a CCD array. But it is the focal length of the camera that determines the scale of the image, namely, the number of angstroms in the spectrum per millimeter on the detector. This scale factor is called the *linear dispersion*.

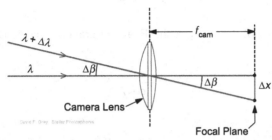

Figure 3.17 The camera in a spectrograph focuses the collimated monochromatic beams from the grating to positions differing by Δx for the difference in angle $\Delta \beta$ corresponding to two wavelengths differing by $\Delta \lambda$. The scale of the image in the focal plane is proportional to the focal length of the camera. When the camera is a mirror, the same diagram holds true if the diagram is folded about the dashed line through the lens.

If two collimated beams having wavelengths λ and $\lambda + \Delta \lambda$ leave the grating at directions differing by a small angle $\Delta \beta$ then, as shown in Figure 3.17, their linear separation in the spectral image is

$$\Delta x \approx f_{cam} \, \Delta \beta,$$

or more generally,

$$\frac{d\beta}{dx} = \frac{1}{f_{cam}},$$

and it follows immediately that the linear dispersion in angstroms per millimeter is given by

$$\frac{d\lambda}{dx} = \frac{d\lambda}{d\beta}\frac{d\beta}{dx}$$

$$= \frac{1}{f_{cam}}\frac{1}{d\beta/d\lambda}$$

$$= \frac{1}{f_{cam}}\frac{d\cos\beta}{n}, \tag{3.13}$$

where $d\beta/d\lambda = n/(d \cos \beta)$ is the angular dispersion for the grating, which we found earlier (Equation 3.6) from the grating equation. (Note the mixed units for $d\lambda/dx$ of Å length in the spectrum over mm length in the focal plane; $10^7 \, \text{Å} = 1 \, \text{mm}$.) "High dispersion" is a small number of angstroms per millimeter and vice versa. High dispersion is typically ~1–2 Å/mm, low dispersion is ~100 Å/mm. Occasionally this statement of dispersion is referred to as "reciprocal dispersion."

Now the question is: what camera focal length should be chosen?

The Camera and Slit Magnification

In the spectrum imaged on the detector, the monochromatic image of the entrance slit dominates the overall resolution of the spectrograph. The angular size of the entrance slit as seen from the collimator (and the grating) is

$$\Delta\alpha = \frac{W'}{f_{\text{coll}}},$$

where W' is the width of the entrance slit. From the grating equation (Equation 3.5), we know that

$$\Delta\beta = -\frac{\cos\alpha}{\cos\beta}\,\Delta\alpha,$$

so

$$\Delta\beta = -\frac{\cos\alpha}{\cos\beta}\frac{W'}{f_{\text{coll}}}$$

is the angular size of the monochromatic image of the entrance slit. Denote the linear width of the slit image in the spectrum by w. Use the geometry of Figure 3.17, but now with the two rays having the same wavelength and coming from the opposite edges of the entrance slit, and with Δx replaced by w, then

$$w = \Delta\beta f_{\text{cam}}$$

$$= -\frac{\cos\alpha f_{\text{cam}}}{\cos\beta f_{\text{coll}}}\,W'. \tag{3.14a}$$

Or we might recast this slightly as

$$\frac{w}{f_{\text{cam}}} = -\frac{\cos\alpha}{\cos\beta}\frac{W'}{f_{\text{coll}}}, \tag{3.14b}$$

to emphasize the proportionality between the angular size of the image, w/f_{cam}, and the angular size of the entrance slit, W'/f_{coll}. The magnification of the width of the entrance slit, w/W', is usually < 1. As mentioned earlier in the chapter, the magnification along the slit is just $f_{\text{cam}}/f_{\text{coll}}$. The negative sign, indicating a reversal of the image, can usually be dropped.

Now we see how to choose f_{cam} and the linear dispersion. In essence, f_{cam} should scale the spectrum so that w spans two or three pixels in the detector to meet the Nyquist sampling requirement. Spreading the spectrum out more than this gives oversampling. A small amount of oversampling is scientifically advantageous in estimating the noise, but is paid for with longer exposure times or brighter limiting magnitudes of the stars that can be observed. Too small an f_{cam} will produce undersampling, information will be lost, and the detector will then limit the spectrograph resolution, a situation that should not normally be allowed.

One other aspect of importance for spectrograph cameras should be noted: their field of view. The range in β can be several degrees, and that is a large field for an optical device. Consequently, wide-angle cameras, usually of the Schmidt design, are used (Lemaitre 2009).

Spectrograph Resolution and the Spectrograph Equation

Now, with the basics in place, we turn to the most important parameter of the spectrograph, its spectral resolution. It specifies the finest details one can see in the spectrum. In the usual case, the width of the entrance-slit image dominates the resolution. Convert the image width, w, to a wavelength interval by combining Equations 3.13 and 3.14,

$$\Delta\lambda = w\frac{d\lambda}{dx}$$

$$= \frac{\cos\alpha f_{cam}}{\cos\beta f_{coll}}\,W'\,\frac{1}{f_{cam}}\frac{d\cos\beta}{n}$$

$$= \cos\alpha\frac{W'}{f_{coll}}\frac{d}{n}. \tag{3.15}$$

In the same vein that Equation 3.5 is the grating equation, this is the "*spectrograph equation.*" It ties together the grating parameters d and n with the spectrograph parameters W' and f_{coll}. We see that the camera focal length has canceled out; the resolution is independent of the camera focal length (presuming the linear dispersion is appropriate for the detector). In general, we say that the resolution is higher (i.e., better) when $\Delta\lambda$ is smaller, and smaller $\Delta\lambda$ is what we want in order to measure the details of spectral lines.

Here is how we make $\Delta\lambda$ small. Equation 3.15 tells us that we have two grating options: to choose a grating with high ruling density (small d) or one blazed to work in high orders (large n). And we have two spectrograph options: choose a small W' or a large f_{coll}. Making W' smaller gives higher resolution, but probably, at least for ground-based observations, at the cost of throwing light away, since W' is usually smaller than the stellar seeing disk. Naturally image slicers are a wise choice at this point. Image slicers recycle the parts of the seeing disk that don't go directly through the slit and would otherwise be lost (as discussed in Chapter 12). Finally there is f_{coll}. Making f_{coll} large is exactly what is done in practice, and it is the reason why coudé spectrographs are large (more in Chapter 12 on this too). Recall how, e.g., Figure 3.1, the beam diverges with the f-ratio of the telescope as it enters the spectrograph. The expansion of the beam is arrested and the beam size is fixed by the collimator. The limit to extending f_{coll} is therefore set by the maximum available grating size ($\lesssim 400$ mm). In some cases mosaics of gratings are used to get around this limitation (e.g., Grundmann *et al.* 1990, Dekker *et al.* 1994, Blasiak & Zheleznyak 2002).

To amplify this point, high-resolution ($\lambda/\Delta\lambda \gtrsim 10^5$) spectroscopy on big telescopes is seriously affected by limited grating size. A big telescope has a correspondingly big effective focal length (or else an equally troublesome small f-ratio). If the size of the telescope is doubled, the stellar seeing disk has twice the diameter, so the spectrograph slit needs to be twice as wide to let in the same fraction of the image. Then to preserve the spectral resolution, the collimator must be twice as far from the entrance slit, so the beam has doubled its diameter by the time it is collimated. With telescope apertures of 3–4 m, four-element mosaic gratings are used to gain sufficient beam size. Excellent atmospheric seeing is an important advantage in this regard.

A quick way to *estimate* the resolving power of a spectrograph is to take λ from the grating equation, divide it by $\Delta\lambda$ from Equation 3.15, assume the Littrow configuration, and get $\lambda/\Delta\lambda \approx 2\tan\phi/(W'/f_{\text{coll}})$.

Although the spectrograph resolution is dominated by the image of the entrance slit, there are other factors that blur the image, most noticeably optical aberrations. Optical aberrations are discussed by Lemaitre *et al.* (1990) and Bland-Hawthorn *et al.* (2017), among many others. The actual instrumental profile of a spectrograph is wider than given by Equation 3.15, and typically roughly Gaussian in shape (example in Figure 12.7).

A well-designed high-resolution spectrograph is a wonderful thing (e.g., Gray 1986, Grundmann *et al.* 1990, Vogt *et al.* 1994, Pilachowski *et al.* 1995, Sheinis *et al.* 2015, Gibson *et al.* 2018). Figure 3.18 gives a taste of the details seen in line profiles with spectral resolving power of ~100 000.

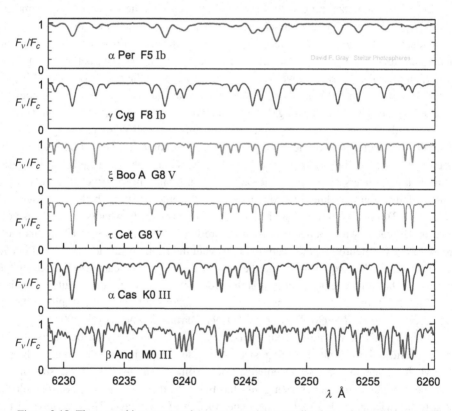

Figure 3.18 These coudé-spectrograph measurements, normalized to unit continuum, have spectral resolution of ~100 000. The noise can just barely be discerned on some of the exposures. The differences in widths and shapes of the spectral lines are easy to see with spectra like these. Data from the Elginfield Observatory.

Echelle Spectrographs

In recent years, cross-dispersed echelle spectrographs have been used effectively for certain types of work (Vogt *et al.* 1994, Lacy *et al.* 2002, Bernstein *et al.* 2003, Barnes *et al.* 2008, Terada *et al.* 2008). The leverage offered by the echelle grating is its high angular dispersion, resulting from working in very high orders, $n \sim 100$, one of the options in Equation 3.15. As a result, the physical size of the spectrograph can be kept small enough to use at the Cassegrain focus, typically attaining modest resolving powers (~40 000). This saves the reflection losses of the relay mirrors needed with a coudé spectrograph. It also saves money since the optics are smaller.

High order implies a large blaze angle. Echelles are often manufactured with integer values of tan ϕ, as in Table 3.1, i.e., an R1 echelle has tan $\phi = 45°$, etc. With such steep angles, the short side of the groove is the working side and the echelle must be used close to Littrow.

Table 3.1 *Echelle R names and their ϕ values (degrees)*

R1	R2	R3	R4	R5	R6
45.0	63.4	71.6	76.0	78.7	80.5

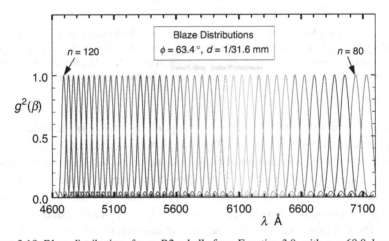

Figure 3.19 Blaze distributions for an R2 echelle from Equation 3.9, with $\alpha = 68.9$ degrees.

Since the wavelength span of the blaze distribution varies inversely as the order number, an echelle spectrum is effectively cut into dozens of small pieces. The blaze distribution for an R2 echelle is shown in Figure 3.19. The overlap of the orders is handled by using a cross disperser, i.e., a low-dispersion grating or prism with its dispersion perpendicular to that of the main grating. This stacks the spectra as in Figure 3.20.

Order   Central λ   A/order

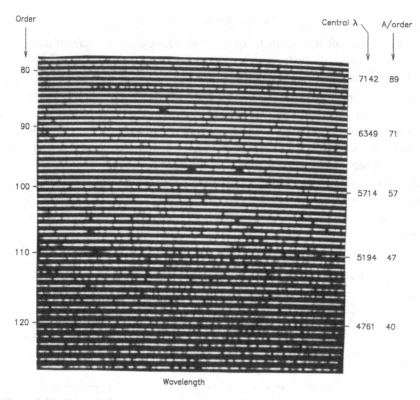

Wavelength

Figure 3.20 This echellogram shows the central portions of 49 orders spanning a range of
wavelength from 4500 Å to 7300 Å. The resolving power is 30 000, and the star (A4, $V =$
12.3) is in the globular cluster M71. The strong line in order 87 is H_α, and the two in order
97 are the Na D lines. Details can be seen in Vogt (1987). Photograph courtesy of S.S. Vogt.

The scientific advantage of the echelle spectrograph lies mainly in the wide wavelength
coverage obtained in a single exposure. This is important for chemical abundance studies,
for instance, where numerous and widely separated lines must be measured. The price paid
is limited accuracy. The narrow blaze distribution causes steep variations in the apparent
continuum with wavelength, making accurate measurements of profiles of strong lines, like
the hydrogen Balmer lines in hotter stars, especially difficult (see Giribaldi *et al.* 2019 for
example). The fact that thousands of angstroms of spectrum are allowed into the spectrograph leads to relatively high levels of scattered light, typically several percent (see
Chapter 12). With low-order gratings, a bandpass filter is used to exclude the light not
being measured, so there is less light to scatter.

There is also the question of how strong the cross dispersion should be. If the adjacent
orders are too close together, spill-over stemming from optical aberrations will contaminate
adjacent orders. But if the orders are spaced well apart, fewer of them will be recorded
because detectors have limited area. The strong variation of brightness across each blaze
makes for a widely ranging signal-to-noise ratio. Echelle data processing is complicated by

the slant of the spectral lines resulting from the cross dispersion and by an instrumental profile that varies with position in the field.

Another area of strength for high-order spectrographs is radial velocity work, in particular, radial velocity variations. The main requirement here is stability, not the integrity of the spectral line profiles (e.g., Robertson *et al.* 2019).

Cross-dispersed echelle spectrographs are powerful tools, and should be used when very wide wavelength coverage is deemed important enough to sacrifice the ultimate in accuracy. Otherwise, lower order spectrographs have the advantage.

Spectra from Interferometers

Fourier spectroscopy, the technique of using an interferometer to measure the spectrum, offers certain advantages for astronomical measurements including high optical throughput and high spectral resolution (see Mertz 1965a, Connes 1970, Bell 1972). Interferometers measure the Fourier transform of the spectrum. Several types of interferometers have been used, notably the Michelson and Fabry–Perot instruments. In the Michelson interferometer shown in Figure 3.21, the path difference between the two arms is x (the mirror moves half this distance). The moveable mirror introduces a phase term $e^{2\pi i x \sigma}$ for wave number $\sigma = 1/\lambda$. For each wavelength, or equivalently δ-function, in the input spectrum, the signal in the exit diaphragm varies sinusoidally with x.

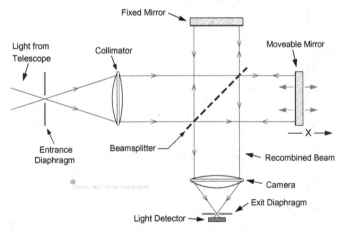

Figure 3.21 A Michelson interferometer uses a beam splitter to create two collimated beams that go to the two mirrors and back. The path length and therefore the phase of the horizontal beam is changed by moving the mirror on the right through the path difference x, which because of the reflection is double the actual displacement of the mirror. The interference pattern, formed when the beams are recombined, is focused on the detector by the camera lens.

Take $F(\sigma)$ to be the full input spectrum and $f(x)$ to be the signal recorded at the detector. If we make these functions symmetric by defining $F(-\sigma) = F(\sigma)$ and $f(-x) = f(x)$, we have a pair of δ-functions for each frequency in $F(\sigma)$. The Fourier transform of the interference pattern, $f(x)$, is then the spectrum of the input light, $F(\sigma)$.

Some of the first useful measurements of infrared stellar spectra were obtained by Mertz (1965b). An example of an interferogram and the resulting spectrum of α Per (F5 Ib) are shown in Figures 3.22 and 3.23. At small path differences, all wavelengths are nearly in phase, and large amplitudes appear in $f(x)$. The phases rapidly become non-coherent as x increases, and the interference between different Fourier components reduces the amplitudes to lower values. There is however a great deal of meaningful structure well away from zero, as illustrated in Figure 3.22.

The maximum path difference, x_{max}, determines the resolution afforded by the interferometer in exact analogy with the width of the diffraction grating, as described by Equation 3.3 and Figure 3.5. That is, the full interferogram, which in principle extends to $x = \infty$, is actually restricted to x_{max}. In the now familiar way, we can view this as multiplication with a box. The instrumental profile is therefore a sinc function with a first zero at $\sigma = 1/x_{max}$. It is common practice to apodize this function to essentially remove the sidelobes. Because it is possible to work over path differences of two meters, interferometers can produce substantially higher resolution than diffraction gratings, the largest of which, as we have seen, is less than half a meter in width. Non-trivial technical difficulties in measuring path positions are overcome with reference laser interferometers to measure the positions of the optical components (e.g., Maillard & Michel 1982).

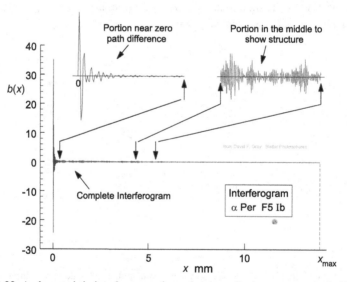

Figure 3.22 A characteristic interferogram shows large amplitudes near zero path difference where the phase differences are still small. The signal out to large path differences, although of smaller amplitude, contains the high Fourier frequency information of the wavelength domain. Two portions are enlarged to show detail. Data courtesy of D.D. Sasselov and J.B. Lester.

In a grating spectrograph, we saw how the spectral resolution could only be kept high by using a small entrance slit (Equation 3.15). In many situations, the seeing disk is larger than the slit and light is lost. An interferometer has no entrance slit; the whole stellar seeing disk can be measured without loss in resolution. This offers a significant efficiency gain over slit spectrographs. On the other hand, image slicers (discussed in Chapter 12) mostly redress this imbalance. Other advantages offered by interferometers include low scattered light, absolute wavelength calibration to $\lesssim 100$ m/s, a well-determined and controllable instrumental profile, "adjustable" resolving power (x_{max}), versatile wavelength coverage, and insensitivity to non-linearities in the detectors. Additional information can be found in the review of Ridgway and Brault (1984).

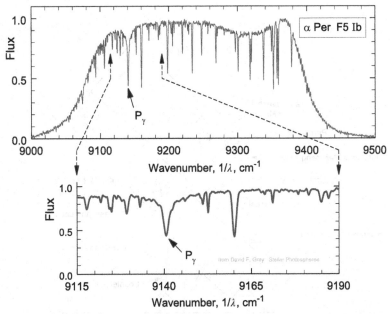

Figure 3.23 The wavelength-domain reconstruction of the interferogram in Figure 3.22 is shown. In the upper portion, the overall shape is the transmission curve of the bandpass filter used to isolate the measured portion of the spectrum. Division by an exposure of a continuous-spectrum lamp is used to remove this shape and normalize to the continuum. The lower part shows an expansion around the Paschen gamma line to better illustrate the details of the line profiles. The spectral resolving power is about 27 000.

Unfortunately, interferometers are not as fast as conventional spectrographs used with multi-element array detectors, nor are they as easy to work with. In the context of stellar photospheres, interferometers do however give us the ultimate in solar atlases with resolving power of ~10^6, signal-to-noise ratios of several hundred, and absolute wavelengths with uncertainties of ~10–100 m/s. These include the solar flux atlases of Kurucz *et al.* (1984), reformatted by Hinkle *et al.* (2000), Wallace *et al.* (2011), and Reiners *et al.* (2016). An interferometer IR atlas of Arcturus (α Boo, K1.5 III) is given by Hinkle *et al.* (1995) and the visible-region counterpart is given by Hinkle *et al.* (2000).

Aspects of Telescopes

Our spectrographs would not be very productive without telescopes to feed in starlight. The telescope's job is to collect and concentrate the starlight so that we have enough photons to get the science done.

The focal ratio of the primary mirror is kept as small as possible to minimize the size of the telescope and the dome housing it. Ratio values in the range of 3–5 are common. The secondary mirror is figured to give the actual working f-ratio, ~10 for Cassegrain foci and ~30 for coudé foci. Long effective focal lengths are needed for coudé operation, as in Figure 3.24.

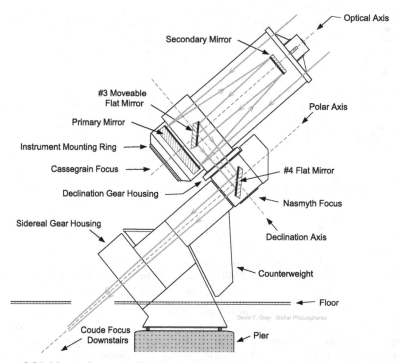

Figure 3.24 Many telescopes allow the beam to be brought to any of several foci. The arrangement for the coudé focus is shown. The #3 flat mirror is placed where the telescope optical axis intersects the declination axis, bringing the beam through the hollow declination axle. The #4 mirror lies at the intersection of the declination and polar axes, and directs the beam through the hollow polar axle to the coudé focus on the floor below the telescope. In this way the coudé focus remains fixed and independent of the direction in which the telescope is pointing. When the Cassegrain or Nasmyth focus is needed, the secondary mirror is changed to give a shorter effective focal length. To use the Cassegrain focus, the #3 flat mirror inside the telescope tube is moved to the side to let the light by.

The effective focal length, F_{eff}, is defined to be the focal ratio at the focus multiplied by the aperture of the telescope, and it is much larger than the physical distance from the primary mirror to the focus. The spectrograph collimator must match this f-ratio. Interchangeable secondary mirrors are used to change the effective focal length and bring the star image to different focal positions, such as Cassegrain versus coudé. The scale in the focal plane of the telescope (with the same idea used in Figure 3.17) is given by

$$S = \frac{1}{F_{\text{eff}}} \text{ radians/mm} \quad \text{or} \quad \frac{206265''}{F_{\text{eff}}} /\text{mm}, \tag{3.16}$$

where F_{eff} is given in millimeters. Typical Cassegrain-focus scales are 10–30″/mm, while at the coudé, values of 2–6″/mm are common. A stellar seeing disk of 2″ would then be ~0.5 mm across at a typical coudé focus.

Light collected by a telescope like the one in Figure 3.24 can be brought to any of the three foci, Coudé, Nasmyth, or Cassegrain, with a change of secondary and #3 mirrors. This allows three different instruments to be mounted at the same time, saving the time and effort of instrument changes.

Aluminum has been the traditional reflecting surface for telescope and spectrograph mirrors, but enhanced and overcoated silver offers throughput advantages. Seasoned aluminum typically has a reflectivity of 85% over much of the visible window, while overcoated silver's reflectivity is about 97%. If one is working at a coudé focus with, say, five mirrors to get the light to the spectrograph slit, and another three mirrors to get the light to the detector, there is potentially a gain of as much as a factor of three. These overcoated surfaces also last two to three times longer than bare aluminum.

Polarization introduced by the telescope is small at the Cassegrain focus but quite significant at the coudé focus. The oblique reflections needed to bring the light to the coudé focus (notice the #3 and #4 mirrors in Figure 3.24) produce about 6 or 7% polarization. How these contributions combine depends on the direction the telescope is pointed, so the net result varies with hour angle and declination.

Fiber optics make an effective alternative to the usual coudé-relay mirrors (e.g., Horton *et al.* 2012). And at the output end, they can be used as an image slicer (Felenbok & Guerin 1988, Furenlid & Cardona 1988). A fiber feed can also bring the starlight to a Cassegrain-type spectrograph placed on a bench instead of having it mounted on the telescope (Hubbard *et al.* 1979). This eliminates flexure and sagging of the spectrograph that always occurs when it is mounted on a telescope, owing to its changing orientation with respect to gravity as the telescope is pointed to different parts of the sky. Fiber optics bring some challenges. Some fibers suffer from transmission inefficiency, especially at shorter wavelengths. Sometimes the f-ratio of the telescope has to be modified (with a lens in the beam) to use the fiber optics effectively, and keeping all rays within the same f-ratio cone at the output-end of the fiber can be difficult.

Seeing is a major consideration for ground-based telescopes. The actual distribution of light in the stellar seeing disk has a modest peak in the center and often shows a large but faint outer halo reaching to several seconds of arc. The details depend on the geographical location, elevation, and the local weather and the stability of the atmosphere. The general distribution of light in seeing disks is discussed by Piccirillo (1973), Young (1974), Diego (1985), and Racine (1989). Physical causes of seeing are reviewed by Coulman (1985). Heat generated by the telescope or its housing can cause poor seeing. Given the importance of the width of the spectrograph entrance slit (Equation 3.15), poor seeing implies losing light, never a good thing, and makes image slicers all the more important. Locating telescopes in sites having good seeing is just as important for spectroscopy as for direct imaging.

Adaptive optics can correct many of the effects of seeing, especially if the seeing is naturally already good. By introducing real-time corrections of seeing distortions in the wave front, images can be sharpened and throughput enhanced. Adaptive optics work better in the infrared than in the visible region. Although these systems are relatively costly, they are becoming vital for high-resolution spectroscopy on giant telescopes (e.g., Beckers *et al.* 2007, Conod *et al.* 2016, Jovanovic *et al.* 2016). Diffraction-limited performance through adaptive optics is anticipated for the Thirty-Meter Telescope (www.tmt.org). If one is able to work at the diffraction limit, a spectrograph that works for a one-meter telescope will work for any size aperture. An alternative to adaptive optics is the emerging technology of photonics. Stemming from the physics of fiber optics, photonic techniques introduce new ways of manipulating and dispersing light that may greatly reduce the size of spectrographs (e.g., Bland-Hawthorn & Kern 2012, Gatkine *et al.* 2019).

Questions and Exercises

1. Sketch qualitative graphs of the diffraction patterns of four hypothetical transmission gratings of infinite width having $b = d/3$, $b = d/2$, $b = 2d/3$, and $b = d$, where b is the grating slit width and d is the slit spacing. Explain what you see.

2. Draw a schematic diagram of a spectrograph using transmission optics for clarity. Show and label all relevant components. Show two rays of starlight coming into your spectrograph converging at the entrance slit with the f-ratio of the telescope. Draw the path of these rays as they pass through the spectrograph, ending at the light detector. Take two wavelengths, say 4000 Å and 6500 Å, and indicate how the rays of light behave after leaving the grating, where the spectrum is formed, and the orientation of the wavelength scale.

3. What do we mean by the term "blazed grating"? How are gratings blazed? Why does blazing not change the angular position of both the diffraction envelope *and* the interference pattern?

4. Suppose you are designing a reflection grating having 300 lines/mm to work in the fifth order at 6000 Å. What facet angle, ϕ, is needed to position the blaze for optimum efficiency? Assume $\alpha = \beta$ for 6000 Å light. For this grating, what is the angular separation of 6000 Å light from 6500 Å light as the beams leave the grating?

5. Calculate the blaze distributions (light versus wavelength) for this grating: dimensions 350×250 mm, 300 lines/mm, blaze angle of 63°, aluminum reflecting surface, low expansion glass substrate. What order would be optimum for observations at 6250 Å? Could this wavelength be observed in more than one order?

6. Define spectral resolving power. Explain what factors determine it for a grating spectrograph and how one goes about optimizing it for stellar observations.

7. Consider a coudé spectrograph with two cameras, one having a focal length of 2080 mm, the other 559 mm. The focal length of the collimator is 6960 mm. Suppose the grating, which has 316 lines/mm, is used in the ninth order at 6250 Å. Assume for convenience that $\alpha = \beta = 60°$. What linear dispersions (in Å/mm) are achieved for these two cameras? If the entrance slit is 200 µm (0.200 mm) wide, what spectral resolution (in Å) can one expect for these two cameras?

8. Can we use normal spectrographs on giant telescopes? Maybe. Convince yourself that telescopes giving diffraction-limited images can use the same spectrograph independent of telescope aperture. One way to proceed is as follows. First remember that the angular size of the diffraction image varies inversely with telescope aperture. Second use the fact that the f-ratio of the collimator is the same as for the telescope, so $f_{coll} = F_{eff}\,B/T$, where F_{eff} is the effective focal length of the telescope and B and T are the spectrograph beam size and telescope aperture. Third remember that the scale (radians/mm) in the focal plane of the telescope is $1/F_{eff}$. Finally, Equation 3.15 tells us that the resolution is determined by the angular size of the entrance slit, W'/f_{coll}. Assume that W' (in mm) is set to the size of the diffraction image (converted to mm) and that the beam size, B, is fixed, i.e., we want to use the same spectrograph.

References

Barnes, S.I., Cottrell, P.L., Albrow, M.D., *et al.* 2008. *SPIE* **7014**, 70140K.

Beckers, J.M., Andersen, T.E., & Owner-Petersen, M. 2007. *Opt Exp* **15**, 1983.

Bell, R.J. 1972. *Introductory Fourier Transform Spectroscopy* (New York: Academic Press).

Bernstein, R., Shectman, S., Gunnels, S., Mochnacki, S., & Athey, A. 2003. *SPIE* **4841**, 1694.

Bland-Hawthorn, J. & Kern, P. 2012. *Phys Today* **65**, 31.

Bland-Hawthorn, J., Kos, J., Betters, C., *et al.* 2017. *Opt Exp* **25**, #15614.

Blasiak, T. & Zheleznyak, S. 2002. *SPIE* **4485**, 370.

Breckinridge, J.B. 1971. *App Opt* **10**, 286.

Connes, P. 1970. *ARAA* **8**, 209.

Conod, U., Blind, N., Wildi, F., & Pepe, F. 2016. *SPIE* **9909**, 990141.

Coulman, C.E. 1985. *ARAA* **23**, 19.

Dekker, H. & D'Odorico, S. 1994. *ESO Messenger* **76**, 16.

Diego, F. 1985. *PASP* **97**, 1209.

Felenbok, P. & Guerin, J. 1988. In *Instrumentation for Ground-Based Optical Astronomy* (New York: Springer-Verlag), L.B. Robinson, ed., p. 260.

Furenlid, I. & Cardona, O. 1988. *PASP* **100**, 1001.

Gatkine, P., Veilleux, S., & Dagenais, M. 2019. *App Sci* **9**, 290.

Gibson, S.R., Howard, A.W., Roy, A., *et al.* 2018. *SPIE* **10702**, 107025.

Giribaldi, R.E., Ubaldo-Melo, M.L., Porto de Mello, G.F., *et al.* 2019. *A&A* **624**, 10.

Gray, D.F. 1986. *Instrumentation and Research Programmes for Small Telescopes*, IAU Symposium 118 (Dordrecht: Reidel), J.B. Hearnshaw & P.L. Cottrell, eds., p. 401.

Griffin, R.F. 1968. *A Photometric Atlas of the Spectrum of Arcturus* (Cambridge: Cambridge Philosophical Society).

Grundmann, W.A., Moore, F.A., & Richardson, E.H. 1990. *SPIE* **1235**, 577.

Harrison, G.R. 1973. *App Opt* **12**, 2039.

Hinkle, K., Wallace, L., & Livingston, W. 1995. *Infrared Atlas of the Arcturus Spectrum, 0.9–5.3 µm* (San Francisco: ASP).

Hinkle, K., Wallace, L., Valenti, J., & Harmer, D. 2000. *Visible and Near Infrared Atlas of the Arcturus Spectrum 3727–9300 Å* (San Francisco; ASP).

Hopkinson, F. & Rittenhouse, D. 1786. *Trans Am Phil Soc* **2**, 201.

Horton, A., Tinney, C.G., Case, S., *et al.* 2012. *SPIE* **4846**, 48463A.

Hubbard, E.N., Angel, J.R.P., & Gresham, M.S. 1979. *ApJ* **229**, 1074.

Ismail, M.A. 1986. *ApSS* **122**, 1.

Jovanovic, N., Schwab, C., Cvetojevic, N., Guyon, O., & Martinache, F. 2016. *PASP* **128**, 121001.

Kurucz, R.L., Furenlid, I., Brault, J., & Testerman, L. 1984. *Solar Flux Atlas from 296 to 1300 nm* (Sunspot, NM: National Solar Observatory).

Lacy, J.H., Richter, M.J., Greathouse, T.K., Jaffe, D.T., & Zhu, Q. 2002. *PASP* **114**, 153.

Lemaitre, G.R. 2009. *Astronomical Optics and Elasticity Theory* (Berlin: Springer), Chapter 4.

Lemaitre, G., Kohler, D., Lacroix, D., Meunier, J.P., & Vin, A. 1990. *A&A* **228**, 546.

Maillard, J.P. & Michel, G. 1982. *Instrumentation for Astronomy with Large Optical Telescopes*, IAU Colloquium 67 (Dordrecht: Reidel), C.M. Humphries, ed., p. 213.

Mertz, L. 1965a. *Transformations in Optics* (New York: Wiley).

Mertz, L. 1965b. *AJ* **70**, 548.

Meyer, C.F. 1934. *The Diffraction of Light, X-Rays, and Material Particles* (Chicago: University of Chicago Press), Appendix F.

Mi, X., Zhang, S., Qi, X., *et al.* 2019. *Opt Exp* **27**, 19448.

Newton, I. 1704. *Opticks*, Printed for W. and J. Innys, Printers to the Royal Society, at the Prince's-Arms in St. Paul's Church-Yard, 2nd ed.

Palmer, C. 2020. *Diffraction Grating Handbook* (Rochester, NY: Richardson Gratings, MKS Instruments, Inc.), 8th ed.

Piccirillo, J. 1973. *PASP* **85**, 278.

Pilachowski, C., Dekker, H., Hinkle, K., *et al.* 1995. *PASP* **107**, 983.

Racine, R. 1989. *PASP* **101**, 437.

Reiners, A., Mrotzek, N., Lemke, U., Hinrichs, J., & Reinsch, K. 2016. *A&A* **587**, 65.

Ridgway, S.T. & Brault, J.W. 1984. *ARAA* **22**, 291.

Robertson, P., *et al.* 2019. *JATIS* i.d. 015003.

Rowland, H.A. 1893. *Phil Mag* **35**, 397.

Sheinis, A.I., *et al.* 2015. *JATIS* **1**, i.d. 035002.

Stroke, G.W. 1967. *Handb Phys* **29**, 426.

Terada, H., Yuji, I., Kobayashi, N., *et al.* 2008. *SPIE* **7014**, 701434.

Vogt, S.S. 1987. *PASP* **99**, 1214.

Vogt, S.S., Allen, S.L., Bigelow, B.C., *et al.* 1994. *SPIE* **2198**, 362.

Wallace, L., Hinkle, K.H., Livingston, W.C., & Davis, S.P. 2011. *ApJS* **195**, 6.

Wood, R.W. 1935. *Phys Rev* **48**, 928.

Young, A.T. 1974. *ApJ* **189**, 587.

4

Light Detectors

Light detectors do the important work of converting the stellar photons into recordable signals. Remarkable and wonderful developments in detectors have occurred in the past few decades, resulting in dramatic increases in speed and improved signal-to-noise ratios. Most notable for stellar-photospheric studies are the silicon array detectors such as the charge-coupled devices, CCDs. The progress in detector development, starting with the human eye, has flowed through photography, photomultiplier tubes, image tubes of various sorts, on to silicon-based arrays. A short history of the conception and early development of CCDs is given by Turriziani (2019).

To make quality observations of stellar spectra, the observer must be cognizant of the basic detector parameters: quantum efficiency, spectral response, linearity, noise, and dark leakage. These properties can be defined for any detector, and you can apply or extend the concepts of this chapter to the detector of your choice. Along with the detector itself, we have electronics and computers. From a computer program, i.e., the control software, signals are sent to digital and logic components that control the detector, issuing integration start and stop pulses, clock sequencing to route the signals through amplifiers and pulse shapers, and to digitize the signal. Then it is back to the computer again for several steps of processing before we get the basic digital tabulation of the spectrum.

First we start with detector sensitivity, expressed in quantum efficiency.

Quantum Efficiency and Spectral Response

Stars are faint. We want the best light detectors we can get, so the first question we ask is how good is the detector at its job? The quantum efficiency tells us. It is defined by

$$q(\lambda) = \frac{\text{number of photons detected}}{\text{number of photons incident}}. \tag{4.1}$$

The closer $q(\lambda)$ is to 1.0, the better. Figure 4.1 shows some comparisons. As wonderful as eyes are, when it comes to quantum efficiency, CCDs win by ~40–50 times, and off the eye's peak-sensitivity wavelength by even more. Astronomical CCDs that are carefully thinned and have an anti-reflection coating can reach quantum efficiencies of ~90%. These coatings can even be customized to modify the spectral response.

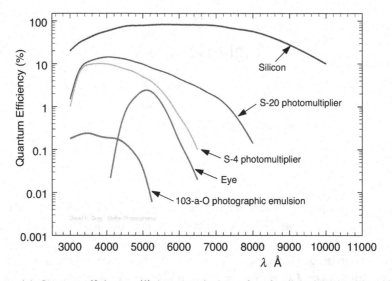

Figure 4.1 Quantum efficiency, $q(\lambda)$, in percent is shown for a few light detectors. The 103-a-O emulsion was used extensively in spectral classification work. The S-4 type cathode was used in the 1P21 photomultiplier as part of the definition of the UBV magnitude system. Silicon detectors not only have the best quantum efficiency, but also valuable capability at the longer wavelengths. The silicon curve is for a thinned CCD and reaches up to ~80%.

Quantum efficiency is a good way to specify the spectral response, but sometimes the responsivity is given instead. It specifies the charge generated by the detector per unit optical energy input. If the incident radiation energy is J and consists of N photons, then

$$J = Nhc/\lambda$$

in which h is Planck's constant, λ is the wavelength of the light, and c is the velocity of light. The charge generated in the detector is

$$Q = q(\lambda)\,eN,$$

where e is the electronic charge. Combining these two equations gives the responsivity:

$$\text{Responsivity} = \frac{Q}{J} = \frac{e\lambda}{hc}\,q(\lambda) = 8.066 \times 10^{-5}\lambda\,q(\lambda) \text{ coulomb/joule} \qquad (4.2)$$

for λ in angstroms.

The measurement of a detector's quantum efficiency is a laboratory job. In a typical situation the detector is compared to a calibrated diode using light from a stabilized source passed through a monochromator. The calibration of the diode reference is done separately, for example by the National Institute of Standards and Technology (NIST). See Coles *et al.* (2017) for calibration in an astronomical context.

So then, silicon detectors, especially thinned and overcoated CCDs, do a remarkably good job at detecting light. Before looking at their other parameters, we take a short look at how these devices work.

Silicon-Diode Arrays

Silicon detectors come in various sizes, often 4096 pixels in each dimension with square pixel sizes in the range of 5–30 μm. Figures 4.2 and 4.3 show two silicon devices. The first is designed specifically for spectroscopy. The silicon is infused with other elements such as phosphorus to improve its conductivity. We concentrate on the charge-coupled device, the CCD, the most common astronomical detector in use.

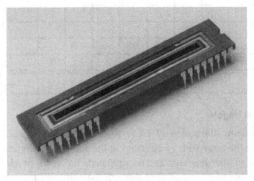

Figure 4.2 An early silicon detector designed for spectroscopic work. Courtesy of EG&G Reticon.

Figure 4.3 The CCD matrix for the Kepler satellite is shown. Each individual CCD is 2.8 by 3.0 cm with 1024 by 1100 pixels. Courtesy of NASA and Ball Aerospace.

Upon being struck by a photon, an electron is freed in a silicon pixel. In effect, each pixel of the array acts like a small capacitor, storing the charges until the integration time is over. Static barrier voltages within the pixel grid keep the electrons penned up in their pixel area. When the integration time is finished, the penning voltages between rows are changed to shift each row down one step, as pictured in Figure 4.4. The charges in the bottom row of pixels are transferred into a serial register and then marched out of the register, pixel by pixel, using a voltage step wave in the on-chip circuitry. As the charge packet from each pixel exits, it is converted to a voltage proportional to the number of charges by means of a small capacitor, $\sim 10^{-13}$ farad, and amplified. Then the next row is down-shifted, and so on until the whole array is read out. There is an error associated with each pixel's signal called the readout noise. A more detailed discussion of CCD construction and operation is given by MacKay (1986).

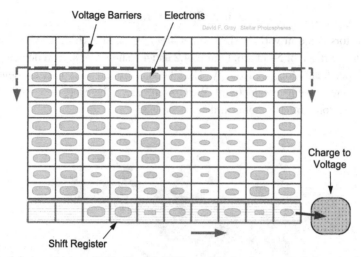

Figure 4.4 A caricature illustration of CCD rows and columns. Each box in the grid represents a pixel in the array with its electrons at the end of an integration. As the voltage barriers (grid lines in the diagram) are manipulated, the rows of electrons are shifted downward. The bottom-most row is filed through the shift register, where the individual pixels are read out serially into a charge-to-voltage converter.

It is also possible to do on-chip binning, where the penning voltages are manipulated to combine the charges from several adjacent pixels before marching them out through the shift register. This speeds up the readout and reduces the noise inherent in the readout process, a technique that is useful when the signal is exceedingly weak and fighting to stay above the noise. Stellar spectroscopy is a natural situation for on-chip binning, since the height of the spectrum usually spans many rows but carries no additional information about the spectrum. Slower readout rates produce less noise during the read process. It takes a few seconds to tens of seconds to read the spectrum out of a 4096^2-pixel CCD. To avoid contamination by additional light impinging on the CCD during readout, a shutter is used to close off the light beam.

The efficiency of charge transfer as rows are moved down during readout is not perfect, with values ranging from ~0.999 50 to 0.999 99 per transfer. After 2048 transfers, for example, 0.999 90 to the 2048th power gives 81%. In effect, some of the charge is left behind and blurs into the following pixels. Slower readout moves more of the charge.

Bias Offset and Dark Leakage

The photon signal recorded by a CCD is on top of a background. If the background is not handled properly, the spectrum can be ruined.

To start with, CCDs are set up to run with a bias level that acts as an artificial floor for the photon signal. The main reason for this is to avoid any negative numbers in the output. While photon count cannot be negative, random noise spikes can. Part of the routine of using a CCD is to record short-exposure dark frames, called bias frames, right before or after the stellar observation. A mean of several bias frames can be used to reduce their noise. The mean bias frame is subtracted from the frame that has recorded light. The exact bias value varies from pixel to pixel, so it is important to take bias frames with the same binning used for the stellar observations.

Second is thermal leakage, also called dark current. Thermal agitation within the crystal lattice frees some electrons in the silicon. When the array is read out, there is no way to distinguish these maverick electrons from the photoelectrons. Solution? Cool the array. At $-70\,°C$, a typical leakage is roughly one electron per second per pixel. Exposure times for stellar spectra can exceed an hour. Fortunately thermal leakage falls exponentially with temperature. Astronomical CCDs are often cooled to about -100 to $-130\,°C$ using liquid nitrogen (which is at $-196\,°C$), forcing the dark leakage down to $\sim 10^{-3}$ electrons per second per pixel. And the temperature must be stable. Leakage is proportional to the exposure time. A long-exposure dark frame needs to be taken to get a good handle on the amount of leakage actually occurring, but an hour dark exposure, for example, can be scaled to the exposure times used for the individual stellar exposures. The exposure-time-scaled leakage frame is subtracted from the stellar frame. Any inaccuracy in determining the leakage causes a zero-level error in the stellar spectrum. The larger the thermal leakage, the larger the uncertainty in the zero level. The depths of strong spectral lines are the most affected by this error. Further, noise from the leakage frame contributes to the overall noise in the final spectrum.

The refrigeration system is much bulkier than the CCD itself and this has to be considered when designing a spectrograph. The usual solution is to place a small pick-off mirror just before the camera focus to direct the beam above or below the optical axis of the spectrograph, where there is plenty of room for the cooling system.

Cooling also affects the spectral response, $q(\lambda)$. Generally there is a decline in red response for cooler temperatures. The effect can be dramatic, amounting to as much as a factor of 10 for a $100\,°C$ change in temperature (Young 1963, Vogt *et al.* 1978). Wide spectral coverage, as in stellar energy distribution measurements, therefore requires temperature stabilization to a fraction of a degree.

The only difference between a bias frame and a dark-leakage frame is the exposure time. Bias frames are short, ~ 1 second, so leakage is nil. Dark leakage frames are long, ~ 1 hour, so the leakage can be confidently seen and measured. It is assumed in most of the material coming next that the bias and dark leakage have been measured and subtracted from the raw signal.

Pixel Variations and Flat Fields

Pixel-to-pixel sensitivity can vary by several percent, even on a very good array. The standard procedure for removing this variation is to divide the stellar exposure by a "flat field" exposure. This is relatively painless, since it involves a short exposure of a continuous spectrum, usually from an incandescent lamp placed in front of the spectrograph entrance slit. The signal-to-noise ratio in the flat field should be made several times higher than the value for the stellar exposure so that very little increase in noise results from the flat-field division.

Fringing can cause problems with the flat-field process. Silicon becomes progressively transparent for $\lambda \gtrsim 7000\,\text{Å}$ and reflections from the front and back of the thin silicon layers interfere. The thinner the silicon, the worse the problem. Fringing can have amplitudes of several percent. The flat-field division does not necessarily remove the fringes and can even enhance them. Fortunately fringing is not always a problem, especially at visible wavelengths and with the thicker silicon chips. More technical information can be found in Malumuth *et al.* (2003) and Howell (2012).

High quantum efficiency led us to focus on CCDs, and it is the number one parameter, but there are other characteristics of importance. Near the top of the list is linearity.

Linearity

When the output of a detector is exactly proportional to the amount of incident radiation, we have a linear detector. Linearity is important because it means that the detector output faithfully represents the stellar spectrum. When linearity is lacking, a response calibration must be made so that the detector signal can be translated into the corresponding amount of light. Not only is this an inconvenience, but it results in loss of accuracy.

Linearity in detectors depends in the first instance on the linear photoelectric effect: one photon, one electron. This is true as long as the energy of the photons exceeds a work function characteristic of the material, but that condition is fulfilled for silicon by all photons in the visible spectral region. Linearity depends in the second instance on the electronics involved in handling the charge produced by the photons. Inside the array, poor charge transfer efficiency is a non-linearity. After the signal leaves the detector, various types of operational amplifiers are used to strengthen and integrate the voltage pulses. Linearity must be maintained at each step to give us what we need.

Digitization of the signal is the final process before the exposure is stored in computer-readable form. Analog-to-digital (A/D) conversion devices are readily available, but their accuracy, linearity, range, and speed can vary widely. Since the final numbers are generated on a scale set by the A/D conversion, we often see output expressed in ADU, i.e., analog-to-digital units. Translating this to numbers of photons requires a determination of the photon-ADU conversion factor for the system as a whole. While it is true that the conversion factor can be made larger or smaller by adjusting amplifier gains, the final value still has to be measured. We will do this later in the chapter.

The basic linearity test of the whole system is to plot ADU output for different exposure times. Obviously the light source must be stable during such measurements, more stable than any non-linearities we might wish to discover. For an exemplary description of silicon detector testing, see Vogt *et al.* (1978). Figure 4.5 shows the results of a linearity test, CCD output versus integration time. Deviations from the least squares line are displayed in the right panel.

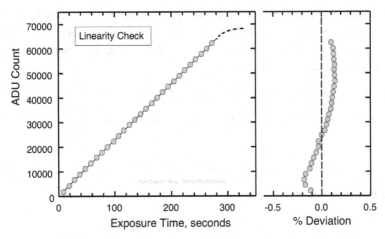

Figure 4.5 CCD output in analog-to-digital units can be plotted for a range of exposure times to check the linearity. For this particular CCD, the deviations from linearity are <0.2%. Eventually, as the exposure is increased, saturation comes in, as indicated by the dashed line at the top in the left panel. Based on Downing *et al.* (2006).

Linearity, as in the left panel of Figure 4.5, will eventually fail as exposure levels are raised to saturation. The pixels run out of electrons and have increasingly less response to incoming light. Putting too much light on the detector could be a problem when measuring calibration lamps, and could even be an issue with stellar spectra when we are pushing toward high signal-to-noise ratios. The maximum photon count before saturation, termed the *well depth*, is important for precise measurements. We return to this property after first considering noise.

Noise in the Measurements

Noise in data is undesirable and unavoidable. There are three basic categories: (1) photon noise in the starlight itself, (2) noise from the sky background, usually negligible for stellar spectroscopy, and (3) equipment noise, mainly from the dark leakage and the readout noise of the CCD. There is no way to remove the statistical noise from a photon stream; it is the theoretical minimum. Photon noise arises from the irregular times of arrival of the photons. The fluctuations in photon rate are described by Bose–Einstein statistics. For a thermal source of characteristic temperature T, the mean-square fluctuation, δN photons, is

$$\delta N = N^{1/2} \left(1 + \frac{1}{e^{h\nu/kT} - 1} \right).$$

(4.3)

Since $h\nu$ is appreciably greater than kT in most cases, we use

$$\delta N = N^{1/2}$$

(4.4)

as an adequate description of photon noise. This same relation holds for stellar photons and for photons coming from the sky background. In fact, since the thermal leakage is measured in the same manner, we apply the same expression to it. Similarly we can express the readout noise in equivalent photon units. Let us bundle together the total combined background value from all three sources, and call it B. If N is the total number of photons of which a number L come from the star, then

$$L = N - B,$$

and the statistical uncertainty in L is the square root of the quadratic sum of the individual errors giving

$$\delta L = \left(\delta N^2 + \delta B^2\right)^{1/2}$$

or, using Equation 4.4,

$$\delta L = (N + B)^{1/2}. \tag{4.5}$$

The signal-to-noise ratio is then

$$s/n = \frac{L}{\delta L} = \frac{N - B}{(N + B)^{1/2}}. \tag{4.6}$$

There are two asymptotic cases. The first is the good case, when $B \ll N$, making N essentially the same as L, and

$$s/n = L^{1/2}. \tag{4.7}$$

This is the domain of pure signal-photon noise and represents the ideal minimum noise case. For example, detecting 10^4 photons gives a s/n of 100 or a 1% error. If the photon flux is constant with time, s/n improves with the square root of the integration time. On the other hand, if we work to the same s/n from star to star, the integration time needs to vary inversely with the brightness of the source; source fainter by 2, get the same L by integrating twice as long. If it took one hour to reach a signal-to-noise ratio of 500 for a 6th magnitude star, it will take a little more than two and one half hours to do the same for a 7th magnitude star.

The opposite regime, where $N \ll B$, implies a small signal on top of a large background – an undesirable situation to be in. Then $N \approx B$ and Equation 4.6 gives

$$s/n = \frac{L}{(2B)^{1/2}}.$$

When stuck in this position, we should strive to increase L by every available means: a larger telescope, better throughput, higher quantum efficiency, and to decrease B.

If we are not in the regime described by Equation 4.7, we need to look more carefully at the background, B. Combining errors in quadrature,

$$B = \delta B^2 = \delta D^2 + \delta S^2 + r^2$$
$$= D + S + r^2,$$

where $\delta D^2 = D$ is the equivalent photon count for the dark leakage, $\delta S^2 = S$ is the photon count for the sky background, and r is the readout noise in photon units. If B is dominated by skylight and/or dark leakage, it scales with the integration time just as L does. If instead B is dominated by the readout noise, it is independent of the integration time.

In any case, while the noise from the dark leakage and the sky may be small, the readout noise is always there, so how do we measure it?

Measuring a Detector's Readout Noise

The overall noise of a CCD system can be very good. Indeed, the intrinsic system noise is frequently limited by the fundamental readout process, so we talk about the "readout noise," meaning the detector noise when thermal leakage is negligible. Just as we use repeated measurements to improve any observation, and we use the root-mean-square scatter of those measurements to estimate the uncertainty, here we use the scatter of repeated short-exposure dark readouts to estimate the readout noise. The difference between two dark frames gives us $2^{1/2}$ times the noise for each individual exposure. We could take many dark frames and look at the distribution of values for each pixel. This would give us a large number of Gaussian distributions, the $1/e$ widths being the readout noise for each pixel. In situations where there are a few problem pixels, this technique can be useful for investigating those few. But it is unlikely that we want to look at millions of individual pixel distributions (e.g., $2048 \times 2048 = 4\,194\,304$). Instead we assume all the pixels have the same noise behavior, which is a good assumption for a CCD since the readout hardware is the same for each of them, and take the root-mean-square of the pixel differences between two exposures. We cannot use just one exposure because then we would be measuring primarily the pixel-to-pixel differences in quantum efficiency. By differencing two exposures, the quantum-efficiency roughness of the array cancels out, and we are essentially using the individual pixel differences to give us the distribution function of the noise. That distribution is usually Gaussian or close to it, and the rms value is the $1/e$ width. Again, since the noise from two exposures enters, the readout noise for the individual exposures is $2^{1/2}$ smaller. So now we have a value for the readout noise in analog-to-digital units.

The translation of ADUs to photons need not be known for much of the scientific work, but one can make an estimate as follows. For sufficiently high light levels, photon noise will dominate and then we know that $s/n = L^{1/2}$, where L is in photon count, as in Equation 4.7. But we can *measure* s/n from repeated observations of a flat-field lamp, and infer from $s/n = L^{1/2}$ the number of photons that must have been detected. We also know the mean light level in ADU from these exposures. The ratio of the inferred photon count to the light in ADUs gives us the conversion factor, the number of photons per ADU. Now we know both the signal and the readout noise in photon count units.

This is all summed up in Figure 4.6. Pairs of lamp exposures are taken over a range of light levels. From them we get the measured values of s/n, which are plotted as a function of mean light level in ADU for each pair of exposures. Equation 4.6, where now the background is the readout noise ($B = \Delta B^2$ is replaced by r^2), is shown for different values of r. The number of photons per ADU is adjusted until the upper part of the curve matches the measurements (left–right shift). Then r is varied in Equation 4.6 to match the lower part of the measurements. We can elaborate on this slightly by writing Equation 4.7 as

$$\log \frac{s}{n} = \frac{1}{2}\log(L_{\mathrm{ADU}}G_0) = \frac{1}{2}\log L_{\mathrm{ADU}} + \frac{1}{2}\log G_0, \qquad (4.8)$$

in which the light count is expressed in ADUs and G_0 is the conversion factor in photons per ADU, called the gain. This is the "Pure Photon Noise" line in Figure 4.6, where G_0 has been adjusted to match the left–right position of the observed points. This process established the value of the gain, $G_0 = 15$ photons/ADU in this particular example. As can be seen from Equation 4.8, when $\log s/n = 0$, $\log G_0 = -\log L_{\mathrm{ADU}}$, shown as the solid dot in the figure. In other words, a quick way to estimate G_0 is to draw a straight line through the photon-noise points and extrapolate it to $\log s/n = 0$. If the photon-noise section is poorly defined, remember, the slope is 1/2.

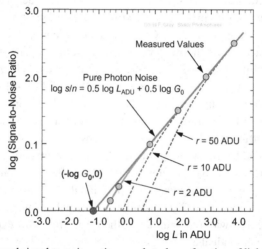

Figure 4.6 Measured signal-to-noise ratios are plotted as a function of light level in analog-to-digital units. Equation 4.6 is plotted as lines for different values of readout noise, r, as labeled. The value for $r = 2$ fits the observations.

A common variation of this plot is shown in Figure 4.7. It is termed a "photon transfer curve." Here the noise itself is plotted against the signal level, so we have essentially Equation 4.8 again but with $\log n_{\mathrm{ADU}}$ on the left and the sign on the last term reversed. Follow the same path: the conversion factor G_0 is adjusted to get the left–right match in the upper portion of the curve and the value for r is adjusted to match the lower portion. The "quick" point, when $\log n = 0$, is shown by the solid dot. For some CCDs, there seems to be a decrease in slope at higher light levels (MacKay 1986, Downing *et al.* 2006, Astier *et al.* 2019), so this kind of calibration should be done with light levels well below the saturation level.

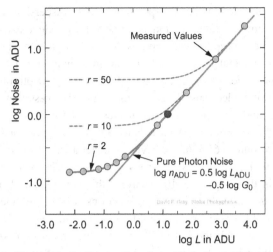

Figure 4.7 Measured noise is plotted as a function of light level. The pure photon noise is used to find the scale factor of photons per ADU. The curve that fits the observations has $r = 2$. This curve approaches the ordinate value of the readout noise.

While it is nice to know the signal-to-noise ratio from Equation 4.6, the real noise in the observations is best determined by comparing repeat exposures of the star. That way all sources of noise are included, readout noise, leakage noise, photon noise, continuum normalization noise, and other source we have not yet recognized.

Well Depth and Dynamic Range

The charge capacity of one pixel is called the well depth or full-well capacity. The well depth depends on the amount of silicon in each pixel; bigger pixels have more potential photoelectrons. If the pixel is exposed to too much light, the well depth is exceeded and the output is capped at its maximum value, and charges begin to leak into adjacent pixels. When the CCD is pushed into saturation, it is fairly obvious. The output numbers fail to scale with the level of exposure and often just stay at a constant value. Fortunately a CCD is not permanently damaged by too much light, although a short dark or low-light exposure to clear errant electrons should be done before resuming scientific work.

Our main concern with well depth is that we be able to collect enough electrons to reach high signal-to-noise ratios. For example, in high-precision spectroscopy, we need to reach high s/n, say 500, to make us sufficiently happy. Then, according to Equation 4.7, we need to detect 250 000 photons. Do real CCDs give us what we need? On the low end, there are CCDs with 9 μm pixels and a well depth of ~85 000, not quite up to what we need. On the high end, pixels range up to 30 μm with well depth of ~10^6, comfortably sufficient. And it is usually a good idea to stay well below saturation. One alternative, if the well depth is too small, is to add together shorter exposures, and this is OK when $B \ll L$. But the extra readout time is taken from your telescope allotment, and there are more numbers to process.

A better solution might be to bin pixels perpendicular to the dispersion. The CCD with only 85 000 well depth is bumped up to the range we hoped for if three rows are binned.

Right along with the concept of well depth is that of the dynamic range, defined by well depth divided by the readout noise. It tells us the maximum signal range the CCD can handle in one exposure. For instance, take a CCD with a well depth of 250 000 electrons and a readout noise of 5 electrons. Then the range in brightness that can be covered is ~50 000. That is big, and it is one of the advantages of CCDs.

In practice the A/D converter plays a role here. Suppose we have an A/D converter with 16 bits. Its maximum count is 65 536. If we wish to cover a well depth of 250 000 electrons, the A/D converter will give us 3.8 electrons per count. The readout noise should be resolved, so if r is in the range of 5–10 electrons, this is workable. If the CCD has well depth 10^6 electrons and a readout noise of 5 electrons, we might choose an 18-bit A/D, giving us a resolution of ~4 electrons, or a 20-bit A/D giving us ~1 electron. If we never wish to work higher than half the full well capacity, then the 18-bit A/D can give us ~2 electron resolution. If we have the money, we choose the 20-bit device or even larger so as to not be limited by numerical resolution.

A Synoptic Review of the Process

At this stage we have covered several aspects of CCDs. It does not hurt to recapitulate the reduction process. The stellar spectrum is the end result of combining four separate exposures: the stellar exposure, the bias frame, the dark or leakage frame, and the flat-file lamp exposure,

$$\text{spectrum} = \frac{(\text{star} - \text{bias}) - E_S(\text{dark} - \text{bias})}{(\text{lamp} - \text{bias})},$$

in which E_S is the ratio of exposure times, stellar/dark. In principle a (dark − bias) is subtracted from the denominator too, but in practice the lamp exposure times are so short that the dark level is negligible.

Quantum Efficiency versus Readout Noise

Although we have considered several parameters for detectors, we are sometimes faced with a choice balancing quantum efficiency against readout noise. The goal is to get the best signal-to-noise ratio in the shortest exposure time. Assuming the dark leakage is small, we alter Equation 4.6 to explicitly including the quantum efficiency, $q(\lambda)$, and the readout noise, $B = r^2$. Also assume that the bias and lamp exposures are done with care so their noise will be much lower than for the stellar exposure. Let l be the number of incident photons, such that the counted number is $L = q(\lambda)l$. Then

$$s/n = \frac{q(\lambda)l}{[q(\lambda)l + r^2]^{1/2}}. \tag{4.9}$$

Figure 4.8 shows how this signal-to-noise behaves for two sample detectors. One of them has a pleasantly high quantum efficiency of 90%, but a not-so-good readout noise of 100 photons. The other has a modest quantum efficiency of 40%, but a more desirable readout noise of 10 photons. Which one is better? The result is interesting. Equation 4.9, as shown in Figure 4.8 for these two case, tells us that the answer depends on the signal-to-noise ratio we need. Faint stars and low s/n profit more from the low readout noise; bright stars and high s/n profit more from the high quantum efficiency.

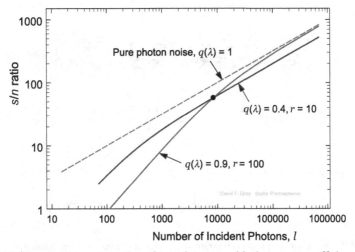

Figure 4.8 The signal-to-noise ratios of two detectors with the quantum efficiencies and readout noise values as labeled are compared. The detector with higher $q(\lambda)$ wins when $s/n > 57$, while the detector with lower $q(\lambda)$ wins when $s/n < 57$.

Cosmic Rays

A word about cosmic rays. They pepper a typical CCD during any exposure. The rate depends on the terrestrial latitude because of the variation in the Earth's magnetic-field shielding, on altitude because many are absorbed by the air, and on shielding in the housing around the CCD. At mid-latitudes, at sea-level, a two-hour exposure can show as many as one per mm^2. If not too much on-chip binning has been done, the larger hits are easy to identify, and interpolation between neighboring pixels can be used to patch over the affected pixel or pixels.

Photomultiplier Tubes

The photomultiplier is the classic pulse-counting detector. They were the detectors used in setting up dependable magnitude systems such as the standard UBV and other filter systems. Although it has desirable properties, including linearity and passable quantum efficiency of ~20%, it remains a single-element detector and therefore is less efficient than an array detector. The efficiency factor can be large, with a potential gain of 4096^2 CCD pixels compared to one for the photomultiplier. And if the CCD has a quantum efficiency of 90%, then it gains another factor of four or more, especially toward longer wavelengths (as per Figure 4.1).

Figure 4.9 shows a photograph of a photomultiplier. It is a vacuum tube with the light-sensitive cathode on the flat end. An electron is emitted from the cathode when a photon is absorbed – the classic photoelectric effect. This electron is multiplied $10^6 - 10^9$ times by electron cascade through a dynode chain that operates in the range of 1–2 kV. Unlike silicon detectors, photomultipliers can be destroyed by excessive illumination.

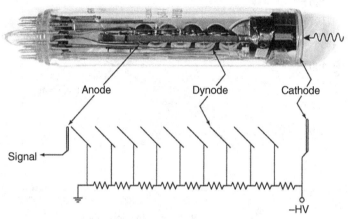

Figure 4.9 This photograph and diagram show the structure of a photomultiplier. Light strikes the cathode on the right, ejecting a photoelectron that is then accelerated by the electrical field of ~100 volts to the first dynode. There two or more electrons are emitted by the impact of the initial electron. The process is repeated over again at each dynode. The overall electron multiplication depends on the number of dynodes and the accelerating voltage. The electrical pulse comes out at the anode and is fed into external amplifiers.

The dynode chain is a near-perfect amplifier (Lallemand 1962, Engstrom 1963), and yields pulses lasting $\sim 10^{-8}$ seconds at the output (anode). These pulses can be counted with appropriate amplifiers and pulse-height discriminators (Morton 1968) up to rates of 10^7 per second, although coincidence corrections need to be applied for the higher count rates. Photomultipliers can also be used in a direct current mode in which the current from the tube is measured instead of individual pulses. More details can be found in Schonkeren (1970) and Young (1974). The same amplification principle is used in electron multiplication CCDs (Ives *et al.* 2008).

Although it has faded from active use, the photomultiplier played an important historical role as an astronomical detector.

The Photographic Plate

The wide use of photography prior to the 1970s is now completely superseded by electronic detectors. The photographic process suffers from non-linear response, low quantum efficiency (see Figure 4.1), and other effects detrimental to precision spectroscopy. Refer to Chapter 4 of the first edition of this book (Gray 1976), Wright *et al.* (1963) or Marx (1988) for details. The one advantage held by photography is the huge number of pixels per exposure. A typical 40 cm by 40 cm photographic plate can have ~10^9 detection elements. Large monolithic CCDs and mosaics of smaller CCDs are gradually gaining on this front.

Still, it is well to remember the legacy of photography in astronomy. Many of the spectral type classifications, such as the MK system, were made using photographic plates.

Questions and Exercises

1. Would orienting the CCD in a spectrograph such that the spectrum is spread across columns, rather than rows, mitigate a poor charge transfer efficiency? How many rows would you expect the height of the spectrum to fill?

2. What effect would there be on the flat-field lamp division if the CCD showed non-linear behavior?

3. What minimum signal-to-noise ratio would you expect in a spectrum where an average of 100 photons have been recorded per resolution element? And for 20 000 photons?

4. Suppose you want to take an exposure of a star's spectrum, but the star is faint, so faint that the dark leakage is as large as the stellar signal. How much have you lost in the signal-to-noise ratio compared to having zero leakage? What happens to the s/n if you now add the readout noise and it too is as large as the stellar signal? What can you suggest to improve the s/n?

5. If the well depth of a certain CCD is 100 000 electrons, how many bits would you recommend having in the A/D converter?

6. Two light detectors have the following parameters: Detector 1, peak quantum efficiency is 0.92, readout noise is 150 equivalent photons; Detector 2, peak quantum efficiency is 0.40, readout noise is one equivalent photon. Assume the dark leakage is zero. Which detector is more effective for faint sources and low signal-to-noise work? Is the opposite one then preferred for bright sources and high-signal-to-noise work? Explain. At what signal-to-noise ratio do these two detectors perform equally well?

References

Astier, P., Antilogus, P., Juramy, C., *et al.* 2019. *A&A* **629**, 36.
Coles, R., Chiang, J., Cinabro, D., *et al.* 2017. *J Instrumentation* **12**, C04014.
Downing, M., Baade, D., Sinclaire, P., Deiries, S., & Christen, F. 2006. *SPIE* **6276**, 627609.
Engstrom, R.W. 1963. *Phototubes and Photocells*, RCA Technical Manual PT-60, pp. 6, 28.
Gray, D.F. 1976. *The Observation and Analysis of Stellar Photospheres* (New York: Wiley), 1st ed., Chapter 4.
Howell, S.B. 2012. *PASP* **124**, 263.
Ives, D., Bezawada, N., Dhillon, V., & Marsh, T. 2008. *SPIE* **7021**, 70210B.

Lallemand, A. 1962. *Astronomical Techniques* (Chicago: University of Chicago Press), W.A. Hiltner, ed., p. 126.

MacKay, C.D. 1986. *ARAA* **24**, 255.

Malumuth, E.M., Hill, R.S., Gull, T., *et al.* 2003. *PASP* **115**, 218.

Marx, S. (ed.) 1988. *Astrophotography* (Berlin: Springer-Verlag).

Morton, G.A. 1968. *App Opt* **7**, 1.

Schonkeren, J.M. 1970. *Photomultipliers, Phillips Application Book* (Eindhoven: Phillips), pp. 6, 28, 116.

Turriziani, S. 2019. AAS Historical Astronomy Division, 29 October edition.

Vogt, S.S., Tull, R.G., & Kelton, P. 1978. *App Opt* **17**, 574.

Wright, K.O., Lee, E.K., Jacobson, T.V., & Greenstein, J.L. 1963. *PDAO Victoria* **12**, 173.

Young, A.T. 1963. *App Opt* **2**, 51.

Young, A.T. 1974. *Methods in Experimental Physics, Volume 12, Part A: Astrophysics, Optical and Infrared* (New York: Academic Press), N. Carleton, ed., p. 1.

5

Radiation Terms and Definitions

The photosphere of a star is filled with light, and some of it escapes to us. To properly describe what we see and also to compute a spectrum from the model photospheres that we will be constructing, we need some concepts and terminology. The two most basic of these are specific intensity and flux.

Specific Intensity: the Cornerstone

Consider the radiating surface depicted in Figure 5.1. This surface might be a real physical boundary or it could be an imaginary surface inside a radiating gas. Denote a small portion of this surface by ΔA through which passes an increment of energy, ΔE_ν, in the direction θ from the normal, contained within the solid angle $\Delta\omega$, in a spectral interval $\Delta\nu$, in a time span Δt. The specific intensity, I_ν, is defined with these parameters as

$$I_\nu = \lim \frac{\Delta E_\nu}{\Delta t\, \Delta A \cos\theta\, \Delta\omega\, \Delta\nu}$$

$$= \frac{dE_\nu}{dt\, dA \cos\theta\, d\omega\, d\nu}. \tag{5.1}$$

The limit is taken as all the quantities diminish toward zero, including ΔE_ν. In this way, we define the specific intensity at a point on the surface. The adjective "specific" in front of the word intensity is often omitted. This same physical quantity goes under the name of "brightness" in radio astronomy. Since I_ν is in the θ direction, it is the projected area, $dA \cos\theta$ that is relevant.

Although the "surface" in Figure 5.1, real or imaginary, is perfectly general, it is often helpful to think of it as the surface of a star with light coming out, i.e., the specific intensity is being emitted. The light direction can also be reversed, and then we can think about the specific intensity being received, for instance by our telescope.

One has a choice of units for specific intensity. Equation 5.1 uses frequency units with $\Delta\nu$ in the denominator. We could equally well define an I_λ by replacing $\Delta\nu$ with a wavelength increment $\Delta\lambda$. The relation between these two spectral distributions is

$$I_\nu\, d\nu = I_\lambda d\lambda, \tag{5.2}$$

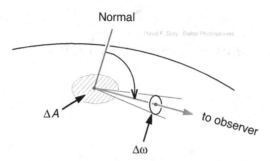

Figure 5.1 The geometrical portion of specific intensity is illustrated here. The increment of area, ΔA, is foreshortened by $\cos \theta$ in the direction of the observer, where θ is the direction of the observer from the surface normal. The increment of solid angle is shown as $\Delta \omega$.

which follows directly from Equation 5.1. The cgs units of I_ν are $\mathrm{erg}/(\mathrm{s\ cm^2\ rad^2\ Hz}) = \mathrm{erg}/(\mathrm{cm^2\ rad^2})$, where Hz (hertz) is used to denote cycles per second. The traditional units of I_λ are $\mathrm{erg}/(\mathrm{s\ cm^2\ rad^2 Å})$. The corresponding mks units are $\mathrm{watt}/(\mathrm{m^2\ rad^2\ Hz})$ and $\mathrm{watt}/(\mathrm{m^2\ rad^2 Å})$, respectively.

The two spectral distributions, I_ν and I_λ, have different shapes for the same spectrum. Characteristic features such as intensity maxima appear at different wavelengths. The solar spectrum, for example, has a maximum in the blue (~5000 Å) for I_λ, but the maximum is in the near infrared (~8800 Å) for I_ν. This curious state of affairs can be understood by looking at the relation between spectral bands for the two cases. Since $c = \lambda \nu$,

$$d\nu = -(c/\lambda^2)\, d\lambda.$$

The minus sign refers to the opposite directions of the coordinates, λ increases, ν decreases. We see that equal intervals in λ correspond to very different intervals in ν across the spectrum. With increasing λ, a constant $d\lambda$ (the I_λ case) corresponds to a smaller and smaller $d\nu$. These smaller $d\nu$ intervals obviously contain diminishing energy compared to the constant $d\nu$ intervals associated with I_ν. Coincidentally, prism and grating spectrographs approximately portray I_ν and I_λ, respectively, since the dispersion varies approximately as λ^{-2} for the prism, and is nearly constant with λ for the grating (Equation 3.6). Most of the time in these pages, we arbitrarily adopt frequency units.

In the computation of model photospheres, which we take up in Chapter 9, I_ν is obtained from the transfer equation, the topic of Chapter 7. Some of the remainder of this chapter and portions of several others are preparation for obtaining I_ν for given boundary conditions and specified geometry.

The mean intensity, J_ν, is defined as the directional average of the specific intensity,

$$J_\nu = \frac{1}{4\pi} \oint I_\nu \, d\omega, \tag{5.3}$$

where the circle on the integral indicates that the integration is done over the whole unit sphere of 4π square radians.

Flux: What Stars Send Us

Flux is a measure of the net energy flow, ΔE_v, across an area ΔA, in a spectral range Δv, and in a time interval Δt as these quantities are taken to the infinitesimal limit. The only directional significance is which direction the energy crosses ΔA. This total net energy, the flux, is the sum of all the ΔE_vs we would need to describe the same radiation using the specific intensity terminology, and we therefore denote it by $\sum \Delta E_v$ and write the definition of flux as

$$\mathfrak{F}_v = \lim \frac{\sum \Delta E_v}{\Delta t \, \Delta A \, \Delta v} = \frac{\oint dE_v}{dt \, dA \, dv}, \tag{5.4}$$

where again the circle on the integral indicates a complete integration over all directions. Notice the lack of θ and solid-angle dependence in Equation 5.4 in contrast to the specific intensity definition in Equation 5.1. We still have Figure 5.1 in mind for dA and θ. Flux can be related to intensity by combining their definitions, which have $dE_v/(dt \, dA \, dv)$ in common, to get

$$\mathfrak{F}_v = \oint I_v \cos \theta \, d\omega. \tag{5.5}$$

This is the fundamental working relation between specific intensity and flux. In theoretical stellar atmospheres, a quantity called "astrophysical flux" is often defined. It is equal to the real flux, $\mathfrak{F}_v$, divided by π. As with specific intensity, we have a choice of frequency or wavelength units.

When the radiation is isotropic, $\mathfrak{F}_v = 0$, that is, equal amounts of radiation flow in both directions across the surface. In this sense the flux is a measure of the flow of radiation and the anisotropy of the radiation field. In particular, if we look at a point on the physical boundary of a radiating sphere like a star, we can write, using Equation 5.5 and ϕ as the azimuthal coordinate,

$$\mathfrak{F}_v = \int_0^{2\pi} \int_0^{\pi} I_v \sin \theta \cos \theta \, d\theta \, d\phi$$

$$= \int_0^{2\pi} \int_0^{\pi/2} I_v \sin \theta \cos \theta \, d\theta \, d\phi$$

$$+ \int_0^{2\pi} \int_{\pi/2}^{\pi} I_v \sin \theta \cos \theta \, d\theta \, d\phi$$

$$= \mathfrak{F}_v^{\text{out}} + \mathfrak{F}_v^{\text{in}}.$$

That is, the full flux is the sum of the flux leaving ΔA plus the flux entering ΔA. At a real stellar surface, the second term is usually zero. If, in addition, there is no azimuthal dependence for $\mathfrak{F}_v$, we have for the spectrum leaving the star

$$\mathfrak{F}_v = 2\pi \int_0^{\pi/2} I_v \sin \theta \cos \theta \, d\theta. \tag{5.6}$$

This is a very important equation. We will use it in many different situations, most particularly to compute spectra.

The simplest configuration occurs when I_ν in Equation 5.6 is independent of direction (over the hemisphere $0 \le \theta \le \pi/2$). Then

$$\mathfrak{F}_\nu = \pi I_\nu, \tag{5.7}$$

since the integral of $\sin \theta \cos \theta$ gives a value of 1/2.

The difference between I_ν and $\mathfrak{F}_\nu$ is worth some consideration. For example, I_ν is independent of distance from the source, whereas $\mathfrak{F}_\nu$ obeys the standard inverse square law. Specific intensity can only be measured directly if we can resolve the radiating surface, otherwise we necessarily measure flux. It is $\mathfrak{F}_\nu$ we measure for most stars, but I_ν we normally measure for the Sun. These differences arise because of the solid angle $\Delta\omega$ appearing in Equation 5.1, but not in Equation 5.4.

To elaborate slightly, consider an angularly large and uniformly bright source at a distance r being measured with the telescope shown in Figure 5.2. Part of the image formed in the focal plane is the small circle marked A_D that represents a resolution element of the detector. The source area A_S matches A_D. The solid angle in which radiation is coming to A_D is $A_D/F^2 = A_S/r^2$. The surface in Figure 5.1 corresponds to the lens. In such an arrangement we are measuring I_ν because it is the radiation per solid angle that the detector records. Now, as r increases, the source diminishes in angular size, but the area A_D is still fully illuminated and the output signal from this detector element remains constant since the source is uniformly bright over its surface. Eventually, as r continues to increase, the whole source will subtend a solid angle equal to and then less than A_D/F^2. When this happens, the source is no longer resolved, and there is a transition from intensity to flux.

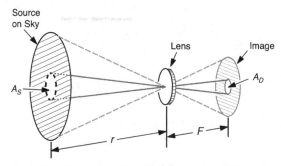

Figure 5.2 A portion, A_S, of the uniformly bright source at a distance r corresponds to a resolution element, A_D, in the image on the detector. The focal length of the lens is F. When A_D is only a portion of the full image, the image is resolved and we measure I_ν. As the distance r increases, the full image of the source becomes smaller until it is completely contained within the one resolution element, A_D. At this r, and for all larger distances, we can only measure the flux, $\mathfrak{F}_\nu$.

The specific intensity and the flux are the two building blocks we use extensively. In special situations the mean intensity, which we have already met, and the K integral, to which we now turn, are useful.

The K Integral and Radiation Pressure

To show the mathematical pattern, reiterate the definition for the mean intensity,

$$J_v = \frac{1}{4\pi} \oint I_v \, d\omega,$$

and for the flux,

$$\mathfrak{F}_v = \oint I_v \cos \theta \, d\omega,$$

and now for the K integral, using the second moment of $\cos \theta$, that is,

$$K_v = \frac{1}{4\pi} \oint I_v \cos^2 \theta \, d\omega. \tag{5.8}$$

Physically this integral is related to the radiation pressure. Electromagnetic radiation has momentum equal to its energy divided by the velocity of light, c. Consider photons transferring momentum to a solid wall in a manner analogous to the usual discussion of particles in kinetic gas theory. The component of momentum normal to the wall taken per unit time and unit area is the pressure,

$$dP_v = \frac{1}{c} \frac{dE_v \cos \theta}{dA \, dt},$$

where θ is the angle of the photon trajectory to the wall (also as in Figure 5.1). Casting this in terms of specific intensity gives

$$dP_v = \frac{I_v}{c} \cos^2 \theta \, d\omega \, dv.$$

The total radiation pressure comes with integration over all directions and over the whole spectrum. Thus

$$P_R = \frac{1}{c} \int_0^\infty \oint I_v \cos^2 \theta \, d\omega \, dv = \frac{4\pi}{c} \int_0^\infty K_v \, dv, \tag{5.9}$$

where the definition of K_v in Equation 5.8 has been used.

A commonly used special case assumes that I_v is independent of direction so that it can be factored out of the directional integration in Equation 5.9. Upon integration of $\cos^2 \theta$, one finds

$$P_R = \frac{4\pi}{3c} \int_0^\infty I_v \, dv. \tag{5.10}$$

In certain applications, a temperature is defined using Equations 5.7 and the Stefan–Boltzmann law (Equation 6.4) to give

$$\pi \int_0^\infty I_v \, dv = \sigma T^4,$$

where $\sigma = 5.6704 \times 10^{-5}$ erg/$(\text{cm}^2 \text{ K}^4 \text{ s})$ or 10^{-8} watt/$(\text{m}^2 \text{ K}^4)$ is the Stefan–Boltzmann constant, and then the radiation pressure can be recast as

$$P_R = \frac{4\sigma}{3c} T^4, \tag{5.11}$$

When Equation 5.11 is used in place of Equation 5.10, T is usually taken to be the kinetic gas temperature. Radiation pressure becomes quite important in the hottest stars. In solar-type and cooler stars, it can be safely neglected (see Table 9.1 for some numerical values).

The Absorption Coefficient and Optical Depth

Next on our list of concepts is the absorption of light. Consider radiation in a specified direction shining on a thin layer of material which itself is so cool that it does not radiate measurably. The intensity of light is found experimentally to be diminished upon passage through a layer of thickness dx by an amount dI_v, where

$$dI_v = -\kappa_v \rho\, I_v\, dx. \tag{5.12}$$

Here ρ is the mass density. The absorption coefficient, κ_v, has units of area per mass, and is therefore a mass absorption coefficient.

There are two physical processes contributing to κ_v. The first is true absorption where the photons are destroyed and the energy thermalized. The second is scattering where the photons are deviated in direction and removed from the solid angle being considered. Consequently, the absorption coefficient is really an extinction coefficient.

The reduction of the radiation as per Equation 5.12 depends on the combination of $\kappa_v \rho$ and dx over some path length L given by

$$\tau_v = \int_0^L \kappa_v \rho\, dx. \tag{5.13}$$

This new quantity, τ_v, is called the *optical depth* (unitless), and we use it often. With optical depth, Equation 5.12 becomes

$$dI_v = -I_v\, d\tau_v.$$

The solution is $I_v = I_v^0 e^{-\tau_v}$, the usual simple exponential extinction law. This is a "passive" situation where no emission occurs and is the simplest example of the radiative transfer equation (the main topic of Chapter 7) and its solution.

The Emission Coefficient and the Source Function

Hot gas in the photosphere will be emitting light. We treat the emission much the same as we did the absorption. An increment of radiation emitted from a layer of thickness dx in a specified direction is defined as

$$dI_v = j_v \rho\, dx. \tag{5.14}$$

Here again ρ is the density, and the emission coefficient, j_v, has units of $\text{erg}/(\text{s rad}^2 \text{ Hz g})$ or $\text{watt}/(\text{rad}^2 \text{ Hz kg})$. As with absorption, the physical processes contributing to j_v are (1) real emission, i.e., the creation of photons, and (2) scattering of photons into the direction being considered.

The ratio of emission to absorption has the same units as I_v, and can be thought of as the specific intensity emitted at some point in a hot gas. It turns out to be convenient to use this ratio when discussing the transfer of radiation, and it is given a special name, the *source function*, and is defined by

$$S_v = j_v/\kappa_v. \tag{5.15}$$

The physics of calculating S_v can sometimes be complicated, especially for spectral lines. Some of these complications are considered briefly in Chapter 13. For now we consider the two extreme cases of pure absorption and pure isotropic scattering as examples of the source function.

The Source Function for Pure Absorption

We have the case of pure absorption when all the absorbed photons are destroyed and all the emitted photons are newly created, with a distribution governed by the physical state of the material. Common usage of the term pure absorption implies the physical state of thermodynamic equilibrium in which all microscopic processes are balanced by their inverse. Emission from a gas in thermodynamic equilibrium is described by the laws of the black-body radiator, the topic of the next chapter. The source function for this case is given by Planck's radiation law,

$$S_v = \frac{2hv^3}{c^2} \frac{1}{e^{hv/kT} - 1}. \tag{5.16}$$

As usual, h is Planck's constant (6.6261×10^{-27} erg s or 10^{-34} J s) and c is the velocity of light. This is the specific intensity emitted by a gas of temperature T. According to this equation, the source function depends only on the frequency of the radiation and the temperature of the material. The black-body source function is often denoted as B_v instead of S_v. We will use this source function extensively in later chapters.

The relation $j_v/\kappa_v = B_v$ is called Kirchhoff's law (Kirchhoff 1860), although the form of B_v, expressed in Equation 5.16, was not discovered until the turn of the following century (Planck 1914).

The Source Function for Pure Isotropic Scattering

In the pure-isotropic-scattering case, all the "emitted" energy is due to photons being scattered into the direction under consideration. We imagine the state of affairs as depicted in Figure 5.3. The contribution dj_v to the emission from the solid angle $d\omega$ is proportional to $d\omega$ and to the absorbed energy $\kappa_v I_v$. It is isotropically re-radiated so the fraction per unit solid angle is $1/4\pi$. In this way we can write

$$dj_v = \frac{\kappa_v I_v d\omega}{4\pi}.$$

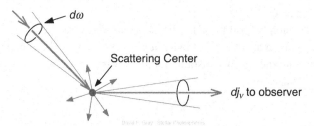

Figure 5.3 Radiation from the solid angle $d\omega$ is scattered isotropically, with dj_v coming toward the observer.

Since j_v and κ_v are both expressed per unit density, the factor of ρ on both sides has been omitted. We obtain all the contributions to j_v by integrating over all solid angles,

$$j_v = \frac{1}{4\pi}\oint \kappa_v I_v \, d\omega.$$

But κ_v is usually independent of ω, so we can write

$$S_v = \frac{j_v}{\kappa_v} = \frac{1}{4\pi}\oint I_v \, d\omega$$

or

$$S_v = J_v,$$

the mean intensity from Equation 5.3. For this case, the source function depends completely on the radiation field and not at all on the local thermal emission.

The Einstein Coefficients

When we deal with spectral lines or bound–bound transitions, we can describe the probabilities for normal spontaneous emission in terms of atomic constants. Consider the spontaneous transition between an upper level, u, and a lower level, l, separated by an energy $h\nu$. We assume the emission is isotropic. Then the probability that the atom will emit its quantum of energy in a solid angle $d\omega$ and in a time dt is $A_{ul} \, d\omega \, dt$. The proportionality constant, A_{ul}, is the Einstein probability coefficient for spontaneous emission. If there are N_u excited atoms per unit volume, the contribution of spontaneous emission to the emission coefficient is

$$j_v\rho = N_u A_{ul} \, h\nu. \tag{5.17}$$

Additional emission, called *stimulated emission*, is induced if a radiation field is present that has photons corresponding to the energy difference between the levels u and l. The induced emission is proportional to the amount of this radiation, and the induced photons show both phase coherence with and go in the same direction as the inducing photons. Because of this, the process is also called negative absorption. The probability for stimulated emission producing a quantum in the solid angle $d\omega$ in the time interval dt is written as $B_{ul}I_v \, d\omega \, dt$, where the directional dependence of the specific intensity becomes very important since it fixes the directional dependence of these new photons. B_{ul} is the Einstein probability coefficient for stimulated emission. You may recognize this as the basic process for lasers.

The true absorption probability is defined in the same way, and the proportionality constant is denoted by B_{lu}. Notice how these Einstein coefficients have been tied to I_v and j_v. Their units are therefore per steradian, and differ by 4π from definitions using energy density.

The mass absorption coefficient for this bound–bound transition can be expressed in terms of these Bs by considering the radiation absorbed per unit path length from the intensity beam I_v, namely, normal absorption less the stimulated emission,

$$\kappa_v \rho I_v = N_l B_{lu} I_v h v - N_u B_{ul} I_v h v,$$

where N_l is the population of the lower level per unit volume. I_v can be canceled in the above equation giving

$$\kappa_v \rho = N_l B_{lu} h v - N_u B_{ul} h v. \tag{5.18}$$

The amount of reduction in absorption from the second term turns out to be only a few percent in the visible spectrum (refer to Table 8.1).

The numerical calculation of the absorption coefficient is the topic of Chapter 8 for continuous absorption and Chapter 11 for line absorption.

Questions and Exercises

1. Experiment with some numerical manipulation of spectra to help understand their nature as a distribution (histogram) of photons. For example, start out by making a graph of Equation 5.16. What units does $B_v = S_v$ have? As with any histogram, one has some choice in the binning interval. What binning interval have you chosen? How would your graph change if the binning interval were widened by a factor of 10? Now use the basic relations in Equation 5.2 to convert I_v to I_λ. Make a graph of I_λ versus λ. Compare this new graph with the I_v graph. Do they peak at the same wavelength? If not, why not? On the I_λ graph, plot I_v as a function of λ. Why are these two plots not the same?
2. Prove that the flux from an isotropically radiating surface (upon which there is no incident radiation from the outside) is related to the specific intensity by $\mathfrak{F}_v = \pi I_v$. Then calculate the flux for a completely isotropic surface (as might be encountered deep inside a star).

References

Kirchhoff, G. 1860. *AnPhys* **185**, 275.
Planck, M. 1914. *The Theory of Heat Radiation* (Philadelphia: P. Blakiston's Son & Co.), translated by M. Masius, p. 168.

6

The Black Body and Its Radiation

Think of a container that is completely closed except for a very small hole in one wall. Any light entering the hole has a very small probability of finding its way out again, and eventually will be absorbed inside the container. The photon's predicament is illustrated in Figure 6.1. We have a perfect absorber – all the light that enters the small hole is absorbed inside. This type of container, or more specifically the hole, is called a black body.

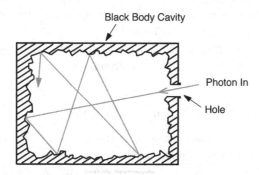

Figure 6.1 A photon enters the container through a small hole and is repeatedly reflected from the interior surface until it is absorbed. The probability of the photon finding its way out can be reduced almost without limit by increasing the size of the container, reducing the size of the hole, and by increasing the absorption of the walls through their blackness, roughness, and irregularity.

The connection of the black body to stellar photospheres starts to fall into place when we reverse the situation and heat the container. Then the walls glow, emitting photons and filling the inside with radiation, similar to conditions inside a stellar photosphere. Each of these photons is reabsorbed inside the container, the loss through the small hole being negligible. Lengthy arguments can be put forward to support what seems obvious to the physically minded, namely that an equilibrium condition exists inside the chamber, provided the temperature of the walls remain constant and uniform, with each physical process balanced by its inverse. A small fraction of the internal radiation leaks out of the hole, and we can measure its spectrum, but this leakage is so small that the thermodynamics of the container is essentially maintained. One might call this a perfect emitter: the emission does not depend on the materials forming the container, nor on the size of the hole (as long as the light can get through and it is small compared to the size of the container). The spectrum is continuous and unpolarized. And the amount of emission is the maximum that can come out of a thermal emitter. It is reproducible and obeys simple laws, as we shall see.

The basic required condition for the black body as an emitting source is that the thermodynamic equilibrium be maintained. At the bottom of a stellar photosphere, the optical depth to the surface is high enough to prevent the escape of most photons. They are reabsorbed not far from where they are emitted. There are nearly as many photons headed into the star as there are headed outward. In other words, the leakage out of the star is negligible, like the leakage out of the hole in the container. Since the photospheric material is then essentially in thermodynamic equilibrium, we expect the radiation laws of a black body to describe the radiation there. Higher photospheric layers deviate increasingly from the black-body case as the leakage becomes more significant. What is different from the container is the gradual drop in temperature outward through the photosphere. So in this case, we think of the equilibrium as being localized in depth, the local black-body emission being characterized by the temperature at each depth. The concept of *local thermodynamic equilibrium*, LTE, is born. We use the LTE concept extensively in later chapters. But there is also a continuous transition from near perfect thermodynamic equilibrium deep in the photosphere to complete non-equilibrium at the top of the atmosphere, termed non-LTE.

In stellar photospheres, when LTE holds, the source function is black-body emission, Planck's law as stated earlier in Equation 5.16. In this chapter, we explore this emission in more detail.

The second intersection of the black body with astronomy is photometric calibration. The black body is the fundamental standard for photometric calibrations, both laboratory and astronomical alike, because of its well-established and reproducible radiation laws.

Observed Relations

The amazing consistency and simplicity of radiation from black bodies was discovered late in the nineteenth century. The observed spectra were found to be continuous, isotropic, and unpolarized. The specific intensity depends only on wavelength and the temperature of the black body. Two fundamental empirical laws were discovered before the full nature of the spectrum was understood. The first of these is a scaling relation, found by Wien (1893), in which the specific intensity can be written in frequency or wavelength units as

$$B_\nu(T) = \nu^3 f\left(\frac{\nu}{T}\right) \text{ or } B_\lambda(T) = \frac{c^4}{\lambda^5} f(c/\lambda T), \tag{6.1}$$

where c denotes the velocity of light and f is a unique function that can be tabulated from the measurements. (Recall the frequency to wavelength translation discussed following Equation 5.2.) If we denote the scaled frequency by $u = \nu/T$, the specific intensity can be written

$$B_\nu(T) = T^3 u^3 f(u) \text{ or } B_\lambda = \frac{1}{c} T^5 u^5 f(u).$$

This helps us see the temperature and frequency dependence more clearly. In particular, the frequency dependence for the frequency-unit case is always the same, $u^3 f(u)$, but the strength scales with the cube of the temperature. The maximum of $u^3 f(u)$ always occurs at the same u since $f(u)$ is a unique function. The observations show the maximum to be at $u = 5.8789 \times 10^{10}$ Hz/K, or

$$\lambda'_{max} T = 5.0994 \times 10^7 \text{ Å K}, \tag{6.2a}$$

where λ'_{max} is the wavelength at which the *frequency* distribution, $B_\nu(T)$, has its maximum. The corresponding relation for $B_\lambda(T)$ shows the amplitude scaling as T^5, while the $u^5 f(u)$ maximum is given by

$$\lambda_{max} T = 2.8978 \times 10^7 \text{ Å K}. \tag{6.2b}$$

The scaling relation in Equation 6.1 is *Wien's law*, although the special cases in Equation 6.2 often go by the same name or sometimes by Wien's displacement law. The fact that the maximum appears at a different place in the spectrum for frequency and wavelength units was discussed near the beginning of Chapter 5. One way to measure the temperature of a black-body-like source is to measure the position of the maximum intensity in the spectrum and apply Equation 6.2. We can also go in the other direction and estimate the spectral position of maximum intensity given the temperature. For example, the solar spectrum is similar to that of a black body with $T = 5772$ K, leading to $\lambda'_{max} \sim 8800$ Å and $\lambda_{max} \sim 5000$ Å.

The second black-body law, the *Stefan–Boltzmann law* (Stefan 1879), says that the total power output is specified purely by the temperature according to

$$\int_0^\infty \mathfrak{F}_\nu \, d\nu = \pi \int_0^\infty B_\nu(T) \, d\nu = \sigma T^4, \tag{6.3}$$

where $\mathfrak{F}_\nu$ is the flux (πB_ν for this isotropic radiator) and σ is the Stefan–Boltzmann constant of value $\sigma = 5.6704 \times 10^{-5}$ erg/$(\text{s cm}^2 \text{ K}^4)$ or $\times 10^{-8}$ watt/$(\text{m}^2 \text{ K}^4)$. The Stefan–Boltzmann relation follows directly from Wien's law. Again let $u = \nu/T$. Then, with $B_\nu(T) = T^3 u^3 f(u)$,

$$\int_0^\infty B_\nu(T) \, d\nu = \int_0^\infty T^3 u^3 f(u) \, T \, du$$

$$= T^4 \int_0^\infty u^3 f(u) \, du$$

$$= \text{constant } T^4. \tag{6.4}$$

As shown in the next section, the constant has a value of σ/π. Thermodynamic derivations of these two laws can be seen in Richtmyer *et al.* (1955), for example.

The shapes of the spectra are illustrated in Figure 6.2. At low frequencies the curves have the form

$$B_\nu(T) = \frac{2kT\nu^2}{c^2} = \frac{2kT}{\lambda^2}, \tag{6.5}$$

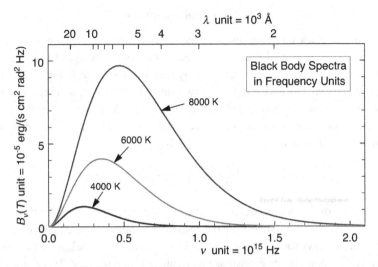

Figure 6.2a The specific intensity emitted by black bodies of three temperatures is shown. The scaling of Wien's law, Equation 6.1, can be seen in the shape of the curves. The area under the curves obeys the Stefan–Boltzmann law, Equation 6.4.

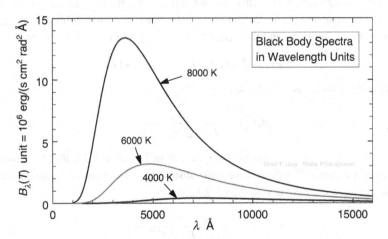

Figure 6.2b The black-body curves have a different shape when using wavelength units instead of frequency units.

called the *Rayleigh–Jeans approximation*, an expression often seen in radio-astronomy work. At high frequencies

$$B_\nu(T) = \text{constant } \nu^3 e^{-\text{constant } \nu/T}.$$ (6.6)

The name *Wien approximation* is associated with this equation.

Planck's Radiation Law

The history of the black body and its fundamental role in starting quantum theory is well known. We omit the historical details here (see Planck 1914) except to point out that any derivation of the relation describing black-body spectra, like those in Figure 6.2, relies on the quantum approach. We derive Planck's law using a two-level atom. (See Born 1957 and Cox & Giuli 1968 for alternative derivations.) The upper level has a population N_u, the lower N_l, and since these atoms are in a container in a state of thermodynamic equilibrium, their populations are related by (see Equation 1.18),

$$\frac{N_u}{N_l} = \frac{g_u}{g_l} e^{-h\nu/kT},$$

where the energy difference between the levels is the difference between their excitation potentials $\chi_u - \chi_l = h\nu$, the energy of the equivalent photon. The equilibrium condition implies that all the ways for the electron to go from u to l must be balanced by return paths.

The number of spontaneous emissions per second per unit solid angle per unit volume is $N_u A_{ul}$, where A_{ul} is the Einstein coefficient defined in the last section of Chapter 5. Similarly, the rate of stimulated emission per unit solid angle and volume is $N_u B_{ul} B_\nu(T)$, and $B_\nu(T)$ must be taken at ν appropriate to the transition. Finally, absorption pumps electrons upward, so for this rate we write $N_l B_{lu} B_\nu(T)$. In addition to radiative transitions we could have collision-induced transitions, but in equilibrium there are as many up as there are down, and they cancel out. And so the equilibrium condition is

$$N_u A_{ul} + N_u B_{ul} B_\nu(T) = N_l B_{lu} B_\nu(T).$$

If we now solve this for $B_\nu(T)$, we find

$$B_\nu(T) = \frac{A_{ul}}{B_{lu}(N_l/N_u) - B_{ul}}, \tag{6.7}$$

which is just the ratio of emission to absorption as in Equation 5.15, that is, Equation 6.7 has the form of a source function. If we also substitute for the population ratio according to the excitation equation above, Equation 6.7 becomes

$$B_\nu(T) = \frac{A_{ul}}{(g_l/g_u) B_{lu} e^{h\nu/kT} - B_{ul}}.$$

But we know that this expression must revert to the Rayleigh–Jeans approximation (Equation 6.5) in the limit of small ν or $h\nu/kT \ll 1$. Therefore we expand the exponential, keeping only the first-order term, and find

$$B_\nu(T) \approx \frac{A_{ul}}{(g_l/g_u)B_{lu} - B_{ul} + (g_l/g_u) B_{lu}h\nu/kT}.$$

This can equal $2kT\nu^2/c^2$ only if

$$B_{ul} = \frac{B_{lu}g_l}{g_u} \quad \text{and} \quad A_{ul} = \frac{2h\nu^3}{c^2}B_{ul}. \tag{6.8}$$

The result of substituting these relations back into Equation 6.7 is

$$B_\nu(T) = \frac{2h\nu^3}{c^2} \frac{1}{e^{h\nu/kT} - 1}. \tag{6.9}$$

This is *Planck's radiation law*. It gives the proper fit to the observed curves in Figure 6.2 and has the correct limiting forms at long and at short wavelengths. It has the same functional form expressed in Equation 6.1. We can also check the maximum in the usual way by setting the derivative of Equation 6.9 equal to zero. Use $x = h\nu/kT$ and obtain

$$x = \frac{3(e^x - 1)}{e^x}.$$

Solution by iteration gives $x = 2.8214$, or $\lambda'_{max} T = 0.50994$ cm K in accord with Equation 6.2.

Integration of Planck's law leads to the Stefan–Boltzmann law. Since we are looking at light leaking out of a small hole and the escaping radiation is isotropic over the outward hemisphere, $\mathfrak{F}_\nu = \pi B_\nu$ according to Equation 5.7. It then follows that

$$\int_0^\infty \mathfrak{F}_\nu \, d\nu = \pi \int_0^\infty \frac{2h\nu^3}{c^2} \frac{1}{e^{h\nu/kT} - 1} \, d\nu = \frac{2\pi h}{c^2} \left(\frac{kT}{h}\right)^4 \int_0^\infty \frac{x^3}{e^x - 1} \, dx,$$

where $x = h\nu/kT$. Evaluating the integral gives $\pi^4/15$ and then

$$\int_0^\infty \mathfrak{F}_\nu \, d\nu = \frac{2\pi^5 k^4}{15h^3 c^2} T^4 = 5.6704 \times 10^{-5} \, T^4 = \sigma T^4.$$

So the observed Stefan–Boltzmann law of Equation 6.4 is retrieved.

Finally, notice that the relations 6.8 are between fundamental atomic constants. These relations must hold independent of the physical situation the atoms are encountering, and therefore are perfectly general relations. We use them again in Chapter 11.

Numerical Values of Black-Body Radiation

Tabulations of black-body functions, such as those of McDonald (1955) or Pivovonsky and Nagel (1961), are useful in some situations, but often it is quicker and more convenient to compute directly from Equation 6.9, especially when $B_\nu(T)$ is being used as the source function in a model photosphere calculation. Accordingly, in frequency units,

$$\begin{aligned}
B_\nu(T) &= \frac{2h\nu^3}{c^2} \frac{1}{e^{h\nu/kT} - 1} \\[2mm]
&= \frac{1.4744998 \times 10^{-47} \nu^3}{e^{4.79924405 \times 10^{-11} \, \nu/T} - 1} \; \text{erg}/\left(\text{s cm}^2 \, \text{rad}^2 \, \text{Hz}\right) \\[2mm]
&= \frac{1.4744998 \times 10^{-50} \, \nu^3}{e^{4.799239 \times 10^{-11} \, \nu/T} - 1} \; \text{W}/\left(\text{m}^2 \, \text{rad}^2 \, \text{Hz}\right) \\[2mm]
&= \frac{3.9728925 \times 10^8}{\lambda^3} \frac{1}{e^{1.438777 \times 10^8/\lambda T} - 1} \; \text{erg}/\left(\text{s cm}^2 \, \text{rad}^2 \, \text{Hz}\right) \; \text{for } \lambda \text{ in Å} \quad (6.10)
\end{aligned}$$

or in wavelength units,

$$B_\lambda(T) = \frac{2hc^2}{\lambda^5} \frac{1}{e^{hc/\lambda kT} - 1}$$

$$= \frac{1.1910432 \times 10^{27}/\lambda^5}{e^{1.438775 \times 10^8/\lambda T} - 1} \; \text{erg}/\left(\text{s cm}^2 \, \text{rad}^2 \, \text{Å}\right) \; \text{for } \lambda \text{ in Å}$$

$$= \frac{1.1910432 \times 10^{24}/\lambda^5}{e^{1.438775 \times 10^8/\lambda T} - 1} \; \text{W}/\left(\text{m}^2 \, \text{rad}^2 \, \text{Å}\right) \; \text{for } \lambda \text{ in Å}, \qquad (6.11)$$

where we have used $h = 6.62607015 \times 10^{-27}$ erg s (or 10^{-34} J s), $k = 1.380649 \times 10^{-16}$ erg/K(10^{-23} J/K), and $c = 2.99792458 \times 10^{10}$ cm/s (10^8 m/s), the velocity of light in a vacuum. If the black body is used in air, c should be decreased from its vacuum value by the index of refraction of air, $n \approx 1.00028$ (see Equations 12.18 and 12.19). Likewise, all wavelengths in air will differ from their vacuum values: $\lambda = \lambda_0/n$, where λ_0 is the vacuum value. The frequency, however, remains unaltered. These differences may not be negligible, amounting to ~0.1% across the visible spectrum or ~0.4 K when T is ~3000 K (Hartmann 2006).

The Black Body as a Radiation Standard

For photometric work, the black-body radiator is used as a primary standard of radiation against which many types of secondary standards are calibrated. The usefulness of the black body in this context stems from the simple unique dependence of the radiation on temperature. However, knowing the temperature and keeping it stable is non-trivial. The intricacies of fixing the temperature scale itself are discussed by Mangum *et al.* (2001) and Pavese and Molinar Min Beciet (2013, especially their Appendix A). A black body suitable for this type of work is built with a cavity in the center surrounded by electrical heating coils. Inside is a small amount of pure metal, typically gold. A small hole allows the radiation to be seen. The apparatus is used by slowly heating the gold until it melts. The system is then allowed to cool to the freezing point of the metal, where there is a thermally stable plateau lasting from a few minutes to about an hour. It is during this interval that the black body is used for photometry. Operation at a metallic freezing point results in good reproducibility and stability. The temperature of freezing gold is 1337.33 K. The black body can be operated at other temperatures by using other metals: platinum freezes at 2045 K; copper at 1357.8 K, silver at 1235 K. The more inert the metal, the less it becomes contaminated with impurities. Alas, the freezing temperature is a function of the purity of the metal. Practical details of black bodies are discussed by Cussen (1982), Latvakoski *et al.* (2010), and Sister *et al.* (2017), for example.

Absolute flux calibration of stars is discussed in Chapter 10. Calibration is necessary if we want to know the real shape of a star's continuous spectrum, also called the energy distribution. But this chore is not easily dispatched. Black bodies are awkward to look at with astronomical telescopes, so secondary standards such as lamps and emitting diodes, calibrated against black bodies, are used instead (e.g., Bless *et al.* 1968, Hayes & Latham 1975, Fraser *et al.* 2007, Kaiser *et al.* 2017). Even then the situation is challenging.

Questions and Exercises

1. Explain what we mean by a black body, and why it is relevant to stellar photospheres. Explicitly address the question of how a closed container, like the one shown in Figure 6.1, can have anything in common with a stellar photosphere that is gaseous and without walls.

2. Starting with Planck's law, Equation 6.9, derive the Rayleigh–Jeans approximation, applicable at long wavelength.

3. The Stefan–Boltzmann law, Equation 6.4, is used to define the effective temperature of a star, as done in Equation 1.2. This in turn is related to the power output and the radius of the star in Equation 1.3. Using these relations, compare the following two stars. The brighter star has 11.3 times the luminosity of the fainter one. The more distant star lies at 21.3 parsecs. The effective temperature of the brighter star is 1.4 times that of the fainter star. Assuming these are main-sequence stars, what is the ratio of their radii? If the lesser star is the Sun, estimate the temperature, radius, luminosity, and mass of the other star. You may wish to consult the data given in Appendix B.

4. How much power actually comes out of a stellar surface? Use the Stefan–Boltzmann law to find the answer for two stars of effective temperature 4000 K and 30 000 K. If these stars were at the distance of the Sun from us, how much power would they give per square meter on Earth?

5. Derive an analytical expression of the temperature of a slowly rotating planet in a circular orbit around a star. Call the effective temperature of the star T_{eff}, the radius of the star R, the radius of the planet a, and the radius of the orbit r. First consider the case where the planet has no atmosphere. Take the planet to be a perfect black body that absorbs and emits isotropically, and that it absorbs all of the stellar radiation impinging upon it. Assume an equilibrium situation exists, e.g., the temperature of the planet is constant. Does the planet's temperature depend on the size of the planet? If not, why not?

   If the star is the Sun and the radius of the orbit is 1 a.u., what temperature might you expect the Earth to have based on your black-body model? How does this compare to the actual temperature of the Earth? What complications might occur if the planet has an atmosphere?

References

Bless, R.C., Code, A.D., & Schroeder, D.J. 1968. *ApJ* **153**, 545.

Born, M. 1957. *Atomic Physics* (New York: Hafner), 6th ed.

Cox, J.P. & Giuli, R.T. 1968. *Stellar Structure* (New York: Gordon and Breach), p. 108.

Cussen, A.J. 1982. *Infrared Sensor Technology*, SPIE 344.

Fraser, G.T., Brown, S.W., Yoon, H.W., Johnson, B.C., & Lykke, K.R. 2007. *SPIE* **6678**, 66780.

Hartmann, J. 2006. *Opt Exp* **14**, 8121.

Hayes, D.S. & Latham, D.W. 1975. *ApJ* **197**, 593.

Kaiser, M.E., *et al.* 2017. *SPIE* **10398**, 1039815.

Latvakoski, H.M., Watson, M., Topham, S., *et al.* 2010. *SPIE* **7739**, 19.

Mangum, B.W., Furukawa, G.T., Kreider, K.G., *et al.* 2001. *JNIST* **106**, 105.

McDonald, J.K. 1955. *Pub Dominion Ap Obs Victoria* **10**, 127.

Pavese, F. & Molinar Min Beciet, G. 2013. *Modern Gas-Based Temperature and Pressure Measurements* (Heidelberg: Springer).

Pivovonsky, M. & Nagel, M.R. 1961. *Tables of Black Body Radiation Functions* (New York: Macmillan).

Planck, M. 1914. *The Theory of Heat Radiation* (Philadelphia: P. Blakiston's Son & Co.), translated by M. Masius, p. 168.

Richtmyer, F.K., Kennard, E.H., & Lauritsen, T. 1955. *Introduction to Modern Physics* (New York: McGraw-Hill), 5th ed.

Sister, I., Leviatan, Y., & Schächter, L. 2017. *Opt Exp* **25**, 589.

Stefan, J. 1879. *Sitzungsberichte der kaiserlichen Akademie der Wissenschaften, Mathematische-Naturwissenschaftliche Classe, II Abteilung,* **79**, 391.

Wien, W. 1893. *Sitzungsberichte der Koniglich Preussischen Akademie der Wissenschaften zu Berlin*, p. 55.

7

Energy Transport in Stellar Photospheres

Light carries most of the energy outward through the stellar photosphere, so radiative transfer is our main focus here. The spectrum we see and measure is the result of this process; it links together the physical parameters of the stellar material and the light. We start by setting up the differential equation describing the flow of radiation along a line through an infinitesimal volume. Then we integrate the equation and move toward shaping it for the geometry of stellar photospheres. Although the integral equation is informative and useful, it is not a direct physical solution of the radiative transfer because the integrand depends on the atomic excitation of the material, which itself depends on the temperature of the material and the radiation field in the material. In applications of theory to real stars, a numerical model of the star's photosphere is formed from which the integrand and then the spectrum of the star can be calculated. This chapter, along with Chapters 8 and 9, develops the tools for this modeling.

Transport of energy by convection is often important below the surface, but rarely carries a significant fraction of the flux in the photosphere. Conduction comes into play only in extreme cases such as white dwarfs. Specific intensity and flux are the terms we use to describe the transport of radiative energy, while the absorption coefficient and optical depth deal with the blocking of radiation. The assumption is that you have already digested Chapter 5.

The Transfer Equation and Its Formal Solution

Consider radiation traveling in a direction s. The change in specific intensity, dI_ν, over an increment of path length, ds, is the sum of the losses, expressed in the absorption coefficient, κ_ν, and gains, expressed in the emission coefficient, j_ν,

$$dI_\nu = -\kappa_\nu \, \rho \, I_\nu \, ds + j_\nu \, \rho \, ds.$$

Now divide the equation by $\kappa_\nu \, \rho \, ds$, which we call $d\tau_\nu$ in accord with Equation 5.13, and write

$$\frac{dI_\nu}{d\tau_\nu} = -I_\nu + \frac{j_\nu}{\kappa_\nu},$$

or using the definition of the source function from Equation 5.15,

$$\frac{dI_v}{d\tau_v} = -I_v + S_v. \tag{7.1}$$

This is the general differential equation of radiative transfer.

Since it is the specific intensity that we ultimately want, integration is needed. To get there, follow a standard integrating-factor technique. Noticing that the variable τ_v appears alone in Equation 7.1, we are led to try a solution of the form

$$I_v(\tau_v) = f(\tau_v)e^{b\tau_v}, \tag{7.2}$$

where $f(\tau_v)$ is a function to be determined. Differentiate this,

$$\frac{dI_v}{d\tau_v} = f(\tau_v)\, b\, e^{b\tau_v} + e^{b\tau_v}\frac{df(\tau_v)}{d\tau_v} = bI_v + e^{b\tau_v}\frac{df(\tau_v)}{d\tau_v}.$$

Then place it back into Equation 7.1,

$$bI_v + e^{b\tau_v}\frac{df(\tau_v)}{d\tau_v} = -I_v + S_v.$$

We can make the first terms on each side of the equation equal by setting $b = -1$, and then equating the second terms gives

$$e^{-\tau_v}\frac{df(\tau_v)}{d\tau_v} = S_v,$$

and integration results in

$$f(\tau_v) = \int_0^{\tau_v} S_v\, e^{t_v}\, dt_v + c_0,$$

in which t_v serves as a dummy variable and c_0 is the integration constant. Then our original Equation 7.2 becomes

$$I_v(\tau_v) = e^{-\tau_v}\int_0^{\tau_v} S_v(t_v)\, e^{t_v}\, dt_v + c_0 e^{-\tau_v}. \tag{7.3}$$

The integration constant c_0 must be $I_v(0)$ as seen by setting $\tau_v = 0$.

It is instructive to rewrite Equation 7.3, bringing the $e^{-\tau_v}$ inside the integral,

$$I_v(\tau_v) = \int_0^{\tau_v} S_v(t_v)\, e^{-(\tau_v - t_v)}\, dt_v + I_v(0)e^{-\tau_v}. \tag{7.4}$$

Figure 7.1 illustrates the line along which radiation is being considered and the positions of the variables for this equation. In particular, at a point τ_v the original intensity, $I_v(0)$, has suffered an exponential extinction $e^{-\tau_v}$, and in a similar way, the intensity generated at any point t_ν, which is $S_\nu(t_\nu)$, undergoes an exponential extinction $e^{-(\tau_\nu - t_\nu)}$ before being summed at the point τ_ν.

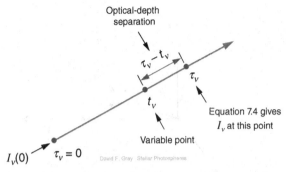

Figure 7.1 Radiation along the line at point τ_ν is composed of the sum of intensities, $S_\nu(t_\nu)$, originating at the points t_ν along the line, but suffering extinction according to the optical-depth separation $\tau_\nu - t_\nu$. Any radiation incident at $\tau_\nu = 0$, called $I_\nu(0)$, suffers extinction by a factor of $e^{-\tau_\nu}$.

Equation 7.4 is the basic integral form of the transfer equation. If we can solve this equation, i.e., compute $I_\nu(\tau_\nu)$, we have the answer we expect from the theory. However, in order to actually do the integration, $S_\nu(\tau_\nu)$ must be a known function. In some situations this is a complicated function. In others it is simple, as shown in the examples of Chapter 5. In the case of local thermodynamic equilibrium (LTE), $S_\nu(\tau_\nu) = B_\nu(T)$, the Planck function, and then we need T as a function of τ_ν to get the solution. For historical reasons, $T(\tau_\nu)$ is called the *temperature distribution*.

The Transfer Equation for Stellar Geometries

Slowly rotating stars are spherical. Equation 7.4 expresses I_ν as a function of optical depth along a line. The first step in transforming to the stellar geometry is to define the optical depth relative to the star along a stellar radius rather than simply along the line of sight. The appropriate spherical coordinates are shown in Figure 7.2. The z axis is chosen toward the observer, and if we measure optical depth along that axis, $d\tau_\nu = \kappa_\nu \rho \, dz$. Thus

$$\frac{dI_\nu}{\kappa_\nu \rho \, dz} = -I_\nu + S_\nu.$$

In general dI_ν / dz in these coordinates can be written

$$\frac{dI_\nu}{dz} = \frac{\partial I_\nu}{\partial r}\frac{dr}{dz} + \frac{\partial I_\nu}{\partial \theta}\frac{d\theta}{dz}, \tag{7.5}$$

where we assume I_ν has no ϕ, i.e., azimuthal, dependence. From the figure we see that

$$dr = \cos\theta \, dz,$$
$$r\,d\theta = -\sin\theta \, dz.$$

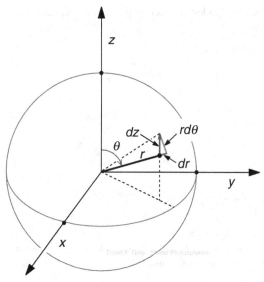

Figure 7.2 The geometry is shown for the conversion of line-of-sight to polar coordinates centered on the star. The z axis is toward the observer.

Then the transfer equation becomes

$$\frac{\partial I_\nu}{\partial r} \frac{\cos\theta}{\kappa_\nu \rho} - \frac{\partial I_\nu}{\partial\theta} \frac{\sin\theta}{\kappa_\nu \rho\, r} = -I_\nu + S_\nu. \tag{7.6}$$

This form of the equation is used in stellar interiors and in the calculation of very thick stellar atmospheres such as those found in supergiants.

Conveniently, the geometrical thickness of the photosphere in most stars is very small compared to the stellar radius. For example, the solar photosphere is ~700 km thick, or ~0.1% of the solar radius (refer to Figure 1.1). Under these conditions, a plane-parallel approximation can be made, namely, θ does not depend upon z so there is no second term in Equation 7.6. Then

$$\cos\theta \frac{dI_\nu}{\kappa_\nu\rho\, dr} = -I_\nu + S_\nu.$$

According to custom, the terminology in this plane-parallel case is changed, with x (as in Figure 7.3, not the x in Figure 7.2) assigned to the depth dimension and the optical depth, τ_ν, increasing *into* the photosphere, i.e., $dx = -dr$. Adopting this convention and writing $d\tau_\nu$ for $\kappa_\nu\rho\, dx$ gives the basic form of the transfer equation used in the central arena of stellar photospheres,

$$\cos\theta \frac{dI_\nu}{d\tau_\nu} = I_\nu - S_\nu. \tag{7.7}$$

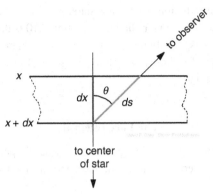

Figure 7.3 In the plane-parallel approximation, the geometry is simpler than in the previous figure. An increment of path length along the line of sight is $ds = -dx \sec \theta$, where dx is the corresponding change in depth into the photosphere.

The optical depth defined in this way is measured along x and not in the direction of the observer, which is at an angle, θ, as shown in Figure 7.3. This amounts to replacing τ_v in Equation 7.1 with $-\tau_v \sec \theta$. The negative sign arises from choosing $dx = -dr$. The integrated form corresponding to Equation 7.4 is then

$$I_v(\tau_v) = -\int_c^{\tau_v} S_v(t_v) \, e^{-(t_v - \tau_v)\sec\theta} \sec\theta \, dt_v. \tag{7.8}$$

The integration limit, c, replaces the $I_v(0)$ integration constant of Equation 7.4 because the boundary conditions are clearly different for radiation going inward ($\theta > 90°$) and coming outward ($\theta < 90°$). In the first case, we start at the outer boundary, where $c = \tau_v = 0$, and work inward. But in the second case, we have radiation at the depth τ_v and deeper until no more radiation can be seen coming out. In other words, when $I_v = I_v^{\text{out}}, c = \tau_v = \infty$. Therefore, the full intensity at the depth τ_v in the photosphere is

$$I_v(\tau_v) = I_v^{\text{out}}(\tau_v) + I_v^{\text{in}}(\tau_v)$$

$$= \int_{\tau_v}^{\infty} S_v \, e^{-(t_v - \tau_v)\sec\theta} \sec\theta \, dt_v - \int_0^{\tau_v} S_v \, e^{-(t_v - \tau_v)\sec\theta} \sec\theta \, dt_v. \tag{7.9}$$

The mathematically minded will usually add that one must require $S_v e^{-t_v}$ go to zero as τ_v goes to infinity to ensure that the first integral exists. Real stars apparently know how to meet this condition without effort.

An important special case of Equation 7.9 occurs at the stellar surface. In this case

$$I_v^{\text{in}} = 0,$$

$$I_v^{\text{out}}(0) = \int_0^{\infty} S_v \, e^{-t_v \sec\theta} \sec\theta \, dt_v, \tag{7.10}$$

where we assume that the radiation from other stars, galaxies, and so forth is completely negligible compared to the star's own radiation. Equation 7.10 is the expression we need to compute the spectrum. It is especially relevant to solar work where intensity measurements can be made as a function of θ. For most stars, we must still integrate I_v^{out} over the star's disk, i.e., we observe the flux.

The Flux Integral

Now we formulate the most important integral in our arsenal. Expand the defining equation for flux (Equation 5.5) in spherical polar coordinates and assume no azimuthal dependence for I_v. Then

$$\mathcal{F}_v = 2\pi \int_0^\pi I_v \cos\theta \sin\theta \, d\theta$$

$$= 2\pi \int_0^{\pi/2} I_v^{\text{out}} \cos\theta \sin\theta \, d\theta + 2\pi \int_{\pi/2}^\pi I_v^{\text{in}} \cos\theta \sin\theta \, d\theta.$$

With the definitions of I_v^{out} and I_v^{in} from Equation 7.9,

$$\mathcal{F}_v = 2\pi \int_0^{\pi/2} \int_{\tau_v}^\infty S_v \, e^{-(t_v - \tau_v)\sec\theta} \sin\theta \, dt_v \, d\theta$$

$$- 2\pi \int_{\pi/2}^\pi \int_0^{\tau_v} S_v \, e^{-(t_v - \tau_v)\sec\theta} \sin\theta \, dt_v \, d\theta,$$

or, if S_v is isotropic,

$$\mathcal{F}_v = 2\pi \int_{\tau_v}^\infty S_v \int_0^{\pi/2} e^{-(t_v - \tau_v)\sec\theta} \sin\theta \, d\theta \, dt_v$$

$$- 2\pi \int_0^{\tau_v} S_v \int_{\pi/2}^\pi e^{-(t_v - \tau_v)\sec\theta} \sin\theta \, d\theta \, dt_v. \tag{7.11}$$

If we now let $w = \sec\theta$ and $x = t_v - \tau_v$, the first inner integral can be written

$$\int_0^{\pi/2} e^{-(t_v - \tau_v)\sec\theta} \sin\theta \, d\theta = \int_1^\infty \frac{e^{-xw}}{w^2} \, dw. \tag{7.12}$$

Integrals of this form are well known. They are called exponential integrals. Their properties are discussed later in this chapter. For now we simply denote them by E_n,

$$E_n(x) = \int_1^\infty \frac{e^{-xw}}{w^n} dw. \tag{7.13}$$

Using this exponential-integral notation, Equation 7.11 becomes

$$\mathfrak{F}_v(\tau_v) = 2\pi \int_{\tau_v}^\infty S_v\, E_2(t_v - \tau_v)\, dt_v - 2\pi \int_0^{\tau_v} S_v\, E_2(\tau_v - t_v)\, dt_v. \tag{7.14}$$

To obtain $E_2(\tau_v - t_v)$ in the second integral, make the substitution $w = -\sec\theta$ and $x = \tau_v - t_v$ in the inner integral of the second term of Equation 7.11. The limit as θ goes from $\pi/2$ to π is approached with negative values of $\cos\theta$, so w goes to ∞, not $-\infty$. Equation 7.14 is the basic relation we seek. It is analogous to Equation 7.9 for intensity. It tells us how much power at any point in the spectrum is flowing through a unit area at the depth τ_v in the photosphere.

The theoretical stellar spectrum sent to us by the star is $\mathfrak{F}_v$ at $\tau_v = 0$, that is,

$$\mathfrak{F}_v(0) = 2\pi \int_0^\infty S_v(t_v)\, E_2(t_v)\, dt_v. \tag{7.15}$$

This says that the surface flux is composed of the sum of the source function at each depth multiplied by an extinction factor, $E_2(t_v)$, appropriate to that depth below the surface, and the sum is taken over all depths contributing a significant amount of radiation at the surface. Notice again that $\mathfrak{F}_v$ is defined per unit area. The total stellar radiation at the frequency v is $4\pi R^2 \mathfrak{F}_v$, where R is the star's radius.

In the process of deriving Equation 7.15, we assumed I_v to be independent of the azimuthal angle ϕ and that S_v and κ_v are isotropic. For many purposes this is perfectly reasonable. In others it is not. In particular, when we deal with Doppler shifts of spectral lines arising from photospheric velocity fields (Chapter 17) or from rotation of the stars (Chapter 18), these assumptions may no longer be true and we are forced to revert back to Equation 5.5, replacing Equation 7.15 with explicit integrations of $I_v(\theta, \phi)$ over the apparent disk of the star. Explicit disk integrations are also required when modeling stars with surface features such as starspots.

The Mean Intensity and the K Integral

In theoretical treatments of radiation transfer, the mean intensity and the K integral are standard tools. We can easily derive expressions for J_v and K_v as functions of τ_v, and obtain expressions similar to Equation 7.15 for flux. The result is

$$J_v(\tau_v) = \frac{1}{2} \int_{\tau_v}^\infty S_v\, E_1(t_v - \tau_v)\, dt_v + \frac{1}{2} \int_0^{\tau_v} S_v\, E_1(\tau_v - t_v)\, dt_v \tag{7.16}$$

and

$$K_\nu(\tau_\nu) = \frac{1}{2} \int_{\tau_\nu}^{\infty} S_\nu E_3(t_\nu - \tau_\nu)\, dt_\nu + \frac{1}{2} \int_{0}^{\tau_\nu} S_\nu\, E_3(\tau_\nu - t_\nu)\, dt_\nu. \tag{7.17}$$

Since we are oriented more toward the practical side, rather than the theoretical, our use of these quantities is minimal. We have now employed three exponential integrals.

Exponential Integrals

Because the second exponential integral, E_2, appears in the flux integral, Equations 7.14 and 7.15, it is the most important one for us. To repeat the equation for the exponential integrals, $E_n(x)$, of the nth order,

$$E_n(x) = \int_{1}^{\infty} \frac{e^{-xw}}{w^n}\, dw. \tag{7.18}$$

The value at the origin is found directly by setting $x = 0$, leaving

$$E_n(0) = \int_{1}^{\infty} \frac{dw}{w^n} = \frac{1}{n-1}.$$

The first three exponential integrals are shown in Figure 7.4, where it can be seen that $E_1(0)$ is infinite, $E_2(0)$ is unity, and $E_3(0)$ is one-half.

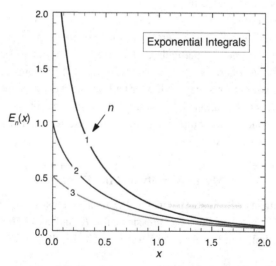

Figure 7.4 Here are plots of the first three exponential integrals, $E_1(x)$, $E_2(x)$, and $E_3(x)$. At $x = 0$, E_1 is unbounded, E_2 is unity, and E_3 is 1/2. At large x, they all converge to e^{-x}/x.

Differentiation leads to a useful relation,

$$\frac{dE_n}{dx} = \int_1^\infty \frac{1}{w^n} \frac{d}{dx} e^{-xw} \, dw = -\int_1^\infty \frac{e^{-xw}}{w^{n-1}} \, dw,$$

or

$$\frac{dE_n}{dx} = -E_{n-1}. \tag{7.19}$$

There is also an important recursive formula,

$$nE_{n+1}(x) = e^{-x} - xE_n(x), \tag{7.20}$$

that is valuable for generating numerical values of $E_{n+1}(x)$ from $E_n(x)$. Abramowitz and Stegun (1964) give the following approximations suitable for computation:

$$E_1(x) = -\ln x - 0.57721566 + 0.99999193x - 0.24991055x^2$$
$$+ 0.05519968x^3 - 0.00976004x^4 + 0.00107857x^5 \quad \text{for } x \le 1$$

and

$$E_1(x) = \frac{x^4 + a_3x^3 + a_2x^2 + a_1x + a_0}{x^4 + b_3x^3 + b_2x^2 + b_1x + b_0} \frac{1}{xe^x} \quad \text{for } x > 1,$$

where

$$a_3 = 8.5733287401, \quad b_3 = 9.5733223454,$$
$$a_2 = 18.059016973, \quad b_2 = 25.6329561486,$$
$$a_1 = 8.6347609825, \quad b_1 = 21.0996530827,$$
$$a_0 = 0.2677737343, \quad b_0 = 3.9584969228.$$

These polynomials fit E_1 with an error of less than 2×10^{-7}. Then from Equation 7.20, the higher order exponential integrals can be generated.

The asymptotic behavior is given by the expansion

$$E_n(x) = \frac{1}{xe^x} \left[1 - \frac{n}{x} + \frac{n(n+1)}{x^2} - \cdots \right]$$

$$\approx \frac{1}{xe^x}.$$

This is a very sharp decline with x, and lends confidence to the convergence of the integral in Equation 7.15. Notice that the asymptotic form is ultimately independent of n.

Computation of the Spectrum

Our foremost goal is to compute the stellar spectrum. We do this by using Equation 7.15 for a range of wavelengths. As we shall see in Chapter 9, the source function is specified as a function of optical depth at a reference wavelength, usually 5000 Å. This optical depth is called τ_0. At this wavelength, Equation 7.15 then gives the surface flux,

$$\mathscr{F}_v = 2\pi \int_0^\infty S_v(\tau_0)E_2(\tau_0)d\tau_0.$$

At other wavelengths, the argument of E_2 and the integration are on the τ_v scales of those other wavelengths, while S_v is on the τ_0 reference scale,

$$\mathscr{F}_v = 2\pi \int_0^\infty S_v(\tau_0)E_2(\tau_v)\, d\tau_v.$$

We reconcile the optical depths from the definition (Equation 5.13), $\rho dx = d\tau_v/\kappa_v = d\tau_0/\kappa_0$ and $\tau_v = \int_0^{\tau_0} \kappa_v/\kappa_0\, d\tau_0$ to give us $\tau_v(\tau_0)$. Then the flux at an arbitrary wavelength is

$$\mathscr{F}_v = 2\pi \int_0^\infty S_v(\tau_0)E_2(\tau_v(\tau_0))\frac{\kappa_v}{\kappa_0}\, d\tau_0, \qquad (7.21)$$

and so our computed spectrum is in hand.

Radiative Equilibrium

Radiative equilibrium is an expression of conservation of energy. Conservation of energy must apply to the flow of energy outward through the stellar photosphere, since there are no sources or sinks of energy within the photosphere. The energy generated in the core of the star is simply flowing outward to the outer boundary. The constraint of radiative equilibrium can be used to derive the theoretical source function mathematically. It is the core issue for the *theoretical* approach to model atmospheres.

The general condition expressing the lack of sources and sinks is that the divergence of the flux be zero everywhere in the photosphere. However, virtually all the calculations we consider are for the plane-parallel geometry. The divergence condition is then

$$\frac{d\mathscr{F}(x)}{dx} = 0 \quad \text{or} \quad \mathscr{F}(x) = \mathscr{F}_0 = \text{constant}, \qquad (7.22)$$

where $\mathscr{F}(x)$ is the total energy flux in erg/(s cm^2) or watt/m^2 as a function of depth. When all the energy is carried by radiation,

$$\mathscr{F}(x) = \int_0^\infty \mathscr{F}_v(x)\, dv = \mathscr{F}_0, \qquad (7.23)$$

with $\mathscr{F}_v$ given in Equation 7.14. Equation 7.22 is the first condition for radiative equilibrium. Although the shape of $\mathscr{F}_v$ can be expected to change very significantly with depth, its integral remains invariant. If convection were to play a significant role, then the convective flux, $\Phi(x)$ would be included,

$$\mathscr{F}_0 = \Phi(x) + \int_0^\infty \mathscr{F}_\nu \, d\nu. \tag{7.24}$$

Milne (1930) sought expressions for radiative equilibrium in order to theoretically solve for the source function. For example, if we take the flux equation, Equation 7.14, and combine it with Equation 7.23, we get Milne's second equation,

$$\int_0^\infty \left[\int_{\tau_\nu}^\infty S_\nu \, E_2(t_\nu - \tau_\nu) \, dt_\nu - \int_0^{\tau_\nu} S_\nu \, E_2(\tau_\nu - t_\nu) \, dt_\nu \right] d\nu = \frac{\mathscr{F}_0}{2\pi}. \tag{7.25}$$

In essence, it says that the solution of the transfer equation is found when an $S_\nu(\tau_\nu)$ is known that satisfies this equation.

To find Milne's first equation, start with the transfer equation in the form

$$\cos\theta \, \frac{dI_\nu}{dx} = \kappa_\nu \rho I_\nu - \kappa_\nu \rho S_\nu \tag{7.26}$$

and integrate over solid angle,

$$\frac{d}{dx} \oint I_\nu \, \cos\theta \, d\omega = \kappa_\nu \rho \oint I_\nu \, d\omega - \kappa_\nu \rho \oint S_\nu \, d\omega,$$

assuming κ_ν is independent of direction. The integral on the left is the flux. The integral in the center is the mean intensity. If S_ν is also independent of direction, the equation becomes

$$\frac{d\mathscr{F}_\nu}{dx} = 4\pi \, \kappa_\nu \rho \, J_\nu - 4\pi \, \kappa_\nu \rho \, S_\nu. \tag{7.27}$$

Then integrate over frequency,

$$\frac{d}{dx} \int_0^\infty \mathscr{F}_\nu \, d\nu = 4\pi\rho \int_0^\infty \kappa_\nu \, J_\nu \, d\nu - 4\pi\rho \int_0^\infty \kappa_\nu \, S_\nu \, d\nu. \tag{7.28}$$

In radiative equilibrium, the left-hand side is zero, leaving the relation

$$\int_0^\infty \kappa_\nu \, J_\nu \, d\nu = \int_0^\infty \kappa_\nu \, S_\nu \, d\nu. \tag{7.29}$$

This is one form of Milne's first equation, but it is often expanded by incorporating the expression for J_ν in Equation 7.16,

$$\int_0^\infty \kappa_\nu \left[\frac{1}{2} \int_{\tau_\nu}^\infty S_\nu \, E_1(t_\nu - \tau_\nu) \, dt_\nu + \frac{1}{2} \int_0^{\tau_\nu} S_\nu \, E_1(\tau_\nu - t_\nu) \, dt_\nu - S_\nu \right] d\nu = 0. \tag{7.30}$$

In this particular form of the radiative equilibrium condition, the value of the flux constant does not appear. As with Equation 7.25, a solution is in hand when we have $S_\nu(\tau_\nu)$ that satisfies this relation.

The third Milne equation is obtained in a similar way. Multiply Equation 7.26 by $\cos\theta$ before integrating over solid angle and frequency with the result

$$\int_0^\infty \frac{dK_\nu}{d\tau_\nu}\, d\nu = \frac{\mathfrak{F}_0}{4\pi}. \tag{7.31}$$

This third Milne relation is generally expanded using the expression for K_ν from Equation 7.17,

$$\int_0^\infty \frac{d}{d\tau_\nu}\left[\frac{1}{2}\int_{\tau_\nu}^\infty S_\nu\, E_3(t_\nu - \tau_\nu)\, dt_\nu + \frac{1}{2}\int_0^{\tau_\nu} S_\nu\, E_3(\tau_\nu - t_\nu)\, dt_\nu\right] d\nu = \frac{\mathfrak{F}_0}{4\pi}. \tag{7.32}$$

The three Milne equations are not independent. The $S_\nu(\tau_\nu)$ that is a solution of one will be the solution of all three.

In the theory of stellar atmospheres, much of the technical effort goes into iteration schemes using Milne's and related equations of radiative equilibrium to find the source function, $S_\nu(\tau_\nu)$. Details can be found in Kourganoff (1963), Mihalas (1978), Kalkofen (1987), and Adelman *et al.* (1996). The "grey case," discussed in the next section, is often a starting point for such iterations.

The Grey Case

One simplification of the equations of radiative equilibrium deserves special comment because of the historic spot it occupies and because it can serve as a starting point in iterative calculations. The simplification is to assume that κ_ν is independent of frequency, hence the name grey. Except for electron scattering, a minor contributor, κ_ν is distinctly non-grey (see Chapter 8). The grey case is, as a consequence, not very realistic.

To proceed, integrate the basic transfer equation (Equation 7.7) over frequency. Denote $\int_0^\infty I_\nu\, d\nu$ by I, $\int_0^\infty S_\nu\, d\nu$ by S, $\int_0^\infty J_\nu\, d\nu$ by J, $\int_0^\infty \mathfrak{F}_\nu\, d\nu$ by $\mathfrak{F}$, and $\int_0^\infty K_\nu\, d\nu$ by K. Then

$$\cos\theta\,\frac{dI}{d\tau} = I - S,$$

where $d\tau = \kappa\rho\, dx$ and κ is the "grey" absorption coefficient. This is a vastly simpler situation because the total radiation is described by a single transfer equation, whereas in the general case we had an infinite number – one for each frequency.

This type of simplification runs through the radiative equilibrium equations and Milne's equations. Thus the equivalents of the radiative equilibrium conditions, Equations 7.23, 7.29, and 7.31, are

$$\mathfrak{F} = \mathfrak{F}_0,$$
$$J = S,$$
$$\frac{dK}{d\tau} = \frac{\mathfrak{F}_0}{4\pi}. \tag{7.33}$$

One of the first solutions was the approximate scheme of Eddington (1926). He assumed hemispherically isotropic outward and inward specific intensity,

$$I(\tau) = I^{out}(\tau) \quad \text{for } 0 \leq \theta \leq \pi/2,$$
$$I(\tau) = I^{in}(\tau) \quad \text{for } \pi/2 \leq \theta \leq \pi.$$

Then I^{out} and I^{in} can be factored out of the spatial integrals, and the mean intensity is

$$J(\tau) = \oint I \frac{d\omega}{4\pi} = \frac{I^{out}}{2} \int_0^{\pi/2} \sin\theta \, d\theta + \frac{I^{in}}{2} \int_{\pi/2}^{\pi} \sin\theta \, d\theta$$

$$= \frac{1}{2} \left[I^{out}(\tau) + I^{in}(\tau) \right].$$

In a similar manner,

$$\mathfrak{F}(\tau) = \pi \left[I^{out}(\tau) - I^{in}(\tau) \right]; \quad \mathfrak{F}(0) = \pi I^{out}(0) = \mathfrak{F}_0$$

and

$$K(\tau) = \frac{1}{6} \left[I^{out}(\tau) + I^{in}(\tau) \right]; \quad K(0) = \frac{1}{6} I^{out}(0) = \frac{1}{6\pi} \mathfrak{F}_0.$$

By Equation 7.33, $S(\tau) = J(\tau)$, giving

$$S(\tau) = J(\tau) = 3K(\tau). \tag{7.34}$$

Then by integrating the third equation in 7.33 and using $K(0) = \mathfrak{F}_0/6\pi$ we have

$$K(\tau) = \frac{\mathfrak{F}_0}{4\pi} \tau + \frac{\mathfrak{F}_0}{6\pi}.$$

Finally, placing this expression for $K(\tau)$ into Equation 7.34, we arrive at the solution, Eddington's approximate solution for the grey case,

$$S(\tau) = \frac{3\mathfrak{F}_0}{4\pi} \left(\tau + \frac{2}{3} \right). \tag{7.35}$$

In this simple case, the source function varies linearly with optical depth. This same behavior is not terribly far from the truth in more rigorous non-grey solutions.

Assuming local thermodynamic equilibrium (LTE), use the frequency-integrated form of Planck's law and write $S(\tau) = (\sigma/\pi)T^4(\tau)$ and $\mathfrak{F}_0 = \sigma T_{eff}^4$, so Equation 7.35 becomes

$$T(\tau) = \left[\frac{3}{4} \left(\tau + \frac{2}{3} \right) \right]^{1/4} T_{eff}. \tag{7.36}$$

Notice that at $\tau = 2/3$ the temperature is equal to the effective temperature, and that $T(\tau)$ scales in proportion to the effective temperature. Equation 7.35 also shows that $\mathfrak{F} = \mathfrak{F}_0 = \pi S(2/3)$, i.e., the surface flux is π times the source function at an optical depth of 2/3, similar to Equation 5.7.

Complete and rigorous solutions of the grey case (e.g., Chandrasekhar 1957) lead to a solution differing only slightly from Equation 7.35. It is usually written in the form

$$S(\tau) = \frac{3 \mathcal{F}_0}{4\pi} [\tau + q(\tau)]$$

or

$$T(\tau) = \left\{ \frac{3}{4} [\tau + q(\tau)] \right\}^{1/4} T_{eff}, \tag{7.37}$$

where $q(\tau)$ is a slowly varying function ranging from 0.577 at $\tau = 0$ to 0.710 at $\tau = \infty$.

Although the grey case is part of history when it comes to interpreting real stellar spectra, it can be useful for thinking through certain situations.

Now let us briefly turn to the convective mode of energy transport.

Convective Energy Transport and Other Factors

The fraction of energy transported by convection in stellar photospheres is tiny. Radiation dominates. Stars on the cool half of the HR diagram, ~F spectral type and cooler, have convective envelopes that extend up to the bottom of the photosphere. For these stars, convection carries just enough energy to slightly lower the temperature gradient in the deepest photospheric layers.

Convection functions through buoyancy. Energy is transported because some rising (or falling) packet of gas possesses an excess (or deficit) of heat compared to its surroundings. Such a cell must have sufficient optical depth across its dimensions to prevent significant radiative energy exchange with the surroundings. The flux carried by convection can be estimated in terms of the average cell characteristics: the temperature excess, ΔT, of the cell over its terminal surroundings, the specific heat at constant pressure, C_p, the cell density, ρ, and the upward velocity, v, of the cell, namely

$$\Phi = \rho C_p \, v \, \Delta T. \tag{7.38}$$

Both rising and sinking cells are involved here with ΔT and v changing sign between the two cases. In the phenomenon of the solar granulation (see Chapter 17), we see the overshoot layers of the convective cell pattern. The momentum of convective cells is not dissipated when they rise into the photosphere, and substantial mass motion exists above those deeper layers that are actually producing the convection. At the bottom of the solar photosphere $\rho \sim 10^{-7}$ g/cm^3. The observed ΔT and v are of the order of 100 K and a few km/s. Using $C_p \sim 10^8$ erg/(gK) for hydrogen, we find $\Phi \sim 2 \times 10^8$ erg/(s cm^2) compared to $\sigma T_{eff}^4 \approx 6 \times 10^{10}$ erg/(s cm^2) carried by radiative flux, confirming convection as a small contributor to the energy flow.

At the same time, convection plays a major role in shaping line profiles through the Doppler shifts and temperature inhomogeneities it causes in the photosphere. Convection also affects stellar atmospheres by generating sound waves. These waves are internally reflected at the outer and inner boundaries of the convection zone, and some resonate in this convective-zone "cavity." They are called P-mode oscillations (P for pressure, the restoring force for sound waves). Both brightness and velocity variations are imprinted on the photosphere, and these can be measured. Their frequencies and amplitudes are used to do seismic sounding inside stars with quite fruitful results. In addition, acoustic waves likely power the lower chromospheres of stars. In a parallel way, convection wiggles the magnetic field lines that extend through the photosphere into the corona. Some of these motions generate Alfvén waves, waves in the magnetic field lines, that propagate energy into coronal layers, raising temperatures to extreme levels on millions of degrees.

Convection is also an efficient mixing mechanism, producing chemical homogeneity. And if the convection zone deepens with stellar evolution, nuclear processed material in the interior can be lifted to the surface, altering the chemical composition of the photosphere. It is worth noting that the timescale for energy to be conveyed through a convective envelope is weeks to months compared to radiation-transfer timescales of 10^5 to 10^6 years.

One more thing: convection interacts with a star's rotation, redistributing angular momentum inside a star, which affects the surface rotation rate that we measure (see Chapter 18). To top it off, when a star has a convective envelope, the rising and falling motions of convection interact with the star's rotation to form a dynamo that generates magnetic fields. These in turn produce surface spots, coronal loops, flares, power for the corona, rigidity and structure in the outer atmosphere, form conduits for escaping mass, and the lever arms for magnetic braking of rotation.

In summary, convection has a profound influence on stars even though it is a trivial player in the conveyance of energy through the photosphere.

Conditions for Convective Flow

A hot convective cell must be buoyant at each level if it is to continue to be driven upward. Since the pressure declines with height, and since the cell can be expected to stay in pressure equilibrium with its surroundings, it expands as it rises. We calculate the density change to see if the expanded cell is still buoyant. The classical Schwarzschild (1906) criterion is obtained by assuming the convective cell behaves adiabatically – no energy leakage whatsoever from the rising cell. Let γ denote the ratio of specific heats and P the total pressure. Then the adiabatic assumption means that $P\rho^{-\gamma} = $ constant. To be buoyant, the density of the cell must decrease more rapidly than the average photospheric density. Thus

$$\frac{1}{\gamma} = \left[\frac{d \log \rho}{d \log P}\right]_{\text{cell}} > \left[\frac{d \log \rho}{d \log P}\right]_{\text{average}}. \tag{7.39}$$

In many calculations, we prefer to use temperature rather than density in these gradients. Using the gas law in the form $P = (\rho/\mu)kT$, where μ is the mean molecular weight, we write

$$\frac{d \log \rho}{d \log P} = 1 + \frac{d \log \mu}{d \log P} - \frac{d \log T}{d \log P}.$$

Placing this into the last term in inequality 7.39 gives

$$\left[\frac{d \log T}{d \log P}\right]_{\text{average}} > 1 - \frac{1}{\gamma} + \frac{d \log \mu}{d \log P} \tag{7.40}$$

as the condition needed for convection.

One way to meet criterion 7.40 is for the gradient on the left side to become large. Large gradients are caused by high opacity, a situation often found in the layers just below the photosphere. Highly opaque layers constrict the outward flow of heat from below. The star is faced with having to expand or, if relation 7.40 is satisfied, it can use convection as a "relief valve."

Convection will also occur if the adiabatic gradient, the right side of relation 7.40, is reduced. In situations where hydrogen and helium do not change their ionization stage and the gases are chemically homogeneous, $d \log \mu / d \log P = 0$; the value of γ for a monatomic gas is 5/3. The Schwarzschild criterion for convection is then $d \log T / d \log P > 0.4$. Polyatomic gases have γ approaching unity as the degrees of freedom increase with the complexity of the molecules. Therefore convection is encouraged in cool stars where molecules are numerous. The value of γ may also be decreased from the monatomic value of 5/3 by the presence of radiation. From the radiation pressure equation, Equation 5.11, and the first law of thermodynamics, one can show (Chandrasekhar 1957, for example) that the radiation trapped in the convective cell obeys an adiabatic gas law with $\gamma = 4/3$.

If we think about moving deeper into the star, the temperature will increase. Should the transition from neutral to ionized hydrogen occur, μ will decrease by about a factor of two. The gradient $d \log \mu / d \log P$ becomes abruptly negative, which according to inequality 7.40, encourages convection. The ionization is essentially a new degree of freedom, and sometimes γ is redefined to accommodate the effect, e.g., Unsöld (1955) and Krishna Swamy (1961). Hydrogen and helium ionization zones are standard features of stellar envelope models.

The Mixing-Length Formulation

The simple standard treatment of energy transported by convection is called the mixing-length formulation. The one free parameter, the mixing length, ℓ, needs to be chosen by guess or empirical calibration, which is not good. Ideally there should be no free parameters. Basically the mixing-length formulation is a dimensional scheme dating back several decades (Biermann 1932, 1938, Öpik 1950, Vitense 1953), and it goes like this. If $\Delta\rho$ is the average difference in density of the convective cell compared to its surroundings, then mechanical energy conservation dictates that $\frac{1}{2}\rho v^2 = g\Delta\rho\ell$, where ℓ is the mixing length, i.e., the distance over which the cell survives, and the other parameters have their usual meaning of density, velocity, and gravitational acceleration. Assuming pressure equilibrium and constancy of the mean molecular weight, $\Delta\rho/\rho = \Delta T/T$, and energy conservation then gives $v = (2g\ell\Delta T/T)^{1/2}$. Combined with Equation 7.38, the convective flux can then be written

$$\Phi = \rho C_p \left(\frac{2g\ell}{T}\right)^{1/2} \Delta T^{3/2}.$$

Expressed in terms of temperature gradients, $\Delta T = \ell \, dT/dx$,

$$\Phi = \rho C_p \left(\frac{2g}{T}\right)^{1/2} \ell^2 \left[\left(\frac{dT}{dx}\right)_{\text{cell}} - \left(\frac{dT}{dx}\right)_{\text{average}}\right]^{3/2}.$$

In stellar interiors calculations, an empirical calibration of ℓ is possible because the radius of the model star depends on ℓ, and values of ~1.5 pressure scale heights are typically found. In the case of stellar photospheres, no such simple calibration exists. Papers exploring the mixing-length approach can be seen in the conference proceedings edited by Spiegel and Zahn (1977). More recent advances use hydrodynamical calculations to calibrate ℓ (e.g., Mosumgaard *et al.* 2018) or to sidestep the mixing-length formulation (e.g., Spada & Demarque 2019). Hydrodynamical modeling of the photospheric overshoot region, that is, the surface granulation, is considered briefly in Chapter 17.

At this point we have in place the definitions and terms we need (Chapter 5), the relevant details of the black-body source function (Chapter 6), and in this chapter the radiative transfer equation has been formulated for a stellar photosphere. Only one thing remains to be covered before we can start investigating model stellar photospheres: the light-stopping properties embodied in the absorption coefficient, κ_ν, the subject of Chapter 8.

Questions and Exercises

1. How good is the plane-parallel approximation? How thick would a photosphere have to be to produce a 1% error in specific intensity?
2. Find an analytical expression for I_ν at the surface of a star when its source function has the form $S_\nu = a + b\tau_\nu$, i.e., express $I_\nu(\tau_\nu = 0)$ in terms of the constants a and b. The answer is the most elementary form for the limb darkening of a star (as in Figures 9.2 and 9.3, for example).
3. In Equation 7.15 we have an expression for computing the spectrum of a star. Within this expression we see the exponential integral E_2. What conditions are necessary to legitimately use Equation 7.15 and when must a more basic expression be used instead? What is that more basic expression? Cite a physical situation in which the assumptions entering Equation 7.15 fail.

References

Abramowitz, M. & Stegun, I.A. (eds.) 1964. *Handbook of Mathematical Functions*, Applied Math Series **55** (Washington, DC: National Bureau of Standards). p. 227.

Adelman, S.J., Kupka, F., & Weiss, W.W. (eds.) 1996. *M.A.S.S. Model Atmospheres and Stellar Spectra. 5th Vienna Workshop*, ASP Conference Series (San Francisco: Astronomical Society of the Pacific).

Biermann, L. 1932. *ZfAp* **5**, 117.

Biermann, L. 1938. *Ast Nach* **264**, 395.

Chandrasekhar, S. 1957. *An Introduction to the Study of Stellar Structure* (New York: Dover), p. 55.

Eddington, A.S. 1926. *The Internal Constitution of the Stars*, 1959 printing (New York: Dover), pp. 322, 325.

Kalkofen, W. (ed.) 1987. *Numerical Radiative Transfer* (Cambridge: Cambridge University Press).

Kourganoff, V. 1963. *Basic Methods in Transfer Problems* (New York: Dover).

Krishna Swamy, K.S. 1961. *ApJ* **134**, 1017.

Mihalas, D. 1978. *Stellar Atmospheres* (San Francisco: W.H. Freeman & Co.), 2nd ed.

Milne. E.A. 1930. *Handbuch der Astrophysik*, **3**, Part 1; reprinted 1966, *Selected Papers on the Transfer of Radiation* (New York: Dover), D.H. Menzel, ed., p. 77.

Mosumgaard, J.R., Ball, W.H., Silva Aguirre, V., Weiss, A., Christensen-Dalsgaard, J. 2018. *MNRAS* **478**, 5650.

Öpik, E.J. 1950. *MNRAS* **110**, 559.

Schwarzschild, K. 1906. *Akad Wiss Goett Nach* **1**, 41.

Spada, F. & Demarque, P. 2019. *MNRAS* **489**, 4712.

Spiegel, E.A. & J.-P. Zahn (eds.) 1977. *Problems of Stellar Convection*, IAU Colloquium 38 (Heidelberg: Springer-Verlag).

Unsöld, A. 1955. *Physik der Sternatmosphären* (Berlin: Springer-Verlag), p. 227.

Vitense, E. 1953. *ZfAp* **32**, 135.

8

The Continuous Absorption Coefficient

As a precursor to calculating the transfer of radiation through a model stellar photosphere, we now look at the continuous absorption coefficient. The wavelength dependence of the continuous absorption coefficient shapes the continuous spectrum emitted by the star: more absorption, less light. The temperature dependence of the absorption also needs to be known, not only to study stars of different temperature, but also because of the temperature variation from the bottom to the top of the photosphere. The strength of spectral lines also depends on the continuous absorption. More continuous absorption means a thinner photosphere with fewer atoms to make spectral lines. Consequently, we must know the continuous absorption coefficient in order to model any part of a stellar spectrum. Our main focus is on absorption in the visible window.

The atomic absorption coefficient, also called the cross section, with units of area *per absorber*, we call α. The absorption coefficient *per neutral hydrogen atom* we denoted by κ without a subscript. We convert κ to the mass absorption coefficient, κ_v in cm^2/g, at the end of the chapter, which is the function needed in the basic flux integral (Equation 7.21),

$$\mathfrak{F}_v = 2\pi \int_0^\infty S(\tau_0) E_2(\tau_v) \frac{\kappa_v}{\kappa_0} \, d\tau_0.$$

From this integral, we compute the continuous spectrum. We use essentially the same expression in Chapter 13 to compute spectral lines by adding the line absorption coefficient, ℓ_v, to κ_v.

The Origins of Continuous Absorption

The total continuous absorption coefficient is the sum of the absorption resulting from many physical processes. The major contributors are considered here one by one. All of them can be classed into one of two categories. The first is an ionization process where a bound–free transition occurs, and the second is a so-called free–free transition, where an electron that is interacting with another particle is accelerated upon absorbing a photon. The remaining possibility is a bound–bound transition which, of course, gives a spectral line. Spectral lines are not included in κ_v, but when there are many overlapped and crowded lines, they act much like the continuous absorption processes. Most of the continuous absorption is due to hydrogen in some form or other. This is a direct result of hydrogen's overwhelming dominance of the chemical composition. Earlier summaries of continuous absorption are given by Vitense (1951), Pecker (1965), and Bode (1967). Some of these are summarized by Allen (1973).

We do not dwell on the atomic theory but instead move directly to the question of how to calculate κ_ν.

The Stimulated Emission Factor

Many of the absorption processes we are about to consider are affected by stimulated emission (recall also the discussion following Equation 5.17). Since stimulated emission amounts to negative absorption, the calculated absorption coefficient has to be reduced accordingly. A simple allowance for stimulated emissions follows from Equations 5.18, 6.7, and the equilibrium excitation process discussed in Chapter 1, namely, the number of stimulated emissions is subtracted from the number of absorptions,

$$\kappa_\nu \rho = N_l B_{lu} h\nu - N_u B_{ul} h\nu$$

$$= N_l B_{lu} h\nu \left(1 - \frac{N_u B_{ul}}{N_l B_{lu}} \right)$$

$$= N_l B_{lu} h\nu \left(1 - e^{-h\nu/kT} \right). \tag{8.1}$$

In other words, the absorption actually produced is lower by the factor $1 - e^{h\nu/kT}$ because of stimulated emissions. While it is true that we have derived Equation 8.1 only for the case of bound–bound absorption, Milne (1924) showed that the same factor also holds with continuous absorption. Table 8.1 shows its size. Most of our continuous absorption coefficients will be reduced by the stimulated emission factor.

Table 8.1 *Stimulated emission factors*

λT (cm K)	T for $\lambda = 5000$ Å	$1 - e^{h\nu/kT}$	% decrease
0.2	4 000	0.999	0.08
0.4	8 000	0.973	2.8
0.6	12 000	0.909	10.0
0.8	16 000	0.834	19.8
1.0	20 000	0.763	31.1

Neutral Hydrogen Bound–Free Absorption

Neutral hydrogen is the major source of absorption for temperatures exceeding about 8000 K. Both the bound–free and the free–free absorption of hydrogen are important. Bound–free transitions, i.e., ionizations, can occur from any level to the continuum. The wavelengths of the spectral lines, the bound–bound transitions, are given by

$$\frac{1}{\lambda} = R\left(\frac{1}{n^2} - \frac{1}{m^2} \right), \tag{8.2}$$

in which $R = 1.09678 \times 10^5$/cm or $R = 1.09678 \times 10^{-3}$/Å is the Rydberg constant for hydrogen and n and m are principal quantum numbers. Fixing n at 2 and letting m run from 3 to infinity gives the Balmer line series at visible wavelengths. To produce continuous absorption, the electron must be lifted beyond the "last" orbit. The minimum energy bound–free case is therefore given by $m = \infty$. In this case, the photon energy is just sufficient to ionize the atom, and we can write $h\nu_n = hc/\lambda_n = hcR/n^2$, where the subscript on ν and λ denotes the minimum-energy value to ionize from the nth level. Photons of wavelength shorter than any given ionization threshold therefore impart a kinetic energy to the freed electron equal to the photon's energy minus hcR/n^2. The excitation potential, χ_n, is defined to be the energy above the ground level, and the continuum is I electronvolts above the ground level (hcR). This is pictured in Figure 8.1. Planck's constant is 6.6261×10^{-27} erg/Hz or 4.1357×10^{-15} eV/Hz. We can then express the excitation potential for the bound levels as

$$\chi_n = I - hcR/n^2 = hcR - hcR/n^2 = 13.598\left(1 - 1/n^2\right) \text{ eV}. \qquad (8.3)$$

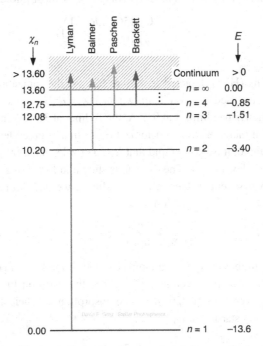

Figure 8.1 This simplified one-dimensional energy-level diagram for hydrogen shows the quantum numbers, n, the binding energy, E, and the excitation potential, χ_n, in electronvolts for the first four levels and the continuum. Transitions from a bound level into the continuum are bound–free absorptions, as shown by the arrows.

The absorption coefficient corresponding to the bound–free transitions of the type shown in Figure 8.1 was original derived by Kramers (1923) and modified by Gaunt (1930). Their result can be written

$$\alpha_{\mathrm{bf}}(\mathrm{H}) = \frac{32}{3^{3/2}} \frac{\pi^2 e^6}{h^3} \frac{R}{n^5 v^3} g_{\mathrm{bf}}$$

$$= \frac{32}{3^{3/2}} \frac{\pi^2 e^6}{h^3 c^3} R \frac{\lambda^3}{n^5} g_{\mathrm{bf}}$$

$$= \alpha_0 g_{\mathrm{bf}} \frac{\lambda^3}{n^5} \ \mathrm{cm}^2/\text{neutral H absorber}, \tag{8.4}$$

with $\alpha_0 = 1.0443 \times 10^{-26}$ for λ in angstroms. In this expression, the electron charge has a value of $e = 4.8032 \times 10^{-10}$ esu, and g_{bf} is the Gaunt factor (not to be confused with the statistical weight g_n), which brings Kramer's original equation into agreement with the quantum mechanical results. It is of the order of unity.

There are two Gaunt factors, one for the bound–free absorption, Equation 8.5, and another for the free–free absorption discussed in the next section (Equation 8.9). Tabulations are given by Karzas and Latter (1961). However, the general expression based on the work of Menzel and Pekeris (1935) is adequate for most purposes,

$$g_{\mathrm{bf}} = 1 - \frac{0.3456}{(\lambda R)^{1/3}} \left(\frac{\lambda R}{n^2} - \frac{1}{2} \right), \tag{8.5}$$

where λR is unitless.

The behavior of $\alpha_{\mathrm{bf}}(\mathrm{H})$ is depicted in Figure 8.2. At the ionization limits, where the kinetic energy of the released electron is nil, we have $\lambda = n^2/R$, so $\alpha_{\mathrm{bf}}(\mathrm{H})$ at these limits is proportional to n. The Lyman limit is at 912 Å, the Balmer at 3646 Å, and so on. This cross section for absorption increases as λ^3 for ionization from any given level. Although the λ^3 dependence is very rapid, it is not so rapid that $\alpha_{\mathrm{bf}}(\mathrm{H})$ for level n becomes negligible when considering $\alpha_{\mathrm{bf}}(\mathrm{H})$ for level $n - 1$. The overlap is shown in Figure 8.2. This means that we must sum $\alpha_{\mathrm{bf}}(\mathrm{H})$ over several values of n. In addition, we have to take into account the populations in each level using Equation 1.18,

$$\frac{N_n}{N} = \frac{g_n}{u(T)} e^{-\chi_n/kT},$$

where N_n/N is the number of hydrogen atoms excited to the level n per neutral hydrogen atom, $g_n = 2n^2$ is the statistical weight, $u(T) = 2$ is the partition function, and χ_n is the excitation potential given by Equation 8.3. The absorption coefficient in cm^2 per neutral H atom for all continua starting at n_0 is then

$$\kappa(\mathrm{H_{bf}}) = \sum_{n_0}^{\infty} \alpha_{\mathrm{bf}}(\mathrm{H}) \frac{N_n}{N}$$

$$= \alpha_0 \sum_{n_0}^{\infty} \frac{\lambda^3}{n^3} g_{\mathrm{bf}} e^{-\chi_n/kT}$$

$$= \alpha_0 \sum_{n_0}^{\infty} \frac{\lambda^3}{n^3} g_{\mathrm{bf}} 10^{-\theta \chi_n}, \tag{8.6}$$

with $\theta = 5040/T$ as usual.

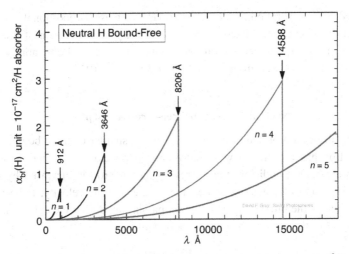

Figure 8.2 The bound–free absorption coefficient per absorber increases with λ^3 for a fixed n, and is proportional to n.

Unsöld (1955) showed that the small contribution of the terms higher than $n_0 + 2$ can be replaced by an integral,

$$\sum_{n_0+3}^{\infty} \frac{1}{n^3} e^{-\chi_n/kT} = -\frac{1}{2} \int_{n_0+3}^{\infty} e^{-\chi_n/kT} \, d\left(1/n^2\right).$$

And according to Equation 8.3, we have $d\chi_n = -I \, d(1/n^2)$ so that

$$\sum_{n_0+3}^{\infty} \frac{1}{n^3} e^{-\chi_n/kT} = \frac{1}{2} \int_{\beta_3}^{I} e^{-\chi_n/kT} \frac{d\chi}{I}$$

$$= \frac{kT}{2I} \left[e^{-\beta_3/kT} - e^{I/kT} \right],$$

where

$$\beta_3 = I \left[1 - \frac{1}{(n_0 + 3)^2} \right].$$

We are justified in neglecting the n dependence of g_{bf} because the integral is small compared to the first three terms that we retained explicitly in the summation. Finally, then Equation 8.6 becomes

$$\kappa(H_{bf}) = \alpha_0 \lambda^3 \left[\sum_{n_0}^{n_0+2} \frac{g_{bf}}{n^3} 10^{-\theta\chi_n} + \frac{\log e}{2\theta I} \left(10^{-\beta_3\theta} - 10^{-I\theta} \right) \right]. \tag{8.7}$$

Here $\log e = 0.43429$ and, again, $\alpha_0 = 1.0449 \times 10^{-26}$ when λ is in angstroms.

In Figure 8.2, the absorption per absorber, $\alpha_{bf}(H)$, increases proportional to n. Now with the inclusion of the excitation factor, $n^2 e^{-\chi_n/kT}$, where χ_n is a function of n according to Equation 8.3, the absorption per neutral hydrogen atom, $\kappa(H_{bf})$, drops rapidly with n. For example, if $\theta = 0.5$, the drop from $n = 1$ to $n = 2$ is 6.4×10^{-5}.

Neutral Hydrogen Free–Free Absorption

The free–free absorption of neutral hydrogen is much smaller than the bound–free. When the free electron has a collision with a proton, its trajectory is altered. If a photon is absorbed during such a collision, the energy of the electron is increased by the photon energy. The strength of the absorption depends on the electron velocity, v, because a slow encounter increases the probability of a photon passing by during the collision. According to Kramers (1923), the atomic absorption coefficient is

$$d\alpha_{ff}(H) = \frac{2}{3^{3/2}} \frac{h^2 e^2 R}{\pi m_e^3} \frac{1}{v^3} \frac{dv}{v}$$

in cm^2 per absorber for the fraction of electrons of mass m_e in the velocity interval v to $v + dv$. The complete free–free absorption per electron is given by integrating this expression over the velocity distribution. In almost every case, it is reasonable to use a Maxwell–Boltzmann speed distribution (Equation 1.12), which gives

$$\alpha_{ff}(H) = \frac{2}{3^{3/2}} \frac{h^2 e^2 R}{\pi m_e^3} \frac{1}{v^3} \int_0^\infty \left(\frac{2}{\pi}\right)^{1/2} \left(\frac{m_e}{kT}\right)^{3/2} v \, e^{-m_e v^2/2kT} \, dv$$

$$= \frac{2}{3^{3/2}} \frac{h^2 e^2 R \lambda^3}{\pi m_e^3 c^3} \left(\frac{2m_e}{\pi kT}\right)^{1/2}. \tag{8.8}$$

The quantum mechanical derivation by Gaunt (1930) leads to nearly the same expression, but modified by the Gaunt correction factor g_{ff} given by Menzel and Perkeris (1935),

$$g_{ff} = 1 + \frac{0.3456}{(\lambda R)^{1/3}} \left(\frac{\lambda kT}{hc} + \frac{1}{2}\right)$$

$$= 1 + \frac{0.3456}{(\lambda R)^{1/3}} \left(\frac{\log e}{\theta \chi_\lambda} + \frac{1}{2}\right), \tag{8.9}$$

where again λR is unitless and $\chi_\lambda = hv = 1.2398 \times 10^4/\lambda$ eV, with λ in angstroms, is the photon energy and $\theta = 5040/T$.

The absorption coefficient in cm^2 per neutral hydrogen atom is proportional to the number density of electrons, N_e, and protons, N_i, giving

$$\kappa(H_{ff}) = \frac{\alpha_{ff}(H) \, g_{ff} \, N_i \, N_e}{N_0}.$$

The number density of neutral hydrogen is denoted by N_0. Using Equation 1.21 to get $N_i N_e / N_0$, and $\alpha_{ff}(H)$ from Equation 8.8, we write

$$\kappa(H_{ff}) = \alpha_0 \lambda^3 g_{ff} \frac{\log e}{2\theta I} 10^{-\theta I}. \tag{8.10}$$

The constant α_0 is as in Equation 8.4 and $I = 13.598$ eV. (Use $I = hcR$ and $R = 2\pi^2 m_e \, e^4 / h^3 c$.)

In stars of B, A, and F spectral types, neutral hydrogen is the dominant continuous absorber. The energy distributions of such stars show strong discontinuities at the ionization edges, e.g., the wavelengths of the ionization limits shown in Figure 8.2. The absorption coefficient of neutral hydrogen is compared to other absorbers in Figure 8.7 below. Other examples of the neutral hydrogen signature can be seen in Figures 10.3, 10.7, and 10.8 in Chapter 10. In cooler stars however, the absorption from neutral hydrogen diminishes and the absorption from the negative hydrogen ion increases.

The Negative Hydrogen Ion

Absorption by the negative hydrogen ion is the major contributor to κ_ν for F stars and cooler. The hydrogen atom is capable of holding a second electron in a bound state because its simple electron–proton structure is highly polarized. The extra electron is bound with 0.7542 eV (Frolov 2015), so all photons with $\lambda \le 16\,440$ Å have sufficient energy to "ionize" the H^- back to the neutral hydrogen atom and a free electron. The extra electrons needed to form H^- come from the ionized metals. The main electron donors are C, Si, Fe, Mg, Ni, Cr, Ca, Na, and K. The effectiveness of H^- therefore depends on the abundances of these elements and their state of ionization. In addition to bound–free absorption, free–free absorption can occur when there is a collision between a (polarized) neutral hydrogen atom and an electron. Following the suggestion by Wildt (1939), calculations as well as laboratory measurements confirmed the negative hydrogen ion as a major source of absorption in the Sun. A short review of the historical development is given by Massey (1969). See also John (1989, 1991, 1992). At temperatures somewhat higher than those in the solar photosphere, not only does the absorption by neutral hydrogen take over, but H^- is ionized to such an extent that it ceases to be a strong absorber. Examples are shown below in Figure 8.7. The H^- absorption also drops off rapidly in the very coolest stars as the metals are no longer ionized and free electrons become scarce.

The bound–free atomic absorption coefficients have been calculated by Geltman (1962), Doughty *et al.* (1966), Krogdahl and Miller (1967), Stewart (1978), and Wishart (1979). The wavelength dependence is shown in Figure 8.3. The maximum is near 8500 Å. The following polynomial fits Wishart's data to better than 0.2% between 2250 Å and 15 000 Å, about five times better than the expected precision of the quantum-mechanical calculations.

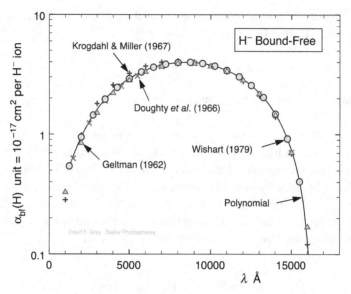

Figure 8.3 The absorption coefficient for the negative hydrogen ion shows a maximum near 8500 Å. There is good agreement among the calculations. The solid line is Equation 8.11.

$$\alpha_{bf}(H^-) = a_0 + a_1\lambda + a_2\lambda^2 + a_3\lambda^3 + a_4\lambda^4 + a_5\lambda^5 + a_6\lambda^6, \qquad (8.11)$$

where $\alpha_{bf}(H^-)$ is in units of 10^{-17} cm^2 per H$^-$ ion; the stimulated emission factor is not included, and the constants have the values

$$a_0 = +0.1199654,$$
$$a_1 = -1.18267 \times 10^{-6},$$
$$a_2 = +2.64243 \times 10^{-7},$$
$$a_3 = -4.40524 \times 10^{-11},$$
$$a_4 = +3.23992 \times 10^{-15},$$
$$a_5 = -1.39568 \times 10^{-19},$$
$$a_6 = +2.78701 \times 10^{-24},$$

when λ is in Å. The H$^-$ ionization is given by Equation 1.19 with $u_0(T) = 1$ and $u_1(T) = 2$ and an electron pressure of P_e, namely,

$$\log \frac{N(H)}{N(H^-)} = -\log P_e - \frac{5040}{T}I + 2.5 \log T + 0.1248.$$

Then, with $I = 0.754$ eV,

$$\kappa\left(H_{bf}^-\right) = 4.158 \times 10^{-10} \, \alpha_{bf} \, P_e \, \theta^{5/2} \, 10^{0.754\,\theta} \qquad (8.12)$$

with units of cm^2 per neutral hydrogen atom.

The free–free component of the H$^-$ absorption becomes dominant in the infrared. Calculations have been made by Geltman (1965), John (1966), Doughty and Frazer (1966), Stilley & Callaway (1970), and Bell and Berrington (1987). The behavior of the absorption coefficient with wavelength and temperature is shown in Figure 8.4. The following polynomials fit the data of Bell and Berrington to typically $\pm 1\%$ for θ in the interval 0.5–2.0 ($T = 10\,000 - 2500$ K) and for λ in the range 2600 to 113 900 Å. In units of cm^2 per neutral hydrogen atom,

$$\kappa\left(H_{ff}^-\right) = \alpha_{ff}(H^-)P_e$$
$$= 10^{-26} P_e \, 10^{f_0 + f_1 \, \log \theta + f_2 \, \log^2 \theta}, \tag{8.13}$$

where

$$f_0 = -2.2763 - 1.6850 \log \lambda + 0.76661 \log^2 \lambda - 0.0533464 \log^3 \lambda$$
$$f_1 = +15.2827 - 9.2846 \log \lambda + 1.99381 \log^2 \lambda - 0.142631 \log^3 \lambda$$
$$f_2 = -197.789 + 190.266 \log \lambda - 67.9775 \log^2 \lambda + 10.6913 \log^3 \lambda - 0.625151 \log^4 \lambda,$$

when λ is in angstroms. Here the stimulated emission factor has been *included* in the polynomial.

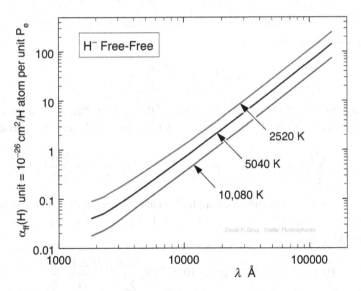

Figure 8.4 The free–free absorption coefficient increases rapidly with wavelength. The values include the stimulated emission factor.

Other Hydrogen Continuous Absorbers

Hydrogen molecules appear in large numbers in cool stars. The H$_2$ molecules are more numerous than atomic hydrogen for stars cooler than mid-M spectral type, and while the molecule itself does not absorb in the visible spectrum, various ions do. The H$_2^+$ absorption has been studied by Bates (1951, 1952), Buckingham *et al.* (1952), Bates *et al.* (1953), Matsushima (1964), and Stancil (1994). It is a significant absorber in the UV, but quite generally is only a few percent of the H^- absorption for $\lambda > 3800\,\text{Å}$. Its absorption can be calculated from

$$\kappa(H_2^+) = \frac{16\pi^4 a_0^5 e^2}{3hc}\,\sigma(\lambda, T)N(H^+) \tag{8.14}$$

cm^2 per neutral hydrogen atom, in which $a_0 = 0.52917 \times 10^{-8}$ cm is the radius of the first Bohr orbit, $\sigma(\lambda, T)$ is the atomic cross section, and $N(H^+)$ is the number of protons per cm^3. Using the formulation given by Bates (1952), with the stimulated emission factor removed, one can separate $\sigma(\lambda, T)$ into λ and T components and write

$$\kappa(H_2^+) = \frac{16\pi^4 a_0^5 e^2}{3hc}\,\sigma_1(\lambda)10^{-U_1\theta}\,N(H^+)$$

$$= 2.51 \times 10^{-42}\,\sigma_1(\lambda)10^{-U_1\theta}\,N(H^+). \tag{8.15}$$

$N(H^+)$ is conveniently calculated from the ionization equation (Equation 1.20) for hydrogen and the electron pressure.

$$N(H^+) = N_0(H)\,\Phi_H/P_e = [N_H - N(H^+)]\Phi_H/P_e,$$

where $N_0(H)$ is the number of neutral hydrogen atoms and N_H is the total number of hydrogen particles. Solving for $N(H^+)$,

$$N(H^+) = N_H \frac{\Phi_H/P_e}{1 + \Phi_H/P_e}.$$

The electron pressure is based on the contribution of the electrons from each of the j elements, but as mentioned above, only C, Si, Fe, Mg, Ni, Cr, Ca, Na, and K need be included,

$$P_e = \sum_C^K N_{ej}kT = \sum_C^K N_j \frac{\Phi_j/P_e}{1 + \Phi_j/P_e}kT = N_H \sum_C^K A_j \frac{\Phi_j/P_e}{1 + \Phi_j/P_e}kT,$$

where the ionization equation has been applied to each element and the abundances relative to hydrogen are $A_j = N_j/N_H$. Solving for N_H and placing it in to the previous equation gives the working equation

$$\kappa(H_2^+) = 3.61 \times 10^{-30}\,\sigma_1(\lambda)\,10^{-U_1\theta}\,\theta P_e \frac{\Phi_H/(1 + \Phi_H/P_e)}{\sum A_j\,\Phi_j/(1 + \Phi_j/P_e)} \tag{8.16}$$

cm^2 per neutral hydrogen atom. In this expression, θ is the usual $5040/T$. Polynomial approximations for $\sigma_1(\lambda)$ and U_1 are

$$\sigma_1(\lambda) = -1040.54 + 1345.71\log\lambda - 547.628\log^2\lambda + 71.9684\log^3\lambda$$

and

$$U_1 = -54.0532 + 32.713\log\lambda - 6.6699\log^2\lambda + 0.4574\log^3\lambda.$$

These approximations match Bates' (1952) data to $\pm 0.3\%$ over the wavelength rage 3800–25 000 Å.

Although there are no stable bound states for H_2^- (Massey 1976), the free–free absorption becomes important in M and cooler stars, equaling a substantial fraction of the H^- absorption. See Somerville (1964), Dalgarno and Lane (1966), and John (1994).

Absorption by Helium

In most stars, helium is the next most abundant element after hydrogen, but its electrons are so tightly bound ($\chi_1 = 19.72$ eV, $I_1 = 24.59$ eV, and $I_2 = 54.42$ eV) that its contribution to κ_ν becomes important only in O and early B spectral types, and primarily in the UV. I_1 corresponds to 504.19 Å, I_2 to 227.82 Å. Absorption edges closer to the visible region come from the higher levels. Because χ_1 is so large, an excited electron will essentially see a net charge of $+1$ below it, and so the absorption will mimic that of hydrogen. Furthermore, an approximate treatment is often acceptable since helium absorption rarely exceeds 10% of the hydrogen absorption. For $n \geq 3$, Ueno (1954) suggested the approximate relation for scaling the hydrogen absorption,

$$\alpha(\text{He}) = 4e^{-10.92/kT} \, \alpha(\text{H}),$$

in which both $\alpha_{bf}(\text{H})$ and $\alpha_{ff}(\text{H})$ would be included in $\alpha(\text{H})$. Detailed treatment of the lower levels can be found in the work of Huang (1948), Underhill (1962), and Vardya (1964).

Because χ_1 is so close to I_1, by the time electrons start populating the excited levels, helium is beginning to ionize. Ionized helium, now a single-electron ion, again behaves much like hydrogen but the energies are scaled up, and the wavelengths scaled down, by a factor of four.

The bound–free absorption of the negative helium ion is generally considered to be negligible because there is only one bound level with an ionization potential of 19 eV. The population of such a level is, under normal conditions, too small to warrant consideration. But the contributions of He^- free–free transitions are marginally significant, increasing slowly toward cooler temperatures and roughly as the square of the wavelength. The free–free absorption can be calculated from the data of McDowell *et al.* (1966) or John (1994); a useful approximation to John's tabulation is

$$\log \alpha\left(\text{He}_{ff}^-\right) = c_0 + c_1 \log \lambda + c_2 \log^2 \lambda + c_3 \log^3 \lambda.$$

The coefficients, with $\theta = 5040/T$, are

$$c_0 = +9.66736 - 71.76242\,\theta + 105.29576\,\theta^2 - 56.49259\,\theta^3 + 10.69206\,\theta^4,$$
$$c_1 = -10.50614 + 48.28802\,\theta - 70.43363\,\theta^2 + 37.80099\,\theta^3 - 7.15445\,\theta^4,$$
$$c_2 = +2.74020 - 10.62144\,\theta + 15.50518\,\theta^2 - 8.33845\,\theta^3 + 1.57960\,\theta^4,$$
$$c_3 = -0.19923 + 0.77485\,\theta - 1.13200\,\theta^2 + 0.60994\,\theta^3 - 0.11564\,\theta^4,$$

giving $a(\mathrm{He_{ff}^-})$ in units of 10^{-26} cm^2 per absorber to a precision of 1% or better between 5000 and 150 000 Å and θ in the range of 0.5–2.0. The stimulated-emission factor has been included in these polynomials.

The absorption coefficient in cm^2 per hydrogen atom is then given by

$$\kappa\left(\mathrm{He_{ff}^-}\right) = 10^{-26}\frac{a\left(\mathrm{He_{ff}^-}\right)A(\mathrm{He})\,P_e}{1+\Phi(\mathrm{He})/P_e}. \tag{8.17}$$

The number abundance of helium relative to hydrogen is $A(\mathrm{He})$, and the factor $1+\Phi(\mathrm{He})/P_e$ in the denominator takes into account the ionization of helium, where $\Phi(\mathrm{He})$ is given by Equation 1.20. As seen in Figure 8.7, this ion is usually a small contributor to the total absorption.

Electron Scattering

Electron scattering is also called Thomson scattering. Free electrons scatter radiation with the same efficiency at all wavelengths. The absorption coefficient is

$$a(e) = \frac{8\pi}{3}\left(\frac{e^2}{m_e c^2}\right)^2$$

$$= 0.6648 \times 10^{-24}\ \mathrm{cm^2/electron}.$$

It has generally been assumed that the intrinsic non-isotropy in the scattering process (phase function of the form $1 + \cos^2\theta$) averages to zero over the symmetrical geometry of the star and many scatterings through the photosphere. We write the absorption per hydrogen atom as

$$\kappa(e) = a(e)\frac{N_e}{H_\mathrm{H}} = a(e)\frac{P_e}{P_\mathrm{H}},$$

where N_H is the number density of hydrogen and P_H is the partial pressure of hydrogen, which is related to the gas and electron pressure in the following way. If N is taken to represent the total number of particles per cubic centimeter, then

$$N = N_e + \sum N_j = N_e + N_\mathrm{H}\sum A_j$$

with N_j particles of the jth element per cubic centimeter, and $A_j = N_j/N_\mathrm{H}$. We solve for N_H and write

$$N_\mathrm{H} = \frac{N - N_e}{\sum A_j}\quad\text{or}\quad P_\mathrm{H} = \frac{P_g - P_e}{\sum A_j},$$

P_g denoting gas pressure, which leads to

$$\kappa(e) = a(e)\frac{P_e}{P_g - P_e}\sum A_j. \tag{8.18}$$

Again, $\kappa(e)$ has units of cm^2 per hydrogen particle. The stimulated emission factor should not be used with $\kappa(e)$.

Electron scattering contributes significant absorption in the photospheres of O and early B stars. Since hydrogen is ionized, $P_e \sim 0.5P_g$ and $\kappa(e) \sim \alpha(e) \sum A_j \sim 7 \times 10^{-25}$ cm^2/hydrogen particle. In cool stars, $P_e \ll P_g$, and then $\kappa(e)/P_e \sim 7 \times 10^{-25}/P_g$. With $P_g \sim 10^3 - 10^5$ dyne/cm^2, electron scattering can be neglected.

Rayleigh scattering, which varies as λ^{-4}, is important in the UV for cool stars, especially when they have significant numbers of molecules in their atmospheres. The reader is referred to Vitense (1951), Gingerich (1964), Vardya (1970), or Allen (1973) for further details. Hayek *et al.* (2011) computed hydrodynamical models for red giants and found Rayleigh scattering to affect most strongly the UV flux, with additional ramifications for granulation characteristics and spectral lines. Rayleigh scattering by hydrogen and helium in hot stars was investigated by Fišák *et al.* (2016) and found to be very small, producing changes in continuum flux of only a few micromagnitudes.

Other Sources of Continuous Absorption

Elements such as carbon, silicon, aluminum, magnesium, and iron produce bound–free absorption that can be important and often dominates in the UV, especially for $\lambda \lesssim 2500$ Å. In fact, the absorption can increase so rapidly toward shorter wavelengths that the atmospheric levels from which the radiation comes lie increasingly above the normal photosphere.

The proper way to handle this type of absorption is to combine well-determined experimental cross sections with the number of absorbers. This can be done in those cases where the cross sections have been measured. Rich (1967), for example, measured silicon (37×10^{-18} cm^2 per absorber for the ground state) and his results were used by Gingerich and Rich (1966) to show that silicon dominates the metal absorption in the 1300–2000 Å region with absorption edges at 1526 Å (ground level ionization) and 1682 Å (first excited level ionization). More recent cross-section results tend to be larger than the earlier ones, e.g., Seaton *et al.* (1994). But in many other cases, less dependable estimates must be used. When a single outer electron is involved in the transition, one might expect the ionization behavior to resemble that of hydrogen (Vitense 1951). A better theoretical approach was given by Burgess and Seaton (1960) called the quantum defect method. It can only be applied to situations of LS coupling. Application has been made by Peach (1967, 1970) and Travis and Matsushima (1968), Dragon and Mutschlecner (1980), among others. The accuracy of the quantum defect method has been considered by Matsushima and Travis (1973). More recent tabulations, based on measured cross sections, are given by Sharp and Burrows (2007). The nature of the metal absorption coefficients for a selection of species is shown in Figure 8.5. Two panels are used to lessen the overlap of curves and not all species of importance are included.

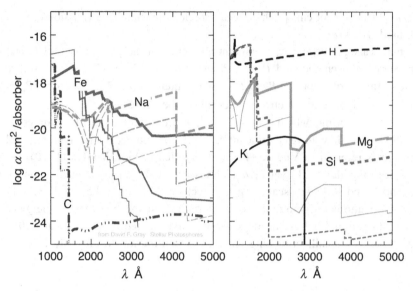

Figure 8.5 Examples of UV absorption by ionization of metals. The ordinate is the same in both panels. Line type signifies the element, as labeled. The upper-most curve in each set is for 6000 K. The next lower curve is for 3652 K. The lowest curve for Fe and Na is for 2500 K. Absorption by the negative hydrogen ion at 2500 K is shown in the right panel. Based on Travis and Matsushima (1968) and Sharp and Burrows (2007).

The strength of the individual components obviously depends on the chemical composition of the star. For the solar composition, these collective absorptions surpass the absorption by H^- for $\lambda < 2500$ Å, as shown by the example in Figure 8.6.

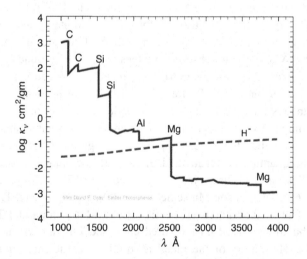

Figure 8.6 One example of the cumulative bound–free absorption of metals at a temperature of 5600 K and log P_e = 1.0. Based on Travis and Matsushima (1968), who used the solar abundances of Goldberg *et al.* (1960).

Several molecular ions are important continuous absorbers in late-type stars. A brief review is given by Vardya (1970). Water vapor absorption has been calculated by Auman (1967). Weck *et al.* (2003) investigate MgH. Sharp and Burrows (2007) calculate absorption coefficients for numerous absorbers expected to be at work in very cool stars and giant planets. Gustafsson *et al.* (2008, Table 1) list the numerous sources they incorporate in their models.

Solid grains and liquids condense in the outer photospheres of very cool stars, from brown dwarfs to red giants. Grains of many types have been included in some models, e.g., Tsuji *et al.* (1996), Allard *et al.* (2001), Ferguson *et al.* (2001), Sharp and Burrows (2007).

Absorption by Lines

Although line absorption is not strictly a part of the continuous absorption, the cumulative effect of many lines nevertheless can behave much like continuous absorption in the upper photosphere. This can be rather important when $T(\tau_v)$ is determined by imposing radiative equilibrium, as emphasized by Kurucz (1979). The problems associated with line absorption stem from the very large number of lines that can arise and the need to know their identification and atomic parameters. Data for millions of lines have been compiled for this purpose (Kurucz & Peytremann 1975, Kurucz 1979, Bell & Tripicco 1995, Gustafsson *et al.* 2008). The *Vienna Atomic Line Database* (VALD, Piskunov *et al.* 1995, Ryabchikova *et al.* 2018) at http://vald.astro.uu.se is very useful in this context. Various techniques have been developed to handle numerous spectral lines, e.g., Gustafsson *et al.* (1975, 2008), Cardona *et al.* (2009).

The amount of absorption by lines depends on temperature, pressure, chemical composition, and, to a lesser extent, on microturbulence (see Chapter 13). It can also vary markedly from one spectral region to the next, although generally there are more atomic lines at shorter wavelengths within the visible window (refer to Figure 10.12). Molecular bands are prevalent in cool stars.

The Total Absorption Coefficient

The final absorption coefficient, that parameter needed in the transfer equation, is now close at hand. We simply add together the various individual absorption coefficients,

$$\kappa_{\text{total}} = \left\{ \left[\kappa(\text{H}_{\text{bf}}) + \kappa(\text{H}_{\text{ff}}) + \kappa(\text{H}_{\text{bf}}^-) + \kappa(\text{H}_2^+) \right] \left(1 - 10^{-\chi_\lambda \theta} \right) + \kappa(\text{H}_{\text{ff}}^-) + \cdots \right\}$$

$$\frac{1}{1 + \Phi(\text{H})/P_e} + \kappa(\text{metals}) + \kappa(\text{He}_{\text{ff}}^-) + \kappa(e) + \cdots. \qquad (8.19)$$

The first few coefficients are reduced by the stimulated emission factor in which, as usual, $\chi_\lambda = 1.2398 \times 10^4/\lambda\,\text{eV}$ when λ is in angstroms and $\theta = 5040/T$; the factor $1/[1 + \Phi(T)/P_e]$ converts the absorption coefficients from normalization per neutral hydrogen atom to per hydrogen particle, thus taking into account the ionization of hydrogen at high temperatures (Equation 1.21). Finally convert to units of cm^2 per gram by dividing by the number of grams per hydrogen particle,

$$\kappa_\nu = \frac{\kappa_{\text{total}}}{\sum A_j m_j}.$$ (8.20)

The mass, m_j, of the jth element in the atmosphere is 1.6606×10^{-24} g times the element's atomic weight. Again, A_j denotes the number abundance relative to hydrogen. A typical value of $\sum A_j m_j$ is $\approx 2 \times 10^{-24}$ g per H particle.

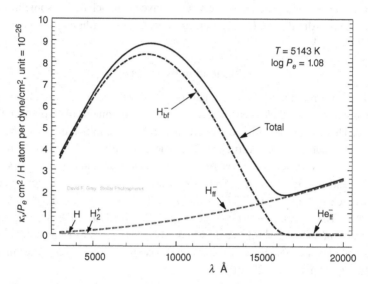

Figure 8.7a Absorption coefficients per unit electron pressure are shown for a relatively cool temperature of 5143 K. The total is dominated by the negative hydrogen ion. The numbers in this and the next three panels are taken from model photospheres at unit optical depth.

Figure 8.7 gives some examples of κ_ν/P_e for selected temperatures. The curves show individual contributions and their total at unit optical depth at 5000 Å in the photospheres of four models. The temperature and electron pressure at that depth are shown in each panel.

The reason for normalizing κ_ν to the electron pressure is that the major contributors for the cooler cases are proportional to P_e. Since P_e increases with surface gravity and with metallicity, these plots are independent of these two basic stellar parameters. In particular, cool stars show a large range in surface gravity and the majority of abundance studies involve cool stars, so normalization by P_e is appropriate. On the other hand, the absorption for neutral hydrogen does not scale with electron pressure, but to be consistent with the other cases, its absorption coefficient has also been normalized to P_e.

The most striking change in moving to higher temperatures is the decline in absorption of the negative hydrogen ion coupled with the rise in absorption of neutral hydrogen. Notice the change in ordinate scales from panel to panel.

In the next chapter, we turn our attention to using these absorption coefficients in the construction of some model photospheres.

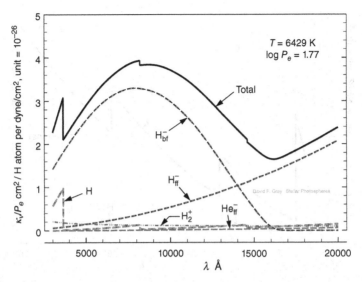

Figure 8.7b Absorption coefficients per unit electron pressure are shown for an intermediate temperature of 6429 K. The total is still dominated by the negative hydrogen ion, but neutral hydrogen is making an appearance.

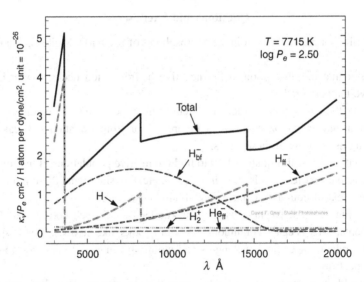

Figure 8.7c Absorption coefficients per unit electron pressure are shown for a temperature of 7715 K. Now neutral hydrogen absorption is comparable to that of the negative hydrogen ion.

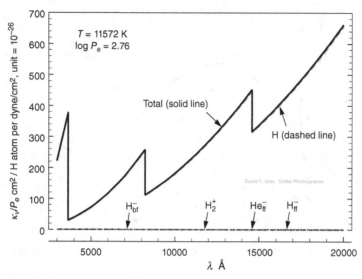

Figure 8.7d Absorption coefficients per unit electron pressure at this high temperature are completely dominated by the absorption by neutral hydrogen. Notice the large increase in the ordinate scale here compared to the previous three panels.

Questions and Exercises

1. What is the dominant absorber in the photospheres of K dwarfs, K giants, A dwarfs, and O stars?

2. Why is it that the absorption of the negative hydrogen ion declines with increasing effective temperature?

3. Why does the absorption coefficient of neutral hydrogen show sharp edges? Using the ionization and excitation energies in Figure 8.1 or from Equation 8.3, calculate the wavelengths of some of these edges.

4. Use Equation 8.3 with Equation 1.19 to find the relative populations of hydrogen atoms excited to the first few levels, e.g., $n = 1, 2, 3$, etc. Consider what these numbers mean in the context of bound–free and bound–bound absorption.

5. Look at Figure 10.3. Explain, using your knowledge of black bodies and absorption coefficients, why Vega's continuous spectrum looks like it does. Be as specific as possible, relating specific wavelengths and atomic parameters to features you identify in the spectrum.

6. How would the conversion to κ_ν in Equation 8.20 be affected if we had a photosphere composed of pure hydrogen? How would κ_{total} be affected?

References

Allard, F., Hauschildt, P.H., Alexander, D.R., Tamanai, A., & Schweitzer, A. 2001. *ApJ* **556**, 357.

Allen, C.W. 1973. *Astrophysical Quantities* (London: Athlone), 3rd ed., p. 99.

Auman, J. Jr., 1967. *ApJS* **14**, 171.

Bates, D.R. 1951. *MNRAS* **111**, 303.

Bates, D.R. 1952. *MNRAS* **112**, 40.

Bates, D.R., Ledsham, K., & Stewart, A.L. 1953. *Phil Trans R Soc London, Ser A*, **246**, 215.

Bell, K.L. & Berrington, K.A. 1987. *J Phys B* **20**, 801.

Bell, R.A. & Tripicco, M.J. 1995. *Astrophysical Applications of Powerful New Databases*, ASP Conference Series **78** (San Francisco: Astronomical Society of the Pacific), S.J. Adelman & W.L. Weise, eds., p. 365.

Bode, G. 1967. *ZfAp* **66**, 170.

Buckingham, R.A., Reid, S., & Spence, R. 1952. *MNRAS* **112**, 382.

Burgess, A. & Seaton, M.J. 1960. *MNRAS* **120**, 121.

Cardona, O., Simonneau, E., & Crivellari, L. 2009. *ApJ* **690**, 1378.

Dalgarno, A. & Lane, N.F. 1966. *ApJ* **145**, 623.

Doughty, N.A. & Frazer, P.A. 1966. *MNRAS* **132**, 267.

Doughty, N.A., Frazer, P.A., & McEachran, R.P. 1966. *MNRAS* **132**, 255.

Dragon, J.N. & Mutschlecner, J.P. 1980. *ApJ* **239**, 1045.

Ferguson, J.W., Alexander, D.R., Allard, F., & Hauschildt, P.H. 2001. *ApJ* **557**, 798.

Fišák, J., Krtička, J., Munzar, D., & Kubát, J. 2016. *A&A* **590**, 95.

Frolov, A.M. 2015. *Eur Phys J D*, **69**, 132.

Gaunt, J.A. 1930. *Phil Trans R Soc London, Ser A*, **229**, 163.

Geltman, S. 1962. *ApJ* **136**, 935.

Geltman, S. 1965. *ApJ* **141**, 376.

Gingerich, O. 1964. *Proceedings First Harvard-Smithsonian Conference on Stellar Atmospheres*, SAOSR 167 (Cambridge, Mass.: Smithsonian Astrophysical Observatory), E.H. Avrett, O. Gingerich, & C.A. Whitney, eds., p. 17.

Gingerich, O. & Rich, J.C. 1966. *AJ* **71**, 161.

Goldberg, L., Müller, E.A., & Aller, L.H. 1960. *ApJS* **5**, 1.

Gustafsson, B., Bell, R.A., Eriksson, K., & Nordlund, Å. 1975. *A&A* **42**, 407.

Gustafsson, B., Edvardsson, B., Eriksson, K., *et al.* 2008. *A&A* **486**, 951.

Hayek, W., Asplund, M., Collet, R., & Nordlund, Å. 2011. *A&A* **529**, 158.

Huang, S.-S. 1948. *ApJ* **108**, 354.

John, T.L. 1966. *MNRAS* **131**, 315.

John, T.L. 1989. *MNRAS* **240**, 1.

John, T.L. 1991. *A&A* **244**, 511.

John, T.L. 1992. *MNRAS* **256**, 69.

John, T.L. 1994. *MNRAS* **269**, 865.

Karzas, W.J. & Latter, R. 1961. *ApJS* **6**, 167.

Kramers, H.A. 1923. *Phil Mag* **46**, 836.

Krogdahl, W.S. & Miller, J.E. 1967. *ApJ* **150**, 273.

Kurucz, R.L. 1979. *ApJS* **40**, 1.

Kurucz, R.L. & Peytremann, E. 1975. *SAOSR* **362**.

Massey, H.S.W. 1969. *Electronic and Ionic Impact Phenomena* (London: Oxford University Press), Vol. 2, p. 1255.

Massey, H.S.W. 1976. *Negative Ions* (Cambridge: Cambridge University Press), 3rd ed.

Matsushima, S. 1964. *Proceedings First Harvard-Smithsonian Conference on Stellar Atmospheres*, SAOSR 167 (Cambridge, Mass.: Smithsonian Astrophysical Observatory), E.H. Avrett, O. Gingerich, & C.A. Whitney, eds., p. 42.

Matsushima, S. & Travis, L.D. 1973. *ApJ* **181**, 387.

McDowell, M.R.C., Williamson, J.H., & Myerscough, V.P. 1966. *ApJ* **144**, 827.

Menzel, D.H. & Pekeris, C.L. 1935. *MNRAS* **96**, 77.

Milne, E.A. 1924. *Phil Mag* **47**, 209.

Peach, G. 1967. *MemRAS* **71**, 1.

Peach, G. 1970. *MemRAS* **73**, 1.

Pecker, J.C. 1965. *ARAA* **3**, 135.

Piskunov, N.E., Kupka, F., Ryabchikova, T.A., Weiss, W.W., & Jeffery, C.S. 1995. *A&AS* **112**, 525.

Rich, J.C. 1967. *ApJ* **148**, 275.

Ryabchikova, T., Pakhomov, Y., & Piskunov, N. 2018. *Galaxies* **6**, 93.

Seaton, M.J., Yan, Y., Mihalas, D., & Pradhan, A.K. 1994. *MNRAS* **266**, 805.

Sharp, C.M. & Burrows, A. 2007. *ApJS* **168**, 140.

Somerville, W.B. 1964. *ApJ* **139**, 192.

Stancil, P.C. 1994. *ApJ* **430**, 360.

Stewart, A.L. 1978. *J Phys B* **11**, 3851.

Stilley, J.L. & Callaway, J. 1970. *ApJ* **160**, 245.

Travis, L.D. & Matsushima, S. 1968. *ApJ* **154**, 689.

Tsuji, T., Ohnaka, K., Aoki, W., & Nakajima, T. 1996. *A&A* **308**, L29.

Ueno, S. 1954. *Kyoto Contr Nr.* **42**.

Underhill, A.B. 1962. *PDAO Victoria* **XI**, 467.

Unsöld, A. 1955. *Physik der Sternatmosphären* (Berlin: Springer-Verlag), 2nd ed., p. 168.

Vardya, M.S. 1964. *ApJS* **8**, 277.

Vardya, M.S. 1970. *ARAA* **8**, 87.

Vitense, E. 1951. *ZfAp* **28**, 81.

Weck, P.F., Schweitzer, A., Stancil, P.C., Hauschildt, P.H., & Kirby, K. 2003. *ApJ* **584**, 459.

Wildt, R. 1939. *ApJ* **89**, 295.

Wishart, A.W. 1979. *MNRAS* **187**, 59P.

9

The Model Photosphere

What is a model photosphere, and why build one? A model is a numerical theory. We start with a few of the basic physical laws and predict a spectrum. Then we compare it with the observed spectrum, which leads us to abandon, adjust, or improve our model, much like we would test any theory. When our model closely reproduces all the available observations, we begin to place some trust in it. We associate properties of the model with the star, such as effective temperature, surface gravity, radius, chemical composition, velocity fields, or rate of rotation. Still, we remain aware of the limitations and imperfections of our models, and work toward improving them.

The material in this chapter is directed toward model photospheres that can be readily constructed. They are certainly simpler than real stars, and yet they do a good job at reproducing stellar spectra. The basic assumptions are:

1. Plane parallel geometry, making all physical variables a function only of depth into the photosphere.
2. Hydrostatic equilibrium, i.e., the photosphere is stable, no accelerations comparable to the surface gravity, no dynamically significant mass loss.
3. Structures such as granulation, starspots, and magnetic fields are secondary and do not prevent the use of mean values of the physical parameters.
4. Local thermodynamic equilibrium (LTE); excitation and ionization behave as described in Chapter 1 and the source function is $B_\nu(T)$, Planck's law.

The model photosphere itself consists of a table of numbers giving the temperature, $T(\tau_0)$, the gas pressure, $P_g(\tau_0)$, the electron pressure, $P_e(\tau_0)$, and the absorption coefficient, $\kappa_\nu(\tau_0)$, as a function of optical depth at $5000\,\text{Å}$, τ_0, for a given chemical composition. We derive these numbers using the physics that connects them. Once the model is in place, we invoke what we already learned in previous chapters to compute the radiation transfer and the spectrum.

As we shall see shortly, the basic difference from one model to the next hinges on three parameters: effective temperature, surface gravity, and metallicity. The effective temperature sets the scale of $T(\tau_0)$. Likewise, the surface gravity sets the scale for $P_g(\tau_0)$ and $P_e(\tau_0)$. The metallicity is the collective abundance of those elements, other than hydrogen and helium, that can be ionized and contribute electrons to the photospheric plasma. The process of assigning a model to a star amounts to adjusting these three parameters to arrive at the best agreement with the observed spectrum.

Our starting point is hydrostatic equilibrium.

The Equation of Hydrostatic Equilibrium

The relation between pressure and optical depth is established from the assumption of hydrostatic equilibrium. Consider the small cylindrical volume of height dx in Figure 9.1. The difference in pressure between the top and bottom of the volume is the weight per unit area of the mass in the volume. The weight is the product of the density, ρ, volume, $dA\,dx$, and surface gravity, g. We let dF represent this weight,

$$dF = \rho g\, dA\, dx.$$

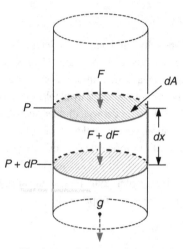

Figure 9.1 A force F is exerted by the overlying gas on the area dA to give a pressure P. The weight of the material in the volume $dA\,dx$ adds a force dF over the increase in depth dx, to give an increase in pressure of dP.

Or, the increment in pressure over the distance dx is

$$dP = dF/dA = \rho g\, dx.$$

With x increasing inward as the pressure does, no negative sign appears in this relation. This is the form of the hydrostatic equation used when geometrical depth, x, is the independent variable, but we mostly use optical depth. Since $d\tau_v = \kappa_v \rho\, dx$, according to Equation 5.13,

$$\frac{dP}{d\tau_v} = \frac{g}{\kappa_v}. \tag{9.1}$$

This is the hydrostatic equation in the form we need.

   The pressure in Equation 9.1 is the total pressure supporting the small volume element. In most stars, the gas pressure, P_g, accounts for the bulk of the total pressure. Only in extreme cases will other pressures be significant compared to P_g. Examples are (1) radiation pressure in very hot stars (Equation 5.11) with

$P_R = (4\sigma/3c)T^4 = 2.52 \times 10^{-15}\,T^4$ dyne/cm^2, (2) magnetic pressure given by $B^2/8\pi$, and (3) turbulence pressure often assumed to be $\frac{1}{2}\rho v^2$, where ρ and v are the density and root-mean-square velocity of the turbulent elements. Numerical examples are given for main-sequence stars in Table 9.1. The magnetic field listed in the table is the value needed to have magnetic pressure equal P_g. Similarly, turbulent elements would have the velocity listed to generate a pressure equal to P_g. In the rest of the chapter these other possible pressures will be ignored and only P_g is retained in the hydrostatic equation.

Table 9.1 *Pressure comparisons*

T_{eff} (K)	Sp	P_g (dyne/cm^2)	P_R (dyne/cm^2)	B (gauss)	v (km/s)
4 000	K5 V	1×10^5	0.6	1585	7.5
8 000	A6 V	1×10^4	10	501	10.6
12 000	B8 V	3×10^3	52	274	13.0
16 000	B3 V	3×10^3	165	274	15.0
20 000	B0 V	5×10^3	403	355	16.7

On our way to finding the run of gas pressure with depth, we need to numerically integrate Equation 9.1. This can be done in a number of ways. The integrand is usually better behaved for numerical work if we write Equation 9.1 as

$$P_g^{1/2}\,dP_g = P_g^{1/2}\frac{g}{\kappa_0}\,d\tau_0,$$

where κ_0 is κ_ν at the reference wavelength of 5000 Å. Integration gives

$$P_g(\tau_0) = \left(\frac{3g}{2}\int_0^{\tau_0}\frac{P_g^{1/2}}{\kappa_0}\,dt_0\right)^{2/3}$$

$$= \left(\frac{3g}{2}\int_{-\infty}^{\log \tau_0}\frac{t_0\,P_g^{1/2}}{\kappa_0\,\log e}\,d\log t_0\right)^{2/3}. \tag{9.2}$$

A logarithmic optical-depth scale gives better numerical precision, which is obvious if the integrand is plotted as a function of depth for a typical case. The procedure is to guess the function $P_g(\tau_0)$, do the numerical integration to get an improved $P_g(\tau_0)$, and continue cycling until convergence. With a reasonable guess at $P_g(\tau_0)$, two to four iterations are typically needed to reach 1% convergence. A poor guess will extend the iteration a cycle or two.

But before we can do this computation, we must know κ_0 as a function of τ_0 since κ_0 appears in the integrand. In Chapter 8 we saw that κ_ν depends on the temperature and the electron pressure, i.e., $\kappa_0 = \kappa_\nu(T, P_e)$ at 5000 Å, so before we can do the integration in Equation 9.2, we must know $T(\tau_0)$ and $P_e(\tau_0)$. Let us first tackle $T(\tau_0)$ and then $P_e(\tau_0)$ in a subsequent section.

The Temperature Distribution in the Solar Photosphere

Our nearest star gives us a special handle on $T(\tau_0)$: the specific-intensity variation across the solar disk. Figure 9.2 shows a real image of the Sun. The drop in brightness from the center to the edge, called the limb, is real, not an artist's rendition to give a spherical look. The effect is called *limb darkening*. It exists because $T(\tau_0)$ declines outward from the deep photosphere to the outermost layers.

Figure 9.2 The solar disk is brighter in the center and darker near the limb. Image courtesy of NASA/GSFC/Solar Dynamics Observatory.

Detailed measurements of solar limb darkening are shown in Figure 9.3. The specific intensity relative to the disk-center value is plotted as a function of $\cos \theta$, where θ has the same meaning it did in Chapter 7 and is shown here in Figure 9.4. There is a very significant wavelength dependence arising from the wavelength dependence of κ_ν. Notice the nearly linear behavior when $\cos \theta$ is used as abscissa. It is $\sin \theta$, however, that

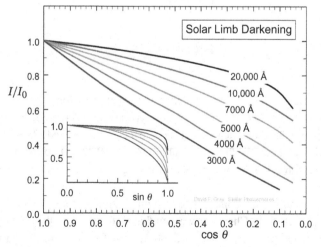

Figure 9.3 Solar limb darkening is shown for several continuum wavelengths. The inset graph shows the variation as a function of position across the disk with the wavelengths in the same order. Based on measurements by Pierce (2000). See also Livingston and Wallace (2003).

corresponds to the apparent fraction of the disk from the center (or the image in Figure 9.2), as shown in the small inset graph.

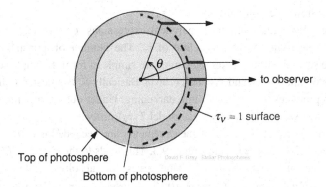

Figure 9.4 This caricature diagram shows how limb darkening comes about. The thickness of the photosphere is enormously exaggerated in order to show the details. The angle θ is zero at disk center and 90° at the limb. Unit optical depth penetration is shown for three limb distances. The dashed curve shows how unit optical depth rises to higher photospheric levels toward the limb.

As a rule of thumb, the characteristic depth to which we see is $\tau_v \sim 1$. In Figure 9.4, three sight lines are shown. The thick line segments indicate the geometrical depth where unit optical depth is reached. We see deepest at disk center, down to relatively hot layers. For sight lines closer to the limb, we see less deep, to less hot layers. Hotter temperatures mean more light. So the rule of thumb qualitatively explains the limb darkening.

To approach it more formally, limb darkening exists because the source function decreases outward. From Equation 7.10, the specific intensity at the $\tau_v = 0$ surface, i.e., the sunlight we see, is

$$I_v(0) = \int_0^\infty S_v \, e^{-\tau_v \, \sec \theta} \sec \theta \, d\tau_v. \tag{9.3}$$

The exponential extinction varies as $\tau_v \sec \theta$, emphasis on the $\sec \theta$, so the position of unit optical depth *along the line of sight* moves upward, i.e., to smaller τ_v, as the line of sight is shifted from the center to the limb, as shown in Figure 9.4. Toward the limb, we see systematically higher photospheric layers, those that are less bright. Suppose we invoke the approximate grey-case solution, where we found a source function with a linear dependence on τ_v (Equation 7.35, here the constant terms are called a and b),

$$S_v = a + b\tau_v.$$

Then from Equation 9.3 we find linear limb darkening, not very different from the observations in Figure 9.3,

$$I_v(0) = a + b \cos \theta.$$

By comparing these last two expressions, we see that when $\cos \theta = \tau_\nu$, the specific intensity on the surface at the position θ equals the source function at the depth τ_ν. This is referred to as the Eddington–Barbier relation. So the grey case orients our thinking and we begin to see how to empirically model $S(\tau_\nu)$.

We can use polynomials or other analytical expressions, or we can set up a purely numerical grid specifying S_ν as a function of τ_ν. The number of adjustable parameters cannot be large since the limb-darkening curves are simple in form. Examples can be found in Mitchell (1959) and Pierce and Waddell (1961). Basically S_ν is adjusted until $I_\nu(0)$ from Equation 9.3 reproduces the observed limb darkening. Under our assumption of LTE, i.e., S_ν is Planck's law, we could equally well model $T(\tau_\nu)$ instead.

On the practical side, two types of measurements are needed. First, the relative limb darkening $I_\nu(\theta)/I_0$ is measured, as in Figure 9.3. This leads to the shape of $T(\tau_\nu)$. Second, the absolute size of I_0 is measured. This gives the absolute scale of $T(\tau_\nu)$ and $T_{\text{eff}}^{\odot} = 5772$ K, as discussed in Chapter 10. This process can be applied to the limb darkening measured at any wavelength. As we shall see a little later in the chapter, the results from any wavelength can be reduced to the standard wavelength, 5000 Å, so then we have $T(\tau_0)$ from measurements at any wavelength.

The rudimentary rule of thumb for the depth of formation, i.e., a sight line penetrates to $\tau_\nu \sim 1$, can be developed more properly by asking how much intensity actually comes from each depth, and that is given by the integrand of Equation 9.3. Again, a logarithmic scale is preferable, so we rewrite Equation 9.3 as

$$I_\nu(0) = \int_{-\infty}^{\infty} S_\nu(\tau_\nu) \, e^{-\tau_\nu \, \sec \theta} \, \tau_\nu \, \sec \theta \, \frac{d \log \tau_\nu}{\log e}. \qquad (9.4)$$

The integrand is called the *contribution function* since it specifies the contribution to $I_\nu(0)$ from each depth. Sample contribution functions are shown in Figure 9.5. These curves are computed using the final $T(\tau_0)$ determined from the limb darkening, but they could also be computed at intermediate stages to assist in the limb-darkening analysis. We see that no light comes from the very highest or the very lowest layers and that there is a simple smooth maximum. Comparing these four examples, we formally confirm that $I_\nu(0)$ originates in higher layers for positions closer to the limb. By virtue of the considerable overlap in depth of these integrands, no fine structure can be discerned.

Limb darkening, as we have now seen, lets us determine the solar temperature distribution from the observations. A number of determinations have also been done theoretically, i.e., by constructing model photospheres with the constraint of flux constancy, as in Equation 7.22 and subsequent discussion. Various determinations of the solar temperature distribution are compared in Figure 9.6.

The models of Pierce and Waddell, Elste, Holweger and Müller, and Maltby depend heavily on empirical information, mainly limb darkening. The other models, Carbon and Gingerich, Gustafsson *et al.*, Bell *et al.*, and Kurucz, are theoretical models based on flux constancy. Good agreement is seen in the layers around unit optical depth, where the

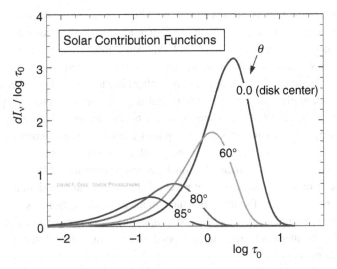

Figure 9.5 Contribution functions (integrands of Equation 9.4) like these tell us about the depth of formation of the surface radiation. The average depth rises toward the limb (larger θ). Negligible contributions come from higher or deeper layers than those shown here. The areas, $I_\nu(0)$, diminish from center to limb, giving the limb darkening. These curves are for $\lambda = 4000\,\text{Å}$. Ordinate unit is $10^{-5}\,\text{erg}/(\text{s cm}^2\,\text{rad}^2\,\text{Hz})$.

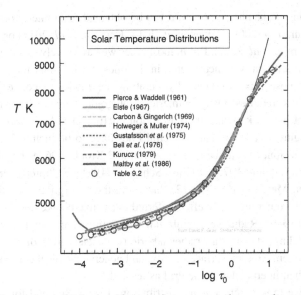

Figure 9.6 Temperature distributions from various sources show good agreement. Solid lines are based on limb-darkening observations. Broken lines are from theoretical determinations. The optical depth scale is for 5000 Å. Circles are from Table 9.2.

observational information is strongest and the theory has the least uncertainty. In the deepest layers, those with $\log \tau_0 > 0$, the uncertain effects of convection enter, leading to some divergence in the models. Pierce and Waddell's model has the radiative gradient. Larger divergence is also seen in the higher layers, $\log \tau_0 \sim -2$ to -4. Here the absorption of spectral lines influences $T(\tau_0)$, and this is challenging to handle theoretically.

Table 9.2 below gives a standard solar model that we use as a reference photosphere. The temperature distribution listed there is shown by the circles in Figure 9.6.

In Figure 9.5, we saw variations in the depth of formation as expressed in contribution functions. Similar variations occur across the profiles of strong spectral lines. Points in the wings of the profile have only small additional absorption contributed from the line, so they are formed deep in the photosphere, not much different from the continuum. By contrast, profile points near the core of a strong line have much more absorption. As a result, they are formed higher in the photosphere where the temperature is lower and less light is emitted. In other words, the shapes of strong lines depend on the temperature distribution, giving us another handle on $T(\tau_0)$.

When we get into the dynamics of the solar photosphere, we should not expect a single $T(\tau_0)$ to properly represent the heterogeneities we see there, in particular the granulation. Temperature differences in the granulation structure amount to several percent (discussed in Chapter 17). Nevertheless, the average temperature distribution is very useful and allows us to understand many aspects of stellar spectra. But how do we get $T(\tau_0)$ for other stars?

Temperature Distributions in Other Stars

Some limited limb-darkening information can be obtained from interferometers for a few stars (e.g., Lacour *et al.* 2008, Baines *et al.* 2014, Kervella *et al.* 2017) or from exoplanet transits (e.g., Morello *et al.* 2017). But in most cases we have to rely on other methods. One is the theoretical path that we encountered in the quest for the solar temperature distribution, namely, impose flux constancy by iterating the model to convergence to attain $d\mathfrak{F}/d\tau_0 = 0$. This requires many numerical integrations over the whole spectrum, $\lambda = 0 \to \infty$. That means knowing the absorption coefficient at "all" wavelengths, a particularly challenging feat when many spectral lines have to be included. A few of many examples of the techniques can be seen in Avrett and Krook (1963a, 1963b), Carbon and Gingerich (1969), Cannon (1973a, 1973b), Scharmer (1981), Kalkofen (1987), Gustafsson *et al.* (2008), and Mészáros *et al.* (2012). Including the absorption from lines lowers the temperatures in the outer layers, an effect referred to as "line blanketing" (Gustafsson *et al.* 1975, Anderson 1989). Radiative equilibrium calculations are complex enough to make customized modeling, star-by-star, rather laborious. Instead, grids of models spanning a range in effective temperature, surface gravity, and general chemical composition are often tabulated. One then interpolates in the grid as needed.

However, if we look at temperature distributions for the Sun and for models of other stars, we see an interesting thing: *they all have nearly the same shape.* The differences are no larger than found between models for the same star found by different people. This leads us to the second and far simpler method of obtaining $T(\tau_0)$. We merely scale a well-established standard temperature distribution, like the Sun's,

$$T(\tau_0) = S_0 \, T_\odot(\tau_0), \tag{9.5}$$

where S_0 is the scaling constant, to get the $T(\tau_0)$ for our target star. The first indication that such a procedure might be possible was seen in the solution of the grey case, where we found (Equation 7.36),

$$T(\tau) = \left[\frac{3}{4}(\tau + q)\right]^{1/4} T_{\text{eff}}.$$

So if we ratio this relation, star to Sun,

$$T(\tau) = \frac{T_{\text{eff}}}{T^\odot_{\text{eff}}} \, T_\odot(\tau),$$

the scaling factor, S_0, in the grey case is the ratio of effective temperatures.

The closeness of $T(\tau_0)$ for non-grey models is illustrated in Figure 9.7. Here $T(\tau_0)$ is plotted for flux-constant models over a wide range of effective temperatures. The "scaled

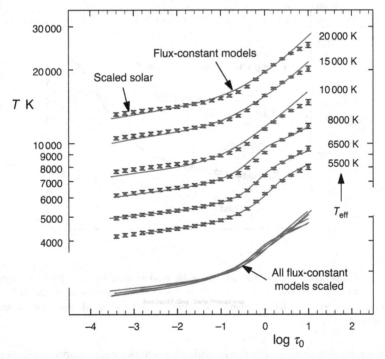

Figure 9.7 Solid lines are for flux-constant models, T_{eff} as shown, $\log g = 4.0$, solar metallicity. Symbol curves are the scaled solar-temperature flux-constant model with error bars based on Figure 9.6. All of the symbol curves are identical except for scaling. The plot at the bottom shows the scatter when these same models are scaled together. All curves based on Kurucz (1979).

solar" case in this plot is the 5800 K model from the same grid of calculations. The ordinate error bars shown for the scaled curves are the estimated uncertainty, taken from the curves in Figure 9.6, which is ~$\pm$1–2%. The differences we see in Figure 9.7 are comparable to the uncertainties in the calculations themselves.

Even changing the surface gravity of the models has only a modest effect on the temperature distributions. Figure 9.8 shows some examples for models of giant stars. The four flux-constant cases, shown by symbols in the figure, differ in the deep layers owing to the way convection is handled. As we saw in Figure 9.5, there is little light leaving the star from such deep layers, so the spectrum is essentially unaffected by such differences. These models also show differences in the higher layers arising from the different treatments of line absorption. The empirical model by Ruland *et al.* (1980) is derived from the profiles of strong spectral lines of β Gem (Pollux K0 III). Their $T(\tau_0)$ has been scaled slightly for the figure to match the Bell *et al.* models. Its main feature is the systematically lower temperatures in the upper layers. And, for comparison, the standard solar model from Table 9.2 is also shown. It agrees remarkably well with the other models. More $T(\tau_0)$ comparisons are given by Gustafsson *et al.* (1975).

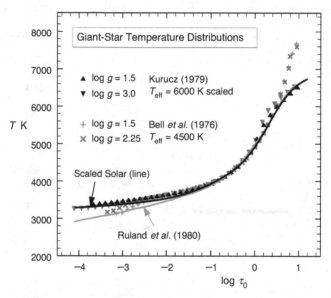

Figure 9.8 Temperature distributions for two giant models from Kurucz (1979) are compared with two from Bell *et al.* (1976). The Kurucz models are scaled down to match those of Bell *et al.* The giant model of Ruland *et al.* (1980) and the standard solar case are also scaled to the Bell *et al.* models.

The upshot is that scaling $T(\tau_0)$ simplifies model building enormously. We take advantage of this in the rest of the chapter. Although there is ample evidence that the effective temperature of a star is given by $T_{eff} = S_0\, T_{eff}^{\odot}$ (e.g., Gray & Kaur 2019), S_0 stands on its own as a specification of how hot a star is. It is the real temperatures specified in $T(\tau_0)$ that determine the spectrum, not an effective temperature. S_0 sets the scale of these temperatures just as T_{eff} does, but it has the advantage of not requiring a flux integration over the entire spectrum. S_0 is found by tuning its value until observations, such as energy distributions or strengths of spectral lines, are matched by the model. A summary of the adopted solar and giant-star temperature distributions is given in Table 9.2. Because these two distributions are nearly identical in deep layers, they produce nearly the same continuous spectrum, but the giant-star distribution produces deeper cores in strong spectral lines.

At this point we assume $T(\tau_0)$ has been determined, and we turn our attention to the calculation of electron pressure, the other variable needed for the calculation of $\kappa_0 = \kappa_\nu(T, P_e)$ at 5000 Å, on our way to solving Equation 9.2. But first a note about the metals that supply electrons for the electron pressure.

Table 9.2 *Standard temperature distributions*

log τ_0	$T_\odot$ solar	T_g giants
−4.0	4305	3810
−3.8	4325	3865
−3.6	4345	3925
−3.4	4370	3985
−3.2	4395	4050
−3.0	4427	4115
−2.8	4455	4185
−2.6	4488	4255
−2.4	4525	4325
−2.2	4565	4405
−2.0	4620	4493
−1.8	4690	4580
−1.6	4775	4670
−1.4	4878	4770
−1.2	4995	4890
−1.0	5135	5025
−0.8	5294	5190
−0.6	5485	5395
−0.4	5720	5680
−0.2	6030	6030
0.0	6430	6440
0.2	6920	6880
0.4	7467	7380
0.6	7955	7940
0.8	8358	8350
1.0	8632	8620
1.2	8810	8780

Metallicity

Metallicity is one of the three basic parameters of a model photosphere. It specifies the photospheric abundances of the metals that can be ionized and contribute electrons for the electron pressure. These elements are C, Si, Fe, Mg, Ni, Cr, Ca, Na, and K under normal circumstances. The usual assumption, which we make here, is that the abundances of all of these elements vary together, so if iron is overabundant by 0.2 dex, so are all the others. By overabundant we mean compared to the solar abundances (see Table 16.1). Since iron has so many spectral lines, it is taken as the cornerstone element and metallicity is symbolically written as

$$[\text{Fe/H}] = \log A(\text{Fe}) - \log A_\odot(\text{Fe}) = \log \frac{A(\text{Fe})}{A_\odot(\text{Fe})}. \tag{9.6}$$

Here $A(\text{Fe})$ is the iron abundance in the star and $A_\odot(\text{Fe})$ is the corresponding abundance in the Sun, both expressed relative to hydrogen by particle numbers. But what is really meant by $A(\text{Fe})$ is the abundances of all the electron donors, not just iron. So replace $A(\text{Fe})$ with the sum of them

$$[\text{Fe/H}] = \log \sum_C^K A_j - \log \sum_C^K A_j^\odot,$$

where the summation index, j, runs from carbon to potassium in the electron-donor list given above. Or

$$\sum_C^K A_j = 10^{[\text{Fe/H}]} \sum_C^K A_j^\odot. \tag{9.7}$$

Thus metallicity is the power of a scaling factor applied to the solar electron donor abundances.

   Most stars in the solar neighborhood (Population I) have $[\text{Fe/H}] \sim 0$ with a range $\sim\pm0.2$ dex. Stars outside this range are called metal rich or metal poor, and most of these lie within a ±0.5 dex span. Very few stars have $[\text{Fe/H}]$ exceeding $+0.5$, while a trickle of extremely metal-poor stars has been found, extending down to $[\text{Fe/H}] \sim -6$. See Chapter 16 for more material on chemical composition. The adopted solar abundances are given in Table 16.1 and Figure 16.6.

   Now on to the electron pressure.

The P_g–P_e–T Relation

The electron pressure, P_e, depends on the temperature, T, and the gas pressure, P_g, which we plan to get by solving Equation 9.2 by iteration. In this section we will find that the determination of P_e also requires its own iteration. Yes, an iteration inside an iteration. For the P_e iteration, proceed as follows. Let the number of ions (denoted by the subscript 1) per unit volume of the jth element be N_{1j} and the number of neutrals (denoted by the subscript 0) per unit volume be N_{0j}. Then the ionization Equation 1.21 for the jth element is

$$\frac{N_{1j}}{N_{0j}} = \frac{\Phi_j(T)}{P_e}.$$

It is $\Phi_j(T) = 1.2020 \times 10^9 (u_{1j}/u_{0j}) \theta^{-5/2} 10^{-\theta I_j}$ that we actually calculate, so we cast the equations in terms of $\Phi_j(T)$. A good approximation is to neglect double ionizations and write $N_{ej} = N_{1j}$ as the number of electrons per unit volume contributed by the *j*th element. Then

$$\frac{\Phi_j(T)}{P_e} = \frac{N_{ej}}{N_{0j}} = \frac{N_{ej}}{N_j - N_{ej}},$$

where the total number of *j*th element particles is $N_j = N_{1j} + N_{0j}$. We solve for N_{ej},

$$N_{ej} = N_j \frac{\Phi_j(T)/P_e}{1 + \Phi_j(T)/P_e}. \tag{9.8}$$

The electron and gas pressures are

$$P_e = \sum_j N_{ej} kT$$

and

$$P_g = \sum_j \left(N_{ej} + N_j \right) kT,$$

which we ratio to get

$$\frac{P_e}{P_g} = \frac{\sum N_{ej} kT}{\sum (N_{ej} + N_j) kT} = \frac{\sum N_j \left[\dfrac{\Phi_j(T)/P_e}{1 + \Phi_j(T)/P_e} \right]}{\sum N_j \left[1 + \dfrac{\Phi_j(T)/P_e}{1 + \Phi_j(T)/P_e} \right]}.$$

The next step is to normalize all N_j values by N_H, the number of hydrogen particles per unit volume, and again denote these relative abundances as $A_j = N_j/N_H$. Then

$$P_e = P_g \frac{\sum A_j \dfrac{\Phi_j(T)/P_e}{1 + \Phi_j(T)/P_e}}{\sum A_j \left[1 + \dfrac{\Phi_j(T)/P_e}{1 + \Phi_j(T)/P_e} \right]}. \tag{9.9}$$

The equation is transcendental in P_e, and is solved by iteration. But the convergence is quicker if we take the upper-most P_e on the right to the other side of the equation and write

$$P_e = \left\{ P_g \frac{\sum A_j \dfrac{\Phi_j(T)}{1 + \Phi_j(T)/P_e}}{\sum A_j \left[1 + \dfrac{\Phi_j(T)/P_e}{1 + \Phi_j(T)/P_e} \right]} \right\}^{1/2}. \tag{9.10}$$

The summation must include H, He, and the metal electron donors, C, Si, Fe, Mg, Ni, Cr, Ca, Na, and K. Since P_e and P_g are both functions of τ_0, Equation 9.10 is solved at each depth, the first cycle of which uses the first guess of $P_g(\tau_0)$. Notice that the A_j and $\Phi_j(T)$ terms are constants for each P_e iteration.

Results of Equation 9.10 for the solar chemical composition are shown in Figure 9.9. Hydrogen dominates at the high temperatures, and when it becomes fully ionized, $P_e \sim P_g/2$, or a bit less if helium is neutral. At cooler temperatures, the constant P_g lines show us the temperature dependence of the ionization of the metals. Of course, P_g is not constant with depth in a stellar photosphere, and to jump ahead to some of the results, Figure 9.9 also shows the loci of three model photospheres of different effective temperatures.

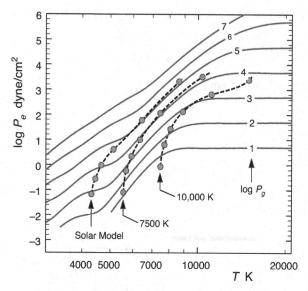

Figure 9.9 Electron pressure is shown as a function of temperature for several values of gas pressure, as labeled, for the solar chemical composition. Dashed lines and circles show the run of parameters with depth in the photospheres for three models, labeled with T_{eff}. The circles are one decade apart and start at $\log \tau_0 = -4$.

The role of metallicity is strongest when hydrogen is neutral and the metals are ionized. In this idealized situation, $\Phi_j(T)/P_e$ is zero for hydrogen and helium and unity for the metals. From Equation 9.9 we then get

$$\frac{P_e}{P_g} = \frac{A_H \times 0 + A_{He} \times 0 + A_C + \cdots + A_K}{A_H + A_{He} + 2A_C + \cdots + 2A_K}.$$

$A_H = 1$ by definition, $A_{He} \sim 0.085$, while the abundances of the metals are smaller by about four orders and can be neglected in the denominator, giving

$$\frac{P_e}{P_g} \approx \frac{1}{(1 + A_{He})} \sum_{C}^{K} A_j. \tag{9.11}$$

Notice that the summation here runs from carbon to potassium in our donor list and does not include hydrogen or helium. For the solar abundances in Table 16.1, the right side of the equation is 3.5×10^{-4}. Electron pressure is much smaller than gas pressure in cool stars. Equation 9.11 is a short-cut way of saying that for cool stars the electron pressure is limited by the availability of donor elements. If the metallicity is [Fe/H] $= -0.5$, for example, we expect the electron pressure to be down by $\sim 10^{-0.5}$. But the idealized case of complete ionization or none at all is only approximately true. For stars hotter than the Sun, hydrogen starts to ionize, while for stars somewhat cooler than the Sun, ionization of the metals is incomplete. Nevertheless, Equation 9.11 is still a useful guide to our thinking. To complicate matters, κ_0 depends on P_e, so the optical depth needed to reach a specified pressure depends on the metallicity. More on this later in the chapter.

Completion of the Model

Now we finally have enough information to integrate Equation 9.2. Remember, the model-specification parameters are the surface gravity, usually in the guise of $\log g$ in cgs units, a metallicity, [Fe/H], and a temperature parameter, T_{eff} or S_0. So let us assume we have $T(\tau_0)$. Then take a first guess at $P_g(\tau_0)$, compute $P_e(\tau_0)$, then $\kappa_0(\tau_0)$ as explained in Chapter 8, and integrate Equation 9.2 to get the new improved estimate of $P_g(\tau_0)$. At the outer boundary, when starting the integration, it is necessary to make some simple trapezoidal integrations for a step or two in $\log \tau_0$. After that, Simpson's rule or some other standard integration formula can be used. Usually a few extra steps are taken at the top of the model to ensure that the pressure at the depths where we need it is independent of how we start the integration. Iteration of Equation 9.2 is continued to convergence, typically $\sim 1\%$ or better, and we then have a complete model for the specified $T(\tau_0)$. If we are using a scaled temperature distribution, *we are finished*. The basic model has been made, we have temperature, gas pressure, electron pressure, and absorption coefficient as a function of our reference optical depth, and we are ready to calculate auxiliary information from the model, such as the geometrical depth and surface flux. Although dozens of iterations on Equation 9.2 may not spark your interest, computers are delighted (if they had feelings) to do these repetitive tasks. Even with a modest computer, it takes much less time to calculate the model than it does to enter the model parameters, S_0, $\log g$, and [Fe/H].

On the other hand, if we are determining $T(\tau_0)$ by the flux constancy condition, we are far from finished and must now calculate $\mathfrak{F}_\nu(\tau_0)$ using Equation 7.14 for many frequencies across the spectrum so that $\int_0^\infty \mathfrak{F}_\nu \, d\nu$ can be properly evaluated as a function of depth. The temperature distribution is then corrected and the model is repeatedly recomputed until flux constancy is achieved.

Table 9.3 lists a model photosphere for the Sun, $S_0 = 1.0$ or $T_{eff} = 5772$ K, $\log g = 4.438$, [Fe/H] $= 0.0$. The first two columns are the same as in Table 9.2. The iterative solution of Equation 9.2 gives the $\log P_g$, $\log P_e$, and $\log \kappa_0/P_e$ columns. These five columns are enough to allow us to compute the spectrum, our main goal. The other

168 9 The Model Photosphere

Table 9.3 *A model solar photosphere*

log τ_0	T	log P_g	log P_e	log κ_0/P_e	log ρ	x
−4.0	4348	2.88	−1.17	−1.22	−8.58	−525
−3.8	4368	3.04	−1.03	−1.23	−8.42	−491
−3.6	4388	3.17	−0.90	−1.24	−8.29	−461
−3.4	4414	3.30	−0.79	−1.25	−8.17	−435
−3.2	4439	3.42	−0.68	−1.26	−8.06	−409
−3.0	4471	3.53	−0.57	−1.28	−7.94	−384
−2.8	4500	3.64	−0.46	−1.29	−7.83	−359
−2.6	4533	3.76	−0.36	−1.30	−7.72	−334
−2.4	4570	3.86	−0.25	−1.32	−7.62	−309
−2.2	4611	3.98	−0.14	−1.34	−7.51	−284
−2.0	4666	4.09	−0.03	−1.36	−7.41	−259
−1.8	4737	4.20	0.09	−1.39	−7.30	−233
−1.6	4823	4.31	0.21	−1.42	−7.20	−207
−1.4	4927	4.42	0.34	−1.46	−7.10	−181
−1.2	5045	4.53	0.47	−1.50	−7.00	−154
−1.0	5186	4.64	0.61	−1.55	−6.90	−126
−0.8	5347	4.75	0.77	−1.60	−6.80	−98
−0.6	5540	4.85	0.95	−1.66	−6.72	−70
−0.4	5777	4.95	1.17	−1.73	−6.64	−43
−0.2	6090	5.03	1.46	−1.81	−6.58	−20
0.0	6494	5.09	1.82	−1.91	−6.55	0
0.2	6989	5.14	2.23	−2.01	−6.53	15
0.4	7542	5.17	2.64	−2.11	−6.54	27
0.6	8035	5.19	2.96	−2.18	−6.54	37
0.8	8442	5.22	3.20	−2.22	−6.54	46
1.0	8718	5.24	3.36	−2.24	−6.53	56
1.2	8898	5.27	3.46	−2.26	−6.51	67

columns are the logarithm of the mass density and the geometrical depth. Tables 9.4 and 9.5 give a few abbreviated reference models for other stars.

Numerous grids of model photospheres have been published. These include Underhill (1968), Bell *et al.* (1976), Kurucz (1979), Wesemael *et al.* (1980), Lester *et al.* (1982), Gustafsson *et al.* (2008), Mészáros *et al.* (2012), Trampedach *et al.* (2013), Bohlin *et al.* (2017), Hainich *et al.* (2019), and many others. Blanco-Cuaresma (2019) gives a critique of grid computations.

Now that we have our basic model photosphere, let us turn to some of the things we can do with it. As we have already noticed, the density and geometrical depth are listed in the tables.

Table 9.4 *Models for dwarfs*

$\log \tau_0$	T	$\log P_g$	$\log P_e$	$\log \kappa_0/P_e$	$\log \rho$	x
		$S_0 = 1.0, \log g = 4.438, [\mathrm{Fe/H}] = -1.0$				
−4.0	4348	3.30	−1.62	−1.22	−8.17	−465
−3.0	4471	3.96	−1.00	−1.27	−7.52	−319
−2.0	4666	4.51	−0.44	−1.35	−6.99	−196
−1.0	5186	4.99	0.38	−1.53	−6.55	−80
0.0	6494	5.28	1.88	−1.90	−6.36	0
1.0	8718	5.37	3.42	−2.25	−6.40	34
		$S_0 = 0.6, \log g = 4.5, [\mathrm{Fe/H}] = 0.0$				
−4.0	2609	3.14	−2.52	−0.09	−8.10	−291
−3.0	2683	3.82	−1.93	−0.16	−7.43	−214
−2.0	2800	4.40	−1.38	−0.27	−6.87	−147
−1.0	3112	4.95	−0.67	−0.52	−6.36	−71
0.0	3896	5.49	0.34	−1.01	−5.93	0
1.0	5231	5.87	1.59	−1.57	−5.68	75
		$S_0 = 1.5, \log g = 4.0, [\mathrm{Fe/H}] = 0.0$				
-4.0	6522	1.22	−0.15	−1.32	−10.43	−2414
−3.0	6707	2.30	0.56	−1.74	−9.36	−1431
−2.0	6999	2.96	1.13	−1.87	−8.71	−804
−1.0	7779	3.45	1.92	−1.95	−8.28	−315
0.0	9741	3.71	2.96	−1.78	−8.19	0
1.0	13077	3.84	3.50	−1.72	−8.37	351

Table 9.5 *Models for giants*

$\log \tau_0$	T	$\log P_g$	$\log P_e$	$\log \kappa_0/P_e$	$\log \rho$	x
		$S_0 = 0.7, \log g = 2.0, [\mathrm{Fe/H}] = 0.0$				
−4.0	2694	1.78	−3.58	−0.17	−9.47	−91959
−3.0	2909	2.45	−2.84	−0.36	−8.84	−66826
−2.0	3176	2.99	−2.14	−0.57	−8.34	−44572
−1.0	3553	3.48	−1.33	−0.81	−7.89	−22393
0.0	4546	3.90	−0.23	−1.31	−7.58	0
1.0	6101	4.30	1.05	−1.80	−7.31	29430
		$S_0 = 0.9, \log g = 2.5, [\mathrm{Fe/H}] = 0.0$				
−4.0	3463	1.80	−2.54	−0.75	−9.56	−40278
−3.0	3740	2.47	−1.80	−0.92	−8.92	−30044
−2.0	4084	3.02	−1.16	−1.10	−8.41	−20896
−1.0	4568	3.57	−0.50	−1.32	−7.91	−10783
0.0	5845	4.07	0.69	−1.74	−7.52	0
1.0	7844	4.22	2.35	−2.08	−7.51	4154

The Geometrical Depth

The geometrical depth is one of those things we are naturally curious about. For most stars, the total geometrical-depth span of the photosphere is tiny compared to the star's radius. Very low surface gravity stars are the exception. Since τ_0 is the independent variable, we can go back to its definition and write $dx = d\tau_0/(\kappa_0\rho)$ or

$$x(\tau_0) = \int_0^{\tau_0} \frac{1}{\kappa_0(t_0)\,\rho(t_0)}\,dt_0. \tag{9.12}$$

The density we calculate from the pressure and temperature using Equation 1.12, namely,

$$\rho = N_H \sum A_j\,m_j \tag{9.13}$$

in which N_H is the number of hydrogen particles/cm^3 and m_j is the atomic weight of the jth element in g/particle (and we are back to the summation without limits, meaning over *all* elements with significant abundances). Then with the total number of particles, $N = N_e + \sum N_j = N_e + N_H \sum A_j$, we have

$$N_H = \frac{N - N_e}{\sum A_j} = \frac{P_g - P_e}{kT \sum A_j}.$$

Placing this N_H into Equation 9.13, ρ into Equation 9.12, and shifting into logarithmic integration, we have

$$x(\tau_0) = \frac{k \sum A_j}{\sum A_j m_j} \int_{-\infty}^{\log \tau_0} \frac{T(t_0)\,t_0}{\kappa_0(t_0)\left[P_g(t_0) - P_e(t_0)\right]} \frac{d\log t_0}{\log e}. \tag{9.14}$$

All of the variables are columns in our model, so numerical integration is straightforward.

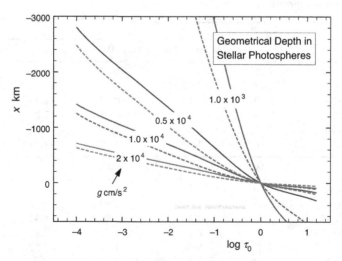

Figure 9.10 Geometrical depth, $x(\tau_0)$, varies nearly linearly with $\log \tau_0$, and the thickness of the photosphere varies inversely with surface gravity, g. Solid curves are for $[\mathrm{Fe/H}] = 0.0$, dashed for $[\mathrm{Fe/H}] = -1.0$.

If surface gravity is changed, the downward compression force on the photospheric gases, i.e., the pressure, is changed by the same amount. With a constant temperature structure, the perfect gas law, Equation 1.10, becomes $PV = $ constant. And in our thin plane-parallel photosphere the volume changes through a change in x. Thus $x(\tau_0)$ is inversely dependent on g. This assumes the overall ionization has not changed. Figure 9.10 shows examples using Equation 9.14. As in Equation 9.12, the zero point for the x scale is taken at log $\tau_0 = 0$.

The density in grams per cubic centimeter is also listed in the tables. We could add other columns to customize the model for other purposes. But let us move on to the main reason we make model photospheres, to get the spectrum.

The Main Goal: Computation of the Spectrum

The computation of the spectrum is the most important use of a model photosphere. We compute the spectrum using Equation 7.15 on a logarithmic scale,

$$\mathfrak{F}_v = 2\pi \int_{-\infty}^{\infty} B_v(\tau_0) \, E_2(\tau_v) \, \frac{\kappa_v(\tau_0) \, \tau_0}{\kappa_0(\tau_0) \log e} \, d\log \tau_0, \tag{9.15}$$

where B_v is the black-body source function. The flux at frequency v requires E_2 on a τ_v scale. We get τ_v from the definition of optical depth,

$$dx = \frac{d\tau_v}{\kappa_v \rho} = \frac{d\tau_0}{\kappa_0 \rho}$$

or

$$\tau_v(\tau_0) = \int_0^{\tau_0} \frac{\kappa_v(t_0)}{\kappa_0(t_0)} \, dt_0 = \int_{-\infty}^{\log \tau_0} \frac{\kappa_v(t_0) \, t_0}{\kappa_0(t_0) \log e} \, d\log t_0. \tag{9.16}$$

Both continua and spectral lines are computed in this same way, using Equation 9.15. Of course when computing lines, the absorption coefficient for the line has to be added to κ_v. The line absorption coefficient is the topic of Chapter 11.

The amount of light contributed to the surface flux as a function of depth in the atmosphere is specified by the integrand of Equation 9.15 (mimicking what we did for specific intensity with Equation 9.4 and Figure 9.5). We call this integrand the *flux contribution function*. It is similar to the contribution function used with I_v, but now includes an integration over the stellar disk. But it still conveys depth-of-formation information and tells us what range of log τ_0 is relevant.

Flux Contribution Functions and the Depth of Formation

Figure 9.11 shows representative flux contribution functions. They tell us where in depth the surface flux originates and where, for example, $T(\tau_0)$ must be defined in order to compute the surface flux with sufficient precision. In particular, the $\pm\infty$ integration limits are easily covered by the $\log \tau_0$ range of our models. This is not always the case. If the absorption is stronger, the depth of formation is higher, and when it is very strong, the formation layers can be above the standard photosphere. The contribution functions in Figure 9.11 indicate that the radiation at 8000 Å is formed slightly higher than the radiation at 5000 Å. This reflects the larger κ_ν at 8000 Å compared to 5000 Å. If these contribution functions are plotted on their own $\log \tau_\nu$ scale instead of the $\log \tau_0$ scale, they all peak near $\log \tau_\nu = 1$.

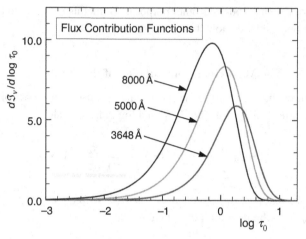

Figure 9.11 Sample flux contribution functions, the integrand of Equation 9.15, for a solar model, showing the depths of formation of the surface flux. The 3648 Å wavelength is just above the Balmer jump. Ordinate unit is 10^{-5} erg/(s cm^2 Hz).

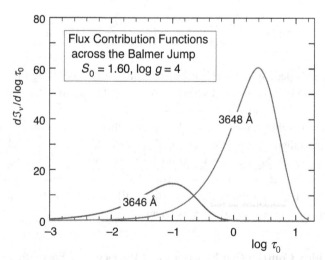

Figure 9.12 In this relatively hot model, the bound–free absorption by neutral hydrogen is strong and varies discontinuously across the Balmer jump, as we saw back in Figure 8.2. The depth of formation of the surface flux responds accordingly. Ordinate unit is 10^{-5} erg/(s cm^2 Hz).

The effect of the κ_ν discontinuity across the Balmer jump is shown for $S_0 = 1.6$ or $T_{\text{eff}} = 9243\,K, \log g = 4, [\text{Fe/H}] = 0$ in Figure 9.12. The stronger absorption on the short-wavelength side raises the depth of formation significantly.

The depth of formation rises slowly when going from cooler to hotter stars. This is illustrated in Figure 9.13. Surface gravity effects are much smaller, being barely noticeable in the hotter models and completely negligible in the cooler ones.

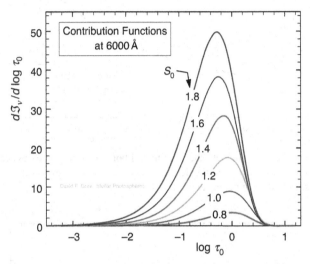

Figure 9.13 Contribution functions at 6000 Å show a modest rise in the depth of formation as the model temperature increases. Ordinate unit is $10^{-5}\,\text{erg}/(\text{s cm}^2\,\text{Hz})$.

Continuous spectra from model photospheres agree reasonably well with the observations. The details of model continua as they relate to real stars are discussed in Chapter 10. Spectral line strengths and profiles for typical lines are also reproduced well by these models; more about this in later chapters. In the rest of this chapter, we explore interesting properties of the models.

Properties of Models: Pressure

The depth dependences of gas pressure and electron pressure are shown for cool models in Figure 9.14 and for hot models in Figure 9.15. In the cool models, the electron pressure is about four orders of magnitude smaller than the gas pressure, as noted in Equation 9.11. Toward deeper layers, the higher temperatures boost the ionization, especially of hydrogen, and hence the electron pressure. Although hydrogen contributes to the cooler models' electron pressure in the deepest layers, it is very far from being completely ionized. Since $\log \tau_0$ is close to a linear x scale (Figure 9.10), the linear behavior of P_g in the upper layers in Figure 9.14 implies a simple exponential increase with a nearly constant scale height. The scale height is eventually driven down in the deeper layers by the rise in temperature.

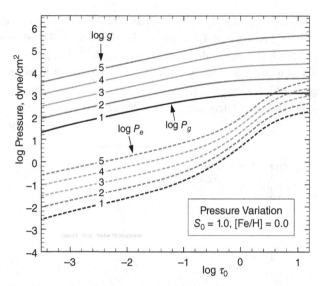

Figure 9.14 For these cool models, the logarithm of both pressures increases nearly linearly with depth in the upper layers. In the deeper layers, gas pressure levels off, while electron pressure rises, as hydrogen begins to ionize.

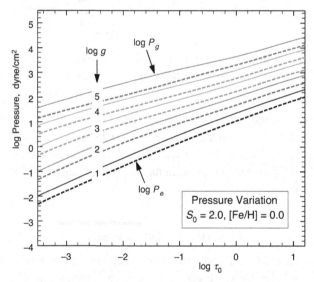

Figure 9.15 In these hot models, the pressures grow nearly linearly with depth, and the electron pressure is close to half the gas pressure owing to the dominance of ionized hydrogen.

By contrast, the hotter models in Figure 9.15 have temperatures high enough to vigorously ionize hydrogen, the electron pressure is dominated by electrons from hydrogen, and P_e approaches $0.5P_g$. The gas pressure is about two orders smaller in the hot model compared to the cool model, a result of κ_0 being much larger in the hot model, so our line of sight does not penetrate as deeply. Here too we see a nearly linear connection between the pressures and $\log \tau_0 \approx x$.

We also see from Figures 9.14 and 9.15 how the increase in gravity compresses the photosphere, increasing both pressures. In the cool-star case, and restricting ourselves to the higher layers where hydrogen is not ionized,

$$P_g \approx g^{0.54} \tag{9.17a}$$

and

$$P_e \approx g^{0.48}. \tag{9.17b}$$

In the deep layers, the exponent in Equation 9.17a rises to ~ 0.65, while the value for Equation 9.17b drops to ~ 0.33.

In the hot-star case in the outer layers,

$$P_g \approx g^{0.85} \tag{9.18a}$$

and

$$P_e \approx g^{0.82}. \tag{9.18b}$$

In the deep layers, both exponents approach 0.5.

Some understanding of these gravity dependences can be obtained by considering Equation 9.2 in its simplest form. First the cool-star case. Here we know that κ_0 is mainly from the negative hydrogen ion and is proportional to P_e. So write

$$dP_g = \frac{g}{\kappa_0} \, d\tau_0 = \frac{g}{P_e(\kappa_0/P_e)} \, d\tau_0, \tag{9.19}$$

and then invoke Equation 9.11 so that P_e is replaced by $P_g \sum_C^K A_j/(1 + A_{\text{He}})$. Thus

$$P_g \, dP_g \approx \frac{g}{\sum\limits_C^K A_j/(1 + A_{\text{He}})} \frac{1}{\kappa_0/P_e} \, d\tau_0$$

and so

$$P_g \approx \left(\frac{2g \, (1 + A_{\text{He}})}{\sum\limits_C^K A_j} \right)^{1/2} \left(\int_0^{\tau_0} \frac{1}{\kappa_0/P_e} \, dt_0 \right)^{1/2}. \tag{9.20}$$

Within the cool-star context, the integral is independent of g, A_{He}, and $\sum_C^K A_j$. In other words, the expected gravity dependence is

$$P_g \approx \text{constant } g^{1/2},$$

which is not far from Equation 9.17a. Using Equation 9.11 again, we get the anticipated cool-star gravity dependence of electron pressure,

$$P_e \approx \text{constant } g^{1/2},$$

in agreement with Equation 9.17b.

In the hot-star case, κ_0 mainly comes from neutral hydrogen bound–free absorption (not proportional to P_e), so the same thought pattern gives

$$P_g \approx \text{constant } g$$

and

$$P_e \approx \text{constant } g,$$

roughly Equation 9.18 in the outer layers.

However, the abundances of the electron donors, $\sum_C^K A_j$, can also be important, as we notice in Equation 9.20.

The Effect of Metallicity on Pressure

Given the fact that metal abundances are about four orders down relative to hydrogen, their contribution to the gas pressure is essentially negligible. Except that optical depth is the independent variable in our formulation and increasing the metals increases the number of free electrons, resulting in a larger continuous absorption coefficient for cool stars (that negative hydrogen absorption again) and a shorter geometrical penetration of our line of sight into the photosphere. Therefore, gas pressure *at a given optical depth* in cool stars must decrease with increasing metal content. Figure 9.16 shows a cool-star example. For [Fe/H] between -4 and 1, the models show that the rate of change with metallicity is nearly constant in the outer layers with

$$P_g \approx \text{constant} \left(\sum_C^K A_j \right)^{-0.4}.$$

Equation 9.20 still applies to this case, so the approximate predicted dependence is

$$P_g \approx \text{constant} \left(\sum_C^K A_j \right)^{-1/2}, \tag{9.21}$$

in passable agreement.

Again with the help of Equations 9.11 and 9.21, we expect

$$P_e \approx \text{constant} \left(\sum_C^K A_j \right)^{1/2}. \tag{9.22}$$

The corresponding detailed relation for the models is shown in Figure 9.17. The metallicity gradient in the figure at $\log \tau_0 = -3$ gives an exponent of 0.4, not far from Equation 9.22.

In the deep layers of Figure 9.17, we see hydrogen taking over as the supplier of electrons and the sensitivity to metallicity is lost. The same is true of hot-star models generally.

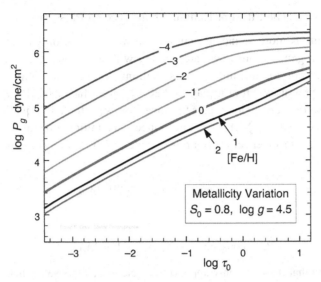

Figure 9.16 Gas pressure as a function of depth shows a strong dependence on metallicity for cool stars.

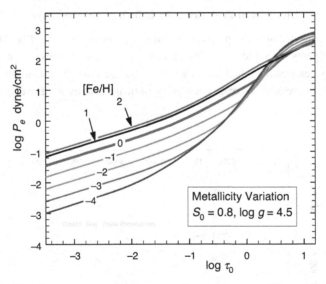

Figure 9.17 The electron pressure in the upper layers increases with metallicity for this cool model.

Notice the similar interplay of gravity and metallicity on the pressures. In later chapters, where the task is to use spectroscopic diagnostics to find log g and [Fe/H], care must be taken to separate these effects. There is one other small factor in front of the integral in Equation 9.20.

The Effect of Helium Abundance on Pressure

Although helium is more abundant than the metals, it is still only a small percentage relative to hydrogen, so unless we are dealing with a very unusual star, helium has little effect. Except for the hottest stars, helium is inert and simply adds weight, increasing the pressures. From Equation 9.20, we get a factor of $(1 + A_{He})^{1/2}$ for P_g, but there is also the conversion factor to move from κ_{total} to κ_ν in Equation 8.20 (or κ_0 in Equation 9.20), namely $\sum A_j \mu_j$, that introduces an additional factor $\approx (1 + 4A_{He})^{1/2}$. The same two factors enter for P_e, but now the sign of the exponent in the first factor is negative. Thus

$$P_g \approx \text{constant} \ (1 + A_{He})^{1/2}(1 + 4A_{He})^{1/2} \tag{9.23}$$

and

$$P_e \approx \text{constant} \ (1 + A_{He})^{-1/2}(1 + 4A_{He})^{1/2}. \tag{9.24}$$

Bearing in mind that these are only approximate relations, suppose the helium abundance were doubled from the solar value of 0.085 to 0.170. The increase in P_g would be ~16%, in P_e, ~8%.

Changes in Pressure with Effective Temperature

Finally, let us consider how the pressure structure changes with S_0 or T_{eff}. Model data are presented in Figures 9.18 and 9.19. The change of pressure at a given optical depth is a useful quantity, especially when we want to understand the behavior of spectral lines (Chapter 13). Gas pressure decreases at a given optical depth as we move to hotter models

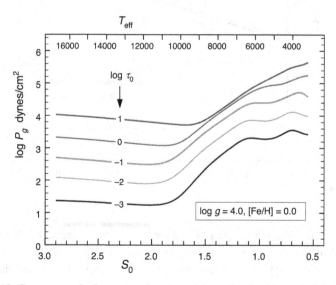

Figure 9.18 Gas pressure is shown as a function of S_0 or T_{eff} for five optical depths.

until hydrogen starts to ionize. We saw back to Figure 8.7 the dramatic increase in opacity in going from the cool to the hot temperatures. Larger opacity means less geometrical penetration to reach a given optical depth, so we see less deeply into the photospheres of hotter stars, and the gas pressure is correspondingly less.

On the other hand, electron pressure increases as we move through the temperatures of the cool stars because of increasing ionization. The rise stops and P_e levels off as ionization of hydrogen takes over. While these relations are by-and-large numerical in nature, it can be useful to approximate the electron pressure for the cooler stars using

$$P_e = \text{constant } 10^{\Psi S_0} = \text{constant } e^{\Omega T}. \tag{9.25}$$

Values of Ψ and Ω are shown in Figure 9.19. These are the slopes of the dashed lines.

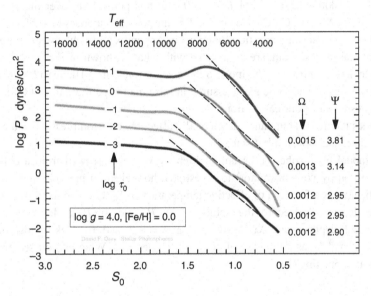

Figure 9.19 Electron pressure is shown as a function of S_0 or T_{eff} for five optical depths.

Comments

When computing a model photosphere for a real star, the basic parameters, S_0 or T_{eff}, $\log g$, and [Fe/H], are adjusted iteratively because a change in one of the three often causes changes in one of the others. To help the process along, we seek out spectral features that have greater and lesser sensitivities to the three parameters. For example, the continuum or energy distribution depends most strongly on the temperature, while certain spectral lines are sensitive to gravity and others are not. These issues are explored in the following chapters. We start in Chapter 10 with the continuum.

Questions and Exercises

1. Sketch qualitative limb-darkening curves for a G8 dwarf as they might appear at 4000 Å, 7000 Å, and at 10 cm. (The 10 cm radio emission comes from the chromosphere and lower corona.) Explain any differences you make in your curves.

2. Explain what we mean by the term "temperature distribution" in the context of stellar photospheres. Why do we need to know a star's temperature distribution when computing a model photosphere? What methods are available for the determination of temperature distributions? Make a sketch of a typical temperature distribution in a stellar photosphere. Be sure to include numerical scales. How is the temperature distribution related to the effective temperature? Does the temperature distribution include a temperature value equal to the effective temperature?

3. Compute the electron pressure at one point in a stellar atmosphere using Equation 9.10. Take $P_g = 1000$ dyne/cm^2 and $T = 5040$ K at this point. Sum over the elements: H, He, C, Si, Fe, Mg, Ni, Cr, Ca, Na, and K. Use the solar abundances in Table 16.1 and the ionization potentials and partition functions in Appendix C. Iterate to get approximately two decimal places. Compare your value with what is shown in Figure 9.9. Which of these elements contributes the most to the electron pressure? The least? Now re-do the calculations using the same gas pressure but raise the temperature to 12 000 K. Ask the same questions of your calculations.

4. Why does the metal abundance dramatically affect the electron pressure in a cool star's photosphere, but not in a hot star's?

5. Use a model photosphere program (written by you or borrowed from a colleague or given to you by your instructor) to investigate how electron pressure and geometrical depth behave as a function of effective temperature or S_0. Use one fixed surface gravity such as $\log g = 4$. Record the values of $\log P_g$ and $\log P_e$ at $\log \tau_0 = 0$ and the geometrical depth span, Δx, between $\log \tau_0 = -2$ and 0. Make plots of P_e/P_g, $\log P_e/P_g$, and Δx as a function of effective temperature. What do you see happening? Why is it happening?

References

Anderson, L.S. 1989. *ApJ* **339**, 558.
Avrett, E.H. & Krook, M. 1963a. *ApJ* **137**, 874.
Avrett, E.H. & Krook, M. 1963b. *JQSRT* **3**, 107.
Baines, E.K., Armstrong, J.T., Schmitt, H.R., *et al.* 2014. *SPIE* **9146**, 91460W.
Bell, R.A., Eriksson, J., Gustafsson, B., & Norlund, Å. 1976. *A&AS* **23**, 37.
Blanco-Cuaresma, S. 2019. *MNRAS* **486**, 2075.
Bohlin, R.C., Mészáros, S., Fleming, S.W., *et al.* 2017. *AJ* **153**, 234.
Cannon, C.J. 1973a. *JQSRT* **13**, 627.
Cannon, C.J. 1973b. *ApJ* **185**, 621.
Carbon, D.F. & Gingerich, O. 1969. *Theory and Observation of Normal Stellar Atmospheres* (Cambridge, Mass.: MIT Press), O. Gingerich, ed., p. 377.
Elste, G.H.E. 1967. *ApJ* **148**, 857.
Gray, D.F. & Kaur, T. 2019. *ApJ* **882**, 148.

Gustafsson, B., Bell, R.A., Eriksson, K., & Nordlund, Å. 1975. *A&A* **42**, 407.

Gustafsson, B., Edvardsson, B., Eriksson, K., *et al.* 2008. *A&A* **486**, 951.

Hainich, R., Ramachandran, V., Shenar, T., *et al.* 2019. *A&A* **621**, 85.

Holweger, H. & Müller, E.A. 1974. *Solar Phys* **39**, 19.

Kalkofen, W. 1987. *Numerical Radiative Transfer* (Cambridge: Cambridge University Press).

Kervella, P., Bigot, L., Gallenne, A., & Thévenin, F. 2017. *A&A* **597**, 137.

Kurucz, R.L. 1979. *ApJS* **40**, 1.

Lacour, S., Meimon, S., Thiébaut, E., *et al.* 2008. *A&A* **485**, 561.

Lester, J.B., Lane, M.C., & Kurucz, R.L. 1982. *ApJ* **260**, 272.

Livingston, W.C. & Wallace, L. 2003. *Solar Phys* **212**, 227.

Maltby, P., Avrett, E.H., Carlsson, M., *et al.* 1986. *ApJ* **306**, 284.

Mészáros, Sz., Allende Prieto, C., Edvardsson, B., *et al.* 2012. *AJ* **144**, 120.

Mitchell, W.E., Jr. 1959. *ApJ* **129**, 93.

Morello, G., Tsiaras, A., Howarth, I.D., & Homeier, D. 2017. *AJ* **154**, 111.

Pierce, A.K. 2000. *Allen's Astrophysical Quantities* (London: Athlone), 4th ed., A.N. Cox, ed., p. 355.

Pierce, A.K. & Waddell, J.H. 1961. *MemRAS* **68**, 89.

Ruland, F.D., Holweger, B.H., Griffin, R., & Griffin, R. 1980. *A&A* **92**, 70.

Scharmer, G.B. 1981. *ApJ* **249**, 720.

Trampedach, R., Asplund, M., Collet, R., Nordlund, Å., & Stein, R.F. 2013. *ApJ* **769**, 18.

Underhill, A.B. 1968. *BAN* **19**, 500.

Wesemael, F., Auer, L.H., van Horn, H.M., & Savedoff, M.P. 1980. *ApJS* **43**, 159.

10

Analysis of Stellar Continua

When we consider a star's continuous spectrum, we are interested in its shape over a broad spectral range, such as over the whole visible window. The continuum shape depends mainly on temperature, so the main reason for measuring it is to determine the star's temperature. The Balmer discontinuity also supplies gravity information for hot stars. We can measure the shape of the continuum using low spectral resolution, i.e., ~10–50 Å. Then faint stars can be reached. That's good. But then the spectral lines are not resolved and the measured fluxes are below the true continuum. That's less than good, but manageable, as we shall see. Spectra observed using low resolution are called *energy distributions*.

A second aspect of continuum studies is to determine the absolute flux reaching Earth from stars. Knowledge of the absolute flux allows us to delve into basic stellar parameters like the sizes of stars and the power output of stars.

The target is to compare continua from model photospheres with the observed ones. We learned in Chapter 9 how to compute the spectrum, so what do we need on the observational front? Basically a relatively simple ultra-low-resolution spectrograph capable of measuring radiation over a wide wavelength range, preferably in all wavelengths simultaneously. With it, an energy distribution is measured in four steps: (1) compare the program star to a standard star, (2) calibrate the shape of the standard star's energy distribution, (3) calibrate the absolute flux at one or more points in the energy distribution of the standard star, and (4) make proper allowance for the spectral lines to get to the true continuum. Most observers will do steps one and four, while relying on specialists to accomplish steps two and three.

The spectrograph is a good place to start.

Ultra-Low-Resolution Spectrographs

An ultra-low-resolution spectrograph is the instrument of choice for energy distribution work. It easily gives the wide wavelength coverage. For example, suppose we have a CCD detector with 1000 pixels, and we wish to cover the 4000 Å to 10 000 Å range. A dispersion of ~6 Å/pixel is easily obtained with a grating working in the first order. Figure 10.1 shows the layout for a low-resolution spectrograph. The usual slit is replaced by an opening large enough to let in the whole seeing disk, but small enough to exclude skylight and to keep the image properly registered on the CCD. A bit of a challenge occurs at the entrance aperture.

First, the size of the seeing disk varies with wavelength. Second, dispersion in the terrestrial atmosphere produces what amounts to a small spectrum, i.e., seeing disks for different wavelengths are not in the same place, but are spread out along a vertical circle from the zenith. In the best arrangement, when the zenith distance exceeds ~45°, the spectrograph is rotated around the optical axis of the telescope so that the spectrograph dispersion is perpendicular to the atmospheric spectrum (equation given below).

The spectral resolution for the conventional slit spectrograph having an entrance slit width W' is given by Equation 3.15, the spectrograph equation,

$$\Delta\lambda = \cos\alpha \frac{W'}{f_{coll}} \frac{d}{n},$$

where α is the angle of incidence on the grating, f_{coll} is the collimator focal length, and d is the grating ruling spacing for the nth order spectrum. In the current case, W' is replaced by the size of the seeing disk. Suppose the seeing disk has an angular size $\Delta\phi$, and the telescope effective focal length is F_{eff}. Then $W' = \Delta\phi\, F_{eff}$, and the resolution is

$$\Delta\lambda = \cos\alpha \frac{F_{eff}}{f_{coll}} \frac{d}{n}\Delta\phi. \tag{10.1}$$

Fluctuations in seeing result in fluctuations in resolution. In favorable cases this is only a few percent and does not directly affect the continuum measurements, since for the most part the continuum hardly changes over a few angstroms. But it does cloud the issue of what spectral lines are included in the $\Delta\lambda$ interval.

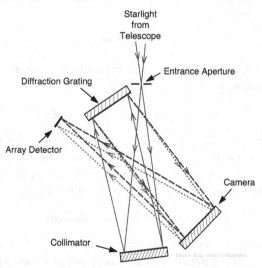

Figure 10.1 The optical layout for a typical low-resolution grating spectrograph suitable for measuring energy distributions.

The collimator and camera focal lengths are chosen in accordance with Equation 3.14, typically with the whole seeing disk scaled to a single pixel, since resolution is not the concern and fainter stars can be observed.

With our spectrograph in hand, we use it to compare our target stars to standard stars.

Observations Using Standard Stars

Standard stars have been set up over a full range of right ascension so that they are available nearby wherever the target star is in the sky. In effect, this means that standard-star energy distributions are known, so all we need to do is differentially compare "our" star with one or more of the standards. How standards are set up is discussed in the next couple of sections. At first glance, it would seem simple enough to record the flux spectrum of our target star and the standard star and ratio them to get the needed result. If the spectrograph is in space on a satellite, you could do just that, but looking through the terrestrial atmosphere from a ground-based observatory is more complicated. The reason is that the air and particles in it cause extinction of the starlight, and the amount of extinction depends on the amount of air we're looking through and on the wavelength. So not only is the starlight dimmed coming through the air, it is dimmed more at shorter wavelengths, changing the shape of the energy distributions, the very thing we want to measure.

The problem is worked around using extinction curves. Multiple observations of at least one star are done over several hours. The rotation of the Earth changes the amount of air the star is seen through. The appropriate geometry has already been considered in Figure 7.3. The plane-parallel approximation is also appropriate for the Earth's atmosphere, and our standard θ angle now measures the angular distance of the star from the zenith. We appropriately rename it Z. Assuming a uniform and constant atmosphere that causes absorption, but no emission (as in the discussion following Equation 5.13), we expect the observed flux to vary as

$$F_v = F_v^0 e^{-\tau_v \sec Z},$$

in which τ_v is the constant optical depth of the full atmosphere toward the zenith and F_v^0 is the flux outside the atmosphere. Or

$$\log F_v = \log F_v^0 - \tau_v \log e \sec Z. \tag{10.2}$$

Figure 10.2 shows one example. The slope, $\tau_v \log e$, is called the extinction coefficient, and the intercept, $\log F_v^0$, is the flux value outside the atmosphere. The zenith distance (for a northern-hemisphere observer) is given by applying the law of cosines to the spherical triangle defined by the zenith, the star, and the north celestial pole,

$$\cos Z = \sin \ell \sin \delta + \cos \ell \sin \delta \cos t,$$

in which ℓ is the observatory latitude, δ is the declination of the star, and t is the absolute value of the hour angle. Essentially the star can be measured as it rises, with data points rising up the extinction curve, reaching its maximum brightness crossing the meridian, and/or as it sets, with data points declining down the curve. An extinction curve is needed for each wavelength we want in the energy distribution.

(Now that we have an equation for the zenith distance, we can use Z to calculate the preferred slit orientation referred to above. The slit needs to be oriented along the vertical circle from the zenith through the star. The rotation angle, called the parallactic angle, p, lies between the hour circle through the star to the vertical circle from the zenith through the star, and is given by applying the law of sines to the same spherical triangle, $\sin p = \sin (90° - \ell) \sin t / \sin Z$.)

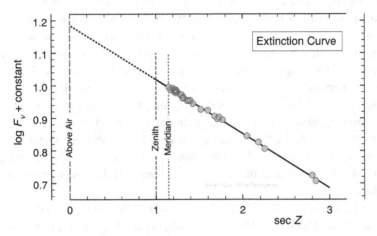

Figure 10.2 Extinction curves like this one are used to correct for the light lost in the terrestrial atmosphere.

The goal is to find the flux ratio, target star to standard star, outside the atmosphere. Rather than using Equation 10.2 to extrapolate to $\sec Z = 0$, it is better to ratio Equation 10.2 for the two stars,

$$\log \frac{F_2}{F_1} = \log \frac{F_2^0}{F_1^0} - \tau_\nu \log e \left(\sec Z_2 - \sec Z_1 \right). \tag{10.3}$$

Here F_2 is F_ν for star 2 and so on. The left side of the equation comes from our measurements at the corresponding zenith distances, and F_2^0/F_1^0 is the ratio we seek. The best case is when the observations are made at the same zenith distance. Then the measured ratio *is* the wanted ratio. The next best case is to cover the same range in zenith distance for both stars. Then the vertical shift of one extinction curve into the other gives us the answer, i.e., we apply Equation 10.3.

Although we do not need to know much about the extinction coefficient to use Equation 10.3, it often has the form (Hayes & Latham 1975, Cochran & Barnes 1981) given by

$$\tau_\nu \log e = A + \frac{B}{\lambda} + \frac{C}{\lambda^4}.$$

If the extinction coefficient has not been measured at all the desired wavelengths, this expression can be used to estimate it from those that have been measured.

Now that we know how to compare two stars, we look at how standard stars are calibrated.

Absolute Calibration of the Shape of Vega's Continuum

Absolute calibration work is not for the faint of heart. The telescope, spectrograph, light detector, and the terrestrial atmosphere imprint their spectral sensitivity and efficiency on the measurements. Ultimately these factors have to be evaluated and removed by measuring a light source with a known spectrum. We have such a source in the black body (Chapter 6). But it is not trivial to measure a black body and a star with the same instruments, although it has been done. Standard lamps sometimes serve an intermediary role. They are calibrated against a black body in the laboratory and then, with relative ease, moved to an observatory site where they can be compared to stars. But they do introduce another layer of uncertainty. Absolute measurements are also more vulnerable to poor sky conditions because Equation 10.2 rather than Equation 10.3 must be used. Absolute calibrations come in two parts. The easier part is to get the absolute shape of an energy distribution, i.e., the real *relative* fluxes as a function of wavelength. The far tougher part is to measure the absolute flux at the standard wavelength of 5556 Å in erg/(s cm^2 Å), the topic of the next section.

Vega (α Lyr, HR 7001, HD 172167, A0 V, $V = 0.03$) has been adopted as the primary photometric standard for the visible window. Why Vega? It is bright, giving good photon statistics with the small telescopes often used in this kind of work. Big telescopes are not designed to point close to the horizon at standard sources not in the sky. Vega has a "clean" spectrum with few spectral lines to complicate exact wavelength registration and it reduces the sensitivity to bandpass size. The star is located at a favorable declination for northern-hemisphere observers, giving a wide range of zenith distances that are so necessary for evaluating the extinction. As a star with a radiative envelope, it is not subject to significant variations arising from starspot transits, magnetic outbursts, or variations during magnetic cycles.

Vega does have some unique characteristics. It is a rapid rotator seen nearly pole-on (Gray 1988, Gulliver *et al.* 1994, Monnier *et al.* 2012). Its projected rotation velocity (see Chapter 18) has been measured at $v \sin i = 23.4 \pm 0.4$ km/s (Gray 1980), 21.8 ± 0.2 (Gulliver *et al.* 1994), 23.0 ± 0.1 km/s or 20.8 ± 0.2 for two different models (Hill *et al.* 2010), and 20.7 ± 0.3 km/s (Yoon *et al.* 2010), while its actual unprojected rotation rate ≈ 200 km/s (Hill *et al.* 2010, Yoon *et al.* 2010). Such rapid rotation leads to a darkening near the equator and an oblate shape. Offshoots include slightly non-standard colors and a higher luminosity compared to most A0 V stars. Vega is also surrounded by a dust shell or ring that enhances the IR emission beyond ~12 μm (Aumann *et al.* 1984, Marsh *et al.* 2006, Müller *et al.* 2010). None of these characteristics fundamentally compromise Vega as a standard source as long as it is not variable. Small flux variations at the $\sim 5 \times 10^{-4}$ level have been reported by Böhm *et al.* (2015). Such small variations are well below the ~1% uncertainty in the calibrations. The secular increase in flux arising from Vega's -14 km/s radial velocity amounts to only 0.04% per century. Unique characteristics make it harder to model the whole energy distribution of Vega, but conforming to a model is not a primary requirement for a standard. Secondary standard stars have been calibrated against Vega, or directly against black bodies in a few cases, as described later.

Techniques and early measurements are discussed and summarized by Code (1960). Additional important information on methods, telescopes, spectrographs, black bodies, standard lamps, and the resulting calibrations of Vega is given by Hayes and Latham (1975), Hayes *et al.* (1975), Tüg *et al.* (1977), Kharitonov *et al.* (1980), and Knyazeva and Kharitonov (1990), for example. Some of the results are compared in Figure 10.3. Hayes (1985) brings together calibrations from six independent investigations and tabulates Vega's mean energy distribution in 25 Å steps continuously from 3300 Å to 10 500 Å, as shown by the line in Figure 10.3. Table 10.1 is based on his summary. The relative fluxes are normalized at 5556 Å. Errors are ~1%.

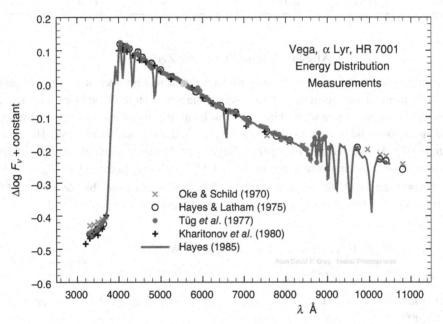

Figure 10.3 A comparison of measurements of the energy distribution of Vega. The agreement is remarkably good. The light dot shows the 5556 Å normalization point.

Technical note: mixed units of frequency flux versus wavelength, as in Figure 10.3, are often convenient and useful, as is the logarithmic ordinate. Sometimes the energy distribution takes on a better shape for certain measurements if F_λ is used instead of F_ν.

Next we move to the determination of the absolute flux at 5556 Å.

Table 10.1 *The energy distribution of Vega*

λ	$\Delta\log F_\lambda$	$\Delta\log F_\nu$	λ	$\Delta\log F_\lambda$	$\Delta\log F_\nu$
3300	−0.011	−0.463	6056	−0.116	−0.041
3400	−0.025	−0.452	6436	−0.194	−0.066
3500	−0.043	−0.444	6800	−0.264	−0.088
3571	−0.048	−0.432	7000	−0.304	−0.103
3636	−0.053	−0.421	7550	−0.407	−0.141
4036	0.404	0.126	7780	−0.449	−0.157
4167	0.362	0.112	8090	−0.498	−0.172
4255	0.339	0.107	8400	−0.554	−0.195
4464	0.279	0.089	8685	−0.549	−0.161
4620	0.236	0.076	8925	−0.572	−0.160
4785	0.186	0.056	9700	−0.677	−0.193
5000	0.133	0.041	10250	−0.761	−0.229
5263	0.068	0.021	10400	−0.781	−0.236
5556	0.000	0.000	10500	−0.796	−0.243
5840	−0.063	−0.020	10800	−0.84	−0.26

Absolute Calibration of the Zero Point

Measuring the absolute flux at 5556 Å, meaning the actual number of photons, requires knowing many factors accurately. These include the temperature of the black body and its stability, the aperture size of the black body or lamp, the distance to the standard source, and the atmospheric extinction. The care required is detailed in, for example, Hayes and Latham (1975) and Tüg *et al.* (1977). Hayes (1985) summarized the values of five determinations, arriving at a value of $F_\lambda = 3.44 \times 10^{-9}$ erg/$(\mathrm{s\,cm^2\,Å})$ or $F_\nu = 3.54 \times 10^{-20}$ erg/$(\mathrm{s\,cm^2\,Hz})$ with an error of ~1.7%. A careful re-evaluation by Mégessier (1995) of the various determinations, placing more emphasis on the direct black-body comparisons, gives

$$\left.\begin{array}{l} F_\lambda = 3.46 \times 10^{-9}\,\mathrm{erg}/(\mathrm{s\,cm^2\,Å}) = 3.46 \times 10^{-12}\,\mathrm{W}/(\mathrm{m^2\,Å}) \\ F_\nu = 3.56 \times 10^{-20}\,\mathrm{erg}/(\mathrm{s\,cm^2\,Hz}) = 3.56 \times 10^{-23}\,\mathrm{W}/(\mathrm{m^2\,Hz}) \end{array}\right\} \text{Vega at 5556 Å.}$$

The estimated error is ~0.7%. Curiously, F_λ is 995 photons/$(\mathrm{s\,cm^2\,Å})$, within error of an even 1000.

Vega has an apparent visual magnitude of $V = 0.03$. If we recast the calibration in terms of a zero-magnitude star, the flux values are 2.8% larger.

$$\left.\begin{array}{l} F_\lambda = 3.56 \times 10^{-9}\,\mathrm{erg}/(\mathrm{s\,cm^2\,Å}) = 3.56 \times 10^{-12}\,\mathrm{W}/(\mathrm{m^2\,Å}) \\ F_\nu = 3.66 \times 10^{-20}\,\mathrm{erg}/(\mathrm{s\,cm^2\,Hz}) = 3.66 \times 10^{-23}\,\mathrm{W}/(\mathrm{m^2\,Hz}) \end{array}\right\} V = 0 \text{ at 5556 Å.}$$

The shape and the zero point for our primary standard are now in place. We can proceed to measure other stars using Equation 10.3. But secondary standards are also handy.

Secondary Standard Stars

When Vega is not visible or is not close enough in the sky to the star we want to measure, it is important to have an alternative. Secondary standard stars are set up by comparing them to Vega or to other secondary standards using Equation 10.3. Candidates for this job are selected with the same criteria mentioned above for Vega, and for their location in the sky. The grid of secondary standards chosen by Oke (1964) is well distributed in right ascension and has generally been used. Tüg (1980a, 1980b) and Taylor (1984) tied additional stars into the grid. The relative errors are in the 1–2% range. Additional stars suitable for use as secondary standards have been compiled by Breger (1976), Kharitonov and Glushneva (1978), Cochran (1981), Hamuy *et al.* (1992), and Glushneva *et al.* (1992).

Fainter standard stars are needed for work with large telescopes. Such standards have been set up, for example, by Stone (1977), Oke and Gunn (1983), Baldwin and Stone (1984), Massey *et al.* (1988), Gutierrez-Moreno *et al.* (1988), Oke (1990), Hickson and Mulrooney (1998), Borisov *et al.* (1998), and Allende Prieto and del Burgo (2016).

Sirius (A1 V, $v \sin i \sim 17$ km/s) and 109 Vir (A0 III, $v \sin i \sim 300$ km/s) have been pressed into service as standards in the infrared over concerns about the dust emission from Vega (e.g., Engelke *et al.* 2010, Bohlin 2014). White dwarf and A star models have been advocated as standards in some cases (e.g., Bohlin *et al.* 2020, Allende Prieto & del Burgo 2016). While the idea of trusting model photospheres to give the correct absolute flux has been around for a long time, and has been used for interesting things (e.g., Gray 1967), we should be careful to not fall into circular reasoning about calibrations. At best, models should not reach beyond the level of secondary standards.

The enormous brightness of the Sun compared to other stars, $\sim 5 \times 10^{10}$ times brighter than Vega, makes it difficult to use as a standard star, but its energy distribution is of considerable interest in its own right.

The Energy Distribution of the Sun

As in many areas of stellar research, the Sun is an anchor-point for radiation measurements. The solar energy distribution has been subjected to extensive measurement. But observing the Sun in integrated light, i.e., flux, is non-trivial because of its half-degree angular size. Incidentally, the term "irradiance" is used for flux in many solar investigations. The task of integrating the light from the solar disk has been accomplished in different ways. One is to replace the focusing mirror in the optical train with a flat mirror, bringing sunlight, but not the solar image, onto the entrance aperture of the measuring system (e.g., Wallace *et al.* 2011). Another is to form a very small image of the Sun such that it matches the entrance aperture (e.g., Beckers 1981, Reiners *et al.* 2016). A clever scheme of using a diffraction image from a pinhole overcomes both the angular size and differential brightness problems (Tüg 1982). A fourth method is to do center-to-limb measurements that relate the disk center to the whole disk (e.g., Neckel & Labs 1984). Some satellite observations use diffusely scattered sunlight (Hilbig *et al.* 2018). A composite of the results of several instruments observing from space, called the Atlas 3 Reference Spectrum, is given by Thuillier *et al.* (2003).

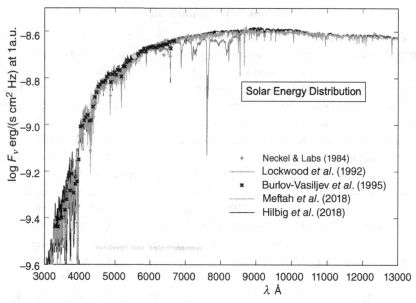

Figure 10.4 A comparison of measurements of the solar flux energy distribution as seen from a distance of one astronomical unit. The light dot is the 5556 Å reference point.

Figure 10.4 shows some of the visual-window observations. Given the fact that these are all independent absolute calibrations, i.e., not just shape determinations, the agreement is very good.

The total power from the Sun, i.e., the integral of the observed energy distribution, is actually usually measured independently of the energy distribution, often with various forms of bolometers (Neckel & Labs 1973, Willson & Hudson 1981, Newkirk 1983, Foukal & Lean 1986, Fröhlich & Lean 2002, Kopp & Lean 2011, Meftah *et al.* 2018). The value adopted here favors Kopp and Lean (2011). The solar power received at the Earth, formerly called the *solar constant*, now more often termed the *total solar irradiance*, and which we will call the *apparent bolometric flux*, is

$$F_{\mathrm{bol}}^{\odot} = \int_{0}^{\infty} F_{\nu}\, d\nu = 1.362 \times 10^{6}\, \frac{\mathrm{erg}}{\mathrm{s\, cm}^{2}} = 1362 \pm 2\, \frac{\mathrm{W}}{\mathrm{m}^{2}}\ \text{at 1 a.u.} \qquad (10.4)$$

At this distance, with one astronomical unit $r = 1.495979 \times 10^{13}$ cm or $\times 10^{11}$ m, the solar luminosity or total power output is

$$L_{\odot} = 4\pi r^{2} F_{\mathrm{bol}}^{\odot} = 3.830 \times 10^{33}\ \mathrm{erg/s}\ \text{or}\ \times 10^{26}\ \mathrm{W}, \qquad (10.5)$$

and using the solar radius of 6.9598×10^{10} cm or $\times 10^{8}$ m, the effective temperature is

$$T_{\mathrm{eff}}^{\odot} = 5772 \pm 2\, K. \qquad (10.6)$$

The solar power reaching Earth varies over the magnetic cycle, as shown in Figure 10.5. The variation is small, but of great interest in the context of solar physics and terrestrial global warming.

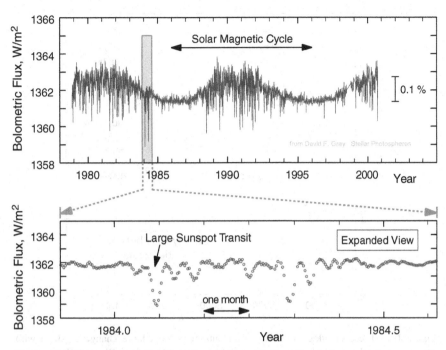

Figure 10.5 Upper panel: Satellite observations of the solar bolometric flux or total irradiance show a variation ~0.1% over the solar magnetic cycle. The Sun is brighter when it is more magnetically active. Lower panel: Expanded timescale to show transits of individual sunspots across the disk as the Sun rotates. Data, see Fröhlich and Lean (1998, 2004); revised mean value as in Equation 10.4.

Now let us turn our attention to what stellar energy distributions look like and what they tell us about stars.

Examples of Stellar Energy Distributions

As expected from basic black-body spectra and what we know empirically from color indices, the energy distributions of stars depend strongly on temperature. This is illustrated in Figure 10.6, where energy distributions are shown for a sample of Hyades cluster stars. The slope does indeed change markedly with spectral type. Also in the hotter stars, we see a well-defined Balmer jump, while for the cooler stars it has dwindled into the noise. This is in line with what we know about the temperature variation of the continuous absorption coefficients shown in Figure 8.7.

Since these five stars are very nearly at the same distance from us, their absolute flux levels show real relative values. That is, we see empirically that the slopes are directly connected to the absolute flux levels. The ability to derive absolute flux by measuring the slope of the Paschen continuum is a powerful tool that we take up again in Chapter 14, where we use it to determine stellar radii.

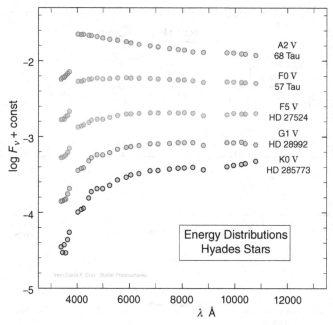

Figure 10.6 These Hyades-star energy distributions show a large change in slope with spectral type and also a large change in absolute flux level. Data from Oke and Conti (1966) updated to Hayes (1985).

The flux shifts to the UV for hotter stars, as shown in Figure 10.7. The slope of the Paschen continuum loses some of its sensitivity to temperature for stars hotter than late B, and UV measurements are better suited for that work. The same happens in reverse for the cool K and M stars, leading one to rely on infrared measurements.

Several compilations of energy distributions have been published. A few examples are Willstrop (1965), Jones (1966), Breger (1976), Glushneva *et al.* (1983), Jacoby *et al.* (1984), Kiehling (1987), Alekseeva *et al.* (1996), Lançon and Wood (2000), Sánchez-Blázquez *et al.* (2006), Falcón-Barroso *et al.* (2011), Stewart *et al.* (2015), and Yan *et al.* (2019). Mean energy distributions as a function of spectral type have been compiled by O'Connell (1973) and Pickles (1998).

Continua from Photospheric Models

The temperature and pressure sensitivity of the continuum can easily be explored using models of the type described in Chapter 9. The flux can be computed at any number of wavelengths using Equation 9.15. Examples are shown in Figure 10.8. Notice how the absolute flux is coupled to the change in slope as the model temperature changes, the effect we noted for the Hyades stars in Figure 10.6. Except for the Balmer jump, the continua show only weak variation with surface gravity. And for the cooler stars, the Balmer jump is obliterated by the numerous spectral lines, so there is zero discernible difference with gravity.

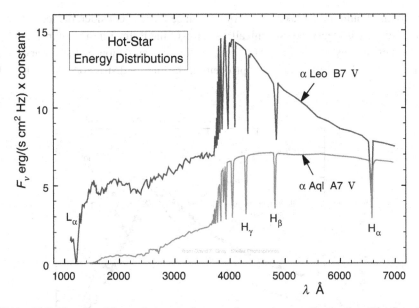

Figure 10.7 The ultraviolet region grows in strength as we move to hotter stars. The flux of these two energy distributions is scaled arbitrarily to facilitate a comparison of their shapes. Based on Code *et al.* (1976).

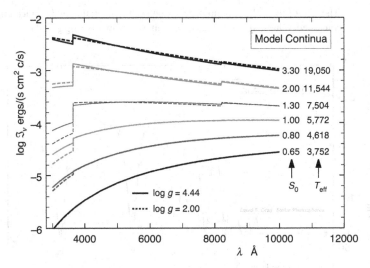

Figure 10.8 Model photosphere continua are shown for several temperatures and two surface gravities. Without spectral lines, metallicity has no effect.

The size of the Balmer discontinuity is a useful gravity index, and it is discussed more fully in Chapter 15. A preview is shown in Figure 10.9. The rise and fall of the Balmer jump corresponds to the temperature range over which neutral hydrogen dominates the continuous opacity (Figure 8.7 again). The good agreement between the models and the observations should not be taken too seriously since there is considerable leeway in how the discontinuity is measured. Specifically, the top of the Balmer jump is wiped out by the Balmer lines as they reach the series limit, so that extrapolation of the observations is required. Figure 10.10 illustrates this problem in the energy distribution of Sirius, where

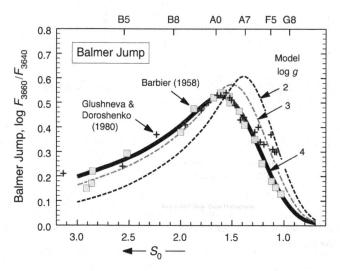

Figure 10.9 The size of the Balmer jump depends on both temperature and surface gravity. Lines show computed values, symbols measured values for dwarfs (log $g \approx 4$).

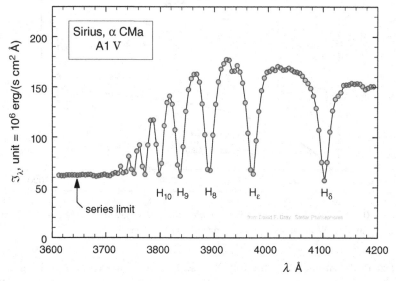

Figure 10.10 When the Balmer jump is strong, so are the Balmer lines. The lines blend together and obliterate the Paschen continuum near the series limit. Resolution is ~5 Å. Ordinate is flux at stellar surface. Based on Panek (1977).

the hydrogen lines overlap shortward of H_δ. Rather than work with the Balmer jump directly, it is better to model simultaneously the Balmer continuum and the Paschen continuum longward of H_δ. Another uncertainty enters through the temperature distribution. We saw back in Figure 9.12 how the flux from opposite sides of the Balmer jump comes from very different depths. If the temperature in the high layers is too high, the Balmer continuum will also be too high or the Balmer jump too small. Similarly as we saw in Chapter 9, the helium abundance can introduce a small additional uncertainty in a derived gravity.

The issue with the absorption by spectral lines is hinted at in Figure 10.10, but it is a hurdle we now confront more generally before models can be matched to energy distributions.

Line Absorption

We want to compare computed model fluxes to observed energy distributions. How spectral lines are handled in this context depends very much on the spectral resolution being used. In the classical situation with a bandpass of ~25–50 Å, the absorption by spectral lines lowers the average measured flux relative to the continuum. If the resolution is ≲1 Å, it may be possible to see the true continuum between the lines. This leads to two conceptual paths we might follow. For the wide bandpass situation, we measure the fraction of the light lost to line absorption and correct the observed energy distribution to find the true continuum. Then we compare this to the model-computed continuum. For the small bandpass situation, we include all the spectral lines directly in the flux calculation and so model the actual observed flux. Each approach has its merits and demerits. The low-resolution path is usually chosen because it is simple and direct, and we will follow it for most of the discussion in this section.

However, the second path is important and deserves a few words. In cases where there are only a few lines, including them in the flux calculation is easy enough to do, although the amount of calculation is increased because several points per line profile are needed. With spectra of cooler stars, where there are hundreds of lines, one is faced with a major computational effort. A more serious issue is the need for atomic data on all the lines in the bandpass: element identification, ionization and excitation potentials, f-values, and damping constants for strong lines (as will become clear in the following chapters). Progress has been made on this front with the Vienna Atomic Line Database (website: vald.astro.uu.se) and the National Institute of Standards and Technology (website: nist.gov/pml/basic-atomic-spectroscopic-data-handbook). Non-atomic information is also needed. As we shall see later, the absorption caused by most lines depends on the small-scale velocity fields characterized by the parameter called the *microturbulence dispersion*, ξ. In many cases, a value of ξ is simply guessed. A poor guess means inaccurate modeling. When there are so many lines that they overlap and no continuum points exist, the only way forward is to compute the flux with all the spectral lines included. Examples can be seen in Figures 16.8, 16.9, 16.10, Fontenla *et al.* 1999, Bean *et al.* (2006), or Siqueira Mello *et al.* (2013).

Back to path one. Measuring line absorption in a 25–50 Å interval requires relatively high-resolution spectroscopic observations, high enough to resolve the spectral lines. Fortunately, we only need the *fraction* of the light lost to line absorption. That means no absolute calibration is needed here, although we still depend on a linear response between light in and signal out from our detector system. Further, the underlying energy-distribution shape is removed by normalizing the local continuum to unity, as shown in the examples in Figure 10.11. The amount of line absorption increases statistically toward shorter wavelengths, with cooler temperature stars, and with higher metallicity, i.e., whenever the lines get more numerous and/or stronger.

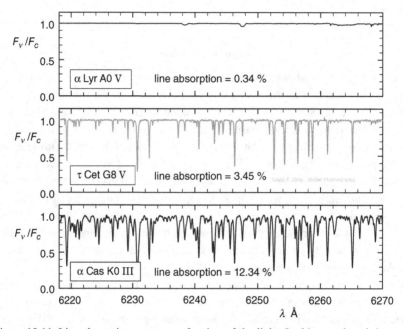

Figure 10.11 Line absorption removes a fraction of the light. In this wavelength interval, the mean flux level is 0.34% below the continuum for α Lyr (upper panel), 3.45% for τ Cet (center panel) and 12.34% below for α Cas (lower panel). Data from Takeda *et al.* (2007) and the Elginfield Observatory.

We can get a feel for the wavelength distribution of lines from Figure 10.12, where the line spectra of Vega, the Sun, and Arcturus are displayed in the three panels. We see that the line absorption is dramatically greater toward shorter wavelengths. When the line-absorption corrections become large, more care must be exercised in their measurement to maintain overall accuracy.

In Figure 10.12a, we see how large the absorption in the terrestrial atmosphere can be. Kim *et al.* (2009) removed the telluric absorption by dividing the Vega spectrum by the spectrum of the rapidly rotating star α Peg (B9 III) from which they removed the stellar lines, a process previously recommended by Bessell (1999).

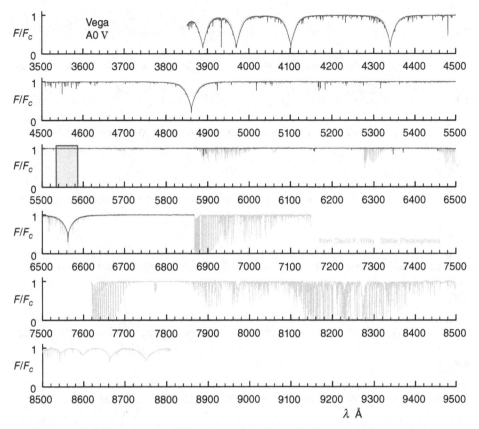

Figure 10.12a The spectrum of Vega shows few lines aside from the obvious very strong hydrogen lines. Two sets of measurements are shown. The darker line shows the spectrum with telluric lines removed (based on Kim *et al.* 2009), while the lighter line has them left in (based on Takeda *et al.* 2007). The rectangle is a 50 Å span centered on 5556 Å.

So we use linear but uncalibrated spectra to measure the line absorption. It can be helpful to compare with previous determinations, but few numbers are published. Some are given by Melbourne (1960), Gray (1967), Oke and Conti (1966), and an extensive listing for β Gem can be found in Ruland *et al.* (1980). In any case, we apply our measured line absorption to the observed energy distributions, and proceed.

Comparison of Model to Stellar Continua

The culmination of our efforts brings model continua to real stellar continua. When the match is successful, we have temperature and occasionally gravity information about the star. Once this basic information is in place, other aspects of the star can be studied, such as its chemical composition and radius, and variations of physical parameters with temperature can be explored. Some examples are instructive. Figure 10.13 shows a fit (shape only) for our primary standard, Vega, where as we saw in the previous section, the line

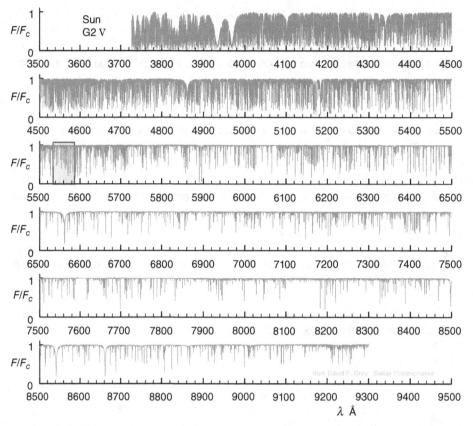

Figure 10.12b The solar spectrum. The number and strength of spectral lines increase dramatically toward shorter wavelengths. The rectangle is a 50 Å span centered on 5556 Å. Data from Hinkle *et al.* (2000).

absorption is almost negligible. The model in the figure is based on the temperature distribution of Dreiling and Bell (1980). In their analysis, Dreiling and Bell found $T_{eff} = 9650 \pm 200$ K and $\log g = 4.0 \pm 0.3$, while a similar modeling by Castelli and Kurucz (1994) gave $T_{eff} = 9550 \pm 100$ K and $\log g = 3.95 \pm 0.05$.

Cooler stars have more line absorption, as we know. The next example in Figure 10.14 is for a cooler star, Procyon. Now the line absorption corrections are significant, and in fact, the greatest uncertainty in this example is the imprecision of the line absorption measurements.

The modeled continuum for the Sun, using $T(\tau_0)$ from Table 9.3, is shown in Figure 10.15. Only a selection of the observations from Figure 10.4 are shown here. Line absorption corrections have been applied to some of the points to give true continuum values. The model fits reasonably well. In this case both the model and the observations are on an absolute scale, so it is not just the shape, but also the absolute flux that is being compared.

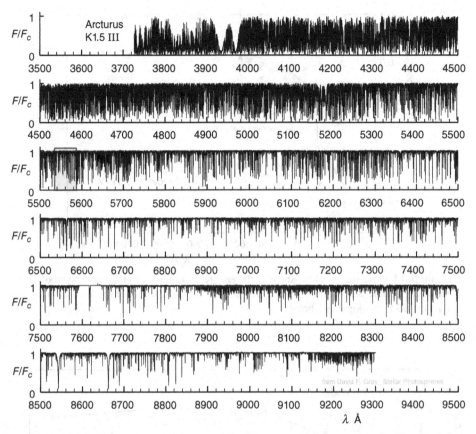

Figure 10.12c The spectrum of Arcturus. The lines are much stronger than for Vega or the Sun. The rectangle is a 50 Å span centered on 5556 Å. Data from Hinkle *et al.* (2000).

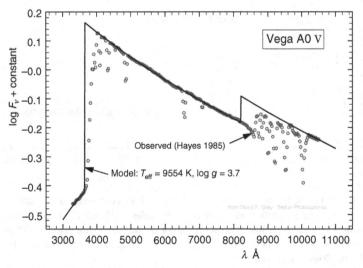

Figure 10.13 The energy distribution of Vega (small circles) is matched by a model having $T_{eff} = 9554$ K and $\log g = 3.7$.

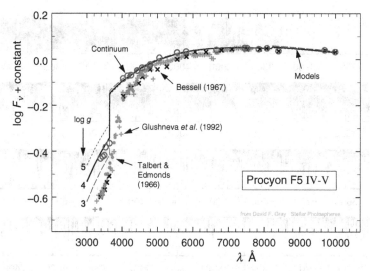

Figure 10.14 Observations of the shape of the energy distribution of Procyon (× symbols), corrected for line absorption (circles), and models (lines). Models have $S_0 = 1.111$ or $T_{\text{eff}} = 6413$ K and log g as labeled.

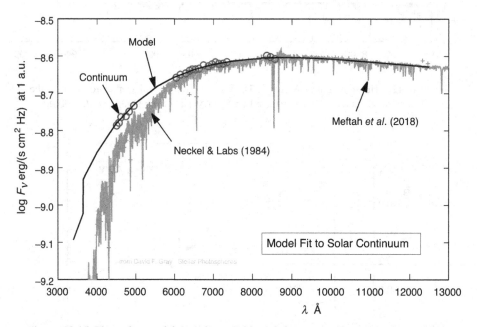

Figure 10.15 The solar model based on Table 9.3 is compared to the observations. Corrections for line absorption are made using the atlas of Hinkle *et al.* (2000) to get the continuum points. Reliable line absorption cannot be measured shortward of ~4500 Å, nor in the ~5000–6000 Å section because there are no continuum points. Line absorption beyond ~10 000 Å is less than ~1%.

As a final illustration of model fitting, let us look back at the Hyades observations from Figure 10.6. The same observations are shown in Figure 10.16, now with line-absorption corrections applied and models matched to the continua. Line absorption cannot be measured for the shortest wavelengths of the coolest two examples because the line blending is so severe that no continuum points exist. The same problem occurs above the Paschen jump. Otherwise, the agreement is satisfactory.

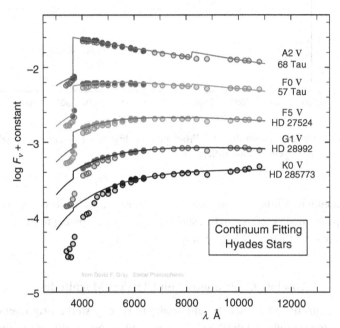

Figure 10.16 Open circles are the observed energy distributions. Filled circles have line-absorption corrections applied. Lines are models fit to the continuum points. Temperature parameters are (S_0/T_{eff}): 1.58/9120, 1.30/7504, 1.12/6465, 1.01/5830, and 0.88/5079.

Temperature Errors with Continuum Fitting

The temperature resolution from fitting stellar continua is good, typically $\pm 1\%$. Two models with a $\sim 2\%$ temperature difference are compared in Figure 10.17. The mismatch with the observations is clearly visible, i.e., the error in matching a model to the observations is of this order or less. In situations where the continuum can be clearly identified in the energy distribution, line absorption can be measured with errors of a fraction of a percent. Uncertainty in line absorption gets worse as the corrections get larger. Once the continuum is lost to overlapping lines, only a lower bound to the true continuum can be estimated.

For stars more distant than ~ 100 pc, interstellar reddening starts to be a problem. The extinction from interstellar dust has close to a $1/\lambda$ dependence. More light is lost to shorter wavelengths, hence the reddening. This is equivalent to changing the slope of the energy distribution and obviously strikes at the heart of continuum fitting. Sometimes reddening corrections are applied, but situations with significant reddening can never be considered on sure ground.

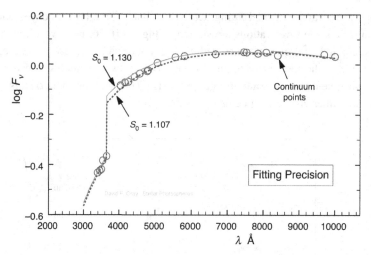

Figure 10.17 The continuum data for Procyon from Figure 10.14 are compared to two models having a 2% temperature difference.

Although continuum fitting to measure stellar temperatures has become commonplace, analysis of Ap (peculiar A) stars can be especially challenging owing to the stratification of chemicals in their photospheres (e.g., Shulyak *et al.* 2013).

Absolute Calibrations with Classical Magnitudes

Magnitude systems have been used traditionally to specify stellar brightness. There are many such systems, such as UBV, RGU, uvby, and so on, as discussed in Chapter 1. The individual magnitudes encompass a much broader wavelength span than typically used for energy distribution measurements and are not intended for such work. The transmission bands are empirically defined by color filters and/or detector response, and through these give a weighted average of the energy distribution across their bandpass. The zero points for magnitudes, i.e., what constitutes a zero magnitude star, are set by agreement and adopted standard stars. But with the absolute calibration of Vega, magnitudes can be placed on a physical scale. This allows us to use magnitudes to infer an absolute zero point when a direct observation from an energy distributions is not available.

The simplest approach here is to start with the effective wavelength of the magnitude's bandpass. For example, the effective wavelength of the V magnitude is 5480 Å. At this wavelength, using the calibrated energy distribution of Vega as discussed above, the absolute flux of a star of magnitude V is

$$\left.\begin{aligned}\log F_\lambda &= -0.4V - 8.441\,\mathrm{erg}/(\mathrm{s\,cm^2\,\mathring{A}}) = -0.4V - 11.441\,\mathrm{W}/(\mathrm{s\,m^2\,\mathring{A}})\\\log F_\nu &= -0.4V - 19.429\,\mathrm{erg}/(\mathrm{s\,cm^2\,Hz}) = -0.4V - 22.429\,\mathrm{W}/(\mathrm{s\,m^2\,Hz})\end{aligned}\right\} \text{ at } 5480\,\mathring{A}.$$

But the effective wavelength depends slightly on the color of the star and that is one of the reasons broad-band magnitudes are inferior to real energy distributions. Cooler stars produce a longer effective wavelength, and the corresponding flux is lower by $\sim 0.018(B - V)$ dex.

A second factor to consider is the line absorption within the V magnitude band. The starlight in the V band includes spectral lines. The true value of the continuum is larger than *V* implies by the fraction of line absorption, similar to what we encountered with line-absorption corrections for energy distributions. But now the situation is more complicated because the line absorption has to be weighted by the response function of the magnitude (i.e., Figure 1.5). For example, the line absorption for Vega multiplied by the V transmission band is 0.41%. The corresponding values for the Sun (G2 V) and Arcturus (K1.5 III) are 7.0% and 13.5%. Such large values suggest that line absorption is best measured explicitly for each program star.

Similar equations can be set up for any effective wavelength in any magnitude system. If this is done for magnitudes covering a very wide range of effective wavelengths, they can be used to map out a rough energy distribution, e.g., Swihart *et al.* (2017).

Bolometric Corrections

In certain situations we might like to estimate the total flux, i.e., power, from a star based on a partial energy distribution or its magnitude and color index. As we already noted in Equation 10.4, the apparent bolometric flux is defined by

$$F_{\text{bol}} = \int_0^\infty F_\nu \, d\nu. \tag{10.7}$$

In general we can measure F_ν over only a part of the full spectrum. The part that is not measured can be inferred from stars that have been more completely measured or, as a last resort, from models. The factor needed to convert the partial measurement to the bolometric one is the *bolometric correction*.

Bolometric flux is a connecting link between stellar atmospheres and stellar interiors. Namely, the total power output of a star, its luminosity, is $L = 4\pi r^2 F_{\text{bol}}$, where r is the distance to the star. Zero interstellar extinction is assumed here. This link is important when making an interior model for a real star or for assigning the star to an evolutionary track.

A second reason to consider bolometric flux is to attach an effective temperature to a star. As in Equation 1.2, the effective temperature is defined in terms of bolometric flux. Bear in mind that the effective temperature is not a physical temperature, but simply a convenient parameter. In particular, it is not central to studies of stellar photospheres the way S_0 and $T(\tau_0)$ are, although it is used when calculating models that invoke flux constancy.

A direct way to define a bolometric correction factor is to write

$$a_{\text{bol}} = \frac{\int_0^\infty F_\nu \, d\nu}{\int_0^\infty W(\nu) F_\nu \, d\nu}, \tag{10.8}$$

in which $W(v)$ is the window function specifying the response of the magnitude being used, like those in Figure 1.5 or 1.6. If we have a_{bol} for the star, we multiply it with our measured denominator to get the bolometric flux. Such factors have been determined in several investigations, such as van Belle *et al.* (2016), where bolometric flux for very cool stars is calibrated against a bandpass at 2.2 μm. Or Buzzoni *et al.* (2010), who look into the metallicity dependence of bolometric corrections in cool stars.

Bolometric corrections are often expressed in magnitudes, and this can lead to some confusion because the zero point is not consistently defined in the literature. Instead of simply converting Equation 10.8 to magnitude units, the bolometric correction (in magnitudes) is defined as

$$BC = -2.5 \log a_{bol} + \text{constant}$$
$$= m_\lambda - m_{bol}, \qquad (10.9)$$

where m_λ is V or a similar bandpass magnitude and m_{bol} is the bolometric magnitude. Physically speaking the constant should be zero, but as with all magnitude systems, it is set arbitrarily. There is additional confusion concerning the sign of BC. The only safe way to navigate this situation is to difference the star's bolometric correction with the solar value specified *in the same system*. Then the constant is eliminated. In the end, check that the bolometric flux is greater than given by any other magnitude (obviously). In other words, the safe relation is

$$m_{bol} - m_{bol}^\odot = \left(m_\lambda - m_\lambda^\odot\right) - (BC - BC_\odot). \qquad (10.10)$$

The Sun has $m_{bol}^\odot = -26.84$ (based on Prša *et al.* 2016) and $m_\lambda^\odot$ for the V magnitude is $m_V^\odot = V_\odot = -26.75$.

For the stellar interiors connection, it follows that the luminosity is $\log(L/L_\odot) = 2\log(r/r_\odot) - 0.4\left(m_{bol} - m_{bol}^\odot\right)$, where r is the distance to the star and $r_\odot = 1/206265$ of a parsec. And for the stellar atmospheres connection, $\log T_{eff}/T_{eff}^\odot = \log S_0 = -0.5\log(\theta/\theta_\odot) - 0.1\left(m_{bol} - m_{bol}^\odot\right)$, where θ is the angular diameter of the star and the solar diameter is $\theta_\odot = 959.8 \pm 0.1$ arcsec.

A sample of bolometric corrections for the V magnitude is shown in Figure 10.18. The smallest corrections have been arbitrarily set to zero. The solar position is shown for $B - V = 0.653$ (Gray 1992, Ramírez *et al.* 2012), $T_{eff}^\odot = 5772$ K, and has a value of $BC_\odot = -0.09$ for this particular curve. A star's BC can then be read from this curve, and Equation 10.10 used to find $m_{bol} - m_{bol}^\odot$ for the star. But specifically, BC and $BC_\odot$ must be taken from the same curve to ensure a consistent zero point.

We see that the corrections are nearly independent of luminosity class so far as they are measured. The steep rise away from the smallest corrections expresses the energy-distribution shift away from the V wavelengths, toward short wavelengths for hot stars and toward long wavelengths for cool stars.

Bolometric corrections can be constructed relative to any magnitude, e.g., Bessel *et al.* (1998).

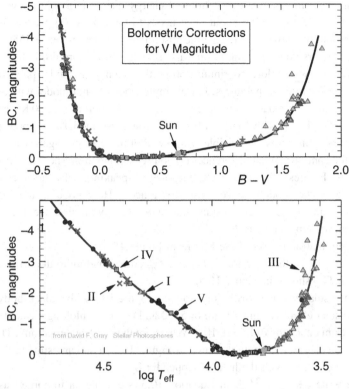

Figure 10.18 Bolometric corrections for the V magnitude are plotted as a function of $B - V$ color index (upper) and log T_{eff} (lower). Symbols denote luminosity class. The smallest corrections for each curve are set to zero. Based on data from Flower (1996).

Our Next Steps

In this chapter we have seen how continua and energy distributions can be used to measure stellar temperature, surface gravity when the Balmer jump is well defined, absolute flux, and luminosity. In the next three chapters, we turn to spectral lines for a wealth of additional information about stellar photospheres. Spectral lines are imprinted on the continuum, but all of the information we need from them comes from the relative flux, i.e., the flux across the profile normalized by the continuum flux at the position of the line (as already done in Figure 10.12). In other words, the energy distribution is removed from the analysis.

Questions and Exercises

1. If one is doing atmospheric extinction corrections, as in Figure 10.2, why not just use Equation 10.2 directly and extrapolate to zero air mass instead of Equation 10.3 as recommended?

2. The people who do absolute calibrations of energy distributions have several difficult problems to overcome. Think about what needs to be done to (a) get the shape of the energy distribution correct and (b) get the absolute flux at the reference 5556 Å position. List all the factors you can think of and discuss why (b) is more difficult than (a).

3. In this exercise, we explore continuum properties. Run your model photosphere program for three or four temperatures. For example, choose a hot model with an effective temperature of ~15 000 K, a middle temperature model at ~7500 K, and a cool model at ~5000 K, all with log g ~ 4.0 or 4.5. For these models, compute the flux at 15–20 wavelengths across the visible window. Plot these values against wavelength to delineate the continua. What differences do you see among these spectra? Compare your results with Figures 10.6 and 10.8. Physically speaking, why does changing the temperature of the models cause these differences? How good is the slope of the Paschen continuum as a temperature indicator? Can you visualize the connection between continuum slope and $B - V$ color index?

4. Estimate the rotation period of the Sun from Figure 10.5.

5. Calculate the size of the variation in $L_\odot$ and $T_{eff}^\odot$ corresponding to the 0.1% solar-cycle variation in $F_{bol}^\odot$ shown in Figure 10.5.

6. Compare the solar energy distribution shown in Figure 10.4 with a black-body spectrum having the solar effective temperature of 5772 K. Do this by plotting a calculated black-body spectrum on a copy of Figure 10.4. Where are the greatest deviations? Do the same for the α Leo energy distribution shown in Figure 10.7. You can estimate the effective temperature of α Leo with the help of Appendix B.

7. When we write $L_\odot = 4\pi r^2 F_{bol}^\odot$ in Equation 10.5, we make an important assumption. Namely, that the $F_{bol}^\odot$ we measure holds for all aspects of view or, equivalently, that the emission is isotropic. But suppose that in magnetic areas of the solar surface, by virtue of magnetic pressure, gas is displaced and the effective opacity is reduced, thus increasing the amount of radiant output. We know that sunspots, a magnetic phenomenon, are localized to latitudes in the range $\sim \pm 5 - 30°$. What does non-isotropic emission imply for $L_\odot$ or other physical parameters? How might we test the solar isotropy? Would limb-darkening measurements along the equator compared to pole-to-pole be useful? Or perhaps a circumpolar satellite probe? Are other stars isotropic emitters? What happens when a star rotates rapidly?

References

Alekseeva, G.A., Arkharov, A.A., Galkin, V.D., *et al.* 1996. *BaltA* **5**, 603.

Allende Prieto, C. & del Burgo, C. 2016. *MNRAS* **455**, 3864.

Auman, H.H., Gillett, F.C., Beichman, C.A., *et al.* 1984. *ApJL* **278**, L23.

Baldwin, J.A. & Stone, R.P.S. 1984. *MNRAS* **206**, 241.

Barbier, D. 1958. *Astrophysik I*, Handbuch der Physik, Vol. **50** (Heidelberg: Springer) S. Flügge, ed., p. 274.

Bean, J.L., Sneden, C., Hauschildt, P.H., Johns-Krull, C.M., & Benedict, G.F. 2006. *ApJ* **652**, 1604.

Beckers, J.M. 1981. *The Sun as a Star*, NASA SP-450 (Washington, DC: CNRS/NASA), S. Jordan, ed., p. 11.

Bessell, M.S. 1967. *ApJL* **149**, L67.

Bessell, M.S. 1999. *PASP* **111**, 765.

Bessell, M.S., Castelli, F., & Plez, B. 1998. *A&A* **333**, 231.

Bohlin, R.C. 2014. *AJ* **147**, 127.

Bohlin, R.C., Hubeny, I., & Rauch, T. 2020. *AJ* **160**, 21

Böhm, T., Holschneider, M., Lignières, F., *et al.* 2015. *A&A* **577**, 64.

Borisov, G.V., Glushneva, I.N., & Shenavrin, V.I. 1998. *A&AT* **17**, 309.

Breger, M. 1976. *ApJS* **32**, 7.

Burlov-Vasiljev, K.A., Gurtovenko, E.A., & Matvejev, Yu.B. 1995. *Solar Phys* **157**, 51.

Buzzoni, A., Patelli, L., Bellazzini, M., Pecci, F.F., & Oliva, E. 2010. *MNRAS* **403**, 1592.

Castelli, F. & Kurucz, R.L. 1994. *A&A* **281**, 817.

Cochran, A.L. 1981. *ApJS* **45**, 83.

Cochran, A.L. & Barnes, T.G., III 1981. *ApJS* **45**, 73.

Code, A.D. 1960. *Stellar Atmospheres* (Chicago: University of Chicago Press), J.L. Greenstein, ed., Chapter 2.

Code, A.D., Davis, J., Bless, R.C., & Hanbury Brown, R. 1976. *ApJ* **203**, 417.

Dreiling, L.A. & Bell, R.A. 1980. *ApJ* **241**, 736.

Engelke, C.W., Price, S.D., & Kraemer, K.E. 2010. *AJ* **140**, 1919.

Falcón-Barroso, J., Sánchez-Blázquez, P., Vazdekis, A., *et al.* 2011. *A&A* **532**, 95.

Flower, P.J. 1996. *ApJ* **469**, 355.

Fontenla, J., White, O.R., Fox, P.A., Avrett, E.H., & Kurucz, R.L. 1999. *ApJ* **518**, 480.

Foukal, P. & Lean, J. 1986. *ApJ* **302**, 826.

Fröhlich, C. & Lean, J. 1998. *Geophys Res Lett* **25**, 4377.

Fröhlich, C. & Lean, J. 2002. *Ast Nach* **323**, 203.

Fröhlich, C. & Lean, J. 2004. *A&AR* **12**, 273.

Glushneva, I.N. & Doroshenko, V.T. 1980. *Soviet Ast* **24**, 421.

Glushneva, I.N. & Ovchinnikov, S.L. 1982. *Soviet Ast* **26**, 548.

Glushneva, I.N., Doroshenko, V.T., Fetisova, T.S., *et al.* 1983. *Trudy Gosud Ast Inst Sternberga* **53**, 50.

Glushneva, I.N., Kharitonov, A.V., Knyazeva, L.N., & Shenavrin, V.I. 1992. *A&AS* **92**, 1.

Gray, D.F. 1967. *AJ* **157**, 92.

Gray, D.F. 1980. *PASP* **92**, 154.

Gray, D.F. 1992. *PASP* **104**, 1035.

Gray, R.O. 1988. *JRASC* **82**, 336.

Gulliver, A.F., Hill, G., & Adelman, S.J. 1994. *ApJ* **429**, 81.

Gutierrez-Moreno, A., Moreno, H., Cortes, G., & Wenderoth, E. 1988. *PASP* **100**, 973.

Hamuy, M., Walker, A.R., Suntzeff, N.B., *et al.* 1992. *PASP* **104**, 533.

Hayes, D.S. 1985. *Calibration of Fundamental Stellar Quantities*, IAU Symposium **111** (Dordrecht: Reidel), D.S. Hayes, L.E. Pasinetti, & A.G.D. Philip, eds., p. 225.

Hayes, D.S. & Latham, D.W. 1975. *ApJ* **197**, 593.

Hayes, D.S., Latham, D.W., & Hayes, S.H. 1975. *ApJ* **197**, 587.

Hickson, P. & Mulrooney, M.K. 1998. *ApJ* **506**, 191.

Hilbig, T., Weber, M., Bramstedt, K., *et al.* 2018. *Solar Phys* **293**, 121.

Hill, G., Gulliver, A.F., & Adelman, S.J. 2010. *ApJ* **712**, 250.

Hinkle, K., Wallace, L., Valenti, J., & Harmer, D. 2000. *Visible and Near Infrared Atlas of the Arcturus Spectrum, 3727–9300 Å* (San Francisco: Astronomical Society of the Pacific).

Jacoby, G.H., Hunter, D.A., & Christian, C.A. 1984. *ApJS* **56**, 257.
Jones, D.H.P. 1966. *IAUS* **24**, 141.
Kharitonov, A.V. & Glushneva, I.N. 1978. *Soviet Ast* **22**, 284.
Kharitonov, A.V., Tereshchenko, V.M., Knyazeva, L.N., & Boiko, P.N. 1980. *Soviet Ast* **24**, 417.
Kiehling, R. 1987. *A&AS* **69**, 465.
Kim, H.-S., Han, I., Valyavin, G., *et al.* 2009. *PASP* **121**, 1065.
Knyazeva, L.N. & Kharitonov, A.V. 1990. *Soviet Ast* **34**, 626.
Kopp, G. & Lean, J.L. 2011. *Geophys Res Lett* **38**, 1706.
Lançon, A. & Wood, P.R. 2000. *A&AS* **146**, 217.
Lockwood, G.W., Tüg, H., & White, N.M. 1992. *ApJ* **390**, 668.
Marsh, K.A., Dowell, C.D., Velusamy, T., Grogan, K., & Beichman, C.A. 2006. *ApJ* **646**, 77.
Massey, P., Strobel, K., Barnes, J.V., & Anderson, E. 1988. *ApJ* **328**, 315.
Meftah, M., Damé, L., Bolsée, D., *et al.* 2018. *A&A* **611**, 1.
Mégessier, C. 1995. *A&A* **296**, 771.
Melbourne, W.G. 1960. *ApJ* **132**, 101.
Monnier, J.D., Che, X., Zhao, M., *et al.* 2012. *ApJL* **761**, 3.
Müller, S., Löhne, T., & Krivov, A.V. 2010. *ApJ* **708**, 1728.
Neckel, H. & Labs, D. 1973. *Problems of Calibration of Absolute Magnitudes and Temperatures of Stars*, IAU Symposium **54** (Dordrecht: Reidel), B. Hauck & B.E. Westerlund, eds., p. 149.
Neckel, H. & Labs, D. 1984. *Solar Phys* **90**, 205.
Newkirk, G., Jr. 1983. *ARAA* **21**, 429.
O'Connell, R.W. 1973. *AJ* **78**, 1074.
Oke, J.B. 1964. *ApJ* **140**, 689.
Oke, J.B. 1990. *AJ* **99**, 1621.
Oke, J.B. & Conti, P.S. 1966. *ApJ* **143**, 134.
Oke, J.B. & Gunn, J.E. 1983. *ApJ* **266**, 713.
Oke, J.B. & Schild, R.E. 1970. *ApJ* **161**, 1015.
Panek, R.J. 1977. *ApJ* **217**, 749.
Pickles, A.J. 1998. *A&A* **92**, 1.
Prša, A., *et al.* 2016. *AJ* **152**, 41.
Ramírez, I., Michel, R., Sefako, R., *et al.* 2012. *ApJ* **752**, 5.
Reiners, A., Mrotzek, N., Lemke, U., Hinrichs, J., & Reinsch, K. 2016. *A&A* **587**, 65.
Ruland, F., Griffin, R., Griffin, R.F., Biehl, D., & Holweger, H. 1980. *A&AS* **42**, 391.
Sánchez-Blázquez, P., Peletier, R.F., Jiménez-Vicente, J., *et al.* 2006. *MNRAS* **371**, 703.
Shulyak, D., Ryabchikova, T., & Kochukhov, O. 2013. *A&A* **551**, 14.
Siqueira Mello, C., Spite, M., Barbuy, B., *et al.* 2013. *A&A* **550**, 122.
Stewart, P.N., Tuthill, P.G., Nicholson, P.D., Sloan, G.C., & Hedman, M.M. 2015. *ApJS* **221**, 30.
Stone, R.P.S. 1977. *ApJ* **218**, 767.
Swihart, S.J., Garcia. V.E., Stassun, K.G., *et al.* 2017. *AJ* **153**, 16.
Talbert, F.D. & Edmonds, F.N., Jr. 1966. *ApJ* **146**, 177.
Takeda, Y., Kawanomoto, S., & Ohishi, N. 2007. *PASJ* **59**, 245.
Taylor, B.J. 1984. *ApJS* **54**, 259.
Thuillier, G., Hersé, M., Labs, D., *et al.* 2003. *Solar Phys* **214**, 1.
Tüg, H. 1980a. *A&A* **82**, 195.
Tüg, H. 1980b. *A&AS* **39**, 67.
Tüg, H. 1982. *A&A* **105**, 395.

Tüg, H., White, N.M., & Lockwood, G.W. 1977. *A&A* **61**, 679.
van Belle, G.T., Creech-Eakman, M.J., & Ruiz-Velasco, A.E. 2016. *AJ* **152**, 16.
Wallace, L., Hinkle, K.H., Livingston, W.C., & Davis, S.P. 2011. *ApJS* **195**, 6.
Willson, R.C. & Hudson, H.S. 1981. *ApJL* **244**, L185.
Willstrop, R.V. 1965. *MNRAS* **69**, 83.
Yan, R., Chen, Y., Lazarz, D., *et al.* 2019. *ApJ* **883**, 175.
Yoon, J., Peterson, D.M., Kurucz, R.L., & Zagarello, R.J. 2010. *ApJ* **708**, 71.

11

The Line Absorption Coefficient

Spectral lines are a trove of valuable information. They tell us about the line-of-sight velocities of stars and about the basic stellar parameters of temperature, surface gravity, chemical composition, and rotation rates. They give us information about photospheric structure, including the temperature distribution, granulation velocities, magnetic fields, and starspots. With so much information stuffed into them, it is not surprising that on the flip side lines are sensitive to many physical parameters. This chapter and the next two move through the apparent maze of physical situations and muster the behavior of spectral lines into useful tools for analysis.

As we did with the continuous spectrum, we start with the absorption coefficient. The main processes for line absorption are (1) natural atomic absorption, (2) pressure broadening, of which there are several, and (3) the ever-present thermal Doppler broadening. The individual shape from each process is convolved with the next to eventually give the final overall line absorption coefficient. One of the remarkable results is that the natural atomic absorption and most the significant pressure broadenings have the same shape, namely the dispersion profile (with the notable exception of hydrogen-line broadening). This leads to a tremendous simplification, as we shall see, since the convolution of dispersion profiles is a new dispersion profile with a half width equal to the sum of the individual half widths of the contributing processes. The thermal broadening, on the other hand, reflects the Maxwellian velocity distribution of the absorbing atoms and ions via the Doppler effect, so its wavelength shape is a Gaussian. The final absorption coefficient is therefore a convolution of a dispersion profile with a Gaussian profile.

The atomic absorption coefficients are denoted by α cm^2/absorber. The mass absorption coefficient for lines is ℓ_v cm^2/g and parallels the continuous absorption coefficient κ_v cm^2/g. But ℓ_v varies rapidly across the width of the spectral line, typically ~1 Å, whereas κ_v varies slowly over tens of angstroms (with the exception of hydrogen ionization edges). The spectrum is again calculated from Equation 7.21, which in this case takes on the form

$$\mathfrak{F}_v = 2\pi \int_0^\infty S(\tau_0) E_2(\tau_v) \frac{\ell_v + \kappa_v}{\kappa_0} \, d\tau_0.$$

We start with the basic line absorption process corresponding to the transition between two atomic levels.

210

Natural Atomic Absorption

The simplest classical model for the interaction of light with an atom is a plane electromagnetic wave interacting with a dipole. This interaction results in absorption or attenuation of the wave. Imagine an incident electromagnetic wave, E, traveling in the x direction. The wave equation for E is

$$\frac{\partial^2 E}{\partial t^2} = v^2 \frac{\partial^2 E}{\partial x^2},$$

where v is the wave velocity. Any function of the form $E = f(x - vt)$ is a solution.

We can allow E to remain general by considering it to be a Fourier composition of sinusoidals. Because the electric field can be linearly decomposed, the manipulations performed on any one Fourier component can be performed on the others as well – the final answer being the sum of the manipulated components. This one Fourier component we write as

$$E = E_0 e^{i2\pi(x - vt)/\lambda} = E_0 e^{i\omega(x/v - t)}, \tag{11.1}$$

where $\omega = 2\pi v/\lambda$ is the angular frequency.

For light, Maxwell's equations tell us (e.g., Stone 1963) that the wave velocity is related to the electrical and magnetic properties of the medium through which it is passing by

$$v = c \left(\frac{\mu_0 \epsilon_0}{\mu \epsilon} \right)^{1/2}.$$

Here ϵ and μ are the permittivity and magnetic permeability in the medium; ϵ_0 and μ_0 are the same for free space. Further, since we are dealing with gases and they have negligible magnetic permeability, $\mu = \mu_0$. We can evaluate ϵ/ϵ_0 by going back to the basic reason why ϵ differs from ϵ_0, which is the charge separation of the dipoles (atoms in the gas) induced by the imposed field, E. The total electrical field is the sum of E and the field of the separated charges. The ratio of this total field to the field of the wave is

$$\frac{\epsilon}{\epsilon_0} = \frac{E + 4\pi Nqz}{E} = 1 + \frac{4\pi Nqz}{E}. \tag{11.2}$$

The number of dipoles per unit volume is N, the size of the dipole charge is q, and their induced separation is z. In most stellar spectra, where we deal with electronic transitions, we replace q with the electron charge, $e = 4.0803 \times 10^{-10}$ esu (in mks units, divide the last term by $4\pi\epsilon_0$ and use $e = 1.602 \times 10^{-19}$ coulombs).

The displacement z can be obtained by considering a single oscillator at $x = 0$ as it is forced into oscillation by the imposed field E. Adding the acceleration, damping, and restoring force terms in that order, we get the usual harmonic oscillator equation with the driving term on the right,

$$\frac{d^2 z}{dt^2} + \gamma \frac{dz}{dt} + \omega_0^2 z = \frac{e}{m_e} E_0 e^{i\omega t}, \tag{11.3}$$

where $m_e = 9.1094 \times 10^{-28}$ grams is the mass of the electron. The damping constant, γ, is of some importance, and is discussed in detail below. The driving wave on the right-hand side of the equation is the same E as in Equation 11.1, but with $x = 0$. The solution is $z = z_0 e^{i\omega t}$. Then $dz/dt = i\omega z$ and $d^2 z/dt^2 = -\omega^2 z$, all of which, when place back into Equation 11.3, gives

$$-\omega^2 z + i\gamma \omega z + \omega_0^2 z = \frac{e}{m} E_0 e^{i\omega t}$$

or

$$z = \frac{e}{m_e} \frac{E_0 e^{i\omega t}}{\omega_0^2 - \omega^2 + i\gamma\omega} = \frac{e}{m_e} \frac{E}{\omega_0^2 - \omega^2 + i\gamma\omega}.$$

The amplitude z reaches its largest value at $\omega = \omega_0$, and so ω_0 is called the resonant frequency. This is the center of the spectral line. For Equation 11.2, where we need z/E, we obtain

$$\frac{\epsilon}{\epsilon_0} = 1 + \frac{4\pi N e^2}{m_e} \frac{1}{\omega_0^2 - \omega^2 + i\gamma\omega}. \tag{11.4}$$

If we further realize that in a gas $\epsilon \approx \epsilon_0$, then we can use the approximation $(1 + \delta)^{1/2} \approx 1 + \frac{1}{2}\delta$ for $\delta \ll 1$, and write

$$\frac{c}{v} = \left(\frac{\epsilon}{\epsilon_0}\right)^{1/2} \approx 1 + \frac{1}{2}\frac{\epsilon}{\epsilon_0} = 1 + \frac{1}{2}\frac{4\pi N e^2}{m_e} \frac{1}{\omega_0^2 - \omega^2 + i\gamma\omega}$$

$$= 1 + \frac{2\pi N e^2}{m_e} \left[\frac{\omega_0^2 - \omega^2}{\left(\omega_0^2 - \omega^2\right) + \gamma^2\omega^2} - i\frac{\gamma\omega}{\left(\omega_0^2 - \omega^2\right) + \gamma^2\omega^2}\right].$$

We recognize the ratio c/v as the index of refraction. In this case, it has an imaginary component, so we write it as $c/v = n + i\boldsymbol{k}$. When $i\boldsymbol{k}$ is combined with the $i\omega x/v$ term in Equation 11.1, a real term is produced, corresponding to exponential extinction, $e^{-\boldsymbol{k}\omega x/c}$. Since the intensity or power of the wave is proportional to EE^*, where the asterisk means complex conjugate, the light extinction can be expressed as

$$I = I_0 e^{-2\boldsymbol{k}\omega x/c} = I_0 e^{-\ell_v \rho x},$$

where standard notation has been used in the last step in which ρ is the density and ℓ_v is the mass absorption coefficient for the line. Upon introducing the full expression for $\boldsymbol{k}$, we have

$$\ell_v \rho = \frac{4\pi N e^2}{m_e c} \frac{\gamma\omega^2}{\left(\omega_0^2 - \omega^2\right)^2 + \gamma^2\omega^2} \quad \text{cm}^2/\text{g}. \tag{11.5}$$

Since this function peaks sharply, giving non-zero values only when $\omega \approx \omega_0$, we write

$$\omega_0^2 - \omega^2 = (\omega_0 - \omega)(\omega_0 + \omega) \approx (\omega_0 - \omega)\,2\omega.$$

In this way, the basic form of the absorption coefficient becomes

$$\ell_\nu \rho = N \frac{2\pi e^2}{m_e c} \frac{\gamma/2}{(\omega - \omega_0)^2 + (\gamma/2)^2}. \tag{11.6}$$

We recognize this as a dispersion profile (Equation 2.8 with $\beta = \gamma/2$); also called a damping profile, a Lorentzian profile, a Cauchy curve, and, by some, the Witch of Agnesi (apparently a mistranslation from Italian, Silver 2017).

If we now consider the absorption coefficient per atom for this natural absorption, α_{nat}, we write for Equation 11.6 $\ell_\nu \rho = N\alpha_{nat}$ and then

$$\alpha_{nat} = \frac{2\pi e^2}{m_e c} \frac{\gamma/2}{(\omega - \omega_0)^2 + (\gamma/2)^2}$$

$$= \frac{e^2}{m_e c} \frac{\gamma/4\pi}{(\nu - \nu_0)^2 + (\gamma/4\pi)^2}$$

$$= \frac{e^2}{m_e c} \frac{\lambda^2}{c} \frac{\gamma\lambda^2/4\pi c}{(\lambda - \lambda_0)^2 + (\gamma\lambda^2/4\pi c)^2} \quad cm^2/absorber. \tag{11.7}$$

The distance from the center of the spectral lines is $\omega - \omega_0$, $\nu - \nu_0$, or $\lambda - \lambda_0$ according to the scale selected. Equation 11.7 is the basic result; the shape of the absorption coefficient for natural atomic broadening is a dispersion profile with its width determined by a damping constant γ. This damping constant is called the "natural" damping, alias "radiation damping," and we will evaluate it shortly. But first we consider a quantum mechanical factor of some importance.

The Oscillator Strength

Direct integration of the dispersion profile in Equation 11.7, or a comparison with the unit area form in Equation 2.8, shows that

$$\int_0^\infty \alpha_{nat}\, d\nu = \frac{\pi e^2}{m_e c}. \tag{11.8}$$

This is the energy per second per atom per square radian absorbed by the total line from the unit I_ν beam. The total energy taken from an I_λ beam is

$$\int_0^\infty \alpha_{nat} d\lambda = \frac{\pi e^2}{m_e c} \frac{\lambda^2}{c}. \tag{11.9}$$

Actual measurements of this total absorption are usually smaller than indicated by Equations 11.8 and 11.9, and by an amount that differs from one spectral line to another. A quantum mechanical treatment leads one to introduce a quantity called the *oscillator strength*, f, such that

$$\int_0^\infty \alpha_{\text{nat}} \, d\nu = \frac{\pi e^2}{m_e c} f. \tag{11.10}$$

The oscillator strength, also called simply the f value, is different for each atomic level and is related to the atomic transition probability B_{lu} from the lower to the upper level (defined near the end of Chapter 5) as follows. The integral in Equation 11.10 must be equivalent to the probability of absorption times the unit energy for the transition, i.e.,

$$\int_0^\infty \alpha_{\text{nat}} \, d\nu = B_{lu} h\nu,$$

which means that

$$f = \frac{m_e c}{\pi e^2} B_{lu} h\nu = 7.484 \times 10^{-7} \frac{B_{lu}}{\lambda} \tag{11.11}$$

for λ in angstroms. Using Equation 6.8, we also see that

$$f = \frac{m_e c^3}{2\pi e^2 \nu^2} \frac{g_u}{g_l} A_{ul} = 1.884 \times 10^{-15} \lambda^2 \frac{g_u}{g_l} A_{ul}, \tag{11.12}$$

where g_u and g_l are the statistical weights of the upper and lower levels, and λ is again in angstroms. Recall in passing that the Einstein probability coefficients defined in Chapter 5 are per unit solid angle and not over the full 4π square radians of a unit sphere, as is sometimes done.

It is also possible to define an f value for emission. The two f values are related by $g_u f_{\text{em}} = g_l f_{\text{abs}}$. Tabulations of measured f values give gf so that no ambiguity arises. Most f values are determined from empirical laboratory measurements, although in less complicated cases calculations can be done. For example, the expression for hydrogen is

$$f = \frac{2^5}{3^{3/2}\pi} \frac{g_{bb}}{l^5 u^3} \left(\frac{1}{l^2} - \frac{1}{u^2} \right)^{-3}, \tag{11.13}$$

where u and l are the quantum numbers of the upper and lower levels as usual, and g_{bb} is the bound–bound Gaunt factor tabulated by Baker and Menzel (1938). The g_{bb} values for the first eight Balmer lines are given to better than 0.35% by $g_{bb} = 0.869 - 3.00/u^3$. Numerically $f = 0.6407$ for H_α, 0.1193 for H_β, 0.0447 for H_γ, and so on. (See also Table 11.4 later in this chapter.)

The Damping Constant for Natural Broadening

The damping constant appearing in Equation 11.7 can be calculated from the classical dipole emission theory by taking a time average of the acceleration (Weisskopf 1932a, Menzel 1961). An equation of the form

$$\frac{dW}{dt} = -\frac{2}{3}\frac{e^2\omega^2}{m_e c^3}\, W = -\gamma_{\text{nat}} W \tag{11.14}$$

results. The solution is $W = W_0 e^{\gamma_{\text{nat}} t}$, where $\gamma_{\text{nat}} = 2e^2\omega^2/(3m_e c^3) = 0.2223 \times 10^{16}/\lambda^2$ for λ in angstroms for both systems of units. The half-half width of α_{nat} is $\gamma_{\text{nat}}/2, \gamma_{\text{nat}}/4\pi$, or $\gamma_{\text{nat}}\lambda^2/(4\pi c)$, depending on which of the forms in Equation 11.7 is used. In the λ form, the half-half width is 0.590×10^{-4} Å for all lines. This classical damping constant is usually smaller by an order of magnitude compared to the observations, and again a quantum-mechanical interpretation must be made to reconcile the difference.

A simple phenomenological way to introduce the quantum aspects is to view the emitted energy W in Equation 11.14 as quantized and write $W = N_u h\nu$, where N_u is the population of the upper level for the transition. Then Equation 11.14 can be rewritten $dN_u/dt = -\gamma_{\text{nat}} N_u$. But we already know from Chapters 5 and 6 that $dN_u/dt = -4\pi A_{ul} N_u$ for each transition starting from the level u, so the total downward rate due to spontaneous emission is

$$\frac{dN_u}{dt} = \sum_{u>l}\frac{dN_{ul}}{dt} = -4\pi N_u \sum_{u>l} A_{ul}.$$

Comparing these equations, we see that

$$\gamma_u = 4\pi \sum_{u>l} A_{ul}. \tag{11.15}$$

The physical meaning of this γ_u can be found by viewing it in terms of the lifetime of the level. Since A_{ul} has units of reciprocal seconds and specifies the probability that the electron will come down from the upper level u in one second, the quantity $\Delta t = 1/(4\pi \sum A_{ul})$ corresponds to the probable time interval the electron will, if left alone, stay in the upper level.

This concept in turn can be coupled to the Heisenberg uncertainty principle. Define the uncertainty of the electron energy in the level u to be ΔW. Then $\Delta W_u \Delta t \gtrsim h/2\pi$ or, replacing Δt with the sum of the Einstein coefficients, $\Delta W_u \gtrsim 2h \sum A_{ul}$. In other words, the energy level u has an intrinsic energy width set by nature at ΔW_u. The lower level shows a similar broadening, ΔW_l, and has a corresponding γ_l. Consider the absorption process depicted in Figure 11.1. The electron leaves the level l from the energy range ΔW_l. The probability of the electron being at some point in this energy band is α_{nat} from Equation 11.7 with $\gamma = \gamma_l$. By a similar token, the probability of the electron ending up at some energy within ΔW_u in the final state is α_{nat} with $\gamma = \gamma_u$. Since any starting energy within the lower level can correspond to the full range in energy in the upper level, we see that the full absorption coefficient is a *convolution* between α_{nat} for the lower state and α_{nat} for the upper state. In Chapter 2 we saw that a convolution between dispersion profiles results in a new dispersion profile having a larger γ given by $\gamma_{\text{nat}} = \gamma_u + \gamma_l$.

But even with the quantum-mechanical version of natural damping, observed widths of stellar spectral lines are always larger than given by the natural atomic damping, sometimes much larger. We go on to discuss some of the other broadening mechanisms.

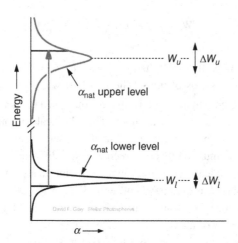

Figure 11.1 The two α_{nat} dispersion profiles show the characteristic widths of the energy levels, ΔW_l for the lower level and ΔW_u for the upper level. Note the break in the ordinate; the energy separation between levels, $W_u - W_l$, is much larger than the level widths. An absorption transition starts somewhere in the lower level with a probability specified by α_{nat} for the lower level and ends somewhere in the upper level with a probability specified by α_{nat} for the upper level.

Pressure Broadening

The term pressure broadening implies a collisional interaction between the atoms that are absorbing the light and other perturbing particles. The other particles can be ions, electrons, or atoms of the same element as the absorbers or another type, or in cool stars, they might be molecules. Early thoughts on pressure broadening are summarized by Michelson (1895). Collisions alter the energies of the atomic levels. The distortion is a function of the separation, R, between the absorber and the perturbing particle, and we expect the upper level to be more strongly altered than the lower level. Figure 11.2 shows a schematic presentation of this situation, where the mean energy of the level is plotted against the distance to the perturber. Transitions near position 1 have nearly the undisturbed or normal energy. As R becomes smaller, the distortions become larger, as shown at positions 2 and 3. Statistically, most encounters are at the larger R values, as suggested by the boxed area.

The altered energy of the transition induced by the collision, i.e., the difference between the curves in Figure 11.2, is often approximated by a power law of the form

$$\Delta W = \mathrm{constant}/R^n \quad \text{or} \quad \Delta \nu = C_n/R^n, \tag{11.16}$$

where n depends on the type of interaction. For example, the dipole coupling force between pairs of atoms of the same kind is proportional to R^{-4}, making ΔW proportional to R^{-3}. Dipole coupling perturbations are called resonance broadening. As a second example, we cite the result of London (1930a, 1930b) who found theoretically that when the perturber is of a different species, the perturbation energy is proportional to R^{-6}, so $n = 6$, called the van der Waals interaction. Table 11.1 gives a summary of the types of interactions that are most important in stars. The interaction constant, C_n, must be measured or calculated for each transition and type of interaction. It is known for relatively few lines.

Table 11.1 *Main types of pressure broadening*

n	Name	Lines affected	Perturber
2	Linear Stark	Hydrogen	Protons and electrons
4	Quadratic Stark	Most lines/hot stars	Ions and electrons
6	Van der Waals	Most lines/cool stars	Neutral hydrogen

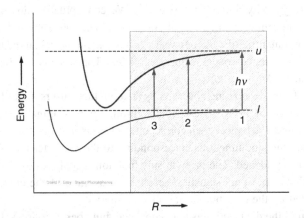

Figure 11.2 Caricature depiction of perturbed energy levels. The unperturbed energies are shown by the dashed lines. Values below the dashed lines are attractive, above are repulsive. The upper level, *u*, is perturbed more than the lower level, *l*. Most transitions occur at the larger *R* values. The difference in perturbed energy in the boxed region is approximated by Equation 11.16.

In most of what we consider, it is assumed that no transitions result from the collisions. This assumption is called the adiabatic assumption. Baranger (1958a, 1958b) has given a generalized theory in which this limitation is not imposed. A review of the progress made in understanding pressure broadening has been given by Van Regemorter (1965). Additional material and references can be obtained from Fuhr *et al.* (1973), Van Regemorter (1973), Griem (1974, 1997), Lesage *et al.* (1999), and other papers referenced below.

A note on terminology: damping from collisions clearly depends on the density and temperature of the gas, and in a stellar photosphere the density varies by orders of magnitude, top to bottom. So even though we may call the pressure damping parameter, γ, the "damping constant," it is anything but constant.

The Impact Approximation

The impact approximation is one of the most useful and versatile tools for dealing with pressure broadening. It is predicated on the abruptness of the collision, hence the name "impact." Abrupt collisions happen when velocities are high. With the relatively high temperatures in stellar photospheres, hydrogen and helium atoms and electrons move at high velocities. Couple this with the modest densities, and the duration of the collisions is small compared to the time between collisions. The impact approximation is also referred to as the interruption or phase-shift approximation.

A simple form of the impact treatment was proposed by Lorentz (1906) in which he assumed the electromagnetic wave was terminated by the impact. This simple treatment was refined under the hands of Lenz (1924), Kallmann and London (1929), and Weisskopf (1932b). In place of a termination of the electromagnetic wave, they showed how a phase change in the wave could occur resulting in broadening. Suppose we have a photon with a time duration Δt, typically about 10^{-9} seconds. We conceptualize this one photon as a product of an infinite sinusoid and a box of duration Δt. The spectrum of the electric field E is then the Fourier transform of a box, Δt sinc $\pi \Delta t (\nu - \nu_0)$, centered on the frequency of the sinusoid, ν_0, and of characteristic width $\Delta \nu = 1/\Delta t$. That means the light, EE^*, has a spectral line of shape Δt^2 sinc^2 $\pi \Delta t (\nu - \nu_0)$.

Now imagine this wave is being absorbed by an atom that suffers multiple impacts. Each impact alters the phase of the wave, essentially cutting up the original wave into pieces, or it terminates the wave. Each piece then corresponds to a box that is a fraction of the original box width, Δt, and the sinc functions corresponding to these shorter boxes are wider, i.e., the spectral line is broadened. The original sinc function is replaced by the sum of several wider sinc functions. The final statistical result is an average of the corresponding sinc^2 functions weighted by the distribution of collision intervals.

The derivation of the distribution is a bit contorted, but goes as follows. Let $P(\Delta t)$ be the probability that there is a collision in the time interval Δt, and $p(\Delta t)$ the probability there is not. Then $P(\Delta t) + p(\Delta t) = 1$ or $p(\Delta t) = 1 - P(\Delta t)$. Now in an increment of time dt_1 the probability of a collision is dt_1/t_0, where t_0 is the average time between collisions. Then the probability of no collision in dt_1 is $p(dt_1) = 1 - dt_1/t_0$. The rules of probability tell us that $p(\Delta t + dt_1) = p(\Delta t)p(dt_1) = p(\Delta t)(1 - dt_1/t_0) = p(\Delta t) - p(\Delta t)dt_1/t_0$. Since the second term is small compared to the first, we can use a Taylor expansion to write $p(\Delta t + dt_1) = p(\Delta t) + (dp/dt)dt_1 + \cdots$. Comparing these last two expressions, we are led to equate dp/dt with $-p(\Delta t)/t_0$ or $dp/p(\Delta t) = -dt/t_0$. Integrating this gives $p(\Delta t) = e^{-\Delta t/t_0}$, where the integration constant is determined by the condition that the maximum of $p(\Delta t)$ is unity. That means $P(\Delta t) = 1 - p(\Delta t) = 1 - e^{-\Delta t/t_0}$, and so $dP(\Delta t) = (d\Delta t/t_0)e^{-\Delta t/t_0}$. This is the probability or weight assigned to each of the sinc^2 functions according to its duration Δt. The statistical shape of the profile is then the weighted average, proportional to

$$\int_0^\infty \Delta t^2 \left(\frac{\sin \pi \Delta t (\nu - \nu_0)}{\pi \Delta t (\nu - \nu_0)} \right)^2 e^{-\Delta t/t_0} \frac{d\Delta t}{t_0}. \tag{11.17}$$

Our mathematical colleagues have figured out how to evaluate this integral (see Petit Bois 1961, for example). The result is

$$\alpha_{\text{impact}} = \frac{\text{constant}}{4\pi^2 (\nu - \nu_0)^2 + (1/t_0)^2}.$$

We recognize this as a dispersion profile, and recast it slightly to look more like Equation 11.7, by taking $\gamma_n = 2/t_0$,

$$\alpha_{\text{impact}} = \text{constant} \frac{\gamma_n/4\pi}{(v - v_0)^2 + (\gamma_n/4\pi)^2}.$$ (11.18)

The damping constant is given a subscript in anticipation of different possible kinds of impacts, as per Equation 11.16, but all of them have the dispersion profile.

In order to use this result in a line profile calculation, we need to evaluate $\gamma_n = 2/t_0$. And to restate a fundamental point, unlike natural damping, impact damping depends on the pressure and temperature, and so is a function of depth in the photosphere. Finding these dependences is next.

Theoretical Evaluation of Impact Damping

We measure the strength of an impact by the size of the phase shift it produces. If we find the phase is shifted by, say, one radian or more, we count it, but if the shift is less, we neglect it. Clearly this is an order of magnitude estimate, for there is no reason to choose one radian versus, say, π radians. The chosen value simply represents what is considered to be a significant phase shift. The frequency change is given by Equation 11.16. The cumulative effect of this change in frequency is the phase shift we seek, namely,

$$\phi = 2\pi \int_0^\infty \Delta v \, dt = 2\pi \int_0^\infty C_n R^{-n} \, dt.$$ (11.19)

By choosing these integration limits, we are temporarily excluding other collisions. The perturber is assumed to move past the atom on a straight line trajectory and maintains a constant velocity, v, as shown in Figure 11.3. The impact parameter, ρ, and the angle, θ, are related by $\rho = R \cos \theta$, so that Equation 11.19 becomes

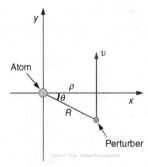

Figure 11.3 The perturber with a relative velocity of v passes the atom at an impact distance of ρ.

$$\phi = 2\pi C_n \int\limits_0^\infty \frac{\cos^n \theta}{\rho^n} \, dt.$$

Using the fact that $v = dy/dt = (\rho/\cos^2 \theta) \, d\theta/dt$ gives $dt = (\rho/v) \, d\theta/\cos^2 \theta$, and so

$$\phi = \frac{2\pi C_n}{v\rho^{n-1}} \int\limits_{-\pi/2}^{\pi/2} \cos^{n-2} \theta \, d\theta. \tag{11.20}$$

The value of the integral is given in Table 11.2 for several values of n.

Table 11.2 *Values of the integral*

n	$\int_{-\pi/2}^{\pi/2} \cos^{n-2} \theta \, d\theta$
2	π
3	2
4	$\pi/2$
5	4/3
6	$3\pi/8$

At this point, we use our definition of a significant phase shift and set $\phi = 1$ radian, thereby defining a limiting impact parameter

$$\rho_1 = \left(\frac{2\pi C_n}{v} \int\limits_{-\pi/2}^{\pi/2} \cos^{n-2} \theta \, d\theta \right)^{1/(n-1)}. \tag{11.21}$$

The subscript 1 reminds us of the 1 radian choice for ϕ. We count only those collisions that have $\phi \geq 1$ radian, or $\rho < \rho_1$. The number of such collisions is given by $\pi\rho_1^2 vN\Delta t_1$, where $\pi\rho_1^2 v$ is the volume swept out per second, Δt_1 is the time interval over which we count the collisions, and N is the number of perturbers per unit volume. If we set $\Delta t_1 = \Delta t_0$, the average time between collisions (as in Equation 11.17), then on average the number of collisions must be one, and $\pi\rho_1^2 vN\Delta t_0 = 1$. In this way

$$\gamma_n = \frac{2}{\Delta t_0} = 2\pi\rho_1^2 vN \tag{11.22}$$

with ρ_1 given by Equation 11.21. The average relative velocity v for the atom of mass m_A and perturber of mass m_p is, according to Equation 1.16,

$$v = \left[\frac{8kT}{\pi} \left(\frac{1}{m_A} + \frac{1}{m_p} \right) \right]^{1/2}. \tag{11.23}$$

The number density of perturbers, N, for Equation 11.22 is given by the model photosphere.

A more general and accurate statement of the damping constant is

$$\gamma_n = N \int_{-\infty}^{\infty} v f(v)\, \sigma_n(v)\, dv, \qquad (11.24)$$

where σ_n is the broadening cross section and $f(v)$ is the velocity distribution of the perturbers, a Gaussian in most cases. The challenge is then to measure experimentally or compute theoretically σ_n. See Peach (1984) and Anstee and O'Mara (1995).

Numerical Values for Impacts with Charged Perturbers

We now obtain some numerical expressions for the quadratic Stark effect. The quadratic Stark effect ($n = 4$) arises from perturbations by charged particles. In a stellar atmosphere these are potentially electrons and ions. From Table 11.2 and Equation 11.21, we have $\rho_1 = (\pi^2 C_4/v)^{1/3}$. Then with N perturbers per unit volume, Equation 11.22 gives us

$$\gamma_4 = 2\pi v N \left(\frac{\pi^2 C_4}{v}\right)^{2/3} \approx 29\, v^{1/3} C_4^{2/3}\, N.$$

A slightly more complicated theory (e.g., Lindholm 1945, Foley 1946, Unsöld 1955) gives

$$\gamma_4 \approx 39\, v^{1/3}\, C_4^{2/3}\, N, \qquad (11.25)$$

which, considering the approximate nature of the calculation, is the same as the previous equation. Nevertheless, to be numerically consistent with material in the literature, we adopt Equation 11.25. If we now take this equation and with the velocities according to Equation 11.23, we could write for the combined perturbations of electrons and ions

$$\gamma_4 \approx 39 C_4^{2/3} N_e \left[\frac{8kT}{\pi}\left(\frac{1}{m_A} + \frac{1}{m_e}\right)\right]^{1/6} + 39 C_4^{2/3} N_{\text{ion}} \left[\frac{8kT}{\pi}\left(\frac{1}{m_A} + \frac{1}{m_{\text{ion}}}\right)\right]^{1/6}. \qquad (11.26)$$

But the masses of the atoms and the ions, m_A and m_{ion}, are so large compared to the electron mass, m_e, that the expression reduces to

$$\gamma_4 \approx 39 C_4^{2/3} N_e \left[\frac{8kT}{\pi m_e}\right]^{1/6}.$$

Expressing the electron number density, N_e, in terms of electron pressure and temperature, the expression becomes

$$\log \gamma_4 \approx 19 + \frac{2}{3}\log C_4 + \log P_e - \frac{5}{6}\log T, \qquad (11.27)$$

where P_e is in dyne/cm^2. We must remember that the constant 19 is quite uncertain.

Values of the interaction constant C_4 are known from laboratory measurements for some lines. Examples are cited in Table 11.3, and additional values can be found in

Table 11.3 *Some Stark interaction constants*

Line	Wavelength	$\log C_4$
Na I D_2	5890 Å	−15.17
Na I D_1	5896 Å	−15.33
Mg I b_2	5173 Å	−14.52
Mg I b_1	5184 Å	−14.52
Mg I	5528 Å	−13.12

Landolt-Börnstein (1950), Griem (1964), Allen (1973), Moity *et al.* (1975), González *et al.* (2002), and Milosavljevic and Djenize (2003). Calculation of C_4 can also be done. Hunger (1960), Mugglestone and O'Mara (1966), Freudenstein and Cooper (1978), Dimitrijevic and Konjevic (1986), Lesage *et al.* (1990), Popovic *et al.* (1999, 2001), and Dimitrijevic *et al.* (2003) give examples of such computations.

Numerical Values for Impacts with Neutral Perturbers

Collisions with neutral perturbers is next. Neutral hydrogen ($T \lesssim 10\,000$ K) is by far the dominant perturber. The classical van der Waals formulation with $n = 6$ is still widely used, although there are recent improvements.

As with the quadratic Stark case, start with Equation 11.22, using ρ_1 given by Equation 11.21 and Table 11.2 with $n = 6$. The result is

$$\gamma_6 = 17 \, v_{\mathrm{H}}^{3/5} \, C_6^{2/5} \, N, \tag{11.28}$$

where, as we did with Equation 11.25, the constant has been increased from 14 to 17. The velocity of hydrogen atoms is v_{H}, given by Equation 11.23, but since hydrogen is so much lighter than most other elements, v_{H} is essentially given by $[8kT/(\pi m_{\mathrm{H}})]^{1/2}$ with m_{H} being the mass of the hydrogen atom. Although the ionization of hydrogen can be taken into account using Equation 1.21, assume here that it is neutral. Further, since hydrogen dominates the chemical composition, the number density of perturbers is adequately given by the gas pressure, $N = P_g/kT$. The result is

$$\log \gamma_6 \approx 20 + \frac{2}{5}\log C_6 + \log P_g - \frac{7}{10}\log T \tag{11.29}$$

for P_g in dyne/cm^2. Unsöld (1955) showed how to evaluate the perturbation energy in an approximate way. The difference between the energies for the two levels of the transition leads to the expression

$$C_6 = 0.3 \times 10^{-30} \left[\frac{1}{(I - \chi_n - \chi_\lambda)^2} - \frac{1}{(I - \chi_n)^2} \right], \tag{11.30}$$

where I is the ionization potential and χ_n is the excitation potential of the lower level in electronvolts. The symbol $\chi_\lambda = h\nu = 1.2398 \times 10^4/\lambda$ electronvolts with λ in angstroms, and is the energy of a photon in the line.

An illustration of the van der Waals and quadratic Stark damping is given for the Na D_2 line in Figure 11.4. Here log C_6 has a value of -31.7. The damping constants are shown as a function of optical depth in a solar-type model. The Stark broadening is much smaller than the van der Waals broadening over most of the photosphere. The increases with depth reflect the depth dependence of P_e (Stark) and P_g (van der Waals) seen in Figure 9.14. This tells us that the dispersion component of the absorption coefficient gets wider with depth.

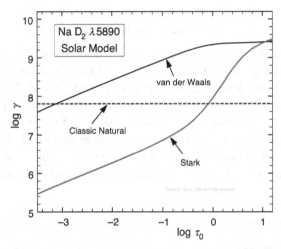

Figure 11.4 Damping constants for the sodium D_2 line are shown as a function of depth in a solar model. Based on Equations 11.27 and 11.29. Although the natural or radiation damping constant is indeed constant, Stark and van der Waals damping increase substantially with depth.

Equations such as 11.27 and 11.29 have been used in a great many line profile calculations. In some cases, the damping constant, notably the γ_6 component, is found to be too small and an empirical "enhancement factor" is used (e.g., Ayres 1977). Numerous papers suggesting improvements have been made, e.g., Blackwell *et al.* (1972) and more recently by Anstee and O'Mara (1991, 1995). Some applications are discussed by Ryan (1998).

In the theory of Anstee and O'Mara (1995), the σ_6 cross sections in Equation 11.24 are calculated and tabulated as a function of *effective* principal quantum numbers, n^*, for the upper and lower levels of neutral species. Barklem *et al.* (1998a) give examples of how these are computed. Armed with these two quantum numbers, one enters their tables to obtain $\sigma_6(\nu_0)$ and p to use in the following expression,

$$\gamma_6 = N_H \frac{4^{p/2}}{\pi} \Gamma\left(\frac{4-p}{2}\right) \left(\frac{\nu}{\nu_0}\right)^{-p} \nu\, \sigma_6(\nu_0),$$

where Γ is the gamma function, $v_0 = 10\,\mathrm{km/s}$, and v is given by Equation 11.23. Additional tables for other atomic configurations are given by Barklem and O'Mara (1997) and Barklem *et al.* (1998b). This formulation reproduces the broadening seen in the wings of stronger solar lines for at least a few significant cases. It is relatively easy to use and can be applied to many lines. More accurate calculations can be done, but on a tedious line-by-line basis. Barklem and O'Mara (1998, 2000) show line-by-line calculations for selected lines of several ions. Tabulated values of γ_6 can also be found in the Vienna Atomic Line Database (VALD: vald.astro.uu.se, Barklem *et al.* 2000a, Ryabchikova *et al.* 2018).

Hydrogen Line Broadening

Hydrogen lines are different from other spectral lines because the atomic structure of hydrogen is particularly subject to the linear Stark effect. The corresponding wavelength shifts can be enormously larger than is typical of "normal" spectral lines, as illustrated in Figure 11.5. It was Struve (1929) who gave convincing arguments that the great widths of the hydrogen lines in early-type stars are due to the linear Stark effect. A decade earlier, Holtsmark (1919) began the studies of theoretical distributions of ions in a plasma and the subsequent broadening caused in removing the degeneracy of the energy levels. The splitting of the energy levels can be expressed as the wavelength shift of the spectral components,

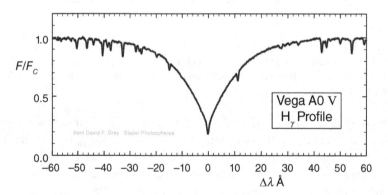

Figure 11.5 The hydrogen lines in hot stars like Vega are about two orders of magnitude wider than typical metal lines. Such lines are mostly "wing" with a small thermal core. Taken at Observatoire de Haute-Provence, resolving power ~16 000, courtesy of F. Royer.

$$\Delta\lambda_j = c_j E, \qquad (11.31)$$

in which E is the microscopic electric field at the hydrogen atom, and c_j is the constant of proportionality for the jth component. Tables of c_j for hydrogen, which are calculated from the quantum numbers, are given by Underhill and Waddell (1959). Historically, E was thought to depend only on the electric fields of ions, which move slowly compared to electrons. One imagined the situation to be static on atomic timescales and to look like Figure 11.6. If we again let R be the separation of some ion from the hydrogen atom, then the field from that ion is $E = e/R^2$, where we have assumed the perturbing ion is singly ionized. This form is consistent with Equation 11.16.

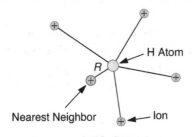

Figure 11.6 The distribution of ions around the absorbing atom gives a non-zero electric field that distorts the energy levels of the atom. The nearest neighbor is at a distance R. The situation is assumed to be static.

Micro-Electric Field Distributions: Static Approximation

The static approximation is the opposite of the impact approximation. Although we are focusing on hydrogen, the discussion in this section is general for all static situations. The relatively slow velocities of ions implies that the configuration does not change significantly during the absorption transition, suggesting the static treatment.

Since the electric field depends on the proximity of (static) ions to the absorbing hydrogen atom, we expect a probability distribution for R. The strongest interaction occurs between the atom and its closest perturber, so the lowest order approximation is given by completely neglecting the lesser fields contributed by the other ions slightly farther away. This approach is know as the *nearest-neighbor approximation*. We take up again the thought pattern that led to Equation 11.17. Let N be the number density of ions. The probability of finding an ion in the shell bounded by R and $R + \Delta R$ is proportional to the number density of ions and the volume of the shell, namely, $4\pi R^2 \Delta R\, N$. We seek that ion that is closest to the atom, so we combine the probability that the ion is *not* in the sphere of radius R, call it $p(R)$, but *is* in the shell of thickness ΔR. The probability, $P(R)$, of both these conditions occurring at the same time is the product of their individual probabilities,

$$P(R) = p(R)\, 4\pi R^2 \Delta R\, N. \tag{11.32}$$

We know that as R increases, $p(R)$ must diminish according to the increased likelihood of encompassing some ion. This leads us to write $p(R + \Delta R) = p(R)\left(1 - 4\pi R^2 \Delta R\, N\right) = p(R) - 4\pi R^2 \Delta R\, p(R)\, N$. But a general Taylor expansion is $p(R + \Delta R) = p(R) + (dp/dR)\Delta R + \cdots$. By comparing these two expressions, we see that $dp(R)/dR = -4\pi R^2 p(R)N$, which upon integration gives $p(R) = p(0)\, e^{-4\pi R^3 N/3}$. But $p(0) = 1$ on physical grounds. Let us normalize R to the mean distance between ions, call it R_0. Then

$N(4\pi/3)R_0^3 = 1$, or $R_0^{-3} = 4\pi N/3$, and the probability of no ion being as close as R can then be rewritten $p(R) = e^{-(R/R_0)^3}$.

Then from Equation 11.32, the probability that the ion is in the shell of thickness ΔR (or dR) is

$$P(R) = p(R)\, 4\pi R^2 \Delta R\, N = 3\frac{R^2}{R_0^3}\, e^{-(R/R_0)^3} \Delta R,$$

where N in Equation 11.32 has been replaced by $\left[(4\pi/3)R_0^3\right]^{-1}$.

This distribution in R can be translated into the distribution of electric fields by using $(R/R_0)^2 = E_0/E$ and $|dR/R_0| = \frac{1}{2}(E_0/E)^{3/2}$, where E_0 is the electric field corresponding to R_0,

$$E_0 = (4\pi/3)^{2/3}\, eN^{2/3}$$
$$= 1.248 \times 10^{-9}\, N^{2/3}. \tag{11.33a}$$

Assuming single ionizations, for each ion there is a free electron, so $N = N_e$, or in terms of photospheric parameters,

$$E_0 = e\left(\frac{4\pi}{3}\frac{P_e}{kT}\right)^{2/3} = 46.72\left(\frac{P_e}{T}\right)^{2/3} \tag{11.33b}$$

for E_0 in esu and N in units of perturbers per cm^3. Thus from $P(R)$,

$$P(E) = \frac{3}{2}\left(\frac{E_0}{E}\right)^{5/2} e^{-(E_0/E)^{3/2}}\frac{dE}{E_0},$$

or using $\beta = E/E_0$,

$$P(\beta) = \frac{3}{2}\beta^{-5/2}e^{-\beta^{-3/2}}\, d\beta. \tag{11.34}$$

This is the simplified distribution based on the nearest-neighbor approximation that explicitly avoids the vector addition of E from several perturbers. A more elaborate calculation that takes into account the full distribution of perturbers and gives a more accurate result for $P(\beta)$ is called the *Holtsmark distribution* (Holtsmark 1919). For example, Underhill and Waddell (1959) give the Holtsmark distribution as the sum

$$P(\beta) = \frac{4}{3\pi}\sum_{j=0}^{\infty}(-1)^j\,\frac{\Gamma[(4j+6)/3]}{(2j+1)!}\beta^{2j+2} \tag{11.35a}$$

in which Γ is the gamma function. Sufficient numerical accuracy is attained if the sum is carried through $j \gtrsim 35$. An alternate expression is

$$P(\beta) = \frac{2\beta}{\pi}\int_0^{\infty} e^{-x^{3/2}}\, x\, \sin\beta x\, dx, \tag{11.35b}$$

which is also evaluated numerically.

Figure 11.7 shows the perturber distributions according to Equations 11.34 and 11.35. The two distributions give the same result for large fields, as one would expect based on the simple argument that strong fields arise from close encounters, in which case one ion dominates and the nearest-neighbor approximation is fulfilled.

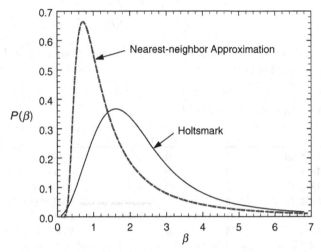

Figure 11.7 The nearest-neighbor and Holtsmark Stark-broadening distributions are compared.

Although using $\beta = E/E_0$ gives a certain compactness to the equations, it is really $\Delta\lambda$ from Equation 11.31 that is needed for the profile. In other words, $\beta = \Delta\lambda_j/(c_j E_0)$, so each jth component has a $P(\beta)$ scaled inversely with its displacement constant c_j.

The Holtsmark Distribution and Hydrogen

The complete hydrogen Stark broadening is the weighted sum of the individual j components (Equation 11.31) with each of them showing a $P(\beta)$ distribution according to Equation 11.35. Following the customary definition (Underhill 1951), we split the total f value between the unshifted Stark component, f_0, and the sum of the shifted components called $f_{\pm}$,

$$f_{\pm} = \frac{1}{l^2}\sum_j f_j = f - f_0,$$

the summation is over all shifted components. The principal quantum number of the lower level is l. The Stark profile is then the sum of the $P(\beta)$'s over all the relevant j values for the transition weighted by their fractional f values,

$$S\left(\frac{\Delta\lambda}{E_0}\right)d\left(\frac{\Delta\lambda}{E_0}\right) = \sum_j \frac{f_j}{\sum f_j} P(\beta)d\beta$$

$$= \frac{1}{l^2 f_{\pm}}\sum_j f_j P\left(\frac{\Delta\lambda}{E_0}\right)d\left(\frac{\Delta\lambda}{c_j E_0}\right)$$

$$= \frac{1}{l^2 f_{\pm}}\sum_j \frac{f_j}{c_j} P\left(\frac{\Delta\lambda}{E_0}\right)d\left(\frac{\Delta\lambda}{E_0}\right). \qquad (11.36)$$

Some sample oscillator strengths and interaction constants from Underhill and Waddell (1959) are given in Tables 11.4 and 11.5.

Table 11.4 *Balmer line oscillator strengths*

Line	λ	u	f_0	$f_{\pm}$	f
H_α	6265.80	3	0.248689	0.392058	0.640742
H_β	4861.33	4	0	0.119321	0.119321
H_γ	4340.47	5	0.007014	0.037656	0.044670
H_δ	4101.74	6	0	0.022093	0.022093
H_ϵ	3970.07	7	0.001310	0.011394	0.012704
H_ζ	3889.05	8	0	0.008036	0.008036

Table 11.5 *Interaction constants and f values for H_α and H_β*

u	j	c_j	f_j	u	j	c_j	f_j
3	1	0.00826	0.3508	4	2	0.00907	0.01456
3	2	0.01653	0.0660	4	4	0.01814	0.08678
3	3	0.02479	0.2087	4	6	0.02720	0.06366
3	4	0.03306	0.1523	4	8	0.03627	0.03654
3	5	0.04132	0.0029	4	10	0.04534	0.03549
3	6	0.04958	0.0033	4	12	0.05441	0.00152
3	8	0.06611	0.0001	4	14	0.06348	0.00010

Figure 11.8 shows the Holtsmark distributions, i.e., the $P(\Delta\lambda/E_0)$ needed in Equation 11.36, for the seven H_α components, each $\Delta\lambda/E_0$ scaled to its c_j, but still with unit area.

The next step is to weight each curve by its oscillator strength, f_j, given in Table 11.5. This leads to Figure 11.9. The f_j values for the last three components

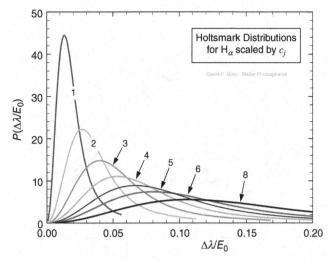

Figure 11.8 Holtsmark distributions for H_α, each labeled with its j value. (There is no $j = 7$ component.) Each curve has unit area.

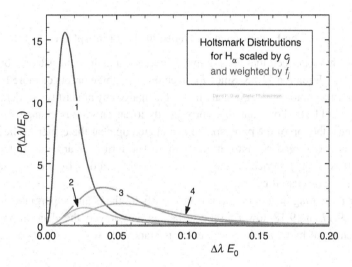

Figure 11.9 Holtsmark distributions for H_α, scaled by c_j and weighted by f_j. The curves for $j = 5, 6$, and 8 have tiny f_j values and their amplitudes are too small to be discerned.

are very small, making their curves imperceptible. These weighted curves are added together to get the sum in Equation 11.36. The full $S(\Delta\lambda/E_0)$ distribution using Equation 11.36 is shown in Figure 11.10 for the first three Balmer lines. The far wing has a $\Delta\lambda^{-5/2}$ dependence.

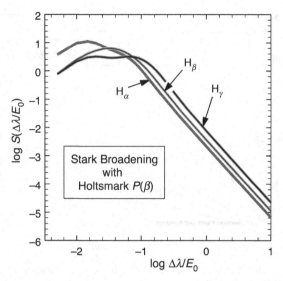

Figure 11.10 The $S(\Delta\lambda/E_0)$ distributions for H_α and H_β using the Holtsmark distribution for $P(\beta)$. Based on Underhill and Waddell (1959). The linear portions have a slope of $-5/2$.

The Dependence on Depth in the Photosphere

Now you will have noticed that the argument in these plots has consistently been $\Delta\lambda/E_0$, with E_0 given by Equation 11.33. Since P_e/T changes by three orders or more from the top to the bottom of a photosphere, the scale factor E_0 changes by at least two orders. Plots like those in Figure 11.10 show that this corresponds to an enormous change in $S(\Delta\lambda/E_0)$. Absorption near the top of the photosphere is bunched up near the center of the line, while in deep layers it is spread out into the far wings. The turn-over around the maximum in Figure 11.10 is usually buried in the line core and overridden by the thermal Doppler broadening to be discussed below.

A similar E_0 scaling occurs in going from dwarfs to giants and supergiants. As we saw in Equations 9.17 and 9.18, lowering the surface gravity lowers the electron pressure. The Stark broadening of hydrogen lines is weaker in stars with lower surface gravity.

Improvements in the $P(\Delta\lambda/E_0)$ Distribution

During the century since Holtsmark's 1919 paper, improvements and new developments have come about. Electron broadening turns out to be important. Early reviews are given by Van Regemorter (1965) and Griem (1974), more recent ones by Demura (2010) and Gigosos (2014). Background papers of interest include Edmonds *et al.* (1967), Griem (1967), Vidal *et al.* (1971, 1973), Stehlé *et al.* (1983, 1988), and Stehlé (1994). Tabulations of $S(\Delta\lambda/E_0)$ are given by Edmonds *et al.* (1967), Vidal *et al.* (1973), Griem (1974), Lemke (1997), Clausset *et al.* (1994), and Stehlé and Hutcheon (1999) among others. The idea with tabulated values is to interpolate to the values of N_e and T in your model photosphere.

Figure 11.11 shows examples of $S(\Delta\lambda/E_0)$ for the first few Balmer lines according to three tabulations. The main differences are in the cores of the lines and here the Doppler broadening blurs them away.

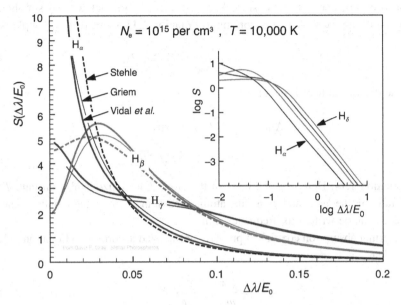

Figure 11.11 Stark profiles, $S(\Delta\lambda/E_0)$, are shown for the first few Balmer lines for $N_e = 10^{15}$ electrons/cm³ and $T = 10\,000$K. The profiles are symmetric about $\Delta\lambda/E_0 = 0$. The inset shows the Vidal *et al.* (1973) functions in logarithmic coordinates.

The Hydrogen Absorption Coefficient

Now that we have the distribution for the shifted components, namely $S(\Delta\lambda/E_0)$, we can write the full absorption coefficient for hydrogen lines as

$$\alpha_H = \frac{\pi e^2}{m_e c}\frac{\lambda^2}{c}\left[\frac{f_\pm S(\Delta\lambda/E_0)}{E_0} + f_0 \delta(\Delta\lambda)\right] \qquad (11.37)$$

in wavelength units. The unshifted component is represented by the delta function, $\delta(\Delta\lambda)$, although $f_0 = 0$ when u and l are both even or odd (as in Table 11.4).

We have now covered the major aspects of pressure broadening. Other wavelength shifts result from the Doppler effect. The thermal component falls in this category, and we now turn our attention to it.

Thermal Broadening

Each atom has a component of velocity along our line of sight to the star due to its thermal motion. Denote this velocity, the radial velocity, by v_R. Then the Doppler shift of the line emitted or absorbed by the atom is given by the usual

$$\frac{\Delta\lambda}{\lambda} = \frac{\Delta\nu}{\nu} = \frac{\upsilon_R}{c}, \tag{11.38}$$

where c is the velocity of light. The distribution of Doppler shifts gives us the shape of the thermal absorption coefficient, but since $\Delta\lambda$ and υ_R are proportional, we may simply use the distribution of atomic velocities according to Equation 1.13, converted to Doppler shifts,

$$N(\lambda)d\lambda = \frac{1}{\sqrt{\pi}\Delta\lambda_0} e^{-(\Delta\lambda/\Delta\lambda_0)^2} \, d\lambda, \tag{11.39}$$

where the dispersion here is denoted by

$$\Delta\lambda_0 = \frac{\lambda}{c}\sqrt{2kT/m} = \frac{\lambda}{c}\upsilon_0$$

$$= 4.301 \times 10^{-7}\lambda\sqrt{T/\mu}\,\text{Å} \tag{11.40}$$

as in Equation 1.14. Here m is the mass of the atom, k is Boltzmann's constant, T is the temperature of the gas, and μ is the atomic mass in atomic mass units. Analogous expressions can be written for frequency units.

The thermal absorption coefficient properly normalized according to Equations 11.9 and 11.10 is then

$$\alpha_{\text{therm}} = \frac{\sqrt{\pi}e^2}{m_e c^2} f \frac{\lambda^2}{\Delta\lambda_0} e^{-(\Delta\lambda/\Delta\lambda_0)^2} \tag{11.41}$$

and

$$\alpha_{\text{therm}} = \frac{\sqrt{\pi}e^2}{m_e c} f \frac{1}{\Delta\nu_0} e^{-(\Delta\nu/\Delta\nu_0)^2}. \tag{11.42}$$

Microturbulence

Small-scale mass motions, where the characteristic dimensions of the moving material are small compared to unit optical depth, have traditionally been dubbed "microturbulence." Velocity fields are well documented in studies of stellar spectral lines, and we take up the subject in more detail in Chapters 16 and 17. In the classical model of microturbulence, one postulates an isotropic Gaussian velocity distribution of dispersion ξ. These small-scale velocities produce Doppler shifts completely analogous to those arising from thermal motions. Consequently, the absorption coefficient for the microturbulent velocities has the Gaussian form of Equations 11.41 and 11.42 with $\upsilon_0 = \xi$ in Equation 11.40.

Combining Line Absorption Coefficients

In the preceding pages, we saw that there are many processes: natural broadening, Stark broadening, van der Waal's broadening, thermal broadening, and microturbulence broadening, all of which take place simultaneously. Consider any two of these absorption coefficients, and imagine the first one to be synthesized by a series of δ-functions. Each of the δ-functions experiences the broadening of the second absorption coefficient. We recognize this as a convolution process. Expand this to include all the above-mentioned processes, and the full absorption coefficient is recognized as a multiple convolution,

$$\alpha_{\text{total}} = \alpha_{\text{nat}} * \alpha_4 * \alpha_6 * \alpha_{\text{therm}} * \alpha_{\text{micro}} \tag{11.43}$$

For normal metal lines, the first three absorption coefficients each have a dispersion profile according to the impact approximation (hydrogen lines are treated separately below). In what is now a familiar process, the first three convolutions result in a new dispersion profile with $\gamma_{\text{total}} = \gamma_{\text{nat}} + \gamma_4 + \gamma_6$. Similarly, the last two absorption coefficients produce a new Gaussian with Doppler dispersion:

$$\Delta\lambda_D = \frac{\lambda}{c}\sqrt{v_0^2 + \xi^2} = \frac{\lambda}{c}\sqrt{\frac{2kT}{m} + \xi^2}$$

or

$$\Delta v_D = \frac{v}{c}\sqrt{v_0^2 + \xi^2} = \frac{v}{c}\sqrt{\frac{2kT}{m} + \xi^2}. \tag{11.44}$$

So with all the dispersion profiles collected together and the two Gaussian profiles now combined, we are left with a single convolution between a dispersion function and a Gaussian, namely,

$$\alpha_{\text{total}} = \left[\frac{\pi e^2}{m_e c}f\frac{\gamma_{\text{total}}/4\pi^2}{\Delta v^2 + (\gamma_{\text{total}}/4\pi)^2}\right] * \left[\frac{1}{\sqrt{\pi}\Delta v_D}e^{-(\Delta v/v_D)^2}\right].$$

It is customary to rationalize the variables by defining

$$a = \frac{\gamma_{\text{total}}}{4\pi}\frac{1}{\Delta v_D} = \frac{\gamma_{\text{total}}}{4\pi}\frac{\lambda^2}{c}\frac{1}{\Delta\lambda_D} \tag{11.45}$$

and

$$u = \Delta v/\Delta v_D = \Delta\lambda/\Delta\lambda_D. \tag{11.46}$$

Then the convolution itself can be written

$$H(u, a) = \int_{-\infty}^{\infty} \frac{\gamma_{\text{total}}/4\pi^2}{(\Delta v - \Delta v_1)^2 + (\gamma_{\text{total}}/4\pi)^2}e^{-(\Delta v_1/\Delta v_D)^2}dv_1$$

$$= \frac{a}{\pi}\int_{-\infty}^{\infty}\frac{1}{(u - u_1)^2 + a^2}e^{-u_1^2}du_1 \tag{11.47}$$

in analogy with Equation 2.21. $H(u, a)$ is called the Hjerting function (Hjerting 1938). Closely allied to the Hjerting function is the Voigt function, $V(u, a) = H(u, a)/(\sqrt{\pi}\Delta v_D)$, which is used equally often. In terms of $H(u, a)$, the absorption coefficient can be written

$$\alpha_{\text{total}} = \frac{\sqrt{\pi}e^2}{m_e c^2} \frac{\lambda^2 f}{\Delta\lambda_D} H(u, a)$$

or

$$\alpha_{\text{total}} = \frac{\sqrt{\pi}e^2}{m_e c} \frac{f}{\Delta\nu_D} H(u, a). \qquad (11.48)$$

One of the simpler ways to calculate $H(u, a)$ is with a series expansion,

$$H(u, a) = H_0(u) + aH_1(u) + a^2 H_2(u) + a^3 H_3(u) + a^4 H_4(u) + \cdots. \qquad (11.49)$$

Interpolation in Table 11.6, taken from Harris (1948), gives the numerical values. The first term, $H_0(u)$, is a Gaussian. In the far wing, beyond the table values, use the asymptotic expressions

$$H_1(u) = \frac{0.56419}{u^2} + \frac{0.846}{u^4}$$

and

$$H_3(u) = -\frac{0.56}{u^4}.$$

Table 11.6a *Hjerting function components*

u	$H_0(u)$	$H_1(u)$	$H_2(u)$	$H_3(u)$	$H_4(u)$
0.0	+1.000000	−1.12838	+1.0000	−0.752	+0.50
0.1	+0.990050	−1.10596	+0.9702	−0.722	+0.48
0.2	+0.960789	−1.04048	+0.8839	−0.637	+0.40
0.3	+0.913931	−0.93703	+0.7494	−0.505	+0.30
0.4	+0.852144	−0.80346	+0.5795	−0.342	+0.17
0.5	+0.778801	−0.64945	+0.3894	−0.165	+0.03
0.6	+0.697676	−0.48582	+0.1953	+0.007	−0.09
0.7	+0.612626	−0.32192	+0.0123	+0.159	−0.20
0.8	+0.527292	−0.16772	−0.1476	+0.280	−0.27
0.9	+0.444858	−0.03012	−0.2758	+0.362	−0.30
1.0	+0.367879	+0.08594	−0.3679	+0.405	−0.31
1.1	+0.298197	+0.17789	−0.4234	+0.411	−0.28
1.2	+0.236928	+0.24537	−0.4454	+0.386	−0.24
1.3	+0.184520	+0.28981	−0.4392	+0.339	−0.18
1.4	+0.140858	+0.31394	−0.4113	+0.280	−0.12
1.5	+0.105399	+0.32130	−0.3689	+0.215	−0.07
1.6	+0.077305	+0.31573	−0.3185	+0.153	−0.02
1.7	+0.055576	+0.30094	−0.2657	+0.097	+0.02
1.8	+0.039164	+0.28027	−0.2146	+0.051	+0.04
1.9	+0.027052	+0.25648	−0.1683	+0.015	+0.05
2.0	+0.0183156	+0.231726	−0.12821	−0.0101	+0.058
2.1	+0.0121552	+0.207528	−0.09505	−0.0265	+0.056
2.2	+0.0079071	+0.184882	−0.06863	−0.0355	+0.051
2.3	+0.0050418	+0.164341	−0.04830	−0.0391	+0.043
2.4	+0.0031511	+0.146128	−0.03315	−0.0389	+0.035

Table 11.6a (*cont.*)

u	$H_0(u)$	$H_1(u)$	$H_2(u)$	$H_3(u)$	$H_4(u)$
2.5	+0.0019305	+0.130236	−0.02220	−0.0363	+0.027
2.6	+0.0011592	+0.116515	−0.01451	−0.0325	+0.020
2.7	+0.0006823	+0.104739	−0.00927	−0.0282	+0.015
2.8	+0.0003937	+0.094653	−0.00578	−0.0239	+0.010
2.9	+0.0002226	+0.086005	−0.00352	−0.0201	+0.007
3.0	+0.0001234	+0.078565	−0.00210	−0.0167	+0.005
3.1	+0.0000671	+0.072129	−0.00122	−0.0138	+0.003
3.2	+0.0000357	+0.066526	−0.00070	−0.0115	+0.002
3.3	+0.0000186	+0.061615	−0.00039	−0.0096	+0.001
3.4	+0.0000095	+0.057281	−0.00021	−0.0080	+0.001
3.5	+0.0000048	+0.053430	−0.00011	−0.0068	0.000
3.6	+0.0000024	+0.049988	−0.00006	−0.0058	0.000
3.7	+0.0000011	+0.046894	−0.00003	−0.0050	0.000
3.8	+0.0000005	+0.044098	−0.00001	−0.0043	0.000
3.9	+0.0000002	+0.041561	−0.00001	−0.0037	0.000

Table 11.6b *Hjerting function components*

u	$H_1(u)$	$H_3(u)$	u	$H_1(u)$	$H_3(u)$
4.0	+0.039250	−0.00329	8.0	+0.0090306	−0.00015
4.2	+0.035195	−0.00257	8.2	+0.0085852	−0.00013
4.4	+0.031762	−0.00205	8.4	+0.0081722	−0.00012
4.6	+0.028824	−0.00166	8.6	+0.0077885	−0.00011
4.8	+0.026288	−0.00137	8.8	+0.0074314	−0.00010
5.0	+0.024081	−0.00113	9.0	+0.0070985	−0.00009
5.2	+0.022146	−0.00095	9.2	+0.0067875	−0.00008
5.4	+0.020441	−0.00080	9.4	+0.0064967	−0.00008
5.6	+0.018929	−0.00068	9.6	+0.0062243	−0.00007
5.8	+0.017582	−0.00059	9.8	+0.0059688	−0.00007
6.0	+0.016375	−0.00051	10.0	+0.0057287	−0.00006
6.2	+0.015291	−0.00044	10.2	+0.0055030	−0.00006
6.4	+0.014312	−0.00038	10.4	+0.0052903	−0.00005
6.6	+0.013426	−0.00034	10.6	+0.0050898	−0.00005
6.8	+0.012620	−0.00030	10.8	+0.0049006	−0.00004
7.0	+0.0118860	−0.00026	11.0	+0.0047217	−0.00004
7.2	+0.0112145	−0.00023	11.2	+0.0045526	−0.00004
7.4	+0.0105990	−0.00021	11.4	+0.0043924	−0.00003
7.6	+0.0100332	−0.00019	11.6	+0.0042405	−0.00003
7.8	+0.0095119	−0.00017	11.8	+0.0040964	−0.00003
8.0	+0.0090306	−0.00015	12.0	+0.0039595	−0.00003

Figure 11.12 shows how the central portion of $H(u, a)$ is dominated by the Gaussian term and how the wings of the dispersion component grow with increasing a. The shaping of stellar line profiles by the line absorption coefficient is illustrated in Figure 11.13.

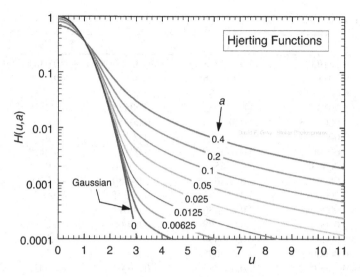

Figure 11.12 Hjerting functions, logarithmic ordinate, are plotted for a wide range of damping parameter, a.

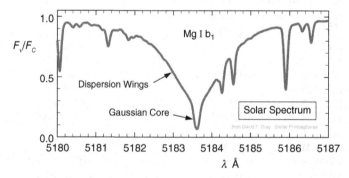

Figure 11.13 A few cases, like this magnesium line in the solar spectrum, clearly show the Gaussian core and the dispersion wings. Notice that the weaker lines, those in the wings of the magnesium line, show only the Gaussian core. Data from Hinkle *et al.* (2000).

In a power expansion like Equation 11.49, a smaller a means faster convergence and smaller error. Generally a tends to be small. For example, if $\gamma_{total} = 10^9$ per second, $T = 6000$ K, at $\lambda = 6000$ Å, and for Fe with $m = 56 \times 1.66 \times 10^{-24}$ grams, we have $a = 0.035$. But the Stark and van der Waals broadening, as well as $\Delta\lambda_D$, increase with depth in the photosphere. Figure 11.14 shows these changes for the magnesium line in Figure 11.13. The rise in Stark and van der Waals damping is similar to what we saw for sodium in Figure 11.4. The inset plot shows $\Delta\lambda_D$ rising slowly from 0.030 to 0.043 in accordance with the slow, square-root, dependence of Equation 11.40 or 11.44. The a damping parameter rises in response to the rise in γ_{total}, ranging from 0.002 to 0.106. The range in $H(u, a)$ can be seen in Figure 11.12, from a near pure Gaussian in high photospheric layers to one with strong damping wings in the deep layers.

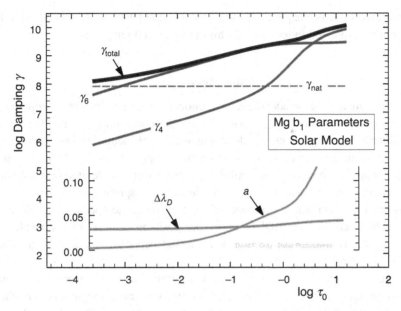

Figure 11.14 As illustrated with the Mg b_1 line, $\lambda 5184$, the Stark and van der Waals damping constants are not constant, but increase with depth, as does $\Delta\lambda_D$ and $a = \gamma\lambda^2/(4\pi c\,\Delta\lambda_D)$. Microturbulence ξ is set at zero for this example.

It is worth emphasizing that the dispersion wings really only show up when the line is strong. The stronger the line, the more dispersion wing one sees, but the relative strength is dictated by the damping constant. When we consider evolved stars, where the surface gravity is two or more orders lower than for dwarfs, pressure broadening is much smaller and therefore so is γ_{total}. At the same time, moving to stars having smaller gravity in the HR diagram often means cooler temperatures and stronger lines. More on this in Chapter 13.

The Special Case of Hydrogen

The general idea of convolutions expressed in Equation 11.43 is also appropriate for hydrogen, but now the Stark component is given by Equation 11.37. Because the linear Stark broadening has neither a Gaussian nor a dispersion shape, it cannot be rolled into $H(u, a)$, but instead is written explicitly as the sum of two convolutions,

$$\alpha_H = \frac{\sqrt{\pi}e^2}{m_e c^2}\frac{\lambda^2}{\Delta\lambda_D}\left[f_{\pm}\frac{S(\lambda)}{E_0} * H(u, a) + f_0 H(u, a)\right], \tag{11.50}$$

where we have now simplified the wavelength notation by writing $S(\lambda)$ for $S(\Delta\lambda/E_0)$. The convolution of $H(u, a)$ with the δ-function in Equation 11.37 gives the $H(u, a)$ in the last term here. Further, in most stars, $S(\lambda)$ is wide compared to $H(u, a)$, so the first convolution essentially returns $S(\Delta\lambda)$. Thus

$$\alpha_H = \frac{\sqrt{\pi}e^2}{m_e c^2}\frac{\lambda^2}{\Delta\lambda_D}\left[f_{\pm}\frac{S(\lambda)}{E_0} + f_0 H(u, a)\right]. \tag{11.51}$$

This simplified form is adequate for most model computations of hydrogen lines. Recall that E_0 can be expressed in terms of electron pressure and temperature.

Other Broadeners

Several other sources of broadening are occasionally relevant. *Hyperfine structure* stems from the interaction of nuclear spin with the angular momentum vector of the rest of the atom, splitting the energy levels of odd atomic-number elements. *Isotopic splitting* occurs with the different nuclear mass of the different isotopes. *Zeeman splitting* requires the absorbers to be in a magnetic field. All three of these turn simple atomic transitions into several, usually closely spaced, transitions. Because the splitting of the original energy levels is small, it amounts to a broadening of the composite spectral line. To properly handle such situations, we need to know the positions and relative strengths of the components. Each component is then treated as an independent line with its own absorption coefficient given by Equation 11.48.

Splitting can alter the shapes of narrow spectral lines and reduce the amount of saturation in strong lines. Wahlgren *et al.* (1997) give an example for hyperfine and isotopic structure of Re II. Other aspects of hyperfine structure are discussed and examples given by Abt (1952), Bely (1966), Holweger and Oertel (1971), Arnesen *et al.* (1982), Bohlender *et al.* (1998), Dworetsky *et al.* (1998), Jomaron *et al.* (1999), Ryabchikova *et al.* (1999), Nielsen *et al.* (2000), Prochaska and McWilliam (2000), Wahlgren *et al.* (2001), Pakhomov *et al.* (2019), and Den Hartog *et al.* (2020). Rare earths show particularly extensive hyperfine structure.

Isotopic splitting is often discussed in the context of nuclear processing of chemicals, and some of its interesting story can be seen in Woolf and Lambert (1999), Nielsen *et al.* (2000), Sneden *et al.* (2002), Aoki *et al.* (2003), and Dolk *et al.* (2003). Zeeman splitting and its polarization are used as a tool for measuring magnetic fields.

The Mass Absorption Coefficient for Lines

In the radiation transfer equation we need $\ell_v = N\alpha/\rho$ cm^2/g, where here N is the number of absorbers per cm^3. Relate the number of absorbers to the number of hydrogen atoms per cm^3, N_H, using

$$N = \frac{N_E}{N_H}\frac{N_l}{N_E}N_H.$$

Here $N_E/N_H = A_E$ is the standard abundance of the element E relative to hydrogen, N_l/N_E is the fraction of the element in the lower level of the transition, and it takes into account the excitation and ionization (Equations 1.19, 1.21, and 1.23). The density is

$$\rho = \sum_j N_j m_j = N_H \sum_j A_j m_j,$$

where the sum is over all the j elements of relative abundances A_j and mass m_j.

Putting these together, the mass absorption coefficient for a typical metal line can now be written in frequency units as

$$\ell_\nu = \frac{\sqrt{\pi}e^2}{m_e c} f \frac{H(u,a)}{\Delta\nu_D} \frac{N_l}{N_E} \frac{A_E}{\sum A_j m_j} \left(1 - 10^{-\chi_\lambda \theta}\right)$$

$$= 1.4974 \times 10^{-2} \frac{A_E f}{\sum A_j m_j} \frac{N_l}{N_E} \frac{H(u,a)}{\Delta\nu_D} \left(1 - 10^{-\chi_\lambda \theta}\right) \qquad (11.52)$$

in units of cm^2 per gram. The stimulated-emission factor is $\left(1 - 10^{-\chi_\lambda \theta}\right)$, $\chi_\lambda = 1.2398 \times 10^4/\lambda$ eV for λ in Å, and $\theta = 5040/T$.

In wavelength units (but keeping the subscript ν on ℓ_ν) the expression is

$$\ell_\nu = 4.995 \times 10^{-21} \frac{A_E f \lambda^2}{\sum A_j m_j} \frac{N_l}{N_E} \frac{H(u,a)}{\Delta\lambda_D} \left(1 - 10^{-\chi_\lambda \theta}\right) \qquad (11.53)$$

with λ and $\Delta\lambda_D$ in Å. The damping parameter a is given by

$$a = \frac{\gamma_{\text{total}}}{4\pi\Delta\nu_D} = 7.96 \times 10^{-2} \frac{\gamma_{\text{total}}}{\Delta\nu_D}$$

or

$$a = \frac{\gamma_{\text{total}}}{4\pi\Delta\lambda_D} \frac{\lambda^2}{c} = 2.65 \times 10^{-20} \frac{\gamma_{\text{total}}\lambda^2}{\Delta\lambda_D} \qquad (11.54)$$

with λ and $\Delta\lambda_D$ in Å.

A similar expression for ℓ_ν can be written for hydrogen lines using Equation 11.51. For discussions of helium lines, see Underhill and Waddell (1959), Griem (1968), Barnard *et al.* (1974), Mihalas *et al.* (1974), Milosavljević and Djeniže (2001), or Pérez *et al.* (2003).

Comments

We can now compute spectral lines with our model photospheres using the version of Equation 7.21 given at the beginning of this chapter. The computed profiles are compared to the observed ones to extract interesting information. Before making such comparisons, we look into how to get observations of spectral lines, the topic of Chapter 12.

Questions and Exercises

1. What is "pressure broadening" in the context of spectral lines? Briefly explain what causes pressure broadening. List the main types of pressure broadening and indicate the perturber for each type.
2. If ϕ in Equation 11.20 is selected to be two radians instead of one, what would be the change in γ_4 and γ_6?

3. Use Equation 11.23 to find the ratio of relative velocities for (a) iron perturbed by an electron, (b) iron perturbed by a proton, (c) iron perturbed by a neutral hydrogen atom, and (d) iron perturbed by a once-ionized magnesium atom. What conclusions can you draw concerning the impacts of these perturbers?

4. Describe the Hjerting function. Explain how we come to need this function in our studies of stellar photospheres. Since the wing portion of the Hjerting function is extremely tiny for weak spectral lines, would it be reasonable to expect weak spectral lines formed in a stellar photosphere to have a Gaussian shape? Support your answer with arguments.

5. Use Figure 11.4 to get an approximate value of γ_{total} for the sodium D_2 line. Then calculate the parameter $a = \lambda^2 \gamma / (4\pi c \, \Delta\lambda_D)$ in the Hjerting function. Is this value small enough so that the series in Equation 11.48 is convergent?

References

Abt, A. 1952. *ApJ* **115**, 199.
Allen, C.W. 1973. *Astrophysical Quantities* (London: Athlone), 3rd ed., pp. 84 & 89.
Anstee, S.D. & O'Mara, B.J. 1991. *MNRAS* **253**, 549.
Anstee, S.D. & O'Mara, B.J. 1995. *MNRAS* **276**, 859.
Aoki, W., Ryan, S.G., Iwamoto, N., *et al.* 2003. *ApJ* **592**, 67.
Arnesen, A., Hallin, R., Nordling, C., *et al.* 1982. *A&A* **106**, 327.
Ayres, T.R. 1977. *ApJ* **213**, 296.
Baker, J.G. & Menzel, D.H. 1938. *ApJ* **88**, 52.
Baranger, M. 1958a. *Phys Rev* **111**, 481, 494.
Baranger, M. 1958b. *Phys Rev* **112**, 855.
Barklem, P.S. & O'Mara, B.J. 1997. *MNRAS* **290**, 102.
Barklem, P.S. & O'Mara, B.J. 1998. *MNRAS* **300**, 863.
Barklem, P.S. & O'Mara, B.J. 2000. *MNRAS* **311**, 535.
Barklem, P.S., Anstee, S.D., & O'Mara, B.J. 1998a. *PASA* **15**, 336.
Barklem, P.S., O'Mara, B.J., & Ross, J.E. 1998b. *MNRAS* **296**, 1057.
Barklem, P.S., Piskunov, N., & O'Mara, B.J. 2000a. *A&AS* **142**, 467.
Barklem, P.S., Piskunov, N., & O'Mara, B.J. 2000b. *A&A* **363**, 1091.
Barnard, A.J., Cooper, J., & Smith, E.W. 1974. *JQSRT* **14**, 1025.
Bely, F. 1966. *Abundance Determinations in Stellar Spectra*, IAU Symposium **26** (London: Academic Press), H. Hubenet, ed., p. 254.
Bergeron, P., Wesemael, F., & Fontaine, G. 1991. *ApJ* **367**, 253.
Blackwell, D.E., Kirby, J.H., & Smith, G. 1972. *MNRAS* **160**, 189.
Bohlender, D.A., Dworetsky, M.M., & Jomaron, C.M. 1998. *ApJ* **504**, 533.
Clausset, F., Stehlé, C., & Artru, M.-C. 1994. *A&A* **287**, 666.
Demura, A.V. 2010. *Int J Spect* **2010**, ID671073.
Den Hartog, E.A., Lawler, J.E., & Roederer, I.U. 2020. *ApJS* **248**, 10.
Dimitrijevic, M.S. & Konjevic, N. 1986. *A&A* **163**, 297.
Dimitrijevic, M.S., Ryabchikova, T., Popovic, L.C., Shulyak, D., & Tsymbal, V. 2003. *A&A* **404**, 1099.
Dolk L., Wahlgren, G.M., & Hubrig, S. 2003. *A&A* **402**, 299.
Dworetsky, M.M., Jomaron, C.M., & Smith, C.A. 1998. *A&A* **333**, 665.
Edmonds, F.N., Jr., Schlüter, H., & Wells, D.C., III 1967. *MemRAS* **71**, 271.

Foley, H.M. 1946. *Phys Rev* **69**, 616.

Freudenstein, S.A. & Cooper, J. 1978. *ApJ* **224**, 1079.

Fuhr, J.R., Wiese, W.L., & Roszman, L.J. 1973. *Bibliography on Atomic Line Shapes and Shifts*, NBSP No. 366.

Gigosos, M.A. 2014. *J Phys D* **47**, 343001.

González, V.R., Aparicio, J.A., del Val, J.A., & Mar, S. 2002. *J Phys B* **35**, 3557.

Griem, H.R. 1964. *Plasma Spectroscopy* (New York: McGraw-Hill).

Griem, H.R. 1967. *ApJ* **147**, 1092.

Griem, H.R. 1968. *ApJ* **154**, 1111.

Griem, H.R. 1974. *Spectral Line Broadening by Plasmas* (New York: Academic Press).

Griem, H.R. 1997. *Principles of Plasma Spectroscopy* (Cambridge: Cambridge University Press).

Harris, D.L., III 1948. *ApJ* **108**, 112.

Hinkle, K., Wallace, L., Valenti, J., & Harmer, D. 2000. *Visible and Near Infrared Atlas of the Arcturus Spectrum, 3727–9300 Å* (San Francisco: Astronomical Society of the Pacific).

Hjerting, F. 1938. *ApJ* **88**, 508.

Holtsmark, P.J. 1919. *Phys Z* **20**, 162.

Holweger, H. & Oertel, K.B. 1971. *A&A* **10**, 434.

Hunger, K. 1960. *ZfAp* **49**, 129.

Jomaron, C.M., Dworetsky, M.M., & Allen, C.S. 1999. *MNRAS* **303**, 555.

Kallmann, H. & London, F. 1929. *Z Phys Chem* **B2**, 207.

Landolt–Börnstein 1950. *Zahlenwerte und Funktionen aus Physik, Chemie, Astronomie, Geophysik und Technik* (Berlin: Springer-Verlag), Vol. **1**, p. 246.

Lemke, M. 1997. *A&AS* **122**, 285.

Lenz, W. 1924. *Z Phys* **25**, 299.

Lesage, A., Lebrun, J.L., & Richou, J. 1990. *ApJ* **360**, 737.

Lesage, A., Konjevic, N., & Fuhr, J.R. 1999. *Spectral Line Shapes 10*, AIP Conference Proceedings, **467**, R.M. Herman, ed., p. 27.

Lindholm, E. 1945. *Ark Mat Ast Fys* **32A**, No. 17.

London, F. 1930a. *Z Phys* **63**, 245.

London, F. 1930b. *Z Phys Chem* **B11**, 222.

Lorentz, H.A. 1906. *Proc Amst Acad* **8**, 591.

Menzel, D.H. 1961. *Mathematical Physics* (New York: Dover).

Michelson, A.A. 1895. *ApJ* **2**, 251.

Mihalas, D., Barnard, A.J., Cooper, J., & Smith, E.W. 1974. *ApJ* **190**, 315.

Milosavljević, V. & Djeniže, S. 2001. *Eur Phys J D* **15**, 99.

Milosavljević, V. & Djeniže, S. 2003. *A&A* **405**, 397.

Moity, J., Pieri, P.E., & Richou, J. 1975. *A&A* **45** 417.

Mugglestone, D. & O'Mara, B.J. 1966. *MNRAS* **132**, 87.

Nielsen, K., Karlsson, H., & Wahlgren, G.M. 2000. *A&A* **363**, 815.

Pakhomov, Yu.V., Ryabchikova, T.A., & Piskunov, N.E. 2019. *Ast Rep* **63**, 1010.

Peach, G. 1984. *J Phys B* **17**, 2599.

Pérez, C., Santamarta, R., de la Rosa, M.I., & Mar, S. 2003. *Eur Phys J D* **27**, 73.

Petit Bois, G. 1961. *Tables of Indefinite Integrals* (New York: Dover), p. 148.

Popovic, L.C., Dimitrijevic, M.S., & Ryabchikova, T. 1999. *A&A* **350**, 719.

Popovic, L.C., Simic, S., Milosavljević, N., & Dimitrijevic, M.S. 2001. *ApJS* **135**, 109.

Prochaska, J.X. & McWilliam, A. 2000. *ApJL* **537**, L57.

Ryabchikova, T., Piskunov, N., Savanov, I., Kupka, F., & Malanushenko, V. 1999. *A&A* **343**, 229.

Ryabchikova, T., Pakhomov, Y., & Piskunov, N. 2018. *Galaxies* **6**, 93.

Ryan, S.G. 1998. *A&A* **331**, 1051.

Silver, D.S. 2017. *American Scientist*, November–December, p. 364.

Sneden, C., Cowan, J.J., Lawler, J.E., *et al.* 2002. *ApJL* **566**, L25.

Stehlé, C. 1994. *A&AS* **104**, 509.

Stehlé, C. & Hutcheon, R. 1999. *A&AS* **140**, 93.

Stehle, C., Mazure, A., Nollez, G., & Feautrier, N. 1983. *A&A* **127**, 263.

Stehlé, C., Mazure, A., Nollez, G., & Feautrier, N. 1988. *A&A* **205**, 368.

Stone, J.M. 1963. *Radiation and Optics* (New York: McGraw-Hill), p. 73.

Struve, O. 1929. *ApJ* **69**, 173.

Underhill, A.B. 1951. *PDAO Victoria* **8**, 385.

Underhill, A.B. & Waddell, J.H. 1959. *NBS Circ* **603**

Unsöld, A. 1955. *Physik der Sternatmosphären* (Berlin: Springer-Verlag), 2nd ed., pp. 326, 331.

Van Regemorter, H. 1965. *ARAA* **3**, 71.

Van Regemorter, H. 1973. *Reports on Astronomy*, Trans IAU **15A** (Dordrecht: Reidel), C. De Jager, ed., p. 155.

Vidal, C.R., Cooper, J., & Smith, E.W. 1971. *JQSRT* **11**, 263.

Vidal, C.R., Cooper, J., & Smith, E.W. 1973. *ApJS* **25**, 37.

Wahlgren, G.M., Sveneric, S.G., Litzen, U., *et al.* 1997. *ApJ* **475**, 380.

Wahlgren, G.M., Brage, T., Brandt, J.C., *et al.* 2001. *ApJ* **551**, 520.

Weisskopf, V. 1932a. *Observatory* **56**, 291.

Weisskopf, V. 1932b. *Z Phys* **75**, 287.

Woolf, V.M. & Lambert, D.L. 1999. *ApJ* **521**, 414.

12

The Measurement of Spectral Lines

In this chapter, we consider the acquisition of spectral line data and the integrity of such data. Several types of measurements can be made of a spectral line: total absorption, profile width, detailed shape and asymmetry, wavelength position, and polarization. Almost all spectral line observations are normalized to the continuum flux, thus setting aside any information about absolute flux and the energy distribution of the star. But in contrast to the continuum measurements dealt with in Chapter 10, where a modest-resolution spectrograph sufficed, spectral line measurements push the limits of spectrograph capability. We often need resolving power $\lambda/\Delta\lambda$ in the range of 100 000 or even higher for accurate profile work. This implies $\Delta\lambda \sim 3$ km/s or ~50 mÅ at 5000 Å. If three detector pixels sample this width, each receives light from only ~16 mÅ of the spectrum! High-resolution spectroscopy is the domain of bright stars and long exposure times. Large aperture telescopes can be just as important for high-resolution spectroscopy as they are for faint galaxy work. High-quality high-resolution measurements also require ultra-stable alignment of optical and mechanical components.

Interferometers are effective in certain situations, but we will not emphasize them here; see Vaughan 1967, Connes 1970, and Ridgway and Brault 1984. Adaptive optics will also not be part of our discussion. Their main thrust is to effectively do away with atmospheric seeing, but they require very specialized telescopes. See Beckers (1993) and Rigaut (2015).

It is assumed that you have digested the spectrograph material in Chapter 3. Different spectrograph designs offer different advantages and disadvantages. Echelle spectrographs used in high grating orders have become popular because of their high efficiency and wide wavelength coverage. They excel in chemical-composition and radial-velocity-variation studies. But the narrow blaze distribution, as in Figure 3.19, is a significant limitation, bringing with it normalization issues and a wide range of photon count across each blaze. Spectrographs used in low grating orders tend to reverse the issues. They have a better, wider blaze distribution but more limited wavelength coverage. Their imaging quality is usually better. Low grating order spectrographs are better at measuring detailed line shapes. We start with the classical coudé spectrograph.

The Coudé Grating Spectrograph

High-resolution grating spectrographs are often rather large, too large to mount on the telescope itself. The coudé focus offers an appropriate solution (see Figure 3.24).

We can understand why the spectrograph is large with the help of Equation 3.15, the spectrograph equation,

$$\Delta\lambda = \frac{W'\,d}{f_{\text{coll}}\,n}\,\cos\alpha. \tag{12.1}$$

Our science requires a small $\Delta\lambda$, but the slit width W' should be large enough to allow most of the stellar seeing disk to enter the spectrograph. For a fixed ratio of ruling spacing to order number, d/n, we have left only f_{coll} as a free parameter. The collimator focal length is then made as large as possible with the important constraint that the size of the collimated beam not exceed the size of the grating. For example, the largest gratings have rulings ~400 mm long. Working at a typical coudé focal ratio of f/30, the collimator would have a focal length of 10.5 m. That makes for a big spectrograph. As larger telescopes are commissioned, there is a need for larger beam sizes. In some cases mosaics of gratings are used to accomplish this.

High-quality gratings are needed, and they must be carefully mounted. Slight imperfections in the mounting cell can cause serious distortions in the instrumental profile. The grating is mounted on a turntable so that it can be rotated to select the wavelength, or to direct the spectrum toward any of several cameras. Bandpass filters are used to exclude light from unused orders or from outside the specific region of interest, thus keeping the contamination of scattered light to a minimum.

Coudé spectrographs are shown in Figures 12.1a and 12.1b. Both spectrographs have an extra mirror to bring the beam into a horizontal orientation. In other cases, coudé

Figure 12.1a The inside of the coudé spectrograph at the Elginfield Observatory. On the left is a short focal length camera Schmidt ($f_{\text{cam}} = 559$ mm). On the right is a long focal length camera ($f_{\text{cam}} = 2089$ mm). The optical axes of the cameras are pointed at the diffraction grating, which is hidden from view behind the camera at the left. The collimator mirror is located on the right outside the field of view.

spectrographs are built around the latitude slant as the beam comes down the polar axis (Figure 3.24). Horizontal coudé rooms are much less expensive to build than those built on the latitude slant, and they are easier to work in. Throughput is increased by using high-reflection coatings, usually silver with dielectric overcoatings. Various technical aspects of coudé spectrographs are discussed by Dunham (1956), Tull (1969, 1972), Richardson *et al.* (1971), Enard (1979), and Gray (1986).

Figure 12.1b The coudé room at the 1.5 m AURA telescope at Cerro Tololo. The grating is visible at the left. Cameras of different focal lengths are mounted on the rectangular beam turret, which is rotated to bring the selected camera into the optical beam. Photo courtesy of Kitt Peak National Observatory.

A simple entrance-slit assembly is shown in Figure 12.2. One slit jaw is moveable to allow adjustment of the slit width. The plane in which the jaws lie is tilted so it is not perpendicular to the optical axis, thereby directing starlight that does not get through the slit to an off-axis guide camera. The image is constrained from moving along the slit by a mask or decker (not shown). Image wander corresponds to spectrum wander on the detector, generally not a good thing.

Incidentally, the atmospheric dispersion discussed early on in Chapter 10 is also relevant here. If the zenith distance is large, make sure the correct portion of this tiny atmospheric spectrum is getting into the spectrograph.

Image slicers should replace the conventional entrance slit whenever the size of the seeing disk exceeds the width of the entrance slit by ~1.5 times. These ingenious devices use multiple reflections to recycle light reflected by the slit jaws, putting it back through the entrance slit. Slicers can improve throughput by a factor of two or three depending on the quality of the seeing, and that is a really good thing. And they also improve the

Figure 12.2 The entrance slit to the coudé spectrograph has a changeable width to accommodate different gratings, cameras, and detectors. The slit is opened wide in this picture so
that it can be seen, but when it is used for observations, it is much narrower, ~0.2 mm.
A cross mask or decker is used in front of the slit to keep the image positioned vertically.

spectrophotometry by scrambling the seeing disk to remove scintillation noise (e.g., Smith
et al. 1987). As a further bonus, slicers give height to the spectrum on the detector, which
ameliorates restrictions of well capacity (Chapter 4) and helps with corrections for cosmic-
ray hits. Image slicers automatically eliminate the need to mask the length of the entrance
slit by supplying a small entrance aperture (often called the slot) comparable in size to the
seeing disk.

The importance of image slicers grows as the new larger telescope apertures outpace
growth in the size of diffraction gratings. It would be shortsighted to think that image
slicers play no role when the seeing is excellent. If the seeing disk is not significantly wider
than the entrance slit, it acts like its own entrance slit, a problem recognized some time ago
by Plaskett (1910). As the small seeing disk wanders around within the real entrance slit,
the spectrum wanders around on the detector. The result is spurious radial velocity
variations and an instrumental profile that is ill defined. We now turn to some of the
specific details.

The Bowen Image Slicer

In this design, a reflection perpendicular to the optical axis distributes the seeing disk across
several thin layers of glass that, by virtue of their geometry, stack their reflections end to
end. The Bowen (1938) slicer is easy to construct (Pierce 1965) and so is relatively
inexpensive. It has two drawbacks. First, the individual segments are not the same distance
from the collimator, so that they cannot all be in exact focus simultaneously. In practice,
this is not serious because the range in distance is small compared to the focal length of the

collimator. Second, the effective height of the composite entrance slit, and hence the height of the spectrum on the detector, varies with the seeing. This makes flat-fielding and related calibrations imperfect. A similar device is describe by Beckert *et al.* (2008).

The Richardson Image Slicer

The Richardson slicer, like the one shown in Figure 12.3, is an amazing invention. It produces a spectrum of a fixed height on the detector, independent of seeing (Richardson 1966), and the spectra from all slices fall on top of each other, on exactly the same area of the detector, with each slice precisely at the same focus, exactly what we want from a slicer. The Richardson slicer is also unique in being able to shepherd valuable starlight around central obstructions along the optical path.

Figure 12.3 A Richardson image slicer like this one replaces the conventional entrance slit. It improves the throughput by getting most of the seeing disk into the spectrograph.

Figure 12.4 shows the optical layout of the Richardson slicer. The recycling optics are two concave mirrors of equal focal length and separated by their radius of curvature. Each mirror is therefore imaged back on itself. The slit mirror of the slicer on the right has the entrance slit to the spectrograph cut in it (in the plane of the figure). This is done by cutting the mirror in half and separating the halves by the desired width of the entrance slit. The slot mirror is also cut in half, but in the opposite orientation. This gap forms the slot, or entrance aperture of the slicer. The telescope is focused on the slit mirror, but a cylindrical lens is placed in front of the slot mirror to bring the beam to a focus sooner in the vertical dimension, compressing the optical beam into a line image in the slot. After entering the slot, the line image diverges in the vertical dimension but is still coming to the original focus on the slit in the horizontal dimension. Therefore another line image is formed on the slit, this time lined up with the slit. The width of this line image is the same as it would be without the slicer, i.e., the width of the seeing disk. The length of the line image depends on the f-ratio of the cylindrical lens and the distance over which the beam has diverged, namely the separation between the slot and slit mirrors. The height of this line image on the slit is imaged into the height of the spectrum on the detector. The distribution of brightness along the line image is that of the telescope beam compressed to one dimension (the illumination plot, center right in the figure). It will taper off rapidly in brightness at the ends because the telescope beam is circular in cross section, and it will be slightly less bright in the center where the compressed image contains the secondary mirror shadow.

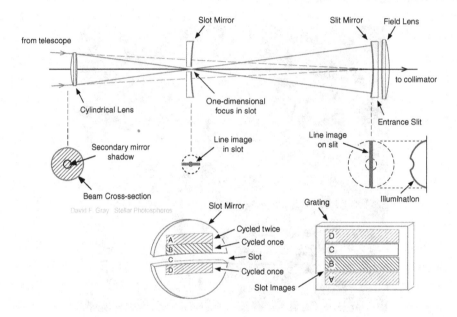

Figure 12.4 The optical layout of the Richardson slicer shows the slot mirror and the slit mirror separated by their (same) radius of curvature, i.e., two focal lengths. Each mirror is split in half to give a central aperture. The slot in the slot mirror is perpendicular to the plane of the diagram. The slit in the slit mirror is in the plane of the diagram and not visible. The cylindrical lens at the top left collapses the beam from the telescope into a line image in the slot. A second line image is formed on the slit. Images (A, B, and D) of the slot (C) are shown on the slot mirror and on the grating (lower right).

In general, the line image on the slit is wider (because of seeing) than the slit; that is why we are using the slicer. A portion of the line image, i.e., one slice, goes through the slit without any reflections from the mirrors. Light that does not go through is recycled by the mirrors as follows. Consider only one side of the seeing disk that is to be recycled. If the optical axes of the slicer's mirrors were along the optical axis of the telescope, the light reflected from the slit mirror would be re-imaged exactly back in the slot and would escape back along the incoming beam from the telescope and be lost. Instead, the one half of the slit mirror has its axis tilted so it is aligned with the top edge of the slot. That small tilt makes the once-recycled light form an image of the slot exactly above the slot (image B in Figure 12.4). Now if the axis of the slot mirror were along the optical axis of the telescope, this light would be reflected back to the slit mirror on the opposite side of the slit, but with no overlap on the slit. So, the slot mirror is twisted to have its axis aligned with the edge of the slit. That makes the recycled light overlap the entrance slit by exactly one slit width, and (hurray!) another slice is now into the spectrograph and on its way to the detector.

The other halves of the slit and slot mirrors are set with their axes on the opposite sides of the slot and slit, respectively, to take care of that part of the seeing disk on the other side of the entrance slit in the original beam (D). So, the slicer has two independently functioning halves.

Light not getting through the slit on the second pass (first recycling) acts like light from the opposite side of the beam from the first pass, except that it is re-imaged on the slot mirror two slot widths up (A), so, after four reflections, it too is recycled. A field lens is placed just beyond the slit inside the spectrograph to image the slot mirror onto the collimator (equivalently, the grating) of the spectrograph.

A typical Richardson slicer produces four slices: one without reflections, two recycled once (two reflections), and one recycled twice (four reflections). Naturally, super-reflecting coatings are used on the mirrors. When the seeing disk is smaller than the slot height, the images of the slot on the grating are illuminated only across the height corresponding to the seeing disk, i.e., the full heights of strips A–D are not filled with light.

It is important to notice that each recycled slice passes through the slit at the same position. Changes in seeing do not alter the position of the spectrum. The height of the spectrum and the distribution of light perpendicular to dispersion are essentially independent of seeing, making for stable spectrophotometry.

As mentioned earlier, the Richardson slicer can also be fabricated to leave a light-less gap between the central images of the slot. This is useful when there is a central obstruction in the beam such as a pick-off mirror. It means that light is not lost from the beam, nor does it contribute to unwanted scattered light, and field errors are not introduced (discussed shortly).

From the front of the spectrograph, we now move to the back, namely cameras.

Spectrograph Cameras

The camera has the important job of imaging the spectrum on the light detector. If the field is wide, which is the case for gratings used in low orders, Schmidt (or occasionally Maksutov) cameras are used (e.g., Bowen 1962, Shulte 1966, Taylor & Schmidt 1975). Aberrations can be reduced and wavelength coverage increased by placing the camera close to the grating. The focus in a Schmidt camera is midway between the center of curvature of the mirror, where the spherical-aberration corrector lens is positioned, and the mirror itself. The focal surface is a sphere concentric to the mirror's sphere. Photographic plates were bent by the plate holder to conform to this surface, but field flatteners are usually needed for the larger electrical detectors, although in principle there is no reason, except cost, that electrical detectors could not be made curved. Examples of cameras are shown in Figure 12.1.

The focal length of the camera is chosen to give the linear dispersion one desires when used with a specified grating and detector (Equation 3.13). Coudé installations often incorporate several cameras to accommodate different resolution requirements. The aperture of a camera depends on the field to be recorded and the collimated beam size. In an optimum design, the camera is large enough to encompass the width of the grating blaze. It is not uncommon for the camera to have an aperture that is a substantial fraction of the telescope aperture.

Although photographic plates were small enough to be placed on-axis inside the camera, most electrical detectors are too bulky and must be located outside the beam. The preferred way to do this is with a small pick-off mirror located in a light-less gap created by a Richardson image slicer. Some folded Schmidt cameras (e.g., Cohen *et al.* 1976) get the detector out of the beam, but require a large flat. Others introduce a central obstruction and the corresponding field error.

Field errors, or non-uniform distribution of light along the spectrum, arise from obstructions in the optical path after dispersion. Since different monochromatic beams go in different directions from the grating, they use different portions of the camera optics, and in particular, any obstruction in the light path blocks different fractions of these different beams. The result is a variation in optical throughput with wavelength across the field. A perfect flat-field lamp division (as discussed in Chapter 4) would remove such field errors. As any instrument user knows, such corrections are never perfect because of unavoidable differences in optical alignment and the distribution of light in the stellar and the lamp beams.

In the end, the camera focuses the spectrum on the detector, perhaps a CCD as discussed in Chapter 4. After removing cosmic-ray spikes and other flaws, processing bias and dark exposures, and dividing by a flat-field lamp, the signal perpendicular to the dispersion is combined to give us a line image, i.e., the amount of signal as a function of position in millimeters along the wavelength coordinate.

The next step is to normalize the spectrum to the continuum.

Continuum Normalization

Standard procedure in dealing with spectral lines is to normalize the continuum to unity. That means no absolute flux calibrations or any other calibrations are needed, only that the detector has a linear response. Division of the stellar exposure by a "flat-field" lamp, as discussed in Chapter 4, removes pixel-to-pixel variations, but there will still remain curvature and slope and possibly irregularities across the field. The observed continuum in spectroscopic observations is shaped not only by the apparent energy distribution of the star, but also by the blaze distribution of the grating (including cross-dispersion optics in echelle spectrographs), the transmission window of order-sorting filters, the transmission and reflection efficiencies of lenses and mirrors along the optical path, and any field and collimation errors. It is impractical to try to evaluate these individual components, and instead the obvious empirical approach is taken. We identify the continuum by looking at the observed spectrum. "Looking" might be with our eyes, but more likely continuum points will be identified through some algorithm that considers high points in specified wavelength intervals. This can be a balancing act between the actual continuum level and noise. The statistical distribution of noise will make the highest points noise excursions that lie above the true continuum. It is a good idea to make a visual check to see if any automated choices are being done satisfactorily.

Generic notations for the spectrum normalized to the continuum are F_v/F_c or F_λ/F_c or just F/F_c since the units cancel to give a unitless ratio, i.e., F_c is in frequency units in the first ratio, wavelength units in the second ratio, and the units of F in the third ratio.

When there are few spectral lines, the continuum is easy to identify and one can use low-order polynomials or splines to represent the continuum for the normalization division. Toward cooler stars, as the density of spectral lines increases, continuum points become more difficult to find. In some cases no continuum can be seen, and the normalization becomes an educated guess. When the spectral lines are broadened by high rotation, relatively large wavelength intervals must be spanned across which there is no continuum information. The same problem arises with wide hydrogen lines. In such cases, the continuum might be estimated from a model photosphere continuum or from some other star that has narrower lines. The problem is accentuated when using high-order echelle spectrographs because of the narrow blazes. Considerable effort has gone into fixing this echelle problem, e.g., Giribaldi *et al.* (2019) and references therein.

After the spectrum is normalized to the continuum, we can explore the spectral lines. But first we review the basics.

The Observed Spectrum versus the True Spectrum

Our observations of spectral lines are always imperfect. The basic processes relating the observed data, $D(\lambda)$, to the true profile, $F(\lambda)$, were discussed in Chapter 2, but are so fundamental that they are resummarized here in Figure 12.5. Here we specify $D(\lambda)$ and $F(\lambda)$ using *line depth*, measured downward from the continuum, so they have maxima at line center. The true profile is convolved with the instrumental profile, $I(\lambda)$, recorded over a finite wavelength range, $W(\lambda)$, and sampled, $Ш(\lambda)$, at the spacing of the detector pixels, $\Delta\lambda$. Or

$$D(\lambda) = F(\lambda) * I(\lambda) \, W(\lambda) \, Ш(\lambda). \qquad (12.2)$$

In the Fourier domain, $i(\sigma)$ filters $f(\sigma)$, the window sinc function, $w(\sigma)$, smooths the transform, removing rapid variations with σ, and the sampling replicates the signal on a spacing of twice the Nyquist frequency, $2\sigma_N = 1/\Delta\lambda$. Or

$$d(\sigma) = f(\sigma) \, i(\sigma) * w(\sigma) * Ш(\sigma). \qquad (12.3)$$

The $W(\lambda)$ window is usually of little concern since it can be made arbitrarily wide by appending continuum values, if the actual observed wavelength span is not already at least a few times larger than the profile being measured.

Aliasing or the overlap of replications of $d(\sigma)$ is avoided (lower right in Figure 12.5) if the highest frequencies contained in $d(\sigma)$ are less than $\sigma_N = 0.5/\Delta\lambda$. This is important. We often have control over the sampling step, $\Delta\lambda$, in the detector we choose, specifically in the size of its pixels. If we choose poorly, the observations could be compromised. Normally we want to profit from the full resolution of the spectrograph, and that means we must choose $\Delta\lambda$ to give a minimum of two samples across the characteristic width of the instrumental profile. Three or even four samples might be better, but that depends on $d(\sigma)$. If $d(\sigma)$ is much narrower than $i(\sigma)$, two is sufficient. In the unexpected case, where $d(\sigma)$ is many times narrower than $i(\sigma)$, a lower resolution spectrograph could be used, yielding the advantage of shorter exposure times and a wider wavelength field.

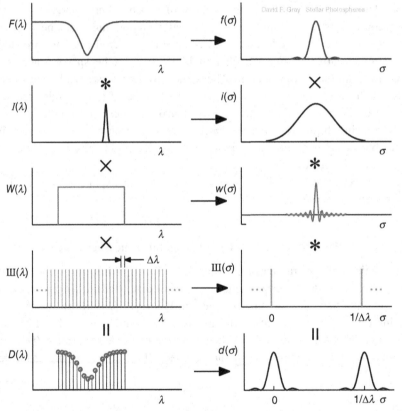

Figure 12.5 The wavelength domain on the left shows how the true spectrum, $F(\lambda)$, is convolved by the instrumental profile, $I(\lambda)$, multiplied by the wavelength window, $W(\lambda)$, and by the sampling Shah function, $\mathrm{III}(\lambda)$, to give the observed data, $D(\lambda)$. On the right we have the corresponding processes in the Fourier domain.

Now we look in detail at the first process in Figure 12.5, the process that defines how much detail our observed spectral lines can reveal.

The Instrumental Profile

Every spectrograph imprints its signature on the shapes of the spectral lines. This signature is the instrumental profile, $I(\lambda)$. It is the profile the spectrograph gives when the input spectrum is a δ-function. When we visualize the stellar spectrum, $F(\lambda)$, as a synthesis of δ-functions with their heights modulated by $F(\lambda)$, as in Figure 12.5, each δ-function gives an $I(\lambda)$ in the output spectrum, but it is shifted to the wavelength of the δ-function. Therefore, the recorded spectrum, $D(\lambda)$, is the sum of these shifted instrumental profiles, our standard convolution,

$$D(\lambda) = \int_{-\infty}^{\infty} I(\lambda - \lambda_0) F(\lambda_0) d\lambda_0$$

$$= I(\lambda) * F(\lambda), \tag{12.4}$$

with zero wavelength chosen to be at the center of the spectral region of interest. We normalize $I(\lambda)$ to unity so the strengths of any features in $F(\lambda)$ remain unaltered. As we know, this kind of convolution is a blurring process, so $I(\lambda)$ is a measure of how much the spectrograph has degraded the true spectrum. To state the obvious, we want $I(\lambda)$ to be significantly narrower than the spectral lines in $F(\lambda)$. If this is not so, the spectral lines take on the shape of $I(\lambda)$ and we are no longer doing high-resolution spectroscopy. (Even in such a disadvantaged situation, some science can still be done, such as chemical analysis and radial velocity determinations, but they would be better done with higher resolution.)

The instrumental profile embodies the combined effects of entrance-slit width, as in Equation 12.1, the grating diffraction width, as in Equation 3.7, the detector pixel size, optical aberrations, and imperfect camera focus. In stellar spectroscopy, the first of these dominates. This is driven by the need to let in as much starlight as possible. Consequently, the blurring effect of $I(\lambda)$ is always part of our observed stellar line profiles. To understand our observations of high-resolution line profiles, we must know $I(\lambda)$. We can measure $I(\lambda)$ using a δ-function spectrum.

Delta-Function Spectra

Very sharp lines, δ-function spectra, are needed to measure the instrumental profile. There are three convenient sources for generating very narrow spectral lines: an isotopic mercury lamp, a stabilized laser, and telluric absorption lines. Lines from these sources are so narrow compared to the instrumental profile of stellar spectrographs that they can be considered to be δ-functions.

A suitable mercury bulb is shown in Figure 12.6. It is simply a sealed glass tube containing a small amount of ^{198}Hg. The mercury is excited to emission using a microwave resonant cavity, a process that produces only minor thermal broadening compared to direct heating or electrical discharge. Air cooling is used for the highest resolution work. The thermal width (Equation 11.40) is 3 mÅ or 0.16 km/s at a room temperature of 300 K. If the instrumental profile is narrower than ~30 mÅ, a Gaussian correction can be applied for the intrinsic width of the mercury line, otherwise none is

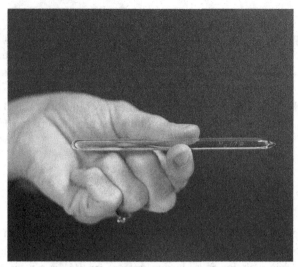

Figure 12.6 A small amount of [198]Hg, too small to see in this picture, is sealed in a glass
tube. The mercury is excited by microwave radiation and cooled by an airflow to keep the
thermal broadening low.

Table 12.1 *[198]Hg spectrum*

λ	Strength	λ	Strength
3125.674	0.72	4046.565	2.14
3131.555	0.52	4077.837	0.17
3131.844	0.63	4339.22	0.01
3341.484	0.12	4347.506	0.03
3650.158	1.51	4358.335	2.04
3654.842	0.56	4916.07	0.004
3663.887	0.09	5460.750	1.00
3663.284	0.39	5769.610	0.18
3704.17	0.002	5789.67	0.004
3906.37	0.006	5790.670	0.16

Adapted from the *American Institute of Physics Handbook*, D.E. Gray, ed., Copyright 1963 by
McGraw-Hill Book Co., used with permission, and from Reader & Sansonetti (1999).

likely to be needed. Table 12.1 lists the strong lines in the [198]Hg visible spectrum. Most
frequently used is λ5461, the mercury green line. The strengths of the other lines are
compared to that of λ5461.

The laser gives a narrow emission line by virtue of the phase coherence of the stimulated
emission. The wavelength spread of an untuned laser encompasses several modes of
oscillation of wavelengths given by $\lambda_n = 2L/n$, where L is the length of the laser's resonant
cavity and n is the integer number of half wavelengths in the spacing L. In a one meter-long
laser, 10–20 modes can lie within the thermal profile. A single mode can be selected by

inserting an etalon into the beam, slightly tilted to the optical axis. One can switch modes by changing the tilt of the etalon, the temperature of the etalon, or by adjusting the length of the laser cavity.

In practice, simple inexpensive lasers without special stabilization and without mode selection can be used as long as the resolving power of the spectrograph does not exceed about 100 000. But with all lasers, the coherence of the beam should be destroyed to reduce diffraction at the entrance slit and to fill the collimator with the proper f-ratio. This can be accomplished be sending the laser light through diffusing screens or by using multiply reflected light (e.g., Griffin 1968).

The telluric absorption lines of molecular oxygen are strong lines with dispersion profiles having half widths of ~13 mÅ (Panofsky 1943). Coupled with the Sun, Moon, or other bright object as a background source, these lines can be suitable δ-functions for measuring the instrumental profile. Panofsky measured several lines near 6300 Å, and his tabulation can be used should allowance need to be made for their finite widths. Griffin (1969a, 1969b, 1969c) used the 6276 Å, 6867 Å, and 7593 Å lines for instrumental profile measurement.

Measurement of the Instrumental Profile

Appropriate optics should be used to make the beam of the δ-function source similar to that of a star's. This is implicit for telluric lines. For other sources, often a lens and diaphragm suffice.

Since the instrumental profile can depend on the focus of the spectrograph, especially for short focal length cameras, $I(\lambda)$ should be checked often, preferably each night stellar observations are taken. If $I(\lambda)$ shows little or no variation, then the measurement cadence can be relaxed. The instrumental profile can vary across the field. Cross-dispersed echelle spectrographs are particularly vulnerable. Checking the profile of the same *stellar* line recorded in adjacent echelle orders can show up field variations in $I(\lambda)$.

The projected slit width, w, measured in millimeters at the spectrum, is independent of the order according to Equation 3.14. Therefore when $I(\lambda)$ is dominated by the slit width (the usual case), it is basically independent of the order. The instrumental profile is converted to a λ scale using the dispersion of the stellar spectrum at the wavelength interest, and it should be normalized to unit area at that wavelength. Ghosts, however, scale in strength as the square of the order, and this should be taken into account if ghosts are significant.

There may also be small wavelength dependences entering through the projection factor of the entrance slit width, $\cos \alpha / \cos \beta$ of Equation 3.14. For this reason, $I(\lambda)$ should be measured with the grating oriented as close as possible to the configuration used for the stellar exposure, even if different orders are used. In extreme cases, a correction may have to be applied. Tests for any wavelength dependence can be made by taking measurements at several different wavelengths, for example, with different lines from Table 12.1.

The top panel of Figure 12.7 shows a spectrograph instrumental profile measured using the mercury green line, $\lambda 5461$. Although the shape is not quite Gaussian, the $1/e$ width is ~50 mÅ or ~2.5 km/s. It is normalized to unit area, which in this example is 1 Å. The lower panel shows the transform or the reduction in Fourier amplitudes this instrumental profile causes.

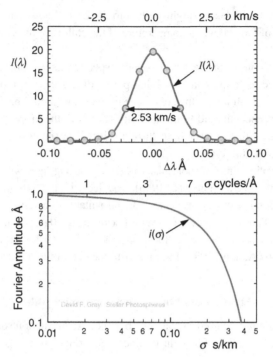

Figure 12.7 Top: a spectrograph instrumental profile measured (points) using the 5461 Å line of ^{198}Hg. The points are 15 μm apart. The resolution is ~60 μm, ~2.5 km/s, or 0.052 Å. Bottom: the Fourier filtering caused by this $I(\lambda)$.

The Instrumental Profile in the Fourier Domain

Denote the Fourier transforms of $D(\lambda)$, $F(\lambda)$, and $I(\lambda)$ in Equation 12.2 by their lower case counterparts, the way we normally do. Then in the Fourier domain, Equation 12.4 gives

$$d(\sigma) = i(\sigma)f(\sigma) \tag{12.5}$$

in which σ is the Fourier frequency measured in cycles per angstrom when the wavelength is expressed in angstroms, or in cycles per kilometer/second or s/km, when the wavelength scale is expressed in velocity units. The function $i(\sigma)$ acts as a filter on $f(\sigma)$, and in the cases of interest for stellar spectroscopy, diminishes monotonically toward higher σ. The lower panel in Figure 12.7 shows the transform of the instrumental profile in the upper panel. Since it is essentially symmetric around $\sigma = 0$, only half of the full profile is shown. Logarithmic coordinates are used to better show the curve at high frequencies.

An example of instrumental blurring of a spectral line is shown in Figure 12.8. In the transform domain, the amplitudes of $f(\sigma)$ are reduced by increasing amounts as σ increases. In the wavelength domain, the core, which has the highest frequencies, is blurred the most. Higher resolution is always advantageous in this context. Higher resolution means a narrower $I(\lambda)$ and a wider $i(\sigma)$, i.e., less filtering, and our observations are closer to reality.

Technical note: remember, in plots like these, $f(\sigma)$ and $d(\sigma)$ are the transforms of the *line depths* of the profiles, so normal absorption lines look mathematically like emission lines. When using a logarithmic scale for amplitudes, use their absolute value, since logarithms of negative numbers are not real numbers.

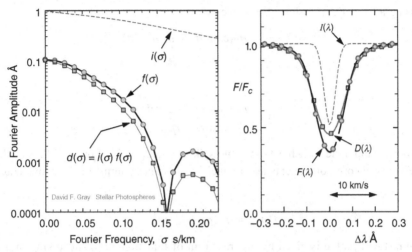

Figure 12.8 Left: transform of the true profile, $f(\sigma)$, is reduced by the transform of the instrumental profile, $i(\sigma)$, down to the transform of the observed profile, $d(\sigma) = i(\sigma)f(\sigma)$. Right: the corresponding profiles in the wavelength domain, $F(\lambda)$ is the true profile, $I(\lambda)$ the instrumental profile arbitrarily scaled in amplitude, and $D(\lambda) = F(\lambda) * I(\lambda)$ is the observed profile. The approximate $1/e$ width of this instrumental profile is 3.3 km/s.

In principle the blurring by $I(\lambda)$ can be removed and $F(\lambda)$ restored, at least in part, by the process to which we now turn our attention.

The Restoration Process

The goal of restoration is to remove the blurring of the instrumental profile and bring the observed line profile as close as possible back to the true one, which is then compared to the profile computed from our model photosphere. The process does not work well with astronomical spectra, but it is nevertheless instructive. There are allied techniques that do work well. The first is to do most of the analysis in the Fourier domain, and we will see numerous examples of this approach in later chapters. The second is to convolve the computed profile with the instrumental profile and then compare the convolution to the observations. Examples of this approach are also given later.

If we temporarily ignore noise in the data, the removal if the instrumental filtering follows directly from Equation 12.5,

$$f(\sigma) = d(\sigma)/i(\sigma), \qquad (12.6)$$

and the restored $F(\lambda)$ is the transform of this ratio. The reconstruction process should only be done below the Nyquist frequency, i.e., for $\sigma \leq \sigma_N = 0.5/\Delta\lambda$, as in Figure 12.5.

We must allow these functions to be complex in order to handle asymmetric profiles. The ratio in Equation 12.6 is therefore a vector division. In rectangular coordinates, with R and I subscripts identifying the real and imaginary components,

$$f(\sigma) = f_R(\sigma) + if_I(\sigma), \tag{12.7}$$

and using a similar expression for $i(\sigma)$ we get

$$f_R(\sigma) = \frac{d_R(\sigma)i_R(\sigma) + d_I(\sigma)i_I(\sigma)}{i_R^2(\sigma) + i_I^2(\sigma)}$$

and

$$f_I(\sigma) = \frac{d_I(\sigma)i_R(\sigma) + d_R(\sigma)i_I(\sigma)}{i_R^2(\sigma) + i_I^2(\sigma)}.$$

The Fourier amplitude, which we use in plots like the left panel in Figure 12.8, is the length of the vector whose components are the real and imaginary terms of the above equation, that is,

$$|f(\sigma)| = \left[f_R^2(\sigma) + f_I^2(\sigma) \right]^{1/2}.$$

So the restored profile is given by the transform of Equation 12.6, but lurking in $d(\sigma)$ is noise, and that noise gets amplified when $i(\sigma)$ is divided out.

Noise and Its Complications

It is difficult or impossible to get the noise in stellar observations low enough to allow restoration to be successful. Figure 12.9 shows the Fourier transform of a 5 Å portion of the observed spectrum of Procyon (F5 IV-V). In the Fourier domain, the signal, $d(\sigma)$ goes from being strong on the left to below the noise on the right. The statistically constant noise level dominates $d(\sigma)$ for log σ higher than about -0.7 dex in the figure. The restoration outlined above amounts to mathematically increasing the Fourier amplitudes of $d(\sigma)$ to undo what the instrumental filtering has done. But the noise as well as the signal is increased by this process. Let $n(\sigma)$ be the noise in the transform. The noise in $i(\sigma)$ is generally negligible compared to the noise in $d(\sigma)$, but if this proves not to be the case, then $n(\sigma)$ includes the noise of both.

The transform like the one in Figure 12.9 can be written

$$d(\sigma) = d_0(\sigma) + n(\sigma), \tag{12.8}$$

in which $d_0(\sigma)$ is the true stellar signal. The restoration is then

$$f_1(\sigma) = \frac{d_0(\sigma)}{i(\sigma)} + \frac{n(\sigma)}{i(\sigma)}.$$

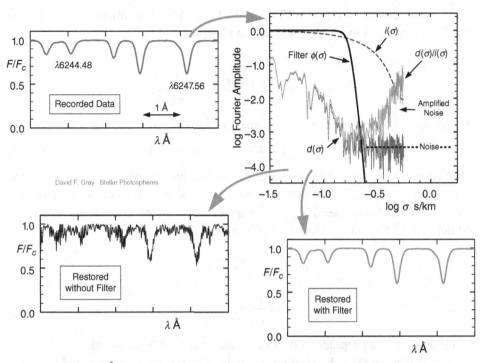

Figure 12.9 A 5 Å portion of the spectrum of Procyon (F5 IV-V) is shown in the upper left. Its transform is shown in the upper right panel in as $d(\sigma)$. For frequencies higher than about -0.7 dex, noise dominates. Dividing out $i(\sigma)$ results in amplified noise, shown by the rising $d(\sigma)/i(\sigma)$ line. Simply returning to the wavelength domain yields the restored (without filter) plot at the lower left, a less-than-satisfactory result. Applying the noise filter, $\phi(\sigma)$, results in a more reasonable result, shown in the lower right panel. The strongest two lines are ~3% deeper after the restoration process.

The first term gives the true signal $f(\sigma) = d_0(\sigma)/i(\sigma)$ back again, but the second is amplified noise, and it increases as $i(\sigma)$ decreases toward higher σ. This is shown in the upper right panel of Figure 12.9. Returning to the wavelength domain with amplified noise like this simply results in a mess, as shown in Figure 12.9 in the panel labeled "Restored without Filter." But because the stellar signal is concentrated toward lower frequencies, while the amplified noise is at higher frequencies, the noise can be reduced with the aid of a low-pass filter.

Fourier Noise Filters

The so-called optimum filter minimizes the square of the errors between the true transform and the filtered-restored transform. Define the filtered-restored transform to be

$$f_2(\sigma) = d(\sigma)\phi(\sigma)/i(\sigma), \tag{12.9}$$

where $d(\sigma)$ is defined by Equation 12.8, and $\phi(\sigma)$ is the filter. The difference between $f(\sigma)$ and $f_2(\sigma)$ is the error in the reconstruction and filtering process,

$$\varepsilon = f(\sigma) - f_2(\sigma)$$
$$= f(\sigma) - \left[\frac{d_0(\sigma)\phi(\sigma)}{i(\sigma)} + \frac{n(\sigma)\phi(\sigma)}{i(\sigma)}\right].$$

And since $d_0(\sigma)/i(\sigma) = f(\sigma)$ this is the same as

$$\varepsilon = f(\sigma)[1 - \phi(\sigma)] - n(\sigma)\phi(\sigma)/i(\sigma). \tag{12.10}$$

The first term shows how the filter introduces an error by not giving exact reconstruction of $f(\sigma)$. The second term shows that the enhanced noise, $n(\sigma)/i(\sigma)$, can be reduced by a properly chosen filter. Specifically, one requires $\int_{-\infty}^{\infty} \varepsilon^2 \, d\sigma$ be a minimum. The result (Middleton 1960, Brault & White 1971, Champeney 1973) is

$$\phi(\sigma) = \frac{d^2(\sigma)}{d^2(\sigma) + n^2(\sigma)}$$
$$= \frac{1}{1 + [n(\sigma)/d(\sigma)]^2}. \tag{12.11}$$

So the optimum $\phi(\sigma)$ depends on the ratio of the noise to observed signal.

In most spectroscopy observations, the noise is statistically constant, and that level can be seen in the high-frequency end of the transform if there is sufficient over-sampling of the data. Such is the case in the transform panel of Figure 12.9, where the mean noise level is labeled "Noise." Call the mean noise level B. Next, for purposes of illustration, and since many spectral lines are roughly speaking Gaussian in shape, approximate $d(\sigma)$ with a Gaussian in the form $A\,e^{-(\pi\beta\sigma)^2} = A\,10^{-b_0^2\sigma^2}$. Then Equation 12.11 becomes

$$\phi(\sigma) = \frac{1}{1 + (B/A)^2 10^{2b_0^2\sigma^2}}. \tag{12.12}$$

The ratio B/A is the mean noise level divided by the zero ordinate of $d(\sigma)$. The fitting parameter, b_0, is chosen according to the Gaussian that most closely matches $d(\sigma)$. Figure 12.10 shows Gaussian models and the corresponding filters as a function of b_0. An observed $d(\sigma)$, scaled to unit ordinate, is matched passably well by the $b_0 = 8$ Gaussian. So for this spectral line, B/A from the observations would be combined with $b_0 = 8$ in Equation 12.12 to give a suitable filter for the restoration.

More generally, when there are several spectral lines, the observed $d(\sigma)$, is used to choose the filter parameters. The panel labeled "Restored with Filter" in the lower right of Figure 12.9 shows the restoration with the noise filter of this form, as shown in the Fourier-domain panel of the figure. This restoration is reasonable, but depends upon (1) the instrumental profile being narrow enough compared to the stellar lines so that only modest

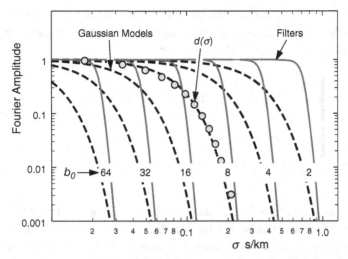

Figure 12.10 Gaussian models for $d(\sigma)$ and the corresponding noise filters are shown as a function of b_0. The transform of an observed line profile is shown by the points. One would choose $b_0 \sim 8$ from this match.

alterations in the stellar signal are involved and (2) the signal-to-noise level of the observations being high, about 1000 for the data in Figure 12.9. In less favorable cases, the amplified noise extends down into the region of stellar signal and the restoration fails or is ambiguous. Too weak a Fourier filter makes the restored profiles noisy and the line depths sensitive to b_0. Too strong a filter reduces the resolution, fundamentally nullifying the restoration process.

Scientifically, there is no advantage in making a restoration over the alternative of proceeding in the opposite direction, that is, convolve the instrumental profile into the calculated spectrum and compare that to the observed spectrum. This alternative avoids the complications of enhanced noise and having to choose a noise filter.

Scattered Light in Spectrographs

Light scattered inside a spectrograph is bad. It distorts the spectral lines and changes their apparent strength. Scattered light comes from optical imperfections, dust on the optical surfaces, striae, bubbles, and granulation in lenses and filters, and from dust in the air. Scattered light can reach several percent, and is not to be casually dismissed.

In principle, scattered light would be included in the instrumental profile if $I(\lambda)$ could be measured over all wavelengths entering the spectrograph. In practice, $I(\lambda)$ is measured over a much narrower interval and the missing far wings of $I(\lambda)$ comprise the scattered light. Suppose we have some spectral line whose profile we wish to measure, centered on $\lambda = 0$. Separate the integration range in Equation 12.4 into three sections: (1) the range over which we actually measure $I(\lambda)$, $\pm \lambda_I$, (2) the wavelength range of light entering the spectrograph, $\pm \lambda_S$, and (3) the formal mathematical extension outside $\pm \lambda_S$ that extends to $\pm\infty$. The last of these is zero (no light outside $\pm \lambda_S$), so we have

$$D(\lambda) = \int_{-\lambda_S}^{-\lambda_I} I(\lambda - \lambda_1) F(\lambda_1) d\lambda_1 + \int_{-\lambda_I}^{\lambda_I} I(\lambda - \lambda_1) F(\lambda_1) d\lambda_1 + \int_{\lambda_I}^{\lambda_S} I(\lambda - \lambda_1) F(\lambda_1) d\lambda_1$$

$$= S(\lambda) + \int_{-\lambda_I}^{\lambda_I} I(\lambda - \lambda_1) \, F(\lambda_1) \, d\lambda_1, \tag{12.13}$$

in which $S(\lambda)$ is the scattered light. And since we are talking about light being scattered, $F(\lambda)$ and $D(\lambda)$ here are fluxes (not line depth). In most cases, $S(\lambda)$ has only a weak wavelength dependence. In which case, Equation 12.13 tells us that adding scattered light is like placing the spectrum on a pedestal or shifting the zero-light level down-ward. If we let $\langle F \rangle$ represent the average flux entering the spectrograph over the full $\pm \lambda_S$ range,

$$S(\lambda) = \int_{-\lambda_S}^{-\lambda_I} I(\lambda - \lambda_1) \, F(\lambda_1) \, d\lambda_1 + \int_{\lambda_I}^{\lambda_S} I(\lambda - \lambda_1) \, F(\lambda_1) \, d\lambda_1$$

$$\approx \langle F \rangle \left[\int_{-\lambda_S}^{-\lambda_I} I(\lambda) \, d\lambda + \int_{\lambda_I}^{\lambda_S} I(\lambda) \, d\lambda \right],$$

where the first integral is the short-wavelength wing and the second is the long-wavelength wing. Since $I(\lambda)$ is, by definition, normalized to unit area, the instrumental profile we measure between $\pm \lambda_I$ has less than unit area according to

$$\int_{-\infty}^{\infty} I(\lambda) \, d\lambda = 1 \approx \frac{S(\lambda)}{\langle F \rangle} + \int_{-\lambda_I}^{\lambda_I} I(\lambda) \, d\lambda. \tag{12.14}$$

This equation expresses the important and interesting result that the measured instrumental profile should be normalized to unity less the fraction of scattered light.

Scattered light in the Fourier transform plane shows up as a spike at $\sigma = 0$. Figure 12.11 shows an extreme example to illustrate the point, where the scattered light amounts to 35%. More commonly, scattered light is less than 5-10%. When this kind of $i(\sigma)$ is multiplied with the transform of $F(\lambda)$, all Fourier amplitudes are reduced by the fraction of scattered light, i.e., the size of the scattered light spike in Figure 12.11, in addition to the normal filtering of $i(\sigma)$. Before we can correct for scattered light, we must be able to measure it.

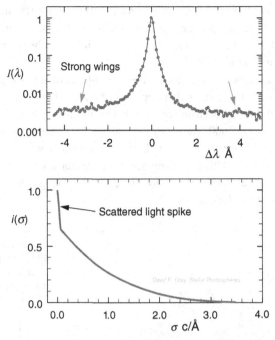

Figure 12.11 An example of a very poor instrumental profile and its transform. The extensive, low-level wings are the scattered light, and they give a δ-function-like spike in the transform at $\sigma = 0$.

Measurement of Scattered Light

No perfect means of measuring scattered light in spectrographs has been devised. But because the scattered light is usually modest, i.e., a few percent, it need not be known with high precision. It is convenient to think of the scattered light being made up of two components: (1) the general scattered light that has no preferential orientation, such as that scattered from dust, and (2) the so-called linearly scattered light that goes in the direction of dispersion and arises from the entrance slit and grating.

The general scattered light can be measured by looking above and below the continuous spectrum formed by an incandescent light bulb (Gray & Evans 1973) or by covering a portion of the entrance slit and measuring the light in the obscured portion of the image (Shane 1932, Conway 1952, Edmonds 1965). By definition, this component of scattering is approximately uniform over the field, so scattered light measured near the spectrum is the same as in the spectrum.

A determination of linear scattering cannot be made so easily because the spectrum itself is in the way. Waddell (1956) placed an absorption cell in front of the entrance slit, and the residual flux in the core of the highly saturated D_1 line was measured. In a similar way, one can use a rejection-band interference filter at the entrance slit to introduce an artificial absorption band into a continuous spectrum. Since the rejection bands are known to be opaque to better than one part in 1000 typically, any light above that level must be due to scattering. An integration of the filter transmission allows one to calculate $\langle F \rangle$, and thus the percentage scattered light. A similar approach was used by Teske (1967) and Goldberg *et al.* (1959) in which a Fabry–Perot interferometer replaced the rejection-band filter.

Remembering that scattered light looks like a zero-level offset, a straightforward test is to look at the cores of strong lines that approach the zero light level, such as the H and K lines of Ca II in the solar spectrum or, even better, in the spectrum of Arcturus (Kurucz *et al.* 1984, Hinkle *et al.* 2000, Reiners *et al.* 2016). If scattered light is significant, the deepest parts will be higher than expected. Spectra with and without scattered light are compared in Figure 12.12. Although there are other weaker lines with deeper cores, their depth will depend on the spectral resolution, whereas the changes in the broad K line are almost independent of resolution.

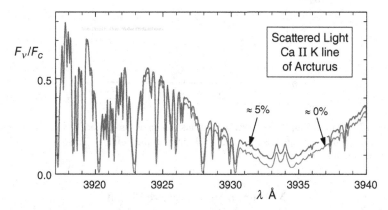

Figure 12.12 The inner profile of the Ca II K line is shown with little to no scattered light compared to ~5% scattered light. Such a difference is easy to see because the core is at low relative flux levels. Based on Hinkle *et al.* (2000).

This check does not say much about scattered light at other wavelengths, and there may also be slight temporal variations in the spectra themselves.

Scattered Light Reduction and Correction

If scattered light can be reduced, its presence can be ameliorated. Keep unwanted light out of the spectrograph: no light leaks from the outside, no "on/off" lights on electronics inside the spectrograph, no fluorescent-type lights that have an afterglow, no clocks or timers with glow-in-the-dark dials. Order-sorting-type filters can be used to restrict the light allowed into the spectrograph to only the wavelength span of interest, i.e., reduce $\pm\lambda_S$. This is one place where gratings used in a single-order gain over cross-dispersed echelles.

Kill off errant light inside the spectrograph. Sometimes offending optics can be identified and cleaned or replaced. Dust can be removed from optical surfaces, for example, with carbon dioxide jets. Dust can be filtered from the air. Baffles can be used to stop light from spreading. Surfaces inside the spectrograph, be they mirror mounts, grating cell, or other hardware, and ceilings and walls in coudé spectrographs or the inside surfaces of smaller spectrographs can all be made non-reflecting, usually with matt black paint. Notice the black dominance in Figure 12.1a.

Corrections for scattered light should be attended to when dealing with the instrumental profile. Perhaps the simplest method is to alter the normalization of the measured $I(\lambda)$ according to Equation 12.14 (Gray 1974). A slightly more laborious approach is to correct the observed spectrum directly (e.g., Teske 1967). Denote the fractional scattered light by $s = S(\lambda)/\langle F \rangle$. We can reasonably take $\langle F \rangle$ to be the average measured spectrum $\langle D \rangle$. Then by calling the observed light D_{obs} and the true value that would exist without scattering D_{true}, we can write

$$D_{obs} = D_{true} - sD_{true} + s\langle D \rangle$$

$$= (1 - s)D_{true} + s\langle D \rangle,$$

where sD_{true} is the light scattered out of the profile and $s\langle D \rangle$ is the light scattered into the profile. Or

$$D_{true} = \frac{D_{obs} - s\langle D \rangle}{1 - s}. \tag{12.15}$$

Using a superscript c to denote continuum values, Equation 12.15 can be written for the continuum as well. Then the true profile in flux is given by D_{true}/D_{true}^c :

$$\frac{D_{true}}{D_{true}^c} = \frac{D_{obs} - s\langle D \rangle}{D_{obs}^c - s\langle D \rangle}$$

$$= \frac{D_{obs}/D_{obs}^c - s\langle D \rangle/D_{obs}^c}{1 - s\langle D \rangle/D_{obs}^c}, \tag{12.16}$$

in which D_{obs}/D_{obs}^c is the observed flux profile. In an uncrowded spectral region, where $\langle D \rangle \approx D_{obs}^c$, this expression becomes

$$\frac{D_{true}}{D_{true}^c} = \frac{D_{obs}/D_{obs}^c - s}{1 - s}. \tag{12.17}$$

Then the correction amounts to raising the zero level by s and renormalizing to unit continuum. Figure 12.13 shows the effect of scattered light on the spectrum. The change in the line cores is most obvious. Superficially this might be mistaken as lower resolution stemming from a wider instrumental profile. With scattered light, however, it is not the higher Fourier frequencies that are preferentially filtered; all frequencies are filtered equally, and the changes are strictly proportional to the line depth at each point. The scattered light "fills in" the line, reducing the contrast, an effect we already noticed with regard to Figure 12.11.

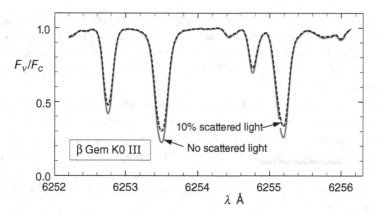

Figure 12.13 Scattered light fills in the line profiles, with the cores being most affected. Elginfield data (No scattered light) with 10% scattered light artificially added.

Determination of the Dispersion and the Wavelength Scale

The values of dispersion and the wavelength scale that goes along with it are central for extracting information from spectral lines. The spectrograph gives us the spectrum as a function of position, x, on the light detector. We need a relation between x and λ. The most direct way to get it is to select a series of spectral lines with known wavelengths spread across the recorded spectrum. Then we measure the x positions of these lines and we have what we need. The spectrum used to get the $\lambda(x)$ relation can be from the star we are studying, if it has suitable lines, or if it doesn't, we can use a totally different star. Or we can turn to spectra from emission-line lamps, telluric absorption spectra, and so on. The $\lambda(x)$ relation can often be represented by a cubic polynomial, making it easy to assign a wavelength to each point in the spectrum, and to calculate the linear dispersion $d\lambda/dx$. This approach is quite adequate for many studies involving profile shapes and the total absorption in the line.

Some items have been left unsaid. First, we need to know the pixel spacing, say δx. From it we generate an x scale, $x = n \, \delta x$, where n is the pixel count starting at 0 and going up to the number of pixels in the direction of dispersion. Second, we need a suitable way to measure the positions of the spectral lines. We might use the positions of the line cores, located by eye or by fitting a vertically oriented parabola or some other such function to the cores. Alternatively, we might use the midpoints of horizontal lines between the sides of the profiles, say halfway up. There are many possibilities. And to measure line positions accurately with any technique the lines should be sharp (more on this later). Third, we need values for the wavelengths for the lines we're using. Compendiums like Moore *et al.* (1966), Hinkle *et al.* (2001), or the National Institute of Standards and Technology web site (physics.nist.gov/asd) are good sources. For accurate specialized studies see Forsberg (1991, Ti I), Litzén *et al.* (1993, Ni I), Nave *et al.* (1994, Fe I), Nave and Sansonetti (2011, several elements), Redman *et al.* (2014, Th), among others. Fourth, if laboratory wavelengths are used with stellar line positions, the stellar spectrum will effectively be shifted to that rest frame, i.e., the radial velocity shifts of the star are lost. Further, since Doppler shifts scale with wavelength, each star's spectrum will be dilated according to its radial velocity, so the $\lambda(x)$ relation can be slightly different for stars having different radial velocities.

In some situations, it is important to distinguish between the vacuum wavelength and the measured wavelength. The usual convention is to give wavenumbers $(1/\lambda)\,\mathrm{cm}^{-1}$ in vacuum, but wavelengths in air a temperature of $15\,^\circ\mathrm{C} = 288.15\,\mathrm{K}$ and a pressure of $760\,\mathrm{mm\ Hg} = 1.01325 \times 10^5\,\mathrm{Pa}$, often denoted by STP (standard temperature and pressure). Most stellar spectrographs are operated in air. Denote the index of refraction of air by $n = c/v$ where c is the velocity of light in vacuum and v is the velocity of light in air. Although n in the visible region is only ~ 1.0003, it produces a very significant shift in wavelength. The air wavelength is related to the vacuum value by $\lambda_{\mathrm{air}} = \lambda_{\mathrm{vac}}/n$, so the wavelength shift is

$$\Delta\lambda = \lambda_{\mathrm{vac}} - \lambda_{\mathrm{air}} = \lambda_{\mathrm{vac}}(1 - 1/n)$$

$$= \lambda_{\mathrm{air}}(n - 1). \tag{12.18}$$

This amounts to 1–$2\,\text{Å}$ in the visible region of the spectrum. In velocity this is ~ 90 km/s.

The wavelength dependence of n can be written (Edlén 1953, 1966, Birch & Downs 1994)

$$10^6(n_{\mathrm{STP}} - 1) = 64.328 + \frac{29498.1}{146 - \lambda^{-2}} + \frac{255.4}{41 - \lambda^{-2}}, \tag{12.19}$$

in which the wavelength is in μm. At $6250\,\text{Å}$, for instance, we find $n = 1.00027662$.

Humidity changes introduce slight variations. Barrell and Sears (1938) give

$$\mathrm{H} = \frac{0.0624 - 0.000680/\lambda^2}{1 + 0.003661\,T}\,V_p,$$

to be subtracted from Equation 12.19, where λ is in μm and V_p is the saturated vapor pressure given by

$$V_p = 4.4748 + 0.3983\,T + 0.0034\,T^2 + 0.0005\,T^3$$

for T in degrees C. At $15\,^\circ\mathrm{C}$ and $6250\,\text{Å} = 0.6250\,\mu\mathrm{m}$, we find $V_p = 12.90\,\mathrm{mm\ Hg}$ and $\mathrm{H} = 0.74$. This is modest compared to $10^6(n_{\mathrm{STP}} - 1) = 276.2$.

In addition, the index of refraction of air is proportional to the air density, the so-called Dale–Gladstone law. At temperatures and pressures other than the standard values, one can use the perfect gas law to write

$$n - 1 = (n_{\mathrm{STP}} - 1)\frac{P/P_{\mathrm{STP}}}{T/T_{\mathrm{STP}}}. \tag{12.20}$$

Here the temperatures are on the Kelvin scale. Temperature and pressure drifts occurring during a night of observations can easily be a few tenths of a degree and one mm of mercury, corresponding to line shifts of -100 and $+100$ m/s, respectively. Seasonal variations can be several times larger. An observatory situated at high altitude will have pressure reductions of ~40% compared to sea level, making for tens of km/s equivalent velocity change.

Wavelength Changes from Barycentric Motion

The positions of spectral lines show apparent variations arising from the motion of the spectrograph relative to the center of mass of the solar system, called the *barycenter*. In situations where velocity or velocity variation is part of the investigation, the barycenter motion is calculated and removed from the observations. Such investigations include measurements of space motions of stars and binary star orbits. The dominant contributors are the orbital motion of the Earth, mean velocity = 29.78 km/s, and the rotation of the Earth, equatorial velocity = 0.465 km/s. Lesser contributions come from the Moon and the planets. Details on calculating the barycentric motion can be found in Stumpff (1979, 1980) and Wright and Eastman (2014).

If your spectrograph is on a satellite orbiting the Earth, the component of the orbital velocity in the direction to the star will also have to be taken into account.

Line Measurements with Low Spectral Resolution

Let us take a step backward and look at low-resolution observations. A good deal of information is not available when spectra have low resolution. Low resolution implies that the instrumental profile dominates the shapes of the spectral lines, so all information about their shapes has been lost. Why might one use low resolution? Low resolution requires less telescope time. Faint objects can be observed. What can we get out of low-resolution spectra? Spectral-type classification is done with low resolution (Chapter 1). Modest precision radial-velocity measurements can be made (see below). Total absorption in the lines can be obtained, and that is sufficient for chemical analyses.

The standard terminology for specifying the total absorption in a line is the *equivalent width*. The total absorption in the spectral line is equivalent to a lightless rectangle with the same area, as in Figure 12.14. Since the vertical extent of the rectangle is from 0.0 to 1.0, the width is numerically the same as the area. The equivalent width is formally defined as

$$W = \int_0^\infty \left(1 - \frac{F_\lambda}{F_c}\right) d\lambda, \tag{12.21}$$

with F_c denoting the flux in the continuum and F_λ the flux in the line profile. Notice that the fluxes can be in either wavelength or frequency units as long as they are the same and cancel. The integration, however, is on a wavelength scale. The equivalent width is expressed in angstroms, not hertz. The wide integration limits are symbolic. In reality the integral covers only one spectral line over the range where the profile can be distinguished from the continuum.

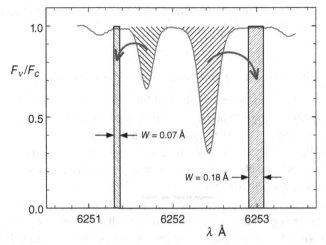

Figure 12.14 This is a short portion of the spectrum of β Her (G7 IIIa). Equivalent widths are shown for two lines. The area or equivalent width, W, of the weaker line is 0.07 Å or 70 mÅ. The stronger line has $W = 0.18$ Å or 180 mÅ. Data from the Elginfield Observatory.

If we delve a little deeper, we realize that we are measuring $D(\lambda) = I(\lambda) * F(\lambda)$, not $F(\lambda)$. Supposedly the convolution does not alter the equivalent width, W, since $I(\lambda)$ has unit area. But $I(\lambda)$ is now very broad, often comparable to or wider than the intrinsic profile $F(\lambda)$. This makes lines blend more with their neighbors, making the measurements less certain. This same effect also results in a lowering of the effective continuum, systematically biasing the equivalent widths. Such issues are considered by Wright (1958, 1966), Helfer *et al.* (1963), Cayrel de Strobel (1967), and Bedell *et al.* (2014). Notice too that scattered light systematically reduces the apparent equivalent widths, as we saw in Figure 12.13.

Integrating the area of a spectral line can be a simple numerical issue if the signal-to-noise ratio is high enough so that the profile is well defined and the line is unblended. Alas, many lines are blended, especially in cooler stars. Then one approach is to manually do blend corrections. A better approach, following Wright (1950, 1966), is to make a series of standard profiles for a range of line strengths using unblended lines or portions of lines. A blended profile can then be fitted into the grid to get its equivalent width. A still better method is to compute a reasonable match to an unblended profile using an appropriate model photosphere, line broadening parameters, and the instrumental profile. Then we can compute a profile for a blended line, using the same parameters, and adjust the equivalent width until the unblended portion is matched. One more thing to remember about blends is that they do *not* behave like random noise. They cannot be "averaged away" by using more lines in the analysis. Blends always work in the same direction, to lower the flux in the profile.

The equivalent widths of weak lines are especially important for chemical-composition studies, as we shall see in Chapter 16. Unfortunately, measuring weak lines with low resolution is not a smooth path to follow. Weak lines are shallow and hard to see clearly unless the signal-to-noise ratio is several hundred or more. Low resolution spreads out weak lines making them even more shallow. The lesson here is to tread carefully and not expect miracles from low-resolution spectra.

Line Broadening and Shape in the Wavelength Domain

Things get more interesting when the spectral lines are well resolved. In cool stars, this means having a spectrograph that yields $\lambda/\Delta\lambda$ in the 100 000 range, or a resolution of ~3 km/s. With resolved lines, we get equivalent widths with as much clarity as the intrinsic spectrum will allow. And far more information begins to flow from the line profiles, information about rotation, photospheric velocity fields, empirical temperature and gravity indices, oscillations, and surface spots.

Historically speaking, the half width was the first rudimentary parameter used to measure line broadening. Simply measure the width halfway down the line. The attractiveness of the half width lies in its ease of measurement and so it was frequently used in initial reconnaissance work. In the past it has been successfully used to measure rotation rates, especially large rotation rates (e.g., Slettebak *et al.* 1975, Abt & Morrell 1995, Fekel 1997). But one number leaves most of the information untapped.

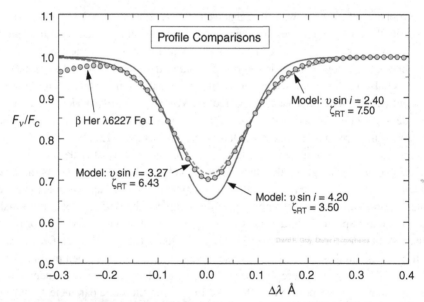

Figure 12.15 Circles: observed profile of Fe I λ6227 for β Her (G7 IIIa) is compared to three model profiles to show how physical information can be extracted from profile shape. The best fit model goes through the observed profile points. The other models have stronger rotation, weaker macroturbulence and weaker rotation, stronger macroturbulence. All three models have the same equivalent width and the same half width as the observed profile. Adapted from Gray (2016).

Consider the profile for β Her in Figure 12.15. The full shape of a line profile contains the combined signals from a multitude of processes. As investigators, we try to unravel these signals and understand the physical processes behind them. The profile in Figure 12.15 has a characteristic shape that is not explained by the Doppler shifts of rotation alone nor by those of macroturbulence alone (large-scale velocity fields in the photosphere), but by a combination of the two. The thermal and instrumental profiles are included in all three models. The amount of broadening is specified by the projected rotation rate, $v \sin i$, and the macroturbulence dispersion, ζ_{RT}. The full meaning of these parameters and the equations on which they are based is discussed in Chapter 17. There is clearly more information to extract from a detailed profile than its half width.

Line Broadening and Shape in the Fourier Domain

A valuable alternative way to view the information contained in profile shapes is to work in the Fourier domain. We have already seen some transforms of spectra earlier in the chapter. Figure 12.16 shows the profiles from Figure 12.15 in the Fourier domain. Recall that only the positive-frequency half is shown and that the transforms are of the profiles' line depths. The zero ordinate of the transform is the equivalent width of the line.

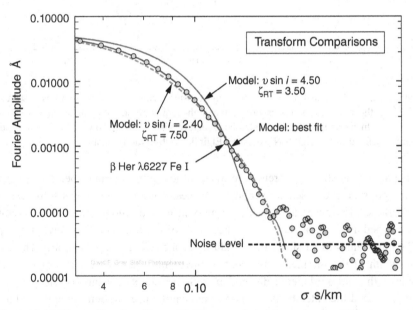

Figure 12.16 The profiles from Figure 12.15 convey their information in a different manner in the Fourier domain. Knowing the white noise level allows any Fourier amplitude to be trusted according to its individual signal-to-noise ratio.

In this type of comparison, no noise filter is involved. The general relation between signal and noise is easy to visualize in the Fourier domain. When the signal-to-noise ratio is improved in the λ domain, the noise level in the σ domain falls in proportion, according to Equations 2.41 or 2.42. With a lower noise level, we can follow a profile's transform to higher frequencies. This is illustrated in Figure 12.17. The first zero and the first sidelobe, revealed clearly in the high S/N exposure, can be important in establishing the nature of line broadening.

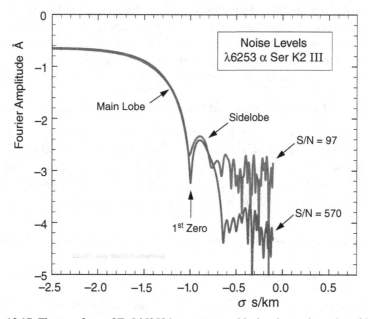

Figure 12.17 The transform of Fe I λ6253 in exposures with signal-to-noise ratios of 97 and 570. As the noise level moves down, the high-frequency structure in the transform is revealed. In particular, the first sidelobe is barely visible in the low S/N exposure, but clearly defined in the high S/N exposure. Elginfield Observatory data.

There are subtle shape differences in line profiles from the same spectrum even when the same physical processes are shaping the lines. Figure 12.18 shows some examples. Most of these differences arise from the amount of saturation. Lines of large strength have absorbed as much as they can in the center of the line, and they have developed a "box" component to their shape that produces a sidelobe in their transform. The characteristic width of the thermal component of broadening, as we saw in Equation 11.44, depends on the microturbulence dispersion, ξ. So it is not surprising that the σ position of the first zero and the sidelobe depend on ξ and can be used to determine the value of ξ. The sharpness of the first zero allows us to find ξ with greater precision than by using the profile.

There are other advantages in the Fourier domain. Important convolutions are involved in shaping line profiles. These turn into products in the Fourier domain, like the instrumental profile product, $d(\sigma) = f(\sigma)\,i(\sigma)$ in Figure 12.7. Products are easier to visualize than convolutions. Signatures of certain physical processes are more readily detected in the σ domain. Important examples are sidelobes like the one shown in Figure 12.18 and sidelobes in the transforms of lines showing strong rotational broadening, i.e., the information is

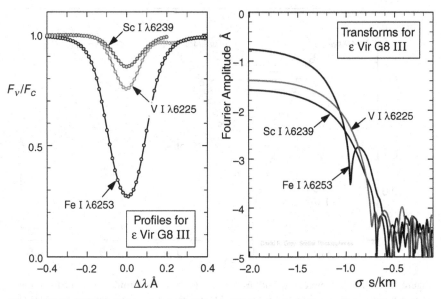

Figure 12.18 Different strength profiles show differences in their transforms. The equivalent widths at 184 mÅ, 42 mÅ, and 27 mÅ. Adapted from Gray (2017).

concentrated into σ bands, making it easier to see above the noise level. We take this up in detail in Chapters 17 and 18.

There is also an important drawback to working in the Fourier domain. Corrections have to be made for any blending with adjacent lines, and blending is common in cool-star spectra. Any distortion caused by a nearby line trespassing onto the target line fundamentally changes the shape we are trying to measure. So we select lines for Fourier analysis that are relatively blend free. Reversing the profile on itself helps guide us in making blend corrections. And comparing the results for several lines can help show up blending problems.

Meanwhile, back in the wavelength domain, we find another interesting analysis tool.

Measurement of Asymmetry

Line profiles in the spectra of F, G, and K stars have more subtle information hidden in them. They are not symmetric. The asymmetries are small and easy to miss. Figure 12.19 shows a line profile on the left. The midpoints of horizontal line segments between the sides of the profile are connected to form the line bisector. Here the levels of the horizontal lines are chosen to correspond to the measured points on the left side, and spline interpolation is used to get the position on the right side. If the profile is symmetric, the line bisector will be a vertical straight line. In this profile plot, you have to look carefully to see that this is not the case. But if the scale of the abscissa is expanded, the shape of the bisector is revealed to be a distorted "C" shape.

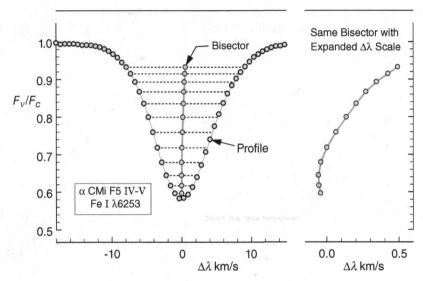

Figure 12.19 The line bisector quantifies the asymmetry of the profile. This line has quite a large asymmetry, but even so, it is hardly visible on the wavelength scale of the profile. The expanded scale on the right allows us to see the shape more clearly. Elginfield Observatory data.

Asymmetries like these arise from velocity fields that are heterogeneous across the surface and that vary with depth in the photosphere. Since we are dealing with velocities, the wavelength scale is expressed in km/s. Decoding bisector shapes is one of the intriguing things we do in Chapter 17.

Photometric error on the profile points, call it δF, translates into velocity errors through the (variable) slope along the profile, $\delta v = \delta F / |dF/d\lambda|$. But each end of the horizontal line segments contributed essentially the same error, raising the total error by $2^{1/2}$, while taking the midpoint reduces it by 2. So the error on a bisector point is

$$\delta v = \frac{1}{\sqrt{2}} \frac{\delta F}{|dF/d\lambda|} \tag{12.22}$$

when $d\lambda$ is in velocity units. As the slope in the core and in the wings goes to zero, the photometric errors increase without bound. This gives a "mushroom"-like shape to the error bars on the bisector points, i.e., increasing errors toward the bottom and toward the top of a bisector (Gray 1983).

Measurement of Line Position

Aside from using a line's wavelength to identify its chemical origin and atomic parameters, the main reason for measuring a line's position is to obtain the Doppler shift and thus the velocity of the emitting gas. Such velocities give us valuable information on the motions of stars in space, orbits of binary stars and extra-solar-system planets, stellar oscillations of various sorts, and mass motions within stellar photospheres. However, what we mean by the position of a spectral line takes on a note of ambiguity when the lines are asymmetric. And there are further complications in cool stars arising from granulation that systematically shifts lines blueward, with larger shifts for weaker lines (again, a topic in Chapter 17).

There are some basic principles. First the obvious. Sharper lines are better for measuring radial velocities. Equation 12.22 formalizes this; the larger $dF/d\lambda$, the smaller the velocity error. This also means that the errors are least through the steepest portion of the profile. Deep lines have larger $dF/d\lambda$ than shallow lines. In a typical late-type star, a line position can be measured with errors of a few m/s. The widths of lines and their intrinsic velocity variability both increase with luminosity, so the smallest errors are for dwarfs.

Next we generally need a reference spectrum with no Doppler shift to give us the absolute wavelength scale. There are several such sources, each with its advantages and disadvantages. These include emission line lamps (Palmer & Engleman 1983, Hinkle *et al.* 2001, Kerber *et al.* 2007), absorption cells (Campbell & Walker 1979, Marcy & Butler 1992, Mahadevan & Ge 2009), telluric lines in the atmosphere above the observatory (Griffin 1973, Kurucz *et al.* 1984, Cochran 1988), telluric lines inside the spectrograph (Gray & Brown 2006), Fabry–Perot interferometers, and laser frequency combs (Molaro *et al.* 2013). These reference spectra are recorded before and after the stellar exposure, or sometimes on top of the stellar exposure.

A fundamental question to ask is what level of precision do we need? If we can work with velocity errors as large as ~0.5 km/s, then many of the issues such as asymmetric lines and granulation blue shifts are small enough to fall under this limit. If we want better than ~0.5 km/s errors, then we ask next if we are measuring velocity *variations*, as in binary-star or extra-solar planet orbits, or absolute velocities. Variations are easier. Asymmetries and granulation shifts are passably stable with time for most stars (less so as we go up in luminosity), so what matters then is how consistent we are with the line-shift measurements. Even here the instability of the intrinsic stellar spectrum ultimately limits the precision at the meter per second level for dwarfs, on up to the kilometer per second level for supergiants. Stellar velocity variations are caused by granulation noise, non-radial oscillations, low-level rotational modulation, and magnetic-cycle variations. Related material is discussed in Chapters 17 and 18. These kinds of variations in the context of detecting extra-solar planets are extensively discussed in the literature, e.g., Dumusque *et al.* (2017), Cabot *et al.* (2021), and references therein.

Naturally, the Doppler shifts of many spectral lines in a star's spectrum are averaged to increase precision. The now-common practice of combining many spectral lines into one measurement by cross-correlating a stellar spectrum with a template was introduced by Griffin (1967) and has subsequently been used in the majority of modern radial-velocity spectrometers. Elaborate spectrographs have been constructed, reaching errors as low as a fraction of one meter per second (e.g., Blackman *et al.* 2020), specifically designed for finding Earth-like planets around other stars. Figure 12.20 shows the radial-velocity variation of a G9 dwarf caused by one of its three planets. The errors on the individual measurements are less than 1 m/s.

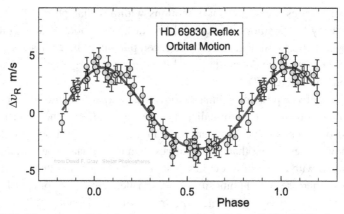

Figure 12.20 The reflex radial-velocity variation of the G9 dwarf, HD 69830, for one of its three planets. This planet has an orbital period of 8.67 days. The amplitude of the variation is 3.5 m/s. Average error is less than 1 m/s. Based on data from Lovis *et al.* (2006).

If we need absolute velocities, then "everything" matters. We are confronted with all of the issues just discussed plus questions of how we define the positions of asymmetrical lines, which lines to include, and how to handle systematic offsets arising from differential blue shifts caused by granulation and gravitational redshifts. In addition, the absolute zero of the wavelength calibration source is needed. Bear in mind that most of these phenomena vary from star to star, potentially introducing velocity differences of hundreds of meters per second. Other matters of technique are discussed by Murdoch and Hearnshaw (1991a, 1991b), Gunn *et al.* (1996), Skuljan *et al.* (2000), Nidever *et al.* (2002), and Borgniet *et al.* (2019); see also Gray and Stevenson (2007). The problem of skylight contamination when dealing with velocity amplitudes below ~1 m/s is discussed by Roy *et al.* (2020).

Now that we have some tools for measuring spectral lines, we move on to consider how they respond to basic physical variables and why they change from one star to the next.

Questions and Exercises

1. How is the instrumental profile of a spectrograph related to the resolving power of the spectrograph?
2. Suppose you are using a Richardson image slicer that illuminates a 5 mm tall section of the entrance slit; what constraints are placed on the spectrograph if the height of the light detector is 0.75 mm?
3. How much does the equivalent width of a line change by the introduction of 5% scattered light?
4. If the transform of a line profile extends out to the Nyquist frequency such that you cannot see the noise level, what might you change in your spectrograph arrangement that would allow you to see the noise level in the Fourier domain?
5. Summarize the advantages you see for analyzing profile shapes in the wavelength domain and in the Fourier domain.

6. Explain how there can be a flux throughput advantage in using a coudé spectrograph compared to a Cassegrain spectrograph. Is this true for all resolving powers? How does seeing affect the discussion?

7. Draw a qualitative log–log plot of the Fourier transform of a typical spectral line, $f(\sigma)$, and show the noise level. Place on your diagram the transform of a typical instrumental profile, $i(\sigma)$. Combine your two functions to form $d(\sigma) = i(\sigma)f(\sigma)$. Now draw in an $i(\sigma)$ that produces twice the degradation, and show the corresponding $d(\sigma)$. If the second $i(\sigma)$ came about by doubling the width of the entrance slit of the spectrograph and that in turn let in twice as much light, would there be an improvement in following $d(\sigma)$ to higher σ?

References

Abt, H.A. & Morrell, N.I. 1995. *ApJS* **99**, 135.

Barrell, H. & Sears, J.E. 1938. *Phil Trans R Soc London* **238**, 1.

Beckers, J. 1993. *ARAA* **31**, 13.

Beckert, E., Strassmeier, K.G., Woche, M., *et al.* 2008. *SPIE* **7018**, 7018J-1.

Bedell, M., Meléndez, J., Bean, J.L., *et al.* 2014. *ApJ* **795**, 23.

Birch, K.P. & Downs, M.J. 1994. *Metro* **31**, 315.

Blackman, R.T., Fischer, D.A., Jurgenson, C.A., *et al.* 2020. *AJ* **159**, 238.

Borgniet, S., Kervella, P., Nardetto, N., *et al.* 2019. *A&A* **631**, 37.

Bowen, I.S. 1938. *ApJ* **88**, 113.

Bowen, I.S. 1962. *Astronomical Techniques* (Chicago: University of Chicago Press), W.A. Hiltner, ed., p. 34.

Brault, J.W. & White, O.R. 1971. *A&A* **13**, 169.

Cabot, S.H.C., Roettenbacher, R.M., Henry, G.W., *et al.* 2021. *AJ* **161**, 26.

Campbell, B. & Walker, G.A.H. 1979. *PASP* **91**, 540.

Cayrel de Strobel, G. 1967. *Transactions of the International Astronomical Union, Reports on Astronomy* (Dordrecht: Reidel), L. Perek, ed., p. 639.

Champeney, D.C. 1973. *Fourier Transforms and Their Physical Applications* (London: Academic Press), p. 121.

Cochran, W.D. 1988. *ApJ* **334**, 349.

Cohen, J.G., Simmons, J.E., & Maney, S. 1976. *PASP* **88**, 966.

Connes, P. 1970. *ARAA* **8**, 209.

Conway, M.T. 1952. *MNRAS* **112**, 55.

Dumusque, X., Borsa, F., Damasso, M., *et al.* 2017. *A&A* **598**, 133.

Dunham, T., Jr. 1956. *Vistas in Astronomy*, Vol. **2** (London: Pergamon), A. Beer, ed., p. 1223.

Edlén, B. 1953. *JOSA* **43**, 339.

Edlén, B. 1966. *Metrologia* **2**, 71.

Edmonds, F.N., Jr. 1965. *ApJ* **142**, 278.

Enard, D. 1979. *ESO Technical Report No. 10*.

Fekel, F.C. 1997. *PASP* **109**, 514.

Forsberg, P. 1991. *Phys Scr* **44**, 446.

Giribaldi, R.E., Ubaldo-Melo, M.L., Porto de Mello, G.F., *et al.* 2019. *A&A* **624**, 10.

Goldberg, L., Mohler, O.C., & Müller, E.A. 1959. *ApJ* **129**, 119.

Gray, D.E. 1963. *American Institute of Physics Handbook* (New York: McGraw-Hill), p. 7.

Gray, D.F. 1974. *PASP* **86**, 526.

Gray, D.F. 1983. *PASP* **95**, 252.

Gray, D.F. 1986. *Instrumentation and Research Programmes for Small Telescopes*, IAUS **118** (Dordrecht: Reidel), J.B. Hearnshaw & P.L. Cottrell, eds., p. 401.

Gray, D.F. 2016. *ApJ* **832**, 68.

Gray, D.F. 2017. *ApJ* **845**, 62.

Gray, D.F. & Brown, K.I.T. 2006. *PASP* **118**, 399.

Gray, D.F. & Evans, J.C. 1973. *ApJ* **182**, 147.

Gray, D.F. & Stevenson, K.B. 2007. *PASP* **119**, 398.

Griffin, R.F. 1967. *ApJ* **148**, 465.

Griffin, R.F. 1968. *A Photometric Atlas of the Spectrum of Arcturus* (Cambridge: Cambridge Philosophical Society).

Griffin, R.F. 1969a. *MNRAS* **143**, 319.

Griffin, R.F. 1969b. *MNRAS* **143**, 349.

Griffin, R.F. 1969c. *MNRAS* **143**, 361.

Griffin, R. 1973. *MNRAS* **162**, 243.

Gunn, A.G., Hall, J.C., Lockwood, G.W., & Doyle, J.G., 1996. *A&A* **305**, 146.

Helfer, H.L., Wallerstein, G., & Greenstein, J.L. 1963. *ApJ* **138**, 97.

Hinkle, K., Wallace, L., Valenti, J., & Harmer, D. 2000. *Visible and Near Infrared Atlas of the Arcturus Spectrum, 3727–9300 Å* (San Francisco: ASP).

Hinkle, K.H., Joyce, R.R., Hedden, A., Wallace, L., & Engleman, R., Jr. 2001. *PASP* **113**, 548.

Kerber, F., Nave, G., Sansonetti, C.J., Bristow, P., & Rosa, M.R. 2007. *ASPC* **364**, 461.

Kurucz, R.L., Furenlid, I., Brault, J., & Testerman, L. 1984. *Solar Flux Atlas from 296 to 1300 nm* (Sunspot: National Solar Observatory).

Litzén, U., Brault, J.W., & Thorne, A.P. 1993. *Phys Scr* **47**, 628.

Lovis, C., Mayor, M., Pepe, F., *et al.* 2006. *Nature* **441**, 305.

Mahadevan, S. & Ge, J. 2009. *ApJ* **692**, 1590.

Marcy, G.W. & Butler, R.P. 1992. *PASP* **104**, 270.

Middleton, D. 1960. *Introduction to Statistical Communication Theory* (New York: McGraw-Hill).

Molaro, P., Esposito, M., Monai, S., *et al.* 2013. *A&A* **560**, 61.

Moore, C.E., Minnaert, M.G.J., & Houtgast, J. 1966. *The Solar Spectrum 2935 Å to 8770 Å*, National Bureau of Standards Monograph 61 (Washington, DC: US Government Printing Office).

Murdoch, K. & Hearnshaw, J.B. 1991a. *ApSS* **186**, 137.

Murdoch, K. & Hearnshaw, J.B. 1991b. *ApSS* **186**, 169.

Nave, G., Johansson, S., Learner, R.C.M., Thorne, A.P., & Brault, J.W. 1994. *ApJS* **94**, 2.

Nave, G. & Sansonetti, C.J. 2011. *JOSAB* **28**, 737.

Nidever, D.L., Marcy, G.W., Butler, R.P., Fischer, D.A., & Vogt, S.S. 2002. *ApJS* **141**, 503.

Palmer, B.A. & Engleman, R., Jr. 1983. *Los Alamos National Laboratory Report LA9615*.

Panofsky, H.A.A. 1943. *ApJ* **97**, 180.

Pierce, A.K. 1965. *PASP* **77**, 216.

Plaskett, J.S. 1910. *JRASC* **4**, 333.

Reader, J. & Sansonetti, C. 1999. *Ultraviolet Atmospheric and Space Remote Sensing: Methods and Instrumentation II, SPIE* **3818.**

Redman, S.L., Nave, G., & Sansonetti, C.J. 2014. *ApJS* **211**, 4.

Reiners, A., Mrotzek, N., Lemke, U., Hinrichs, J., & Reinsch, K. 2016. *A&A* **587**, 65.

Richardson, E.H. 1966. *PASP* **78**, 436.

Richardson, E.H., Brealey, G.A., & Dancey, R. 1971. *PDAO Victoria* **XIV**, 1.

Ridgway, S.T. & Brault, J.W. 1984. *ARAA* **22**, 291.

Rigaut, F. 2015. *PASP* **127**, 1197.

Roy, A., Halverson, S., Mahadevan, S., *et al.* 2020. *AJ* **159**, 161.

Shane, C.D. 1932. *Lick Obs Bull* **16**, 76.

Shulte, D.H. 1966. *App Opt* **5**, 457.

Skuljan, J., Hearnshaw, J.B., & Cottrell, P.L. 2000. *PASP* **112**, 966.

Slettebak, A., Collins, G.W. II, Boyce, P.B., White, N.M., & Parkinson, T.D. 1975. *ApJS* **29**, 137.

Smith, M.A., Graves, J.E., Jaksha, D.B., Plymate, C.L., & Ramsey, L.W. 1987. *PASP* **99**, 654.

Stumpff, P. 1979. *A&A* **78**, 229.

Stumpff, P. 1980. *ApJS* **41**, 1.

Taylor, D.J. & Schmidt, E.G. 1975. *PASP* **87**, 357.

Teske, R.G. 1967. *PASP* **79**, 110.

Tull, R.G. 1969. *Sky and Telescope* **38**, 156.

Tull, R.G. 1972. *Auxiliary Instrumentation for Large Telescopes*, Proceedings of ESO/ CERN Conference, Geneva, S. Laustsen & A. Reiz, eds., p. 259.

Vaughan, A.H., Jr. 1967. *ARAA* **5**, 139.

Waddell, J. 1956. *An Empirical Determination of the Turbulence Field in the Solar Photosphere Based on a Study of Weak Fraunhofer Lines*, PhD thesis, University of Michigan.

Wright, J.T. & Eastman, J.D. 2014. *PASP* **126**, 838.

Wright, K.O. 1950. *PDAO Victoria* **8**, 281.

Wright, K.O. 1958. *Trans IAU* **10**, 558.

Wright, K.O. 1966. *Abundance Determinations in Stellar Spectra*, IAUS **26** (London: Academic Press), H. Hubenet, ed., p. 15.

13

The Behavior of Spectral Lines

Absorption lines in stellar spectra show differences in shape and strength according to the physical conditions in the photospheres in which they originate. Complex and intriguing interactions arise in the interplay of spectral lines with temperature, pressure, chemical composition, radiation, velocity fields, and sometimes magnetic fields. It can be challenging to rigorously model these interactions. On the other hand, the panoramic view of spectral-line behavior can be understood in relatively simple terms, as presented in this chapter.

The most fundamental point to keep in mind is that the strength of a spectral line depends on the number of absorbers producing that line. Thus the atomic level populations are of primary interest. But the number of absorbers also accumulates along the path length through the photosphere. If the *continuous* absorption is strong, the path length will be short and vice versa. In effect, the depth of the photosphere changes with the amount of continuous absorption. Because of this, it is the *ratio* of the line absorption to the continuous absorption, ℓ_v/κ_v, that determines line strength.

One of the goals of our studies is to understand the physical processes behind the various line profiles and line strengths shown by stars. Another is to use that understanding, sometimes coupled with empirical information, to interpret line behavior, enabling us to measure fundamental properties of stars and stellar photospheres. This chapter is the first step along this path. Here we focus on radiation transfer and the formation of the thermal profile, which includes the damping components. In later chapters we look into application of these tools and the reshaping of line profiles by photospheric velocity fields and rotation.

Most of the behaviors of spectral lines can be understood within the framework of local thermodynamic equilibrium (LTE). There is one temperature at each photospheric depth given in $T(\tau_0)$. Using it, we can rely on our standard equations for calculating excitation and ionization, and we can use the black-body source function, $S_v(T) = B_v(T)$.

Why Do Stellar Spectra Have Absorption Lines?

Within the wavelength span of a spectral line, the absorption of the line, ℓ_v, increases the absorption from the continuous absorption alone to $\ell_v + \kappa_v$, which means that our line of sight does not penetrate as deeply into the photosphere. So within the line, we see light from the higher layers that are cooler and emit less light; less light means an absorption line. Consider the Eddington approximation, Equation 7.35, that relates the source function to the flux, here recast slightly as

$$S_\nu(\tau_\nu) = \frac{3}{4\pi} \mathfrak{F}_\nu(0) \left(\tau_\nu + \frac{2}{3} \right),$$

where $\mathfrak{F}_\nu(0)$ is the surface flux we see. In particular, when $\tau_\nu + 2/3 = 4\pi/3$ or $\tau_\nu = (4\pi - 2)/3 \approx 3.5$, call it τ_1, we see that the source function and the surface flux are equal,

$$S_\nu(\tau_1) = \mathfrak{F}_\nu(0). \tag{13.1}$$

Across a line profile, ℓ_ν changes, being larger toward the center of the line. The condition $\tau_\nu = \tau_1$ is true higher up in the atmosphere for ν nearer line center, and holds for progressively deeper layers for frequencies farther from the line center. Consequently, there is a mapping according to Equation 13.1 between S_ν as a function of τ_ν and $\mathfrak{F}_\nu(0)$ as a function of position across the profile. This is shown in Figure 13.1. Because the source function decreases outward, an absorption line is formed.

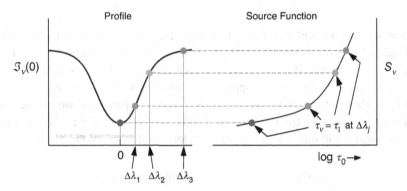

Figure 13.1 There is a mapping between the source function and the line profile. τ_ν takes on the value of τ_1 at $\Delta\lambda_j = 0, \Delta\lambda_1, \Delta\lambda_2, \Delta\lambda_3$, corresponding to the dots. The decline of the source function outward through the photosphere causes absorption lines.

This general idea carries over to the full proper flux integration, where the simplistic τ_1 is replaced by contribution functions.

The Flux Integral for Lines

The absorption in the line, ℓ_ν, is added to the absorption in the continuum, κ_ν, but otherwise we use the standard flux integral given in Equation 7.15 with the understanding that we know $T(\tau_0)$ as discussed in Chapter 9. Thus $d\tau_\nu = (\ell_\nu + \kappa_\nu)\rho dx$ and $d\tau_0 = \kappa_0 \rho dx$, where the subscript 0 denotes a standard wavelength, usually 5000 Å, and the τ_ν scale is given by

$$\tau_v = \int_0^{\tau_0} \frac{\ell_v + \kappa_v}{\kappa_0} \, dt_0$$

$$= \int_{-\infty}^{\log \tau_0} \frac{\ell_v + \kappa_v}{\kappa_0} t_0 \frac{d\log t_0}{\log e}. \tag{13.2}$$

The absorption coefficients are known according to the discussions in Chapters 8 and 11. We then compute the surface flux across the line profile,

$$\mathfrak{F}_v(0) = 2\pi \int_0^{\infty} S_v(\tau_v) \, E_2(\tau_v) \, d\tau_v$$

$$= 2\pi \int_0^{\infty} B_v(T) \, E_2(\tau_v) \frac{d\tau_v}{d\tau_0} d\tau_0$$

$$= 2\pi \int_{-\infty}^{\infty} B_v(T) \, E_2(\tau_v) \frac{\ell_v + \kappa_v}{\kappa_0} \tau_0 \frac{d\log \tau_0}{\log e}. \tag{13.3}$$

The integrand, i.e., the contribution function, plays the same role it did with the continuum flux (Figures 9.11 or 9.13 for example). It shows us the photospheric layers where the surface flux is generated. But of course with the rapid variation of absorption across the line, there is a corresponding rapid response in the contribution functions. Figure 13.2 gives an example. These integrands replace the τ_1 used above. In particular, we see that the flux for a profile point comes from a range in depths, not just a single τ_1. We also see the important migration to higher layers for points closer to the core of the line.

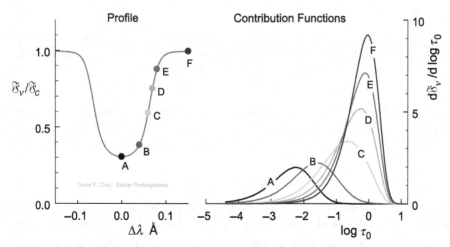

Figure 13.2 Contribution functions on the right are shown for the profile points on the left. The model has $S_0 = 1.0$, $\log g = 4.0$, and solar chemical composition. The atomic parameters are for an iron line at 6000 Å. Unit ordinate for contribution functions is 10^{-5} erg/(s cm^2 Hz).

The Behavior of Weak Lines

In predicting and understanding changes in spectral lines, it is often useful to consider the simplest case, namely how weak lines behave. Taking the lead from Equation 13.2, write for the full optical depth

$$d\tau_v = (\ell_v + \kappa_v)\,\rho\,dx$$

and for the line and continuum optical depths separately

$$d\tau_\ell = \ell_v\,\rho\,dx$$
$$d\tau_c = \kappa_v\,\rho\,dx.$$

Then $\tau_v = \tau_\ell + \tau_c$, and in the spirit of Equation 13.1, $\tau_\ell \approx (\ell_v/\kappa_v)\tau_c$, and the line depth is

$$\frac{\mathfrak{F}_c - \mathfrak{F}_v}{\mathfrak{F}_c} \approx \frac{S_v(\tau_c = \tau_1) - S_v(\tau_v = \tau_1)}{\mathfrak{F}_c},$$

where, as usual, $\mathfrak{F}_c$ is the continuum flux and $\mathfrak{F}_v$ is the flux in the line. Since the line is weak, we can expand the second term in the numerator using a Taylor series,

$$S_v(\tau_v = \tau_1) = S_v(\tau_c = \tau_1) + \frac{dS_v}{d\tau_c}(\tau_c - \tau_v) + \cdots.$$

In this way, the line depth becomes

$$\frac{\mathfrak{F}_c - \mathfrak{F}_v}{\mathfrak{F}_c} \approx \frac{\tau_\ell}{\mathfrak{F}_c}\frac{dS_v}{d\tau_c} = \frac{\ell_v}{\kappa_v}\frac{\tau_c}{\mathfrak{F}_c}\frac{dS_v}{d\tau_c}$$

$$\approx \text{constant}\,\frac{\ell_v}{\kappa_v}, \tag{13.4}$$

in which the constant contains all the factors that are invariant across the line profile. This expression shows us two important things.

1. The profile mimics the shape of ℓ_v as long as the line is not so strong that saturation effects cause distortions. This is especially true of weak lines that are formed completely in deep layers over a relatively narrow range of depth. Note that this is just for the thermal profile. Doppler broadening from macroturbulence and rotation dominate the *shape* of actual stellar profiles, but do not change the total absorption caused by the line. As we know from Chapter 11, ℓ_v for weak lines is a Gaussian.
2. The strength of the line depends on the ratio of line to continuous absorption. For weak lines, the whole profile scales with this ratio, which means that the equivalent width scales the same way, i.e., $W = \text{constant}\,\ell_v/\kappa_v$.

This second point is used extensively later in the chapter. While we naturally expect a stronger ℓ_v to lead to a stronger line, it is equally true that a weaker κ_v also leads to a stronger line. In essence, the strength of the continuous absorption sets the optical thickness of the photosphere. For a fixed ℓ_v, weaker continuous absorption means a deeper photosphere and a greater path length over which the line can absorb light and grow stronger.

Since the contribution functions for lines of intermediate strength and stronger extend over a large range in the photosphere, as in Figure 13.2, it is more-or-less meaningless to attempt to specify a single depth of formation for an equivalent width or for the line as a whole.

Above the Photosphere: the Behavior of Very Strong Lines

The relation between the source function and the shapes of line profiles goes beyond the photosphere proper. There are a few spectral lines that are so strong in some stars that their cores are formed above the photosphere. The most prominent of these are the H and K lines of ionized calcium in cool stars. And a very interesting and surprising thing happens: the cores show emission reversals, like those illustrated in Figure 13.3 (and in Figure 12.12).

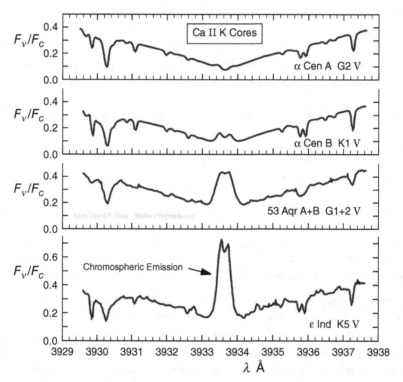

Figure 13.3 Cores of very strong lines, like these Ca II K lines, are so opaque in their cores that our line of sight does not penetrate beyond the chromosphere. Only the central portion of these lines is shown. The emission in the cores shows that the source function is larger in the chromosphere than in the photosphere. The resolving power is ~33 000. Based on Pasquini *et al.* (1988).

The absorption increases monotonically from the wings to the very center of the line. We see higher and higher layers from the wings to the line center. Following the thought pattern for Figure 13.1, we conclude that for these stars the source function *rises* coming out of the photosphere. There is a temperature inversion! The chromosphere, the layers just above the photosphere, are hotter than the photosphere. This is contrary to the normal expectation of energy flow from hotter layers inside the star to cooler interstellar space. Figuring out this puzzle has occupied many minds, and the full solution is still not in hand.

As the sketch in Figure 1.1 pointed out, the temperature not only rises through the chromosphere, but it gets much hotter further out in the corona. Something other than normal heat flow is needed to power these thin outer layers. Stronger chromospheric emission, like that of ε Ind in Figure 13.3, is associated with stronger magnetic fields and often used as a proxy for such fields.

Before exploring the dependence of spectral lines on chemical composition, temperature, and pressure, we consider aspects of the more general case, where LTE is not assumed.

A More General Approach

How do we proceed without LTE? The two pillars of LTE are collisional excitation, allowing us to calculate level populations with Equation 1.19 and the black-body source function. And they are connected. Let us start with the source function.

Recall from Equation 5.15 that the source function is $S_\nu = j_\nu/\kappa_\nu$. Generalize the emission in Equation 5.17 to include a frequency dependence, $\psi(\nu)$, for the spontaneous emission coefficient, A_{ul},

$$j_\nu \rho = N_u \, A_{ul} \, \psi(\nu) \, h\nu,$$

where we use our standard notation of u for the upper level and l for the lower level of the transition. In a similar fashion, let $\phi(\nu)$ be the frequency distribution of the absorption and stimulated emission, such that

$$\kappa_\nu \rho = N_l \, B_{lu} \, \phi(\nu) \, h\nu.$$

Both $\psi(\nu)$ and $\phi(\nu)$ are normalized to unit area. Dividing these expressions and including stimulated emission gives us the source function:

$$S_\nu = \frac{j_\nu}{\kappa_\nu} = \frac{N_u \, A_{ul} \, \psi(\nu)}{N_l \, B_{lu} \, \phi(\nu) - N_u \, B_{ul} \, \phi(\nu)}.$$

If we now bring in Equations 6.8, namely, $B_{ul} = B_{lu} g_l/g_u$ and $A_{ul} = (2h\nu^3/c^2)B_{ul}$, the source function can be written

$$S_\nu = \frac{2h\nu^3}{c^2} \frac{1}{(N_l/N_u)(g_u/g_l) - 1} \frac{\psi(\nu)}{\phi(\nu)}. \tag{13.5}$$

This is getting close to the black-body source function, but it shows us how the population ratio, N_l/N_u, is needed, and also the dependence on the ratio of the emission to absorption profiles.

This can be a complicated situation because in order to calculate the radiation field, we need S_ν, but S_ν depends on the radiation field through its effect in populating the two levels. Collisions also affect the level populations. In fact, when collisions completely dominate the excitation process, the cyclic coupling between the radiation and the source function is broken and we are back to the LTE case (Equation 1.18), where the population ratio is given by

$$\frac{N_u}{N_l} = \frac{g_u}{g_l} e^{-h\nu/kT}.$$

Furthermore, if we have an equilibrium situation, then for each emitted photon there must be an identical photon that is absorbed – in other words, $\chi(\nu) = \phi(\nu)$. Then we are also back to $S_\nu = B_\nu$.

But when collisions do not do all the excitation, we need to calculate the ratio of level populations for all relevant levels, taking the radiative excitation into account. Regrettably, realistic non-LTE calculations are complex, consisting of solving many simultaneous equations, using known transition probabilities, photoionization cross sections, and collision cross sections for each level connected to the transition of interest. For example, in their non-LTE study of iron, Thévenin and Idiart (1999) included 5560 radiative transitions in their calculations. Rate equations are set up for the relevant atomic levels. By invoking the condition of a steady state, the equations are solved simultaneously, usually by iteration. See Jefferies (1968), Athay (1972), Rutten (1988), Nissen and Gustafsson (2018) for a broad perspective; Collet *et al.* (2005), Korotin *et al.* (2018), Reggiani *et al.* (2019), Osorio *et al.* (2020), Amarsi *et al.* (2020), among many others for recent applications.

Much of the focus of non-LTE calculations has been on chemical abundances. In many cases, the more complete calculations produce modest difference, ~0.05 dex in deduced abundances. But each situation is unique, and as seen in some of the references just cited, large differences can sometimes occur. We expect LTE to falter when dealing with the cores of strong lines, since they are formed high in the photosphere, where the collision rates are low and the radiation field is becoming highly non-isotropic. Usually the source function in the high layers falls below the black-body case, causing deeper cores in the lines. This can be imitated by LTE with lower temperatures in these layers.

We continue to use LTE models, judging them to be a close approximation in the majority of cases. This is especially true when modeling the shapes of line profiles, since the small inadequacies in the thermal profile are overridden by the reshaping imposed by macroturbulence and rotation.

Variation of Spectral Lines with Basic Physical Parameters

In the following material, we investigate why spectral lines look different in different stars. The term "variation" means the systematic differences in lines, primarily their strengths, as we move through sequences of stars ranging in, for example, chemical composition, effective temperature, and surface gravity.

The strength of the *thermal component* of a spectral line determines the strength of the line. A line may be reshaped by various Doppler-shift distributions, but its equivalent width remains the same. As lines change strength for whatever reason, they all go through the same patterns of behavior. The main factor is the number of absorbers. Therefore excitation and ionization are in the forefront. Secondary factors include the thermal and microturbulence broadening in $\Delta\lambda_D$ in Equation 11.44. And if the line grows strong enough for the wings to develop, the damping parameters come into play (Equation 11.45). The chemical abundance of the absorbing species plays a foundational role, of course, but for most of the stars we see, this variation is relatively small, especially compared to temperature variations. But, let us start there.

The Abundance Dependence and the Curve of Growth

Chemical composition is one of the defining characteristics of a star. We might well expect the strengths of spectral lines to depend on the chemical composition, since more of a chemical implies more absorbers and more absorbers implies stronger lines for that species. As we have seen, excitation and ionization play crucial roles in determining how many absorbers materialize from the element. But for now, we assume these factors are sufficiently favorable to make the line visible, and hold them constant while we look into the variation with the chemical abundance factor $A_E = N_E/N_H$. We saw in the expression for ℓ_ν, Equation 11.52 or 11.53, that ℓ_ν depends on the product of the oscillator strength, f, with A_E. In other words, it is actually the product fA_E that controls the line strength. We will use this fact to our advantage later. For now, we focus on A_E.

Figure 13.4 shows how an iron line grows in strength as A_E is increased. We call this the line's *curve of growth*. Each and every spectral line has a curve of growth, and all curves of growth look similar to the one in Figure 13.4. Notice that the ordinate is W/λ. Division by λ is customary because in many applications of the curve, the equivalent widths of different

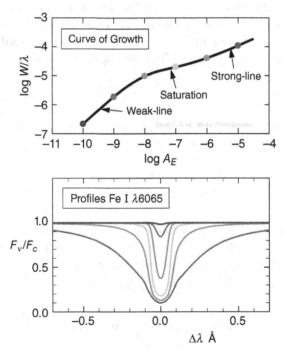

Figure 13.4 Both the equivalent width (upper panel) and the profile (lower panel) strengthen with increased chemical abundance, A_E. The dots on the curve of growth correspond to the profile in the lower panel. The model has $S_0 = 0.85$, $\log g = 4.5$ cm/s^2, and solar chemical composition.

lines are placed on the same graph, and their Ws scale with the Doppler effect. Division by λ removes this scaling.

We visualize three phases of line growth: weak, saturation, and strong, as labeled in Figure 13.4. For a weak line, the equivalent width is nearly proportional to the abundance, $W \approx$ constant A_E or $\log W \approx$ constant $+ \log A_E$. This follows from Equation 13.4 applied to weak lines; each point in a weak line profile varies as ℓ_ν / κ_ν, and ℓ_ν is proportional to A_E, making W proportional to A_E. As A_E increases, the line saturates in the core – it cannot absorb any more there. The only way it can increase its equivalent width and enter the strong phase is by growing wider, which it does by growing wings as the dispersion component of ℓ_ν becomes significant. How soon the wings develop hinges on the size of the damping constant, γ_{total}. In reality, there is a continuous transition from the weak to the strong case.

There is also a continuous transition in the shape of the thermal profile with increasing A_E. In the weak-line section, the Doppler core dominates. Its width is set by $\Delta\lambda_D$ so the core *depth* grows in proportion to A_E. As saturation approaches, the core bottoms out, not to zero, but to the flux coming from the upper-most photosphere where the temperature reaches its minimum. The equivalent width "stalls" during this process, showing only small change with increasing A_E. The saturation portion of the curve of growth would approach a constant W if the dispersion component were not there to form wings. The growth of the wings is a bit more complex. The inner wings experience saturation, as can be seen in Figure 13.4 by comparing the inner-wing depth with the core depth of the profiles where saturation becomes obvious. Meanwhile, the far wing is behaving like a weak line, growing without saturation. This tells us that the growth in W for a strong line is slower than for a weak line.

The approximate growth in strong-line equivalent width can be estimated by saying that the wing is so strong the core can be ignored. Then the dispersion component of ℓ_ν dominates and

$$\ell_\nu \approx \frac{\text{constant } \gamma_{\text{total}} A_E}{\Delta\lambda^2}.$$

Next place this ℓ_ν into Equation 13.4 and realize that if we neglect depth variations in these terms, they can be factored out, leading to

$$\frac{W}{2} \approx \int_{\Delta\lambda_c}^{\infty} \frac{\text{constant } \gamma_{\text{total}} A_E}{\Delta\lambda^2} \, d\Delta\lambda,$$

in which the radiation transfer is invariant with A_E and is contained in the "constant," and $\Delta\lambda_c$ is of the order of the width of the core that we exclude because its contribution is small compared to that of the wings (and to avoid the artificial singularity we introduced at $\Delta\lambda = 0$). Substituting $u^2 = \Delta\lambda^2 / (\text{constant } \gamma_{\text{total}} A_E)$ we get

$$W \approx 2 \sqrt{\text{constant } \gamma_{\text{total}} A_E} \int_{\Delta\lambda_c}^{\infty} \frac{1}{u^2} \, du \approx \text{constant} \sqrt{\gamma_{\text{total}} A_E}, \qquad (13.6)$$

where we have used the fact that the integral is some fixed number, since the arbitrary $\Delta\lambda_c$ is independent of A_E. We see that the equivalent width of a strong line varies approximately as the square root of the abundance or $\log W \approx$ constant $+ 0.5 \log A_E$.

The Abundance Dependence and Metallicity

The simple curve of growth discussed in the last section assumes that only the abundance of the species forming the line changes, while the rest of the photosphere remains unchanged. A more likely scenario is that all metal species change together. The metallicity, [Fe/H], the measure of the abundance of electron donors (Equation 9.6), is a general way of specifying photospheric abundances for our purposes. Specifically in cool stars, metallicity affects the electron pressure, which in turn can affect the ionization balance and the strength of the negative-hydrogen-ion absorption, both of which can affect the line strength. In hot stars, where the electrons come from the ionization of hydrogen and the continuous absorption is from neutral hydrogen, the role of metallicity is nil.

Consider iron in a solar temperature star as an example. Most of the iron is ionized, so the number density of ions, $N_1 \approx$ constant, and then the ionization equation (Equation 1.21),

$$\frac{N_1}{N_0} = \frac{\Phi(T)}{P_e},$$

tells us that the number of absorbers, N_0, for a line of Fe I is proportional to the electron pressure. Similarly, κ_ν, dominated by the bound–free negative hydrogen ion, is proportional to P_e (Equation 8.12). Therefore ℓ_ν/κ_ν is independent of P_e and neutral iron curves of growth are insensitive to metallicity.

The same reasoning tells us that increasing metallicity will weaken Fe II lines according to the increase in κ_ν. Figure 13.5 shows an example for Fe II.

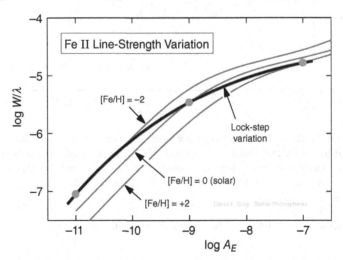

Figure 13.5 The change in equivalent width for pressure-sensitive lines, like this Fe II line, depends on the metallicity when iron is mainly ionized. The model has $S_0 = 1.0, \log g = 4.4$, and metallicity as labeled. The line has $\chi = 2.5\,\text{eV}$ and $\zeta = 1.5\,\text{km/s}$. A large range in metallicity is used to exaggerate the effect.

As a final abundance study, consider the strong sodium D_2 line in Figure 13.6. The upper panel shows the change when the model metallicity is held constant while the sodium abundance is varied over ± 1 dex, i.e., A_{Na} spans a factor of 10. This, of course, changes the number of line absorbers so ℓ_ν changes by the same factor. The middle panel reverses the situation with the sodium abundance held constant while the metallicity is run over a factor of 10. Now the number of line absorbers is constant and the change comes from κ_ν through the electron pressure and its coupling to the abundance of the electron donors. The bottom panel puts the two abundance changes together. This is the most likely real scenario.

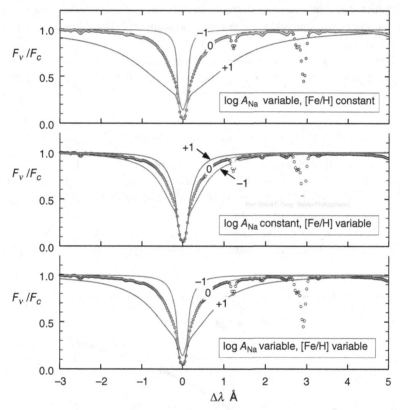

Figure 13.6 Variations in the Na I D_2 line with chemical changes for a solar model. Curve labels are log abundance changes. Top panel: Na abundance alone is changed. Middle panel: metallicity abundance alone is changed. Bottom panel: both Na and metallicity abundances are changed by the same amount. The points show the actual solar spectrum from Hinkle *et al.* (2000).

Next we look at weak lines and how they change with temperature.

Temperature Dependence for Weak Metal Lines

Temperature is the variable that most strongly controls the line strength. This sensitivity arises from the strong temperature dependence in the excitation–ionization process.

Figure 13.7 shows the behavior of typical weak metal lines with changing effective temperature. This is the concept upon which the basic spectral classification sequence is understood. The rise and fall sections are assigned "case numbers." All lines show a similar behavior.

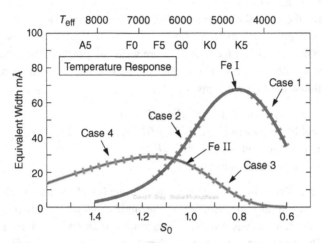

Figure 13.7 Equivalent width variations of Fe I $\lambda6246$ ($\chi = 3.60$ eV) and Fe II 6247 ($\chi = 3.89$ eV) with temperature. The regions for the four cases are shown. Note the reversal of the temperature coordinate to be in accord with the HR diagram.

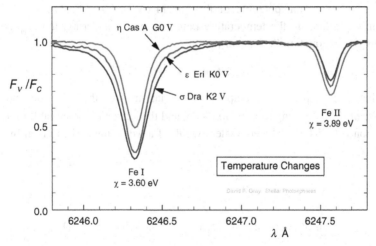

Figure 13.8 Observed line profiles illustrating the behavior shown in Figure 13.7. The Fe I line is in the Case 2 section. The Fe II line is in the Case 3 section. Based on Elginfield Observatory data.

The peak values move to lower temperatures with lower excitation potential. When χ is as small as a few tenths of an electronvolt, as it is for many V I lines, Case 2 extends to low temperatures and Case 1 is essentially gone. And if a curve should rise to larger equivalent widths, where the line experiences saturation, the peak of the curve will be flattened, according to the curve-of-growth behavior.

Observed profiles for Fe I $\lambda6246$ and Fe II $\lambda6247$ are shown in Figure 13.8. Note the position of these three spectral types in Figure 13.7. We see that the Fe I line lies in Case 2, while the Fe II line lies in Case 3.

We can follow the general reasons for these variations by considering the ratio of line to continuous absorption, ℓ_v/κ_v, for weak lines, according to Equation 13.4 simplified to the one-point depth of formation. Since the lines are weak, all points in the profile follow this ratio. Therefore, so does the equivalent width, W. The rise and fall portions of the curves in Figure 13.7 fall into the following cases.

Case 1: neutral line, element neutral, $dW/dT \gtrsim 0$.
Case 2: neutral line, element ionized, $dW/dT < 0$.
Case 3: ionic line, element neutral, $dW/dT > 0$.
Case 4: ionic line, element ionized, $dW/dT \lesssim 0$.

When discussing dW/dT, limit the changes so that the line does not switch cases.

For Case 1, the number of absorbers is the number of neutral atoms, and since the element is mainly neutral, this number is essentially independent of temperature, as in Figure 1.11. The excitation, on the other hand, increases exponentially, according to Equation 1.19, so the line absorption for a weak line varies with temperature as

$$\ell_v = \text{constant } e^{-\chi/kT},$$

where χ is the excitation potential of the line in eV and $k = 8.6173 \times 10^{-5}$ eV/K is Boltzmann's constant. In the temperature range we are considering, the H_{bf}^{-} continuous absorption gives (Equation 8.12 again),

$$\kappa_v = \text{constant } P_e \, T^{-5/2} \, e^{0.754/kT}.$$

Recall that the appropriate temperature variation for the electron pressure is $P_e \approx \text{constant } e^{\Omega T}$ according to Equation 9.25, and the ionization potential for the negative hydrogen ion is 0.754 eV. The equivalent width of a weak line is then given by

$$W = \text{constant } \frac{\ell_v}{\kappa_v} \approx \text{constant } T^{5/2} \, e^{-(\chi+0.754)/kT - \Omega T}$$

or

$$\ln W \approx \text{constant} + 2.5 \ln T - (\chi + 0.754)/kT - \Omega T$$

and

$$\frac{d \ln W}{dT} \approx \frac{2.5}{T} + \frac{\chi + 0.754}{kT^2} - \Omega.$$

Convert to common logs to get

$$\frac{d \log W}{dT} \approx \log e \left(\frac{2.5}{T} + \frac{\chi + 0.754}{kT^2} - \Omega \right). \qquad (13.7)$$

The center term represents the increase in excitation (χ) coupled with the decrease in numbers of H^- ions (0.754 eV). When χ is a few electronvolts, this term dominates and the slope is positive. When χ is near zero, the slope drops to near zero, the situation for V I lines noted above. Case 1 in Figure 13.7 has $\chi = 3.60$ eV and so has a positive slope.

In Case 2, we follow a neutral line to higher temperature stars, where the species is ionized. Unlike the number of neutral atoms, the number of ions varies strongly with temperature. We take this into account in ℓ_ν using the ionization equation, Equation 1.20, with ionization potential I,

$$\frac{N_1}{N_0} \approx \text{constant } T^{5/2} e^{-I/kT} / P_e.$$

Then, using the same remaining factors for excitation and for κ_ν,

$$\frac{d \log W}{dT} \approx \log e \left(\frac{\chi + 0.754 - I}{kT^2} \right). \tag{13.8}$$

Here the ionization dominates (I) and the slope is negative, as seen in Figure 13.7, Case 2.

Case 3, ionic line, element neutral and we are back to cooler stars. Then ℓ_ν/κ_ν leads us to

$$\frac{d \log W}{dT} \approx \log e \left(\frac{5}{T} + \frac{\chi + 0.754 + I}{T^2} - 2\Omega \right). \tag{13.9}$$

Again the central term carries the day and the slope is positive.

Case 4, ionic line, element ionized, and

$$\frac{d \log W}{dT} \approx \log e \left(\frac{2.5}{T} + \frac{\chi + 0.754}{kT^2} - \Omega \right), \tag{13.10}$$

the same equation as Case 1, but now working at higher temperatures. For modest temperatures of ~6000 K, the positive terms only slightly outweigh the electron-pressure term (Ω), making for a slope near zero. But for higher temperatures, the positive terms shrink and the slope is negative. Notice that the slope goes negative because of the increase of the electron pressure, not because of the second ionization of iron.

As expected, and as can be seen from the previous few equations, the curves peak at higher temperatures for higher excitation potentials. Higher excitation potentials also produce weaker lines for the same f value.

These kinds of relations are informative albeit somewhat simplistic. There is, of course, not one temperature in a photosphere, but a full $T(\tau_0)$. The levels of excitation and ionization can vary markedly from the top to the bottom of the photosphere, and there is no sharp transition across ionization stages at particular effective temperatures. With these provisos, we continue on.

Temperature Dependence for Strong Metal Lines

When lines grow out of the weak category, the changes with temperature described in Cases 1–4 above are slowed down owing to saturation. Strong lines have wings, by definition, and we can follow the strength of the far *wings* by looking at ℓ_ν/κ_ν, but not the core and therefore not the equivalent width. But understanding the temperature dependence of strong-line wings is good. Wings on strong lines are relatively easy to measure and use as a diagnostic tool. Figure 13.9 shows some samples of temperature differences for the strong sodium D lines.

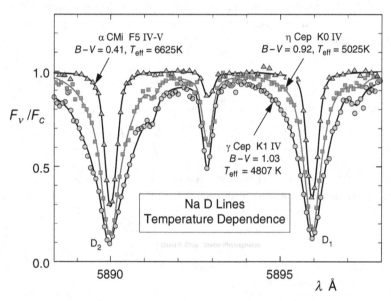

Figure 13.9 The strong sodium D lines grow primarily in width with declining temperatures. Ancient Elginfield Observatory data, *circa* 1970. Resolution 0.27 Å.

For the wings of strong lines, ℓ_ν has a dispersion-profile shape and is proportional to γ_{total}. The Stark and van der Waals components of γ_{total} depend on temperature (and on pressure, which we address later), as in Equations 11.27 and 11.28, i.e.,

$$\gamma_{total} = \text{constant } T^{-5/6} + \text{constant } T^{-7/10} + \gamma_{nat},$$

where the first temperature dependence is for the quadratic Stark term and the second for the van der Waals term. For example, take Case 1 (neutral line, element neutral) and include γ_{total},

$$\frac{\ell_\nu}{\kappa_\nu} \approx \text{constant } \gamma_{total} \, T^{\frac{5}{2}} \, e^{-\frac{\chi+0.754}{kT}-\Omega T}.$$

Incorporating the temperature dependence for γ_{total} leads to an awkward expression, but if we take a case like the one shown in Figure 11.4, the van der Waals $T^{-7/10}$ term is the important one, and then

$$\ell_\nu/\kappa_\nu \approx \text{constant } T^{-7/10} \, T^{5/2} \, e^{-(\chi+0.754)/kT-\Omega T} = \text{constant } T^{1.8} \, e^{-(\chi+0.754)/kT-\Omega T}.$$

Thus the temperature sensitivity of the wing has been slightly reduced compared to the weak-line Case 1.

Temperature Dependence for Hydrogen Lines

Hydrogen Balmer lines are strong in hot stars. They have an atomic absorption coefficient that is temperature sensitive through the Stark effect, as we saw in Chapter 11. But the main temperature-caused variation is from the rise in excitation, causing the lines to strengthen, followed by a rise in ionization, causing the lines to weaken. Figure 13.10 shows examples of the change in H_γ profiles with temperature, while Figure 13.11 shows the equivalent width behavior. The relatively high excitation of the Balmer series (10.2 eV) causes the excitation growth to continue to quite high temperatures so that the maximum occurs at $T_{\text{eff}} \approx 8000-10\,000$ K. At these temperatures, most of the other elements are fully ionized, and their lines are mainly in the UV. A diagram similar to Figure 13.11 for Paschen lines is given by Frémat *et al.* (1996).

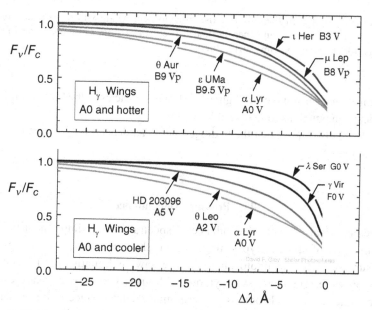

Figure 13.10 The observed profiles of H_γ decline in strength from ~A0 V for both hotter (top) and cooler (bottom) stars. Based on Gray and Evans (1973).

On the cool side of the peaks in Figure 13.11, hydrogen is mainly neutral, and since the lines come from the neutral atoms, this is similar to Case 1 for metal lines. On the hot side of the peaks, hydrogen is ionized, and we have a situation similar to Case 2, but with κ_ν coming mainly from bound–free hydrogen absorption. We might also recall that the Stark broadening functions scale with the mean electric micro-field (Equation 11.33), $E_0 = \text{constant }(P_e/T)^{2/3} = \text{constant }N_e^{2/3}$, so that E_0 has some similarities to a damping constant. This additional modest temperature dependence can be added to the rough description found using the ℓ_ν/κ_ν ratio. Although this ratio gives qualitative agreement

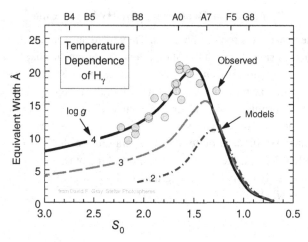

Figure 13.11 Hydrogen lines show a rise in equivalent width with temperature for cool stars, reach a maximum, and then decline. Observations from Gray and Evans (1973); models from Carbon and Gingerich (1969).

for the temperature sensitivities in Figure 13.11, the intricacies of the Stark profiles point us to numerical computations to get the full picture.

Figure 13.11 also shows that the hydrogen lines can depend on pressure (log g). So do some metal lines.

The Pressure Dependence

The basic empirical pressure dependence of spectral lines is implicit in the luminosity classification of spectral types discussed in Chapter 1. Since surface gravity sets the pressure scale in the photosphere, as we saw in Equations 9.17 and 9.18 for example, pressure effects are synonymous with gravity effects. Pressure effects seen in spectral lines are useful tools for determining stellar surface gravities, as detailed in Chapter 15. Pressure effects are inherently weaker than temperature effects. On the other hand, stellar temperatures range over only about a factor of 10, whereas surface gravities range over 4–5 orders. In contrast to temperature dependences, which affect all spectral lines, pressure effects influence only some lines, leaving most unaltered.

Pressure effects are visible in three ways. The first is caused by a change in the ratio of line absorbers to the continuous opacity through the electron pressure and the ionization equilibrium. This process enhances the lines of ions in high-luminosity stars. The second arises in the pressure sensitivity of the damping constant for strong lines. Generally, the lower collisional damping in high-luminosity stars means weaker damping wings. The third is the pressure dependence of the linear Stark broadening in hydrogen, again leading to weaker lines in higher luminosity stars.

Pressure Effects for Weak Metal Lines

Consider a normal metal line such as the case illustrated in Figure 13.12. The gravity sensitivity is small for Fe I and large for Fe II. For these kinds of lines in cool stars, we can state the following rules:

Rule 1: Lines where most of the element is in the next higher ionization stage are not sensitive to pressure changes, e.g., neutral line, element ionized, have $dW/dg \approx 0$.

Rule 2: Lines where most of the element is in the same ionization stage weaken with larger pressure, e.g., neutral line, element neutral or ionic line, element ionized, have $dW/dg < 0$.

Rule 3: Lines where most of the element is in the next lower ionization stage are pressure sensitive, becoming weaker with larger pressure, e.g., ionic line, element neutral, have $dW/dg < 0$.

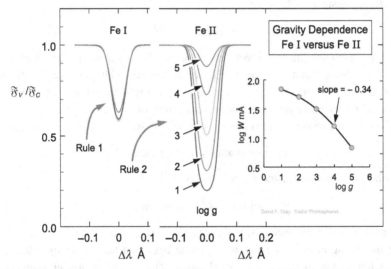

Figure 13.12 Gravity dependence for Fe I (left) compared to Fe II (center) for five values of log g. The Fe I profiles have the same color coding as the Fe II profiles, but they show so little change that they mostly lie on top of each other. The change in Fe II equivalent width is shown in the inset graph at the right. The model has $S_0 = 1.0$ and solar composition. The excitation potential is $\chi = 2.5$ eV and the microturbulence dispersion is $\xi = 1.5$ km/s for both lines.

We gain insight into these rules by again considering the $\ell_\nu / \kappa_\nu \approx$ constant W, just as we did for the temperature dependences. Assume the ionization stage is not altered by the change in pressure so that the line-forming species does not switch rules. Write the ionization equation (Equation 1.21) in the form

$$\frac{N_{r+1}}{N_r} = \frac{\Phi(T)}{P_e}, \tag{13.11}$$

where $\Phi(T)$ includes all the terms not dependent on pressure and is a constant in this discussion.

For Rule 1, the line is formed by an element in the rth ionization stage, while N_{r+1} is approximately equal to the total number of the element, i.e., is constant, making the line absorption coefficient for our line

$$\ell_v = \text{constant } N_r \approx \text{constant } P_e.$$

For cool stars, where the negative hydrogen ion dominates, $\kappa_v = \text{constant } P_e$ too, so ℓ_v/κ_v and W are independent of pressure, as stated in Rule 1.

For Rule 2, the line originates from the r th ionization stage and the element is mainly in that stage, $N_r \approx N_{\text{total}} = \text{constant}$. Then

$$W \approx \text{constant } \ell_v/\kappa_v \approx \text{constant}/P_e.$$

If we invoke Equation 9.17b as it applies to deep layers, the layers in which weak lines are formed, $P_e \approx \text{constant } g^{0.33}$, we find

$$W \approx \text{constant } g^{-0.33}, \tag{13.12}$$

which is close to the more correct and variable value shown in the inset graph in Figure 13.12. Notice that the line in this case declines with pressure not because ℓ_v weakens, but because the H_{bf}^- continuum opacity strengthens, being proportional to P_e. In effect, the photosphere becomes thinner as κ_v increases.

For Rule 3, our line comes from the $r + 1$ ionization state, while $N_r \approx \text{constant}$, making

$$\ell_v = \text{constant } N_{r+1} \approx \text{constant}/P_e.$$

Then with $\kappa_v = \text{constant } P_e$, and $P_e \approx \text{constant } g^{0.33}$,

$$W \approx \text{constant}/P_e^2 \approx \text{constant } g^{-0.66}.$$

The actual gravity dependence is slightly weaker, with an exponent ≈ -0.51.

With increasing effective temperature, neutral hydrogen takes over from the negative hydrogen ion as the major continuous absorber. Then κ_v is no longer proportional to electron pressure and the rules have to be changed accordingly.

Pressure Effects for Strong Metal Lines

When lines are strong, their cores are saturated and the three rules formulated in the previous section have to be augmented with the pressure sensitivity of the damping constants. The broad wings of the Mg I b lines are illustrated in Figure 13.13. On the main sequence, γ_{total} is dominated by the pressure-dependent broadening, but with decreasing surface gravity, this interaction decreases, making for narrower lines.

We get the sense of the pressure variation by noting that the line absorption coefficient, ℓ_v, is proportional to γ_{total} when the dispersion component is strong. The Stark damping, γ_4, is proportional to the electron pressure (Equation 11.27), the van der Waals damping, γ_6, is

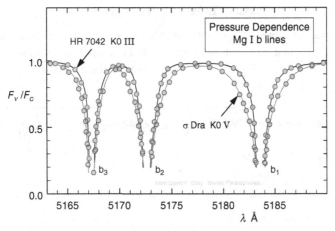

Figure 13.13 The wings of the magnesium b lines are stronger in dwarfs than in giants. Based on Cayrel de Strobel (1969) with permission from MIT Press.

proportional to the gas pressure (Equation 11.29), and the natural damping, γ_{nat}, is independent of pressure. So in cool stars, where $\kappa_v = $ constant P_e, we have out in the wings, where there is no saturation,

$$\frac{\ell_v}{\kappa_v} \approx \text{constant} \frac{N \gamma_{\text{total}}}{P_e} = \text{constant } N \frac{\phi_4(T)P_e + \phi_6(T)P_g + \gamma_{nat}}{P_e},$$

in which N is the number of absorbers, and $\phi_4(T)$ and $\phi_6(T)$ represent the damping-constant factors that do not depend on pressure. When most of the element is ionized and the line comes from the neutral species, $N = N_r = $ constant P_e from Equation 13.11, canceling the P_e in the denominator. Then, using the approximations of Equation 9.17,

$$\ell_v/\kappa_v \approx \text{constant } g^{1/2} + \gamma_{\text{nat}}. \tag{13.13}$$

For the Na I D lines in solar-temperature stars, γ_{nat} for the most part can be neglected (Figure 11.4), leaving a modest positive gravity dependence. For the Mg I 8806 Å line in similar stars, the natural damping is exceptionally large, and there is little gravity dependence (Edvardsson 1988, Ruck & Smith 1993).

On the other hand, if the line comes from the ionized species, then Equation 13.11 tells us that $N = N_{r+1}$, which is essentially independent of P_e. Since the gravity dependence of P_g and P_e are similar (Equation 9.17 again), we are left with

$$\ell_v/\kappa_v \approx \text{constant } \gamma_{\text{nat}}/P_e \approx \text{constant } g^{-1/2}. \tag{13.14}$$

The Ca II H and K lines, for example, show an inverse gravity dependence for stars above the main sequence (Lutz *et al.* 1973).

Equivalent widths show weaker gravity dependences owing to the saturation in and near the core of the line.

Pressure Effects for Hydrogen Lines

Hydrogen Balmer lines are pressure sensitive when they are strong (like the observations in
Figure 13.11), and that behavior has been used for luminosity classification since the
inception of spectral types. Figure 13.14 shows some of these early data.

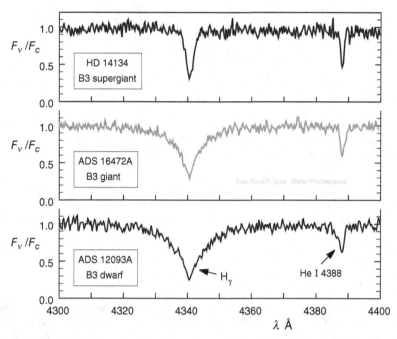

Figure 13.14 Classic photographic observations of H_γ made by Petrie (1953) showing the
gravity dependence.

We probe the pressure variation with the now usual ℓ_ν/κ_ν ratio. On the cool side of the peak,
hydrogen is neutral so the number of absorbers is independent of pressure. But the E_0 factor gives

$$\ell_\nu \approx \text{constant } E_0 = \text{constant } P_e^{2/3}.$$

The bound–free negative hydrogen ion gives most of κ_ν, and it is proportional to P_e. Thus

$$\ell_\nu/\kappa_\nu \approx \text{constant } P_e^{-1/3} \approx \text{constant } g^{-1/6}, \qquad (13.15)$$

where the electron pressure dependence on gravity is taken from Equation 9.18b for the
deep layers where many of the hydrogen lines are formed. This is a relatively weak gravity
dependence, but it can be discerned in Figure 13.11, where the curves are seen to vary
inversely with gravity.

On the hot side of the peaks in Figure 13.11, the ionization Equation 13.11, tells us that the
number of neutral hydrogen atoms is approximately proportional to the electron pressure. But
neutral hydrogen also supplies most of the continuous opacity, so this dependence on
electron pressure cancels out. The remaining pressure factor in ℓ_ν is E_0. Thus

$$\ell_\nu/\kappa_\nu \approx \text{constant } P_e^{2/3} \approx \text{constant } g^{1/3}. \tag{13.16}$$

This increase in line strength with gravity is what we see in Figure 13.11.

In still hotter stars, electron scattering becomes an important part of κ_ν. From Equation 8.18,

$$\kappa_\nu \approx \kappa_e = \frac{\alpha(e)\, P_e \sum A_j}{P_g - P_e} \approx \text{constant},$$

since at these temperatures $P_g \approx 2P_e$. This brings back into ℓ_ν/κ_ν the previously canceled electron pressure from the ionization equation. Then Equation 13.15 becomes

$$\ell_\nu/\kappa_\nu \approx \text{constant } P_e^{5/3} \approx \text{constant } g^{5/6}.$$

It is well to remember that these rough estimates of temperature and gravity dependences are meant to gain understanding, not for numerical analysis. A proper analysis of observations requires careful and full numerical modeling.

Additional Factors

Although the basic behavior of spectral lines has been mapped out, there are many additional factors that influence lines. Some of these are fully covered in the following chapters. There are deviations from LTE that can alter line strengths, especially the cores of lines (which are not covered here). The saturation portion of the curve of growth is subject to ambiguity between the

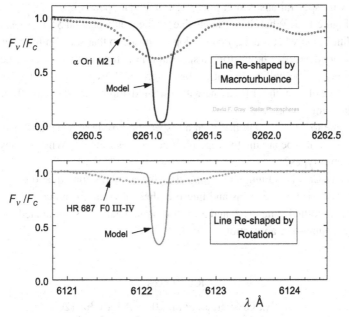

Figure 13.15 Upper panel: Ti I $\lambda6261$ is reshaped by photospheric velocity fields, called macroturbulence. Lower panel: Ca I $\lambda6122$ is reshaped by rotation. Elginfield Observatory data.

temperature distribution and the microturbulence dispersion (Chapter 16). And most dramatically, line profiles are reshaped by photospheric velocity fields (Chapter 17) and by rotation of the star (Chapter 18), as the preview in Chapter 12 has already indicated. Figure 13.15 refreshes the point. Line profiles, not just line strengths, give us the important information.

Onward

The remaining chapters forge ahead with applications of the tools we have developed. How do we apply these tools to measure stellar physics? What can we find out? What uncertainties are involved? Real stars are sending us light. It is our job to scrutinize it.

Questions and Exercises

1. Go to one of the spectral atlases, such as Griffin (1968) or Hinkle *et al.* (2000). Locate the following types of spectral lines:
   (a) a weak line
   (b) a saturated line with weak wings
   (c) a line showing damping wings
   (d) a line in the Balmer series
   (e) a line with emission in its core
   (f) a blended line
   (g) a sodium line.
2. Explain why spectral absorption lines exist in stellar spectra.
3. Look at Figure 13.1. What change in the line profile would occur if $S_v(\tau_0)$ were a horizontal straight line? And what if $S_v(\tau_0)$ had a slope opposite to that show in the figure?
4. What factors determine how strong a spectral line will be? What factors determine how wide a spectral line will be? What factors determine how deep a spectral line will be?
5. Draw a curve of growth for (a) a line with a weak damping constant and (b) a line with a strong damping constant.
6. Is it possible to have a weak line with a dispersion profile in a stellar spectrum?
7. At what spectral type are the hydrogen Balmer lines strongest? Why are they weaker in other spectral types?
8. For a line in Case1 of Figure 13.7, calculate the numbers for Equation 13.7. Compare the numerical size of the terms and interpret their physical meaning.
9. Derive the modified Rules 1, 2, and 3 for hot stars, where κ_v is dominated by the neutral hydrogen bound–free absorption.

References

Amarsi, A.M., Grevesse, N., Grumer, J., *et al.* 2020. *A&A* **636**, 120.
Athay, R.G. 1972. *Radiation Transport in Spectral Lines* (Dordrecht: Reidel).

Carbon, D.F. & Gingerich, O. 1969. *Theory and Observation of Normal Stellar Atmospheres*, Third Harvard-Smithsonian Conference on Stellar Atmospheres (Cambridge, Mass.: MIT Press), O. Gingerich, ed., p. 377.

Cayrel de Strobel, G. 1969. *Theory and Observation of Normal Stellar Atmospheres*, Third Harvard-Smithsonian Conference on Stellar Atmospheres (Cambridge, Mass.: MIT Press), O. Gingerich, ed., p. 35.

Collet, R., Asplund, M., & Thévenin, F. 2005. *A&A* **442**, 643.

Edvardsson, B. 1988. *A&A* **190**, 148.

Frémat, Y. Houziaux, L., & Andrillat, Y. 1996. *MNRAS* **279**, 25.

Gray, D.F. & Evans, J.C. 1973. *ApJ* **182**, 147.

Griffin, R.F. 1968. *A Photometric Atlas of the Spectrum of Arcturus* (Cambridge: Cambridge Philosophical Society).

Hinkle, K., Wallace, L., Valenti, J., & Harmer, D. 2000. *Visible and Near Infrared Atlas of the Arcturus Spectrum, 3727–9300 Å* (San Francisco: Astronomical Society of the Pacific).

Jefferies, J.T. 1968. *Spectral Line Formation* (Waltham: Blaisdell).

Korotin, S.A., Andrievsky, S.M., & Zhukova, A.V. 2018. *MNRAS* **480**, 965.

Lutz, T.E., Furenlid, I, & Lutz, J.H. 1973. *ApJ* **184**, 787.

Nissen, P.E. & Gustafsson, B. 2018. *A&AR* **26**, 6.

Osorio, J., Allende Prieto, C., Hubeny, I., Mészáros, Sz., & Shetrone, M. 2020. *A&A* **637**, 80.

Pasquini, L., Pallavicini, R., & Pakull, M. 1988. *A&A* **191**, 253.

Petrie, R.M. 1953. *PDAO Victoria* **9**, 251.

Reggiani, H., Amarsi, A.M., Lind, K., *et al.* 2019. *A&A* **627**, 177.

Ruck, M.J. & Smith, G. 1993. *A&A* **277**, 165.

Rutten, R.J. 1988. *Physics of Formation of FeII Lines outside LTE*, IAUC 94 (Dordrecht: Reidel), R. Viotti *et al.* eds., p. 185.

Thévenin, F. & Idiart, T.P. 1999. *ApJ* **521**, 753.

14

The Measurement of Stellar Radii and Temperatures

Two of the first things we ask about stars is how big are they and how hot are they. From the scientific perspective, size and temperature are intimately connected. Suppose we denote the flux emitted from the stellar surface by $\mathfrak{F}_\nu$ and the flux we measure at the Earth by F_ν. Next, consider two concentric spheres. The inner sphere represents the star of radius R, the outer one has a radius r, which is the distance from us to the star. Then by conservation of energy (assuming no interstellar extinction), we have

$$4\pi r^2 F_\nu = 4\pi R^2 \mathfrak{F}_\nu. \tag{14.1}$$

We can relate this equation to the effective temperature (Equation 1.2) by integrating over frequency,

$$\int_0^\infty F_\nu \, d\nu = (R/r)^2 \int_0^\infty \mathfrak{F}_\nu \, d\nu = \theta_R^2 \, \sigma T_{\text{eff}}^4, \tag{14.2}$$

where θ_R is the angular radius of the star. The most basic determination of a star's effective temperature uses this relation. F_ν is measured over a wide enough spectral interval to fix $\int_0^\infty F_\nu \, d\nu$, the angular radius is measured using techniques to be described shortly, and T_{eff} follows.

Stars are far away. That makes their angular sizes very small. One of the larger ones is for the early K giant Arcturus with a value of $0.01071''$ or 10.71 milli-arcseconds (mas). The very largest values are for a few nearby M giants and supergiants, approaching 50 mas. Direct measurement of such small angles can be done using interferometers, lunar occultations, and eclipsing binary stars, yielding radii for a few hundred stars. But most stars are too small in angular size to be measured with these techniques. Photometric methods, on the other hand, can be used to measure stars of any angular size. They are in fact the only methods applicable to the majority of stars, and photometric methods often employ model photospheres.

When measurement is impractical or the effort is not warranted, methods of inference can be used in the form of "calibrations" of radius and effective temperature with spectral type, color indices, or other convenient empirical parameters.

The linear radius, R, comes simply by combining the measured angular size with the distance. If we use the conventional units of parsecs (3.0857×10^{13} km) for distance,

304

seconds of arc for the angular radius, θ_R, or diameter, θ_D, and solar radii for R ($R_\odot = 6.961 \times 10^5$ km) then

$$R = 215.0\,\theta_R\,r$$

$$= 107.5\,\theta_D\,r. \tag{14.3}$$

The most commonly available distances come from parallax measurements.

The concept of a stellar radius is predicated on the thinness of the photosphere compared to the distance from the center of the star to the photosphere. Since the depth of formation is nearly the same for all continuum radiation in the visible region of the spectrum, the visible radius is well defined. The same statement cannot be made of the full electromagnetic spectrum. The Sun, for example, is larger at some radio wavelengths because the "surface" of emission is in the corona. Similarly the radii of cool supergiants depend on the wavelength, being larger when the star is viewed in opaque wavelength regions such as the strong molecular bands. Still, for most stars, the radius is sufficiently well defined.

Effective temperature is a more limited concept. It is essentially a shorthand for the total power per unit area coming from an average surface element of the star, i.e., $\int_0^\infty \mathfrak{F}_\nu d\nu = \sigma T_{\text{eff}}^4$. We immediately run into conceptual difficulties if the star is non-spherical or has a surface brightness that varies with position on the star, for example if the star has starspots, chemical patches, or shows gravity darkening due to rapid rotation. Then T_{eff} depends on the orientation of the star to the line of sight and could vary with rotation phase. As discussed in Chapter 9, the effective temperature does not even correspond to some unique physical temperature in the photosphere. Rather it is the temperature distribution, $T(\tau_0)$, that determines the spectral characteristics. In many situations, the scale factor, S_0, in Equation 9.5 is a very serviceable alternative to T_{eff}.

We put these caveats aside and spend the first part of the chapter considering stellar radii, followed by a discussion of stellar temperatures. Since radii and temperatures are so closely related, expect to see techniques used in the first part of the chapter come up again in the second part.

The Solar Radius

The precise radius of the Sun is of considerable interest for solar physics, but also since we normally express stellar radii in solar units. As we saw in Figure 1.1 and again in Chapter 9, the photosphere of the Sun extends over a few hundred kilometers. One hundred kilometers at one astronomical unit subtends $0.14''$, so defining the solar radius to better than $\sim 0.1''$ requires some thinking. Further, when observations are made from the Earth's surface, terrestrial seeing becomes a significant impediment. Hill *et al.* (1975) review some of the earlier definitions of the solar limb and present one of their own. Chollet and Sinceac (1999) discuss their determination and summarize earlier radius values, which range over nearly a full arcsecond. For a more historical perspective, see Vaquero *et al.* (2016) and Rozelot *et al.* (2018). Recent measurements give $\theta_R = 959.7'' \pm 0.1$ (Sofia *et al.* 2013) and $959.86'' \pm 0.06''$ (Meftah *et al.* 2018). These translate into a linear radius of

696043 ± 145 km and 696159 ± 87 km. Real variations on the scale of 0.1″ may exist (Sofia *et al.* 2013, Rozelot *et al.* 2018).

The Sun is remarkably round. Oblateness measurements show the equatorial radius to be larger (due to rotation) than the polar radius by only 7.9 mas or 5.7 km (Meftah *et al.* 2015)!

Radii from Model Photospheres

Photometric techniques are powerful and widely applicable. Starting with the fundamental relation expressed in Equation 14.1, write

$$\theta_R = R/r = (F_v/\mathfrak{F}_v)^{1/2} \tag{14.4}$$

radians, or

$$R = 4.434 \times 10^7 \, r \, (F_v/\mathfrak{F}_v)^{1/2} \tag{14.5}$$

for R in units of the solar radius and the distance r in parsecs. Photometric observations are used to determine the absolute apparent monochromatic flux, F_v, and a model photosphere is used to compute the absolute flux at the star's surface, $\mathfrak{F}_v$, using Equation 7.21 or 9.15. The use of absolute fluxes to determine radii in this way was first introduced by Gray (1967). Notice that the exponent of 1/2 means that only half the percentage error in the flux ratio appears in the radius.

Suppose we check the solar case where we already know the answer. There are several measurements of the solar energy distribution, also called the irradiance, as we saw in Figure 10.4. We can use any convenient wavelength, since the ratio $F_v/\mathfrak{F}_v$ is independent of wavelength if the correct model is chosen. Take the stellar standard-calibration wavelength of 5556 Å, for example. The measured flux is uncertain by 1–2%, but an average value is $F_v = 1.953 \times 10^{-9}$ erg/(s cm² Hz) including spectral lines. In this region, line absorption is 6.0 ± 0.5%, so the continuum flux is 1.06 times larger, or 2.070×10^{-9} erg/(s cm² Hz). Using the model for the Sun in Table 9.2, we compute $\mathfrak{F}_v = 9.613 \times 10^{-5}$ erg/(s cm² Hz), giving $\theta_R = 957.2″$. The observed solar radius is 959.7″ (Sofia *et al.* 2013) or 959.9″ (Meftah *et al.* 2018), so our values is within 0.3%, better than might be expected from the 1–2% uncertainty in F_v alone. To a certain extent this is circular reasoning since the absolute scale of the solar $T(\tau_0)$ is set using the absolute flux measurements of the Sun, but it does show how the method works and that everything is internally consistent.

In practice, we want to apply the method to other stars. The key here is to determine the proper model for the star. In Chapter 9, we found that three parameters were needed to select a model photosphere: a temperature parameter such as S_0, a surface gravity, and a metallicity. But we also found in Chapter 10 (e.g., Figure 10.8) that the Paschen continuum is mainly dependent on temperature and only weakly dependent on gravity and metallicity. So one path to the right model is the shape of the Paschen continuum, using approximate values for gravity and metallicity.

Choosing a Model Using the Paschen Continuum

The model photosphere implicitly relates the slope of the Paschen continuum to the absolute flux, as we saw, for example, in Figure 10.8. Figure 14.1 illustrates the selection of a model using the Paschen continuum. The observed energy distribution of Vega (Table 10.1) is placed within the temperature grid of model continua, using the slope of the Paschen continuum to decide its position. Then the logarithmic flux difference can be read from the two ordinates, yielding $F_\nu/\mathfrak{F}_\nu$ for Equation 14.4. In this case, the ordinate difference is -16.182 dex, and $\theta_R = 1.67$ mas follows from Equation 14.4. (Or adjust S_0 until the observed Paschen continuum is matched by the model, and then take $F_\nu/\mathfrak{F}_\nu$ from any convenient point or selection of points, since it is now the same for all points except for observational noise.)

In the process of assigning a model to the star, we have implicitly determined the star's temperature, but it is not used in the radius calculation, nor is any connection made to the concept of effective temperature or bolometric correction. No reliance is placed on other spectral regions and the bolometric flux is simply not needed. The method does rely on the absolute calibration – both the shape and the absolute flux at 5556 Å. On the down side, line absorption in cooler stars must be allowed for, as in Figures 10.14 or 10.16, which implies that some spectroscopic observations are needed in addition to the basic measurements of the energy distribution across the Paschen continuum.

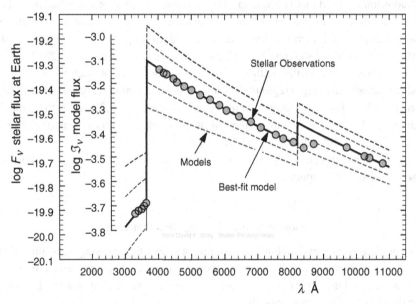

Figure 14.1 The radius of Vega is found to be $\theta_R = 1.67$ mas using the Paschen continuum. Based on Figure 10.13. Best-fit model has $T_{\text{eff}} = 9554$ K, $\log g = 3.7$, [Fe/H] $= 0.0$. Other models shown are in 5% temperature steps.

This technique has been applied mainly to nearby stars where interstellar dust is not a problem (Gray 1967, 1968a, Parsons 1970, Shipman 1972, 1979, Scargle & Strecker 1979, Shipman & Sass 1980, Perrin & Karoji 1987, Adelman & Rayle 2000, Allende Prieto & del

Burgo 2016). For an analysis of Vega incorporating the Balmer continuum, see Lane and Lester (1984). Interstellar dust will redden the continuum, making the star appear cooler than it is, and it will dim the apparent magnitude of the star reducing F_ν. The potential problem of interstellar dust can be mitigated, but not removed, by using spectral lines, which are unaffected by it, but first we consider the integrated form of Equation 14.2.

The Bolometric Technique

One commonly used approach for determining radii is to integrate Equation 14.2,

$$\theta_R = \left(\int_0^\infty F_\nu \, d\nu \middle/ \int_0^\infty \mathfrak{F}_\nu \, d\nu \right)^{1/2} = \left(F_{\text{bol}} / \sigma T_{\text{eff}}^4 \right)^{1/2}, \qquad (14.6)$$

where F_{bol} is the apparent bolometric flux and T_{eff} is the effective temperature by definition. As we have seen, it can be quite difficult to actually measure the bolometric flux. As a consequence, F_{bol} is often estimated by augmenting a few measured points with predictions from model photospheres or energy distributions from standard stars. But the same incomplete measurements or extensive model computations are used to get T_{eff}. In essence, superfluous unmeasured flux is added to both the numerator and the denominator of the flux ratio just to satisfy the formal definitions of bolometric flux and effective temperature. Some of the techniques and complications are discussed in Leggett *et al.* (1986), Bell and Gustafsson (1989), Blackwell and Lynas-Gray (1994), and Boyajian *et al.* (2012).

A more direct route is the method discussed in the previous section, which can be extended to other wavelength regions if desired. It is important to realize that when a model spectrum has been properly matched to the observations, the $F_\nu / \mathfrak{F}_\nu$ ratio is the same for all wavelengths. The effective temperature, if it is wanted, follows from the integrated spectrum, just as it did with the bolometric approach.

Problems of incomplete energy distributions are also avoided by using spectral lines to choose the model photosphere.

Choosing a Model Using Spectral Lines

Now we look into the powerful but more complicated situation of using metal lines to select the model. With metal lines, all three model parameters: temperature, surface gravity, and metallicity, come into play. Only the equivalent widths are needed, not the line profiles, easing the observational requirements.

We are blessed with numerous spectral lines from the metals, especially for the F, G, and K stars. Many of these lines convey the same information. This built-in redundancy can be used to reduce the errors of the analysis. Still, a careful selection of lines is needed to arrive at a unique solution. The electron pressure, P_e, increases with both higher log g and higher [Fe/H], so lines sensitive to P_e respond to changes in these two variables in a similar manner. Figure 14.2 illustrates the abundance variations (relative to the Sun) for measured equivalent widths of V I, Ni I, Fe I, and Fe II lines. These lines are all from the iron group and so no differential abundance variations are expected. Thus metallicity is taken to be a measure of the abundances of all the metals.

In the temperature range of G and early K stars, the V I lines are very temperature sensitive, but insensitive to electron pressure. Two or three such lines of unequal strength can be used to determine the microturbulence dispersion, ζ. The Ni I and Fe I lines show small temperature sensitivity and modest dependence on P_e, while Fe II lines show strong dependence on both temperature and P_e. The intersection of the Fe I curve with the V I curve is one solution. The intersection of the Ni I curve with the V I curve is another solution. In principle, the two should be the same, so their difference is indicative of the precision of the solution. The ambiguity between gravity and metallicity is resolved by requiring the metallicity of the model to agree with the average abundance for the two solutions (ordinates of the two intersection points), typically to within ~0.05 dex.

The solution in Figure 14.2 gives all three model parameters, S_0, log g, and [Fe/H]. For this model we then calculate $\mathfrak{F}_\nu$ and the radius follows from Equation 14.4, as before. Although reddening from interstellar dust has been avoided by using spectral lines, any extinction will still dim F_ν, forcing us to make a correction for it. Interstellar extinction is small to negligible within ~100 pc of the Sun (Bailey *et al.* 2010).

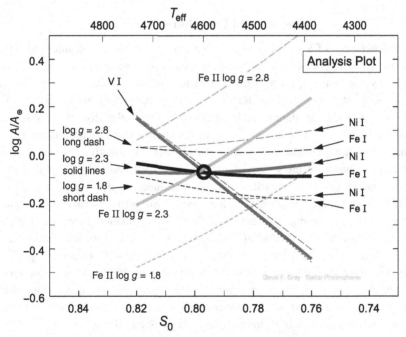

Figure 14.2 Different spectral lines respond to different physical parameters. Each line shows the chemical abundance relative to the Sun, $A/A_\odot$, needed to match an observed equivalent width, as a function of temperature (S_0 scale factor on the bottom, $T_{\text{eff}} = S_0 \times 5772$ K on the top). Three sets of curves are for three log g values: long dash, log $g = 2.8$; solid lines, log $g = 2.3$; short dash, log $g = 1.8$. The solution is indicated by the circle. Elginfield Observatory data for α UMa, K0 II-III. Based on Gray and Kaur (2019).

Notice again how associating a photospheric model with the star not only gives us the radius, but also the temperature, and in cases like this one, the surface gravity and the metallicity as well.

Absolute Flux Is Needed for Radii

We have now seen how the Paschen continuum or equivalent widths of selected spectral lines can be used to choose a model and derive a radius. However, there are many other spectral characteristics that can be used to match a model to a real star, and some of these will be discussed in material yet to come. But the one critical element that is needed to find the radius is an *absolute flux calibration* so that F_ν can be used in Equation 14.4. This point distinguishes choosing a model for a radius calculation versus other purposes such as finding a star's temperature, which can be done without knowledge of the absolute flux.

We now consider methods for determining stellar dimensions that are nearly independent of photospheric models. From the perspective of stellar photospheres, these offer valuable checks on the model results. From the perspective of stellar physics, these other methods give us independent determinations of basic stellar properties.

Stellar Radii Using Interferometers

Angular dimensions of a selection of stars have been measured with interferometers. To understand the basic principle, think of the apparent disk of a star as a hole in a screen (i.e., the dark sky); we can expect a diffraction pattern to be associated with it. We saw in Chapter 3 with diffraction gratings, for example, how the diffraction pattern is the Fourier transform of the aperture intensity distribution. If the stellar disk were a square, the transform would be a sinc x function in both directions, with x being the interferometer baseline. But with the circular disk of stars, the transform is slightly modified in shape and is circularly symmetric. More specifically, if the stellar disk showed no limb darkening, its transform would be $J_1(x)/x$, where the Bessel function $J_1(x)$ has a first zero at $x = 1.22\lambda/\theta_D$ (compared to λ/θ_D for the sinc x). So interferometers are devices for measuring these transforms, also called visibility or correlation functions. Figure 14.3 shows interferometer observations of θ Cyg, an F3 dwarf, and the theoretical transform fit to the observations. Interferometric measurements are closely related to speckle photometry (Gezari *et al.* 1972, Lynds *et al.* 1976, Blazit *et al.* 1977) and adaptive optics, but the longer baselines of true interferometers offer a huge advantage in resolving the small angular sizes of stellar disks. See Jennison (1961) for good introductory treatment. A comprehensive summary of techniques is given by Saha (2002).

Historically, two basic interferometric techniques have proved useful for measuring angular diameters. One is to combine the beams from two apertures prior to detection of the light. Interferometers used this way are called phase-coherent or amplitude interferometers. The original Michelson stellar interferometer (Michelson & Pease 1921) and more recent two-telescope or full array versions of it are of this type (Labeyrie 1975, Faucherre

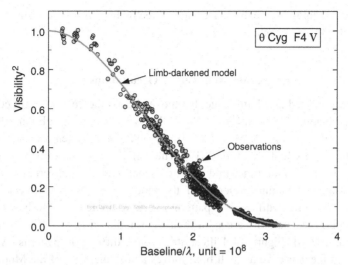

Figure 14.3 The Fourier transform or visibility curve (squared) for interferometer observations of θ Cyg is shown. The angular diameter of the star is 0.573 ± 0.009 mas. The maximum baseline was 250 m or ≈ 3.6 × 10⁸ wavelengths. Based on Figure 1 of White *et al.* (2013).

et al. 1983, Armstrong *et al.* 1998). The second technique correlates the light from two apertures after they are detected, i.e., converted from light to electronic signals. The intensity interferometer is the example for this class (Hanbury Brown *et al.* 1967, 1974, Dravins *et al.* 2015, Le Bohec & Holder 2006). Both techniques have advantages. The intensity interferometer can have larger separation of its apertures, reaching to smaller θ_R, but requires many more photons compared to the amplitude interferometer.

Historically this resulted in cool giants and supergiants being measured with amplitude interferometers and hot stars being the targets of the intensity interferometer. Michelson's original device yielded only seven angular diameters (Pease 1931), all for K and M stars above the main sequence. The maximum separation of the beams was 7.3 meters, limiting the resolution to 17 mas or larger. The intensity interferometer used an aperture separation up to 188 meters, corresponding to an angular measurement of 0.4 mas. The angular diameters of 32 stars are given by Hanbury Brown *et al.* (1974). But Michelson-type interferometers have now reached baselines of 330 m (ten Brummelaar *et al.* 2005), and routinely reach sub-milli-arcsecond diameters. See, for example, Nordgren *et al.* (2001), van Belle *et al.* (2002), Mozurkewich *et al.* (2003), Boyajian *et al.* (2009, 2012), Hutter *et al.* (2016), and Baines *et al.* (2018). These results have now become the foundation of direct radius observations.

If the transform is not measured beyond the first zero, a correction for limb darkening must be made (usually based on model photosphere calculations), and this correction increases the radius over the uniform-disk case by 2–10% depending on the wavelength (see Figure 9.3). Baselines of some interferometers are now large enough to measure transforms to frequencies past the first zero, and when the first sidelobe can be mapped out, the limb darkening is essentially measured directly. In some cases, surface features can even be resolved (e.g., Montargès *et al.* 2016).

Radii from Lunar Occultations

The Moon can be used as a knife edge that moves across the beam of light coming from the star (MacMahon 1908, Whitford 1939, Williams 1939, Nather & Evans 1970, Nather & McCants 1970, Jennings & McGruder 1999), and recently even asteroid occultations have been recorded (Benbow *et al.* 2019). The diffraction pattern of the occultation is recorded as a function of time, and the time coordinate is converted to an angular coordinate using the angular velocity of the moon or asteroid relative to the Earth. Examples of theoretical diffraction patterns were computed by Morbey (1972), and comparable observed patterns are published by Schmidtke *et al.* (1986) and Richichi *et al.* (1998a, 1998b). Figure 14.4 illustrates typical diffraction patterns. As the finite angular size of the star (like a small box) moves across the edge of the Moon, we get a classic convolution of the star's angular dimension with the knife-edge diffraction pattern, i.e., the observed pattern is blurred, reducing the amplitudes of the fringes. Modeling of the blurring yields the angular size of the star.

The horizontal diffraction scale is proportional to the wavelength, so if starlight is detected over too wide a wavelength band, the oscillations for different wavelengths get out of phase and mimic the blurring we wish to detect. In a parallel manner, the "shadow" of the lunar limb seen in the starlight crosses different strips of the telescope aperture at different times, again resulting in a blurring of the pattern. To complicate matters one more step, the timescale of the occultation depends on the slope of the lunar limb where the starlight crosses it. Fast photometric equipment is needed with millisecond integration times, since the complete occultation covers a time span of ~0.5 second. There is precious little time to collect the photons needed to delineate the diffraction pattern, while increases

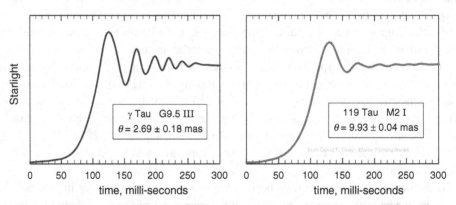

Figure 14.4 Here are two examples of diffraction patterns seen during lunar occultations. The plots show the brightness of the star as a function of time. The larger the angular diameter of the star being occulted, the smaller the fringe contrast. Notice the smaller fringe amplitudes in the pattern for 119 Tau compared to that of γ Tau. Adapted from Richichi *et al.* 1998b.

in optical bandpass or telescope size (which would help get the photon count up) have the detrimental effects just noted.

There are other limitations to this method. We cannot move the Moon to a given position in the sky. Rather, we are obliged to wait and take what is offered. The Moon passes over a limited region of the sky, mapping out its half-degree-wide path across the sky. The regression of the nodes of the lunar orbit coupled with the 5° inclination of the orbit shifts the lunar path relative to the background stars so that over the saros period of 18.6 years a 10° swath around the ecliptic is covered. The potential wait of 18.6 years requires a sizeable endowment of patience. Not to be stopped by such limitations, occultations have now yielded many angular diameters. Early results were obtained for α Sco (Cousins & Guelke 1953, Evans *et al.* 1953), with more recent work on the star by Ridgway *et al.* (1982) and Richichi and Lisi (1990). Numerous investigations encompassing a wide range of stars have subsequently been done. Most are for evolved G, K, and M stars. A sampling is Beavers *et al.* (1982), Radick *et al.* (1982), Tej and Chandrasekhar (2000), Richichi and Calamai (2001), Baug and Chandrasekhar (2012), Dyachenko *et al.* (2018). Aldebaran (α Tau, K5 III) has one of the better-determined values of 20.58 ± 0.03 mas (Richichi & Roccatagliata 2005, Richichi *et al.* 2017). The majority of occultation results have much larger uncertainties. A summary of occultation and interferometer angular radii is given by Duvert (2016).

Radii from Eclipsing Binary

Eclipsing binaries are unique in yielding their size without needing to know their distance. The scale of the system comes through the orbital-motion Doppler shifts that yield radial velocity in absolute units of km/s. The total duration of the eclipse combined with the duration of ingress and egress is combined with the relative orbital velocities to give the radii times the sine of the angle of inclination, but the inclination is ~90°, since these systems are eclipsing. Introductory and procedural information can be found in Roy and Clarke (1977) and Irwin (1962) or Kopal (1990). The results are reviewed by Harris *et al.* (1963), Snowden and Koch (1969), Lacy (1977), Barnes *et al.* (1978), Popper (1980), and Andersen (1991). Typical formal errors are ~5% with a range of 1% to 11% for dwarfs. For evolved stars the errors are about twice as large. A particularly well-determined case is for WW Aur, consisting of two A-type stars with a period of 2.5 days (Southworth *et al.* 2005). The radii are 1.927 ± 0.011 and 1.841 ± 0.011 $R_\odot$. A plethora of papers have been published in recent years, e.g., Hełminiak *et al.* (2018) and Suchomska *et al.* (2019).

Some care should be exercised. The orbital dimensions should be compared to the size of the stars to assess the possibility of tidal interaction and mass exchange. Binaries showing small separations, i.e., a semimajor axis less than about five stellar radii, usually have anomalous spectra which can lead to radial velocity errors. At the same time, binaries with large separation have small radial velocity variations, and the spectral lines of the two stars are difficult or impossible to separate, which means the orbital analysis is correspondingly difficult or impossible to perform. As we shall see below (Figure 14.5), radius determinations from selected eclipsing binaries agree well with those found using other methods.

Radii from Non-Radial Oscillations

Although stars can oscillate in many different ways, it is the pressure-mode oscillations that connect most directly with our quest for radii. Background information can be found in Brown *et al.* (1991), Kjeldsen and Bedding (1995), Arentoft *et al.* (2008), Beck *et al.* (2015). These oscillations are subtle in their observational signals. Radial velocity variations are typically a few m/s and less (e.g., Kjeldsen *et al.* 2008, Beck *et al.* 2015). Brightness variations range from milli-magnitudes to micro-magnitudes (e.g., Appourchaux *et al.* 2008, Michel *et al.* 2009), and line-profile variations are ~0.05% of the continuum (Hekker & Aerts 2010), placing stringent demands on the observational equipment and technique. Further, most stellar targets must be observed more-or-less continuously over many periods, and these range from a few minutes for dwarfs to tens of days for high-luminosity stars, during which the equipment must remain stable. Generally, larger stars show lower frequencies and larger amplitudes, as one would expect from basic physical principles. The Fourier transforms of the time series for these oscillations show a series of peaks corresponding to the various resonant states, modulated by an envelope centered on ν_{max}, as illustrated in Figure 14.5. (Note: although we have used σ to denote Fourier frequency in all previous discussions, the ν_{max} notation is so universally used in the literature that we use it here.) Other examples can be seen in Bedding *et al.* (2007), Appourchaux *et al.* (2008), and Huber *et al.* (2009, 2012, 2014).

The frequency interval between consecutive peaks of the same spherical-harmonic degree and one step in radial order is approximately constant, with a spacing $\Delta\nu$, called

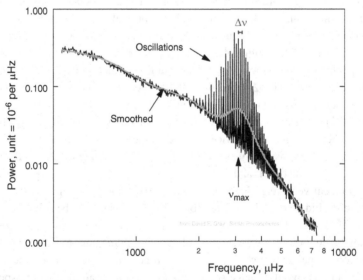

Figure 14.5 An example of the power spectrum of oscillations from brightness variations. These data are from a 700-day span of solar observations. The general drop of the curve results from the behavior of granulation and supergranulation. The peaks labeled "oscillations" are the resonant frequencies with spacing $\Delta\nu = 135$ μHz (every other peak). The maximum is at frequency $\nu_{max} = 3090$ μHz. Based on Michel *et al.* (2009).

the large-frequency separation, shown in Figure 14.5. This frequency step is approximately related to the mean stellar density,

$$\Delta v / \Delta v_\odot \approx \sqrt{\rho/\rho_\odot} = \sqrt{\mathfrak{M}/R^3} \qquad (14.7)$$

with $\Delta v_\odot = 135.1 \pm 0.1$ µHz, ρ and $\rho_\odot$ denoting the mean density of the star and the Sun, and $\mathfrak{M}$ and R are the star's mass and radius in solar units. Equation 14.7 is the first of two so-called scaling relations. The second is

$$v_{max}/v_{max}^\odot = \left(\mathfrak{M}/R^2\right)/\sqrt{S_0} = (g/g_\odot)/\sqrt{S_0}, \qquad (14.8)$$

in which the mass, $\mathfrak{M}$, and the radius, R, are in solar units, $v_{max}^\odot = 3090 \pm 30$ µHz (Huber *et al.* 2009), $g_\odot$ is the solar surface gravity, and $S_0 = T_{eff}/T_{eff}^\odot$ is our usual temperature parameter. In favorable cases Δv and v_{max} can be measured to a few percent or less. Combining Equations 14.7 and 14.8 gives the radius,

$$R = \left(v_{max}/v_{max}^\odot\right)\sqrt{S_0}/(\Delta v/\Delta v_\odot)^2 \qquad (14.9)$$

There is also an empirical relation between Δv and v_{max} found by Stello *et al.* (2009) (see also Campante *et al.* 2011), namely, $\Delta v = 0.263\, v_{max}^{0.77}$. Using this, either of the variables v_{max} or Δv could be exchanged if both were not known from the observations. Scaling relations more generally are considered by Huber *et al.* (2011). Huber *et al.* (2012) and White *et al.* (2013) showed how radii computed from oscillations agree to ~4% with those from interferometers.

Summary Plot of Radii

Figure 14.6 summarizes radii determined with the various methods. It has an appearance similar to an HR diagram. The lower bound, shown by the line, represents the zero-age main sequence (ZAMS). The values for the very largest stars, at the top right of the diagram, are often uncertain and variable (Montargès *et al.* 2016, 2017; Joyce *et al.* 2020).

Early stages of evolution lift stars above the zero-age line, so the scatter right above the ZAMS is real. We see a general decline in main-sequence radii toward cooler temperatures. There is a faster decrease near $B - V \approx 0.3$, a result of entering the domain where convective transport helps cool the stellar envelope, resulting in radii that are smaller than would be the case with radiative transport alone. This accelerated drop in radius continues as the convection zone grows in depth with declining effective temperature. Values for the zero-age curve are tabulated in Appendix B.

Further evolution moves stars from their starting point on the main sequence toward the middle of the diagram, through the subgiant domain, and on into the giant region. More massive stars follow through to become supergiants. We can use the diagram to infer a radius for a star not measured directly. Estimates for the dwarfs are the most precise, but if the luminosity class is known from a spectral classification, or can be found from the absolute magnitude through the HR diagram, the radius can be inferred, at least approximately, for evolved stars too.

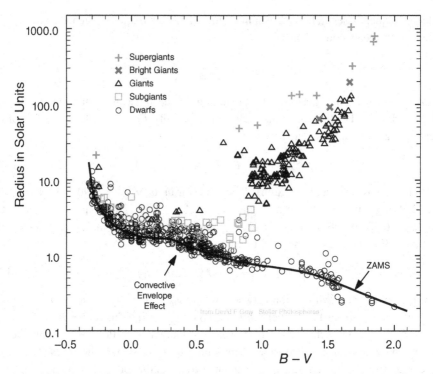

Figure 14.6 Measured stellar radii are shown as a function of $B - V$ color index. Parallaxes are from the Hipparcos (van Leeuwen 2007) or Gaia (Gaia Collaboration 2018) tabulations. The continuous line represents the zero-age main sequence (ZAMS); it is tabulated in Appendix B. Data sources: Gray (1968a updated), Perrin & Karoji (1987), Baines *et al.* (2010, 2016, 2018), Boyajian *et al.* (2013), Eker *et al.* (2014), von Braun *et al.* (2014), Gray & Kaur (2019).

Radii of pre-main sequence stars have been measured by Rhode *et al.* (2001), and radii of white dwarfs are discussed by Provencal *et al.* (1998). Many of the published values are catalogued (Pasinetti Francassini *et al.* 2001).

We now turn our attention to the basic indirect method for determining stellar radii.

The Surface-Brightness Method

The surface-brightness method is an empirical way of replacing the calculated model flux with an observational proxy. The general idea was first suggested long ago by Pickering (1880) and used by Russell (1920) and Hertzsprung (1922), and was brought to fruition by Wesselink (1969) and Wesselink *et al.* (1972). Start with Equation 14.4 and solve for the surface flux,

$$\mathfrak{F}_v = F_v/\theta_R^2.$$

In this method, we use this equation in both directions. First, using angular radii determined from any of the previously discussed methods, along with the observed apparent flux, F_v, we calculate the value of $\mathfrak{F}_v$ from the equation. It represents the *empirical* value of the flux at the stellar surface. The relation can be used at any wavelength. For example, we might choose the visual magnitude flux, in which case we find the visual-band flux at the stellar surface. If we convert to the customary magnitude units, the equation becomes

$$\mathcal{S}_V = m_V + 5 \log \theta_R, \qquad (14.10)$$

where $\mathcal{S}_V = -2.5 \log \mathfrak{F}_v$ is traditionally called the *surface brightness* and $m_V = -2.5 \log F_V$ is the apparent visual magnitude. For each star with a known angular radius, we get a value of $\mathcal{S}_V$. Then we plot $\mathcal{S}_V$ as a function of some convenient parameter such as a color index. Figure 14.7 shows what this might look like. There is a reasonably tight relation. For $B - V$ smaller than 0.5, a cubic polynomial is shown. Between $B - V = 0.5$ and 1.5, a quadratic polynomial fits sufficiently well. Beyond $B - V \approx 1.5$, the scatter becomes too large and this color index should not be used for this job.

So now we can use Equation 14.10 in the opposite direction by entering the curve in Figure14.7 with the color index of a star whose radius is unknown. We find $\mathcal{S}_V$ and use Equation 14.10 to get θ_R. Notice that any units can be used for $\mathcal{S}_V$ as long as consistency is maintained in the backward and forward processing. Milli-arcseconds are used in Figure 14.7.

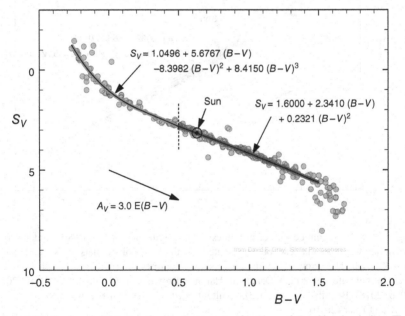

Figure 14.7 Surface brightness, $\mathcal{S}_V$, is shown as a function of $B - V$ color index. This surface-brightness relation should only be used for $B - V$ less than ~1.5. The polynomials switch at the dashed line. Interstellar reddening has almost no effect since it moves points nearly parallel to the relation, as shown by the arrow below the curve. Data from Hanbury Brown *et al.* (1974), Baines *et al.* (2010, 2016, 2018), Boyajian *et al.* (2013), von Braun *et al.* (2014), and Gray and Kaur (2019).

Some caveats go along with this scheme. Do all luminosity classes follow the same curve? Yes, they do, within observational error (see for example Figures 5 and 7 in Mozurkewich *et al.* 2003). Does metallicity have any effect? Metallicity has no measurable effect on the ordinate, but it does slightly alter $B - V$. A simple correction applied to the observed values tightens up the relation,

$$(B - V)_0 = (B - V)_{obs} - 0.13 \, [\text{Fe/H}] \times (B - V). \tag{14.11}$$

This brings the observed $(B - V)_{obs}$ to its value for $[\text{Fe/H}] = 0$. As we saw in Figure 10.12, there is less line absorption at longer wavelengths, so the metallicity dependence can be reduced by working at longer wavelengths. The color index $V - K$ is often used. Figure 14.8 shows the corresponding surface brightness plot. These relations are simple to use and can give inferred radii to a few percent. The surface-brightness method is used with numerous color indices by Boyajian *et al.* (2014).

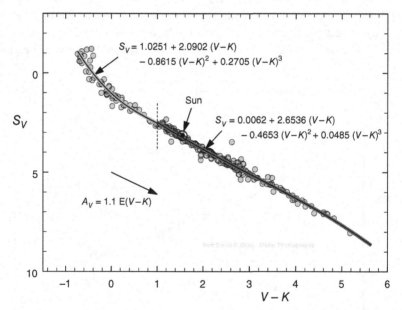

Figure 14.8 V-magnitude surface brightness, $\mathcal{S}_V$, as a function of $V - K$ color index shows a smooth variation that can be represented by the two polynomials above and below $V - K = 1.0$. The arrow below the curve shows the direction a point moves if the star is seen through interstellar dust. Data from Hanbury Brown *et al.* (1974), Baines *et al.* (2010, 2016, 2018), Boyajian *et al.* (2013), Challouf *et al.* (2014), von Braun *et al.* (2014), and Gray and Kaur (2019).

Barnes *et al.* (1978) and Popper (1980) use the bolometric version of the surface-brightness method, but that brings in the undesirable uncertainties of bolometric corrections, introducing unnecessary complications.

At this point in the chapter we turn to the determination of stellar temperatures, while the other factors of surface gravity and metallicity are in the background.

Stellar Classification by Temperature

Casting our thoughts back to Chapter 1, we remember that the basic letters of the spectral classification scheme are temperature indicators based on spectral lines. Likewise color indices express the variation of the energy distributions with temperature. Assigning temperatures to these classifications requires a calibration that uses an ensemble of stars whose temperatures have been measured. It turns out that several of the more sophisticated temperature indices we will consider shortly also need this kind of calibration. So we need to actually measure temperatures for a large enough sample to establish the calibrations.

When we want to investigate a star, we are not likely to make much progress until we pin down its basic physical parameters, with temperature leading the list.

Measured Temperatures

Reflect for a moment; the temperature varies by about a factor of two from the bottom to the top of a stellar photosphere (e.g., Figures 9.6 or 9.8), so there is no one temperature for a star, but rather a temperature distribution, $T(\tau_0)$. Since the shapes of temperature distributions show only small differences, it makes good sense to use the scale factor, S_0, from Equation 9.5,

$$S_0 = T(\tau_0)/T_\odot(\tau_0),$$

as a measure of a star's temperature. Then the temperature is scaled to the Sun's just the way radius and mass normally are. On the other hand, tradition gives *effective* temperature the favored status. It can be measured through Equation 14.2, or

$$T_{\text{eff}} = \left(\int_0^\infty F_\nu \, d\nu/\sigma \right)^{1/4} \Big/ \theta_R^{1/2}, \qquad (14.12)$$

where the integral is the bolometric flux. The angular radius, θ_R, is obtained from one of the non-photometric methods described above, i.e., interferometers, occultations, eclipsing binaries, or oscillations. The absolute apparent flux, F_ν, is supposed to be measured over the entire spectrum. But in fact, almost never is it actually measured in its entirety. Model photospheres or template spectra (e.g., Pickles 1998) are used to fill in gaps between the measurements (e.g., von Braun *et al.* 2014) or to supply bolometric corrections. Well-determined cases have bolometric-flux errors of ~5% (Mozurkewich *et al.* 2003, Baines *et al.* 2018). On the positive side, errors on this front are reduced because of the 1/4 power in Equation 14.12. When all sources of uncertainty are included, temperature errors are ~60 K for the F-G-K range and rise significantly for hotter and cooler stars.

Since radii found from interferometers, occultations, eclipsing binaries, and oscillations are almost independent of model photospheres, they form the cornerstone through Equation 14.12 against which results from other methods can be compared or calibrated.

Stellar Temperatures from Model Photospheres

Associated with each model photosphere is some temperature parameter. If we compute the bolometric flux, we get the effective temperature for the model. If instead the model uses a scaled $T(\tau_0)$, then it has an S_0 value. But the crucial step, as it was when finding radii, is to assign an appropriate model to a target star. Any number of observational parameters can be used, including continuum slope, continuum discontinuities, spectral line strengths, and line-depth ratios. Each has its advantages and disadvantages, and the instrumentation required ranges widely. Those parameters with broad wavelength coverage are susceptible to the complications of interstellar reddening. It is well to remember that three parameters characterize a model photosphere: a temperature parameter, surface gravity, and metallicity. In the current discussion, we emphasize the temperature parameter.

An important advantage of model photospheres is that, unlike the measurements of the effective temperature using Equation 14.12, no absolute fluxes are needed to get temperatures. Relative fluxes or relative line strengths allow us to select the appropriate model, and its temperature is then assigned to the star.

The Energy Distribution Thermometer

We are already familiar with the temperature variation of the energy distributions. The Paschen continuum is one of the best temperature indicators. The depth of formation for the Paschen continuum is characterized by $\tau_0 \sim 1$, so the temperatures in the deep layers of the photosphere are being measured. Several examples were given in Chapter 10, e.g., Figures 10.13, 10.14, 10.16, and 10.17. Another is given here in Figure 14.9. In essence, the model's temperature parameter is adjusted until the shape of the observed Paschen continuum is matched by the model's.

The more rapid the change in slope of the Paschen continuum with temperature, the better the thermometer we have. We find from Figure 10.16, for example, that percentage changes are larger for cooler stars. If the continuum slope can be measured to 2%, a typical value, temperature can be pinned down to ~1% for cool stars and ~5% for stars above 10 000 K. The loss of sensitivity for higher temperatures occurs rather abruptly. We expect to lose leverage when the peak of the star's energy distribution lies significantly shortward of the Paschen continuum. These hot stars are best tackled with UV Lyman-continuum observations from satellites, rockets, and balloons. A similar issue arises toward cooler stars as the energy distribution moves out of the visible window. The situation for these stars is further compromised by the large line absorption that can make it difficult to even find continuum points. The role of line absorption is reviewed in Figure 14.9. At the short-wavelength end of the Paschen continuum, the line absorption corrections become large, hard to measure, and often underestimated.

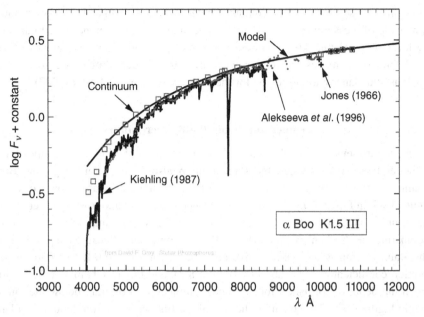

Figure 14.9 Three measurements of the energy distribution show good agreement. The squares show the continuum after line-absorption corrections are applied to the observed energy distribution, but the corrections are imperfect, especially at shorter wavelengths. The model is matched to the shape of the continuum giving $S_0 = 0.710$ ($T_{eff} = 4098$). $\log g = 1.7$, $[Fe/H] = -0.54$ are assumed, but have minor effect.

Paschen continua and energy distributions more generally have been used in a large number of investigations, including, for instance, those by Wolff *et al.* (1968), Schild *et al.* (1971), Blackwell *et al.* (1975), Tur *et al.* (1995), Aznar Cuadrado and Jeffery (2001), Adelman *et al.* (2002), Goraya (2006), Stępień and Lipski (2008), Hoard *et al.* (2010), Rauch *et al.* (2010), Ramírez & Allende Prieto (2011), and Ribas *et al.* (2017). In some cases, the models include the spectral lines, so line-absorption corrections are averted, or rather traded for the uncertainties in handling the line computations.

Synthetic Color Indices

A totally different way to use energy distributions as a temperature indicator is to compute color indices. Computed color indices are then compared directly to the observed ones to select a model for a star, and the model's temperature is assigned to the star. These are ambitious computations in which all the spectral lines are included in the calculation of the flux. The computed energy distribution is multiplied by the transmission function of the magnitude bandpass (as in Equation 1.4). The computation of broad-band color indices such as $B - V$ or $R - I$ is exemplified by Buser and Kurucz (1978), Bell and Gustafsson (1989), Bessell *et al.* (1998), Krawchuk *et al.* (2000), Houdashelt *et al.* (2000), and Knyazeva and Kharitonov (2000). Intermediate-band indices such as $b - y$ have also been used in this way (e.g., Philip & Relyea 1979, Lester *et al.* 1986, Balona 1994).

In favorable cases, errors of the computed index can be as small as 1–2%, comparable to the precision of measurement ≈ 0.01 magnitude, and fixing the temperature to about this same level of precision. Uncertainties in the bandpass functions and the handling of spectral lines often makes the real errors a few times larger. Problems associated with anomalous abundances are discussed for the case of Am stars by Lester (1987).

Temperature from the Balmer Jump

The temperature sensitivity of the Balmer jump is useful in pinning down temperatures of stars hotter than ~F0, although the Balmer jump also depends on log g. The Balmer discontinuity, technically given by the ratio of flux to the longer and shorter of 3647 Å, is plotted back in Figure 10.9. Since the confluence of Balmer lines (refer to Figure 10.10) makes it impractical to measure the flux immediately above the jump, in practice one ends up comparing the Paschen continuum to the Balmer continuum. Because the physical cause of the Balmer jump is the onset of hydrogen being ionized from the second level, the absorption coefficient changes abruptly (Figure 8.2), and this results in significant differences in the depth of formation for flux above compared to the flux below the jump, as shown in Figure 9.12. Consequently, the size of the Balmer jump from model calculations depends on the gradient of the temperature distribution. However, differences in temperature distributions are unlikely to be significant among real B and A stars, and it also helps that most of them are on or close to the main sequence, since evolutionary changes move them rapidly to cooler temperatures. In other words, the range in log g is also small.

There are several ways to specify the Balmer jump. Figure 14.10 shows an empirical approach introduced by Barbier and Chalonge (1941) and developed by Chalonge and

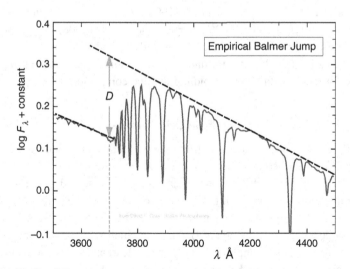

Figure 14.10 The difference between the extrapolated Paschen continuum and the Balmer continuum at 3700 Å is defined as D in the Barbier–Chalonge–Divan system. Based on Zorec *et al.* (2009).

Divan (1952). The observed Paschen continuum is extrapolated down to 3700 Å, where the discontinuity parameter is defined by $D = \log\left(F_\lambda^P / F_\lambda^B\right)$ with P and B denoting the Paschen and the Balmer fluxes at 3700 Å. D is then independently calibrated against T_{eff}.

A more physical tactic is to directly model the same observations. Figure 14.11 shows an example using the same observed spectrum. The temperature resolution is ~100 K.

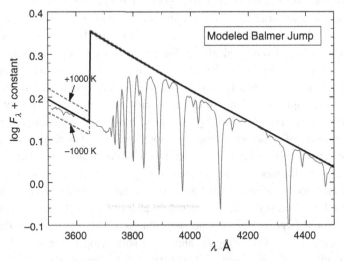

Figure 14.11 The spectrum from the previous figure is now explicitly modeled. The solid-line model has $T_{\text{eff}} = 17\,197$ K, $\log g = 4.0$, [Fe/H] = 0.0. The other two models differ by ±1000 K.

Other examples of temperature measurement using the Balmer discontinuity are found in the work of Chalonge and Divan (1952), Gray (1968b), Singh (1985), Sokolov (1995), Hovhannessian (2004), Zorec *et al.* (2009), and Aidelman *et al.* (2012). Uncertainty in temperatures using the Balmer jump for A stars and hotter is a few percent. Except in extreme cases, the chemical composition has a small effect for these hot stars since hydrogen is ionized and completely dominates the electron pressure. One of the extreme cases is peculiar A stars (Ap stars), where the Balmer jump can be badly damaged by other absorbers (Shulyak *et al.* 2010). The impact of anomalous and extreme CNO abundances on model Balmer jumps has been investigated by Auer and Demarque (1977). If we move to stars as cool as late F, normal chemical composition plays a role, but the Balmer jump becomes too small and is virtually obliterated by the numerous metal lines.

Since the measurements of the Balmer jump focus on a narrow span of wavelength, interstellar reddening is a minor issue. The same is true for temperature measurements based on spectral lines, to which we now turn our attention.

Temperature from the Hydrogen Lines

The basic concept in using spectral lines to find temperature is similar to finding the spectral type of a star, but now we use quantitative measurements. First the hydrogen lines.

In B and A stars the hydrogen lines are tens of angstroms wide and dominate the visible spectrum. Examples of the temperature sensitivity of H_γ were shown in Figures 13.10 and 13.11. Other Balmer lines show a similar behavior. As seen in Figure 13.11, a simultaneous solution for gravity and temperature is needed for stars hotter than about late A. On the other hand, for stars cooler than late A, the rapid change in strength with temperature and the tiny gravity dependence make these lines good temperature indicators on their own. It is challenging to measure a true equivalent width for a broad hydrogen line, so the advantages of profile fitting should be used whenever possible. Figure 14.12 shows a small grid of models compared to the observed H_α profile for an early G star. Cayrel *et al.* (1985) estimated temperature resolution to be a few tens of degrees in this kind of analysis.

Additional uncertainty stems from different theoretical formulations of the hydrogen-line broadening (Chapter 11). Some of the intricacies are explored by Cowley and Castelli (2002). Asymmetries, especially in Lyman lines, have been uncovered (Pelisoli *et al.* 2015). Inadequacies of underlying model photospheres are also significant. In stars ~F5 and later, convective transport cools the deep layers of the photosphere, but it is in exactly these layers that the Balmer lines are formed. It is to be expected then that the theoretical treatment of convection affects the resulting model profiles and the stellar temperatures deduced from them. Ludwig *et al.* (2009) and Amarsi *et al.* (2018) explored this issue and found a T_{eff} error as large as 300 K, which is perhaps a worst-case scenario. Barklem *et al.* (2002) found errors of ~100 K using H_γ and H_β in solar-temperature stars. Similar uncertainty was found in the detailed study of two A9 dwarfs by Fossati *et al.* (2011). Soderblom (1986) compared H_α profiles of the Sun with those of the binary components of α Cen to get their effective temperatures: 5770 ± 20 K for the primary and 5350 ± 50 K for the secondary. Small errors of precision were also found by Cayrel *et al.* (2011) in their analysis of several cool stars using H_α. The results from Balmer lines and iron lines were

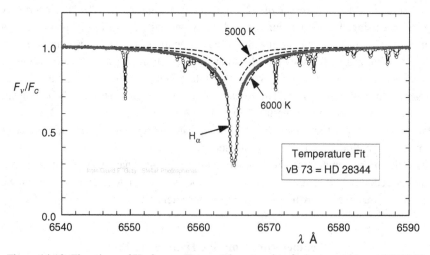

Figure 14.12 The wings of H_α for van Bueren 73 are used to find a temperature of 5900 K. The thick solid line is the adopted model. Resolving power ~33 000, Canada-France-Hawaii coudé spectrograph. Based on Cayrel *et al.* (1985).

put together by Niemczura *et al.* (2017) to find the temperatures of 50 mostly A and F stars with errors of ~100–200 K. There is a tendency for temperatures derived from H_α to be too low by ~50–100 K (Giribaldi *et al.* 2019). The chemical history of the galaxy was investigated by del Peloso *et al.* (2005) using model temperatures determined with the aid of H_α profiles.

For those stars hotter than late A, the temperature sensitivity is about two or three times worse than for the cooler stars, but good results have still been obtained. Details of applications can be seen in Adelman *et al.* (2002), for example. The effective temperature and other basic parameters of the white dwarf, Sirius B, were measured by Barstow *et al.* (2005) using Balmer lines. Frémat *et al.* (1996) found good agreement between models and observations of the Paschen lines, while Holberg *et al.* (1986) used Lyman-α profiles to find temperatures of white dwarfs. Catanzaro *et al.* (2004) pinned down T_{eff} and log g for the chemically peculiar star HR 6000 using H_γ and H_δ, and they explore how extreme variations from normal chemical composition can play a role.

Hydrogen lines are a good addition to the temperature-measurement repertoire. Metal lines are even more valuable. Most metal lines are much narrower than hydrogen lines, making continuum rectification a negligible problem. We now review their capabilities.

Metal Lines as Temperature Indicators: Equivalent Widths

Equivalent width is our basic measure of line strength. To use metal lines to measure temperature, it is mainly a matter of selecting the appropriate lines, those with high temperature sensitivity. But because a simultaneous solution for temperature, gravity, and metallicity is normal for metal lines, it is also valuable to seek out lines with different sensitivities to these parameters. Figures such as 13.7, 13.8, and 13.9 help us with this task, and Figure 14.2 illustrates one way to do the analysis, where we have enlisted lines with the required different sensitivities. The best leverage on temperature comes when we have lines with opposite dW/dT derivatives. Next best are lines having very different size derivatives even if of the same sign. In solar-type stars, the lines of neutral metals are useful because they are insensitive to pressure, holding log g at bay (e.g., Figure 13.12). In hotter stars, lines of ionized metals serve the same purpose. Weak to modest strength metal lines are typically formed in deep photospheric layers, and these are the best ones to use. Weak lines largely bypass any uncertainties in $T(\tau_0)$ and the microturbulence dispersion. With cooperative lines, the temperature can be found to ~± 50–60 K for F, G, and K stars.

Since measuring equivalent widths can be difficult when the overlap of lines becomes common, as is often the case in cool stars, a practical alternative is to measure instead the core depths of spectral lines.

Temperatures from Line-Depth Ratios

With high signal-to-noise data, the core depths of spectral lines can be precisely measured. Pairing up lines having very different temperature sensitivities and ratioing their core depths makes for a good thermometer. Line depths, being measured downward from the continuum, are independent of zero-level errors such as those arising from scattered light in the spectrograph or CCD leakage. Figure 14.13 shows one suitable pair of lines and the variation with temperature. There are many such pairs in the spectra of cool stars; a few are shown in Figure 14.14. The Doppler broadening by macroturbulence and rotation (Chapters 17 and 18) is the same for all weak lines and is also the same when the lines being ratioed are of similar strength. In other words, their influence cancels out when the line depths are ratioed. Line blends are a problem only if they occur right in the line cores, which is a distinct improvement over the situation for equivalent widths. Line-depth ratios can yield resolution on the temperature coordinate better than 5 K in favorable cases, which corresponds to one-hundredth of a spectral class or about 0.002 magnitudes in $B - V$ (Gray 1989, Gray & Brown 2001). Such resolution is 10 or 20 times greater than the accuracy with which the absolute temperature scale itself can be established.

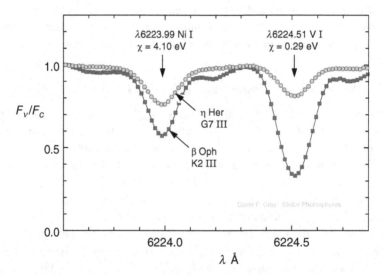

Figure 14.13 The vanadium line on the right shows much stronger variation with tempera-ture than the nickel line on the left. Line depths are measured from the continuum ($F_\nu/F_c = 1$) down to the center of the line core. The V I to Ni I line-depth ratio is 0.787 for η Her and 1.567 for β Oph. Elginfield Observatory data.

Line-depth ratios can be computed directly from model photospheres, in which case the model's temperature can be assigned directly to the star. The impediment to this approach is that the line profiles must be modeled in full, especially the broadening by macroturbulence and rotation. As a result, most applications of line-depth ratios have relied on empirical external calibrations using stars with known temperatures. Figure 14.15 illustrates one type of calibration curve. Since lines change strength with temperature, line-depth ratios are necessarily restricted to a finite range over which they can be used before one of the components becomes too weak or too strong. The ratio in

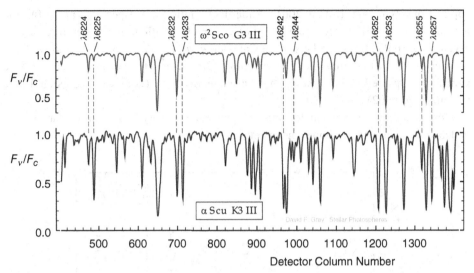

Figure 14.14 Pairs of lines suitable for line-depth ratios are marked. The warmer star is on top, ω^2 Sco, a G3 giant. The cooler star is on the bottom, α Scu, a K3 giant. The line in each pair showing the larger change is from vanadium with typical excitation potential of 0.3 eV. The temperature-stable lines are mainly iron with excitation potentials of ~2–4 eV; $\lambda 6224$ is a nickel line, as noted in the previous figure.

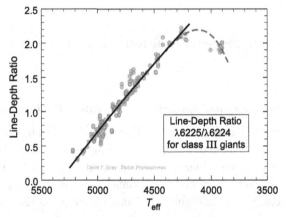

Figure 14.15 A calibration curve for the V I $\lambda 6225$ to Ni I $\lambda 6224$ line-depth ratio. The slope gives the sensitivity of the ratio. The scatter arises mainly from temperature errors. For temperatures below ~4200 K, the curve turns over and temperature becomes double valued. Based on data from Gray and Brown (2001).

Figure 14.15 shows good temperature sensitivity for temperatures from ~5300 K down to ~4200 K. Cooler than 4200 K the curve turns over, making line-depth ratios near 2.0 double valued. Line-depth ratios, when the ratio differs significantly from unity, also depend slightly on metallicity and surface gravity, complicating their use in some applications (Gray & Brown 2001). Line-depth ratios are especially powerful for measuring small differences between stars or small variations in temperature of any given star. Applications can be seen in Catalano *et al.* (2002), Kovtyukh *et al.* (2006), Kim *et al.* (2012), Ramm (2015), and Gray (2017), and to the Sun where ~1.5 K variations appeared to be seen during the magnetic cycle (Gray & Livingston 1997, Caccin *et al.* 2002).

Most stars show no variation down to ~1 K, but a few do vary for reasons that are not yet understood. Figure 14.16 shows one example.

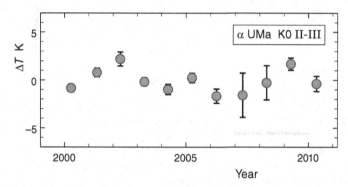

Figure 14.16 The variation of the mean of 10 line-depth ratios converted to temperature units is shown as a function of time for α UMa. Based on Gray (2018).

As you would expect, temperature has been measured for a great number of stars. A summary is valuable.

A Summary and Empirical Temperature Calibrations

Figure 14.17 summarizes some of the better-determined temperatures. In one sense, this is, like Figure 14.6, a statement of the "results." Temperatures range from ~50 000 K or ~10 times the solar value on downward. The data in Figure 14.17 stop with M stars. There are cooler stars and a more-or-less continuous transition to large planets, but this takes us too far afield. We see in Figure 14.17 that the temperatures of most stars follow a close-knit relation. Stars of all luminosity classes follow the same curve. That brings us to a second use of the figure, namely as a calibration. We can infer a star's temperature without much work by using proxy parameters such as a spectral type or a color index. Of course, a great deal of work goes into the fundamental measurements of stellar temperatures as we have seen in the preceding pages. Over the years numerous temperature calibrations have been published (e.g., Philip & Relyea 1979, Ridgway *et al.* 1980, Böhm-Vitense 1981, Lane & Lester 1984, Flower 1996). Many of these were based on relatively few stars and have been superseded as better observations have become available.

From $B - V = 0.0$ and cooler, the temperature scatter is ~0.02 dex or ~5%. Assuming this scatter is largely observational noise, and taking a typical error of $B - V \sim 0.015$ magnitudes, T_{eff} is predictable with an uncertainty of ~1%. This is probably an optimistic estimate.

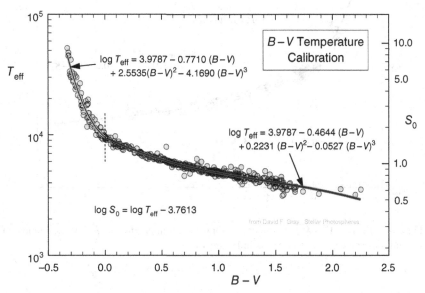

Figure 14.17 Effective temperature varies smoothly with $B - V$ color index. Polynomial fits and their equations are shown. The S_0 temperature parameter is given by the same polynomials less a 3.7613 dex offset. Based on Flower (1996), Baines *et al.* (2010, 2016, 2018), Boyajian *et al.* (2013), von Braun *et al.* (2014), Gray and Kaur (2019). Appendix B tabulates these relations.

Because these types of color indices lose leverage for hot stars, there is an upswing in the curves toward the left, which means that a small error in the color index translates into a large temperature error. Even worse, the fundamental temperatures for hot stars are less well determined, and there can be issues with interstellar extinction, since on average the hotter stars are farther away, or, in some cases, have dust shells around. All this is to say that the hot-star curves should only be used to get a rough idea of the temperature. Scatter on the hot-star portion of the calibration is ~10–15%.

Calibrations like this can be made with virtually any color index. The $V - K$ index is popular, and Figure 14.18 shows the calibration for it. (Careful though; errors on K are often quite large.) Although there are small differences, these plots are basically similar to each other. Boyajian *et al.* (2013) give plots for 33 different color indices!

Caution is always appropriate when using methods of inference like these temperature calibrations. Checking the consistency with different color indices and with spectral type is good practice. Appendix B and Figure 1.5 can be useful. And a literature search might be informative if the star has been studied previously.

In the next chapter, we move on to determining photospheric pressure and log g. Often we will find that radius and temperature are prerequisites for such determinations.

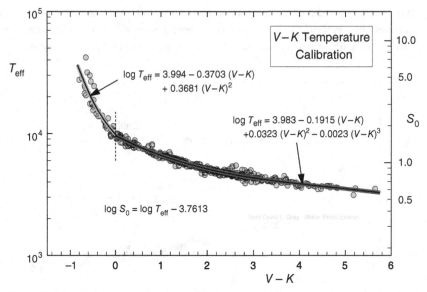

Figure 14.18 The $V - K$ temperature calibration is similar to the $B - V$ calibration in general appearance. Data sources as in Figure 14.17.

Questions and Exercises

1. Optical interferometers are beginning to allow us to resolve the surfaces of some of the closer and larger stars. Estimate the angular diameters of some of the nearby stars. What stellar-atmosphere information is likely to come from such interferometer observations?

2. Derive the temperature and radius of Vega using the shape of the continuum given in Table 10.1 with the absolute calibration given in Chapter 10, $F_v = 3.56 \times 10^{-20}$ erg/(s cm^2 Hz) at $\lambda 5556$. Plot this spectrum using $\log F_v$ as ordinate and wavelength as abscissa. Next fit Vega's continuum, i.e., energy distribution, into a grid of continuous spectra where $\log \mathfrak{F}_v$ is plotted as a function of wavelength. The model grid might be computed by you or taken from the literature, or you can get a rough grid from Figure 10.8. When the comparison is complete, you should get something that resembles Figure 14.1. The temperature comes by associating the best-fit model's temperature with the star. What temperature do you find? Find the difference between ordinate scales; it gives $\log F_v/\mathfrak{F}_v$ to use with Equations 14.4 and 14.5. Look up the parallax in the SIMBAD database and compute the linear radius of Vega. How do your answers compare to what is given in Figure 14.1? How close is the agreement with the numbers in Appendix B? Check your value against Figure 14.6.

3. Explain the limitations inherent in determining stellar radii using the lunar occultation technique. Do the same for eclipsing binaries.

4. Use the surface brightness relation in Figure 14.7 to find the radii of Vega and ε Cyg. Use the magnitudes from the SIMBAD database. Estimate the temperature of Vega from the relation given in Figure 14.14. How do these numbers compare to what you found in Exercise 2 above and with Gray and Kaur (2019)?

5. Rapidly rotating stars are oblate spheroids, not spheres, and the temperature is generally lower at the equator compared to the poles. Ponder the meaning of radius and effective temperature for such stars. What mistakes would we make in this context if we saw a rapid rotator in the pole-on aspect of view and assumed it was a slow rotator?

6. In cool supergiants, the depth of formation varies greatly with wavelength, often by a significant fraction of the star's size. Does it make any sense to try to define a radius for such stars? What do we do with Equation 14.2? Can we still define an effective temperature?

7. Is it more difficult to determine effective temperatures for O stars than for G stars? Explain.

References

Adelman, S.J. & Rayle, K.E. 2000. *A&A* **355**, 308.
Adelman, S.J., Pintado, O.I., Nieva, F., Rayle, K.E., & Sanders, S.E., Jr. 2002. *A&A* **392**, 1031.
Aidelman, Y., Cidale, L.S., Zorec, J., & Arias, M.L. 2012. *A&A* **544**, 64.
Alekseeva, G.A., Arkharov, A.A., Galkin, V.D., *et al.* 1996. *BaltA* **5**, 603.
Allende Prieto, C. & del Burgo, D. 2016. *MNRAS* **455**, 3864.
Aller, L.H. 1963. *The Atmospheres of the Sun and Stars* (New York: Ronald), 2nd ed., p. 379.
Amarsi, A.M., Nordlander, T., Barklem, P.S., *et al.* 2018. *A&A* **615**, 139.
Andersen, J. 1991. *A&AR* **3**, 91.
Appourchaux, T., *et al.* 2008. *A&A* **488**, 705.
Arentoft, T., *et al.* 2008. *ApJ* **687**, 1180.
Armstrong, J.T., Mozurkewich, D., Rickard, L.J., *et al.* 1998. *ApJ* **496**, 550.
Auer, L.H. & Demarque, P. 1977. *ApJ* **216**, 791.
Aznar Cuadrado, R. & Jeffery, C.S. 2001. *A&A* **368**, 994.
Bailey, J., Lucas, P.W., & Hough, J.H. 2010. *MNRAS* **405**, 2570.
Baines, E.K., Döllinger, M.P., Cusano, F., *et al.* 2010. *ApJ* **710**, 1365.
Baines, E.K., Döllinger, M.P., Guenther, E.W., *et al.* 2016. *AJ* **152**, 66.
Baines, E.K., Armstrong, J.T., Schmitt, H.R., *et al.* 2018. *AJ* **155**, 30.
Balona, L.A. 1994. *MNRAS* **268**, 119.
Barbier, D. & Chalonge, D. 1941. *AnAp* **4**, 30.
Barklem, P.S., Stempels, H.C., Allende Prieto, C., *et al.* 2002. *A&A* **385**, 951.
Barnes, T.G., Evans, D.S., & Moffett, T.J. 1978. *MNRAS* **183**, 285.
Barstow, M.A., Bond, H.E., Holberg, J.B., *et al.* 2005. *MNRAS* **362**, 1134.
Baug, T. & Chandrasekhar, T. 2012. *MNRAS* **419**, 866.
Beavers, W.I., Cadmus, R.R., Jr., & Eitter, J.J 1982. *AJ* **87**, 818.
Beck, P.G., *et al.* 2015. *A&A* **573**, 138.
Bedding, T.R., *et al.* 2007. *ApJ* **663**, 1315.
Bell, R.A. & Gustafsson, B. 1989. *MNRAS* **236**, 653.

Benbow, W., *et al.* 2019. *Nature Ast* **3**, 511.
Bessell, M.S., Castelli, F., & Plez, B. 1998. *A&A* **333**, 231.
Blackwell, D.E. & Lynas-Gray, A.E. 1994. *A&A* **282**, 899.
Blackwell, D.E., Ellis, R.S., Ibbetson, P.A., Petford, A.D., & Willis, R.B. 1975. *MNRAS* **171**, 425.
Blazit, A., Bonneau, D., Josse, M., *et al.* 1977. *ApJL* **217**, L55.
Böhm-Vitense, E. 1981. *ARAA* **19**, 295.
Boyajian, T.S., McAlister, H.A., Cantrell, J.R., *et al.* 2009. *ApJ* **691**, 1243.
Boyajian, T.S., McAlister, H.A., van Belle, G., *et al.* 2012. *ApJ* **746**, 101.
Boyajian, T.S., von Braun, K., van Belle, G., *et al.* 2013. *ApJ* **771**, 40.
Boyajian, T.S., van Belle, G., & von Braun, K. 2014. *AJ* **147**, 47.
Brown, T.M., Gilliland, R.L., Noyes, R.W., & Ramsey, L.W. 1991. *ApJ* **368**, 599.
Buser, R. & Kurucz, R.L. 1978. *A&A* **70**, 555.
Caccin, B., Penza, V., & Gomez, M.T. 2002. *A&A* **386**, 286.
Campante, T.L., *et al.* 2011. *A&A* **534**, 6.
Catalano, S., Biazzo, K., Frasca, A., & Marilli, E. 2002. *A&A* **394**, 1009.
Catanzaro, G., Leone, F., & Dall, T.H. 2004. *A&A* **425**, 641.
Cayrel, R., Cayrel de Strobel, G., & Campbell, B. 1985. *A&A* **146**, 249.
Cayrel, R., van't Veer-Menneret, C., Allard, N. F., Stehlé, C. 2011. *A&A* **531**, 83.
Challouf, M., Nardetto, N., Mourard, D., *et al.* 2014. *A&A* **570**, 104.
Chalonge, D. & Divan, L. 1952. *AnAp* **15**, 201.
Chollet, F. & Sinceac, V. 1999. *A&AS* **139**, 219.
Cousins, A.W.J. & Guelke, R. 1953. *MNRAS* **113**, 776.
Cowley, C.R. & Castelli, F. 2002. *A&A* **387**, 595.
del Peloso, E.L., da Silva, L., & Porto de Mello, G.F. 2005. *A&A* **434**, 275.
Dravins, D., Lagadec, T., & Nuñez, P.D. 2015. *A&A* **580**, 99.
Dreiling, L.A. & Bell, R.A. 1980. *ApJ* **241**, 736.
Duvert, G. 2016. VizieR Online Data Catalogue: II/345. https://cdsarc.unistra.fr/viz-bin/cat/II/345.
Dyachenko, V., Richichi, A., Balega, Yu, *et al.* 2018. *MNRAS* **478**, 5683.
Eker, Z., Bilir, S., Soydugan, F., *et al.* 2014. *PASA* **31**, 24.
Evans, D.S., Heydenrych, J.C.R., & van Wyk, J.D.N. 1953. *MNRAS* **113**, 781.
Faucherre, M., Bonneau, D., Koechlin, L., & Vakili, F. 1983. *A&A* **120**, 263.
Flower, P.J. 1996. *ApJ* **469**, 355.
Fossati, L., Ryabchikova, T., Shulyak, D.V., *et al.* 2011. *MNRAS* **417**, 495.
Frémat, Y., Houziaux, L., & Andrillat, Y. 1996. *MNRAS* **279**, 25.
Gaia Collaboration 2018. *A&A* **616**, 1.
Gezari, D.Y., Labeyrie, A., & Stachnik, R.V. 1972. *ApJL* **173**, L1.
Giribaldi, R.E., Ubaldo-Melo, M.L., Porto de Mello, G.F., *et al.* 2019. *A&A* **624**, 10.
Goraya, P.S. 2006. *ApSS* **306**, 93.
Gray, D.F. 1967. *ApJ* **149**, 317.
Gray, D.F. 1968a. *AJ* **73**, 769.
Gray, D.F. 1968b. *ApJL* **153**, L113.
Gray, D.F. 1989. *ApJ* **347**, 1021.
Gray, D.F. 2017. *ApJ* **845**, 62.
Gray, D.F. 2018. *ApJ* **869**, 81.
Gray, D.F. & Brown, K. 2001. *PASP* **113**, 723.
Gray, D.F. & Kaur, T. 2019. *ApJ* **882**, 148.
Gray, D.F. & Livingston, W.C. 1997. *ApJ* **474**, 802.
Hanbury Brown, R., Davis, J., Allen, L.R., & Rome, J.M 1967. *MNRAS* **137**, 393.

Hanbury Brown, R., Davis, J., & Allen, L.R. 1974. *MNRAS* **167**, 121.

Harris, D.L., III, Strand, K.A., and Worley, C.E. 1963. *Basic Astronomical Data* (Chicago: University of Chicago Press), K.A. Strand, ed., p. 273.

Hekker, S. & Aerts, C. 2010. *A&A* **515**, 43.

Hełminiak, K.G., Tokovinin, A., Niemczura, E., *et al.* 2018. *A&A* **622**, 114.

Hertzsprung, E. 1922. *Ann Sterrwacht Leiden* 14.

Hill, H.A., Stebbins, R.T., & Oleson, J.R. 1975. *ApJ* **200**, 484.

Hoard, D.W., Howell, S.B., & Stencel, R.E. 2010. *ApJ* **714**, 549.

Holberg, J.B., Wesemael, F., & Basile, J. 1986. *ApJ* **306**, 62.

Houdashelt, M.L., Bell, R.A., & Sweigert, A.V. 2000. *AJ* **119**, 1448.

Hovhannessian, R. Kh. 2004. *Ap* **47**, 499 (translated from *Astrofizika* **47**, 589).

Huber, D., Stello, D., Bedding, T.R., *et al.* 2009. *Comm Asteroseismology* **160**, 74.

Huber, D., *et al.* 2011. *ApJ* **743**, 143.

Huber, D., *et al.* 2012. *ApJ* **760**, 32.

Huber, D., *et al.* 2014. *ApJS* **212**, 2.

Hutter, D.J., Zavala, R.T., Tycner, C., *et al.* 2016. *ApJS* **227**, 4.

Irwin, J.B. 1962. *Astronomical Techniques* (Chicago: University of Chicago Press), W.A. Hiltner, ed., p. 584.

Jennings, J.K. & McGruder, C.H., III 1999. *AJ* **118**, 3061.

Jennison, R.C. 1961. *Fourier Transforms and Convolutions for the Experimentalist* (New York: Pergamon).

Jones, D.H.P. 1966. *IAUS* **24**, 141.

Joyce, M., Leung, S.-C., Molnár, L., *et al.* 2020. *ApJ* **902**, 63.

Kiehling, R. 1987. *A&AS* **69**, 465.

Kim, C., Moon, B.-K., & Lee, I.-H. 2012. *J Korean AS* **45**, 25.

Kjeldsen, H. & Bedding, T.R. 1995. *A&A* **293**, 87.

Kjeldsen, H., Bedding, T.R., Arentoft, T., *et al.* 2008. *ApJ* **682**, 1370.

Knyazeva, L.N. & Kharitonov, A.V. 2000. *ARep* **44**, 548.

Kopal, Z. 1990. *Mathematical Theory of Stellar Eclipses* (Dordrecht: Kluwer).

Kovtyukh, V.V., Soubiran, C., Bienaymé, O., Mishenina, T.V., & Belik, S. I. 2006. *MNRAS* **371**, 879.

Krawchuk, C.A.P., Dawson, P.C., & De Robertis, M.M. 2000. *AJ* **119**, 1956.

Labeyrie, A. 1975. *ApJL* **196**, L71.

Lacy, C.H. 1977. *ApJS* **34**, 479.

Lane, M.C. & Lester, J.B. 1984. *ApJ* **281**, 723.

Le Bohec, S. & Holder, J. 2006. *ApJ* **649**, 399.

Leggett, S.K., Mountain, C.M., Selby, M.J., *et al.* 1986. *A&A* **159**, 217.

Lester, J.B. 1987. *MNRAS* **227**, 135.

Lester, J.B., Gray, R.O., & Kurucz, R.L. 1986. *ApJS* **61**, 509.

Ludwig, H.G., Behara, N.T., Steffen, M., & Bonifacio, P. 2009. *A&A* **502**, L1.

Lynds, C.R., Worden, S.P., & Harvey, J.W. 1976. *ApJ* **207**, 174.

MacMahon, P.A. 1908. *MNRAS* **69**, 126.

Meftah, M., Irbah, A., Hauchecorne, A., *et al.* 2015. *Solar Phys* **290**, 673.

Meftah, M., Corbard, T., Hauchecorne, A., *et al.* 2018. *A&A* **616**, 64.

Michel, E., Samadi, R., Baudin, F., *et al.* 2009. *A&A* **495**, 979.

Michelson, A.A. & Pease, F.G. 1921. *ApJ* **53**, 249.

Montargès, M., Kervella, P., Perrin, G., *et al.* 2016. *A&A* **588**, 130.

Montargès, M., Chiavassa, A., Kervella, P., *et al.* 2017. *A&A* **605**, 108.

Morbey, C.L. 1972. *PDAO Victoria* **14**, 45.

Mozurkewich, D., Armstrong, J.T., Hindsley, R.B., *et al.* 2003. *AJ* **126**, 2502.

Nather, R.E. & Evans, D.S. 1970. *AJ* **75**, 575.

Nather, R.E. & McCants, M.M. 1970. *AJ* **75**, 963.

Niemczura, E., Polińska, M., Murphy, S.J., *et al.* 2017. *MNRAS* **470**, 2870.

Nordgren, T.E., Sudol, J.J., & Mozurkewich, D. 2001. *AJ* **122**, 2707.

Parsons, S.B. 1970. *ApJ* **159**, 951.

Pasinetti Francassini, L.E., Pastori, L., Covino, S., & Pozzi, A. 2001. *A&A* **367**, 521.

Pease, F.G. 1931. *Ergebn d exakt Naturwiss Berlin* **10**, 84.

Pelisoli, I., Santos, M.G., & Kepler, S.O. 2015. *MNRAS* **448**, 2332.

Perrin, M.-N., & Karoji, H. 1987. *A&A* **172**, 235.

Philip, A.G.D. & Relyea, L.J. 1979. *AJ* **84**, 1743.

Pickering, E.C. 1880. *Proc Am Acad Arts Sci* **16**, 1.

Pickles, A.J. 1998. *PASP* **110**, 863.

Popper, D.M. 1980. *ARAA* **18**, 115.

Provencal, J.L., Shipman, H.L., Høg, E., & Thejll, P. 1998. *ApJ* **494**, 759.

Radick, R.R., Africano, J.L., Schneeberger, T.J., Tyson, E.T., & Rogers, W.F. 1982. *AJ* **87**, 885.

Ramírez, I. & Allende Prieto, C. 2011. *ApJ* **743**, 135.

Ramm, D.J. 2015. *MNRAS* **449**, 4428.

Rauch, T., Orio, M., Gonzales-Riestra, R., *et al.* 2010. *ApJ* **717**, 363.

Rhode, K.L., Herbst, W., & Mathieu, R.D. 2001. *AJ* **122**, 3258.

Ribas, I., Gregg, M.D., Boyajian, T.S., & Bolmont, E. 2017. *A&A* **603**, 58.

Richichi, A. & Calamai, G. 2001. *A&A* **380**, 526.

Richichi, A. & Lisi, F. 1990. *A&A* **230**, 355.

Richichi, A. & Roccatagliata, V. 2005. *A&A* **433**, 305.

Richichi, A., Ragland, S., & Fabbroni, L. 1998a. *A&A* **330**, 578.

Richichi, A., Ragland, S., Stecklum, B., & Leinert, Ch. 1998b. *A&A* **338**, 527.

Richichi, A., Dyachenko, V., Pandey, A.K., *et al.* 2017. *MNRAS* **464**, 231.

Ridgway, S.T., Joyce, R.R., White, N.M., & Wing, R.F. 1980. *ApJ* **235**, 126.

Ridgway, S.T., Jacoby, G.H., Joyce, R.R., Siegel, M.J., & Wells, D.C. 1982. *AJ* **87**, 808.

Roy, A.E. & Clarke, D. 1977. *Astronomy: Structure of the Universe* (Bristol: Adam Hilger), p. 121.

Rozelot, J.P., Kosovichev, A.G., & Kilcik, A. 2018. *Sun and Geosphere* **13**, 63.

Russell, H.N. 1920. *PASP* **32**, 307.

Saha, S. 2002. *RMPhys* **74**, 551.

Scargle, J.D. & Strecker, D.W. 1979. *ApJ* **228**, 838.

Schild, R., Peterson, D.M., & Oke, J.B. 1971. *ApJ* **166**, 95.

Schmidtke, P.C., Africano, J.L., Jacoby, G.H., Joyce, R.R., & Ridgway, S.T. 1986. *AJ* **91**, 961.

Shipman, H.L. 1972. *ApJ* **177**, 723.

Shipman, H.L. 1979. *ApJ* **228**, 240.

Shipman, H.L. & Sass, C.A. 1980. *ApJ* **235**, 177.

Shulyak, D., Ryabchikova, T., Kildiyarova, R., & Kochukhov, O. 2010. *A&A* **520**, 46.

Singh, M. 1985. *ApSS* **113**, 325.

Snowden, M.S. & Koch, R.H. 1969. *ApJ* **156**, 667.

Soderblom, D.R. 1986. *A&A* **158**, 273.

Sofia, S., Girard, T.M., Sofia, U.J., *et al.* 2013. *MNRAS* **436**, 2151.

Sokolov, N.A. 1995. *A&AS* **110**, 553.

Southworth, J., Smalley, B., Maxted, P.F.L., Claret, A., & Etzel, P.B. 2005. *MNRAS* **363**, 529.

Stello, D., Chaplin, W.J., Basu, S., Elsworth, Y., & Bedding, T.R. 2009. *MNRAS* **400**, L80.

Stępień, K. & Lipski, Ł. 2008. *Cont Ast Obs Skalnaté Pleso* **38**, 353.

Suchomska, K., Graczyk, D., Pietrzyński, G., *et al.* 2019. *A&A* **621**, 93.

Tej, A. & Chandrasekhar, T. 2000. *MNRAS* **317**, 687.

ten Brummelaar, T.A. McAlister, A., Ridgway, S.T., *et al.* 2005. *ApJ* **628**, 453.

Tur, N.S., Goraya, P.S., & Sharma, S.D. 1995. *PASP* **107**, 730.

van Belle, G.T., Thompson, R.R., & Creech-Eakman, M.J. 2002. *AJ* **124**, 1706.

van Leeuwen, F. 2007. *A&A* **474**, 653.

Vaquero, J.M., Gallego, M.C., Ruiz-Lorenzo, J.J., *et al.* 2016. *Solar Phys* **291**, 1599.

von Braun, K., Boyajian, T.S., van Belle, G.T., *et al.* 2014. *MNRAS* **483**, 2413.

Wesselink, A.J. 1969. *MNRAS* **144**, 297.

Wesselink, A.J., Paranya, K., & DeVorkin, K. 1972. *A&AS* **7**, 257.

White, T.R., Huber, D., Maestro, V., *et al.* 2013. *MNRAS* **433**, 1262.

Whitford, A.E. 1939. *ApJ* **89**, 472.

Williams, J.D. 1939. *ApJ* **89**, 467.

Wolff, S.C., Kuhi, L.V., & Hayes, D. 1968. *ApJ* **152**, 871.

Zorec, J., Cidale, L., Arias, M.L., *et al.* 2009. *A&A* **501**, 297.

15

The Measurement of Surface Gravity

In this chapter, we view things from the perspective of surface gravity, the second basic parameter of a stellar photosphere, temperature being the first, metallicity the third. Surface gravity determines the scale of the gas pressure and, along with the chemical composition, the scale of the electron pressure, as we saw in Chapter 9. Generally speaking, stars higher in the HR diagram have lower gravity, and by a large factor, so that we usually use log g when working with the numbers. Traditionally, cgs units are used (cm/s^2). In these units, log g ranges from ~4.5 for dwarfs to ~1 for supergiants. The large gravity span comes about from the combination of a modest range in mass, a large range in radius, and the R^2 factor in $g = G\mathfrak{M}/R^2$.

At our disposal are several spectroscopic tools for measuring a star's log g value, but remember, these will also depend on the temperature, and in cases where hydrogen is not ionized, also the metallicity through the abundances of the electron donors.

One important offshoot of determining a surface gravity is to combine it with a radius determination to find the mass. Mass is the preeminent parameter determining a star's characteristics, so this is a nice result. Alas, the errors on this path are usually large, partly because of relatively large uncertainty in measured gravity values, and partly from the radius errors that enter with double weight.

First in our toolbox is the continuum.

The Continuum as a Gravity Indicator

The sensitivity of the visible continuum to surface gravity was illustrated in Figure 10.8. The Balmer jump is the only continuum feature sufficiently sensitive to gravity to make a useful tool, but the behavior of the Balmer jump is complex. Figure 15.1 shows the Balmer jump as a function of log g. Each curve is for a constant temperature. Clearly, if one is going to enter this plot with a measured Balmer jump in the hope of deriving log g, knowing the temperature is mandatory. But several of the curves around $S_0 \sim 1.5$ give two possible answers. Which one is correct then requires a preliminary estimate of the surface gravity. In many situations, the absolute magnitude or spectral luminosity class can be a sufficient guide. In practice, the energy distribution is usually matched with a model continuum, and fitting the Balmer jump is an integral part of the process, so one is not actually using Figure 15.1 directly.

Instead, from Figure 15.1, we can understand the leverage the Balmer jump brings to finding log g. Simply put, the greater the slope of the curve, the better the leverage. On the cool side of $S_0 \sim 1.3$, the steep slope means that if the Balmer jump is measured to ~5% or 0.02 dex, log g is found to within ~0.2 dex or a factor of ~1.5, assuming no temperature error. However, the Balmer jump toward cooler stars is compromised by the multitude of metal lines that appear already in the late F stars, i.e., $S_0 \sim 1.1$ or ~6500 K (Auer & Demarque 1977, Lane & Lester 1984). The Balmer jump in the solar spectrum is measured only with great difficulty (Houtgast 1968), and unfortunately the metallicity affects the size of the Balmer discontinuity in the cooler stars.

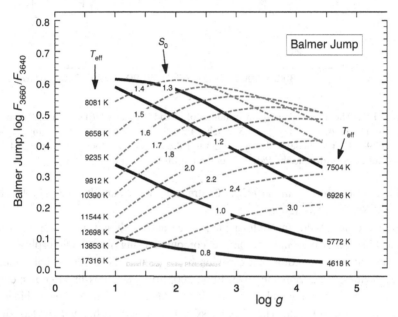

Figure 15.1 The size of the Balmer jump, log F_{3660}/F_{3640}, is shown as a function of surface gravity for constant temperatures as labeled with S_0 and T_{eff}. The actual values depend on $T(\tau_0)$.

But in hotter stars, the metallicity is not a factor. For stars hotter than $S_0 \sim 1.5$, the curves in Figure 15.1 show that better leverage is found for the low-g stars. The slope declines to less-useful levels for larger gravities and for hotter temperatures.

An explicit example is given in Figure 15.2. Three models are compared to the observed energy distribution of an A5 star. The error in matching the observations is $\delta\log g \sim 0.1$ dex. The small sensitivity of the Paschen continuum to gravity requires only a small adjustment of the optimum temperature for each log g value. As noted earlier, the absolute size of the model Balmer jump also depends on $T(\tau_0)$; steeper gradients produce bigger jumps. Notice the inverse changes of the Paschen jump compared to the Balmer jump.

Numerous investigators have incorporated the Balmer jump as a gravity indicator. Examples can be seen in the papers of Bless (1960), Newell *et al.* (1969), Dreiling and

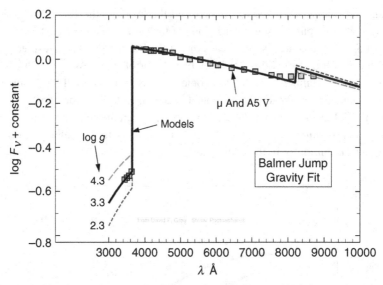

Figure 15.2 The observed energy distribution of μ And (Breger 1976) is compared to three models. The model with log $g = 3.3$ matches the observed Balmer jump. For this model, $T_{eff} \approx 7930$ K. The model with log $g = 2.3$ is 1.8% cooler, the model with log $g = 4.3$ is 2.4% hotter, according to the Paschen-continuum fit.

Bell (1980), Bergeron *et al.* (1992a), Oestreicher and Schmidt-Kaler (1999), Bessell (2007), and Shokry *et al.* (2018).

Photometric indices for the Balmer jump, such as $U - B$ and $u - y$, can be used, but require calibration. Color indices can be computed using model photospheres, and a theoretical grid in a two-color plot can then be used with the observed color indices to estimate gravities (e.g., Zabriskie 1977, Bond 2005). It is also possible to do an empirical calibration using known g values from binary stars, although for many binaries the photometry can be difficult. The Geneva system has been calibrated by Kunzli *et al.* (1997). Shipman and Sass (1980) found the mean surface gravity of white dwarfs to be $\sim 10^8$ cm/s^2 using this technique. Horizontal-branch stars were measured by Hayes and Philip (1988).

The Gravity Dependence of Hydrogen Lines

Pressure sensitivity of the wings of hydrogen lines has been used as one of the classical luminosity indicators in early-type stars (Morgan *et al.* 1943, Abt *et al.* 1968). Empirical calibrations were done by Petrie (1953), Petrie and Lee (1965), and Millward and Walker (1985), for example. Attempts at quantitative interpretation began early in the twentieth century with the work of Hulburt (1924), Russell and Stewart (1924), and especially Elvey and Struve (1930). The introduction of model photospheres was fostered by Underhill (1950). In the years following, the development of our understanding of the hydrogen-line broadening has led to many revisions and readjustments of the deduced numbers.

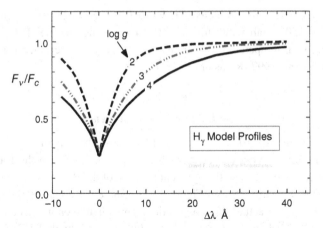

Figure 15.3 Hydrogen line profiles are stronger in stars having higher surface gravity owing to the response of the Stark broadening from the increased pressure. Model temperature is $S_0 = 1.7$ or $T_{\text{eff}} = 9812$ K.

The gravity dependence rapidly diminishes with temperature below about 9000 K (see Figures 13.11 and 13.14; see also González Delgado & Leitherer 1999). But for hotter stars, the sensitivity is much better. The calculated hydrogen-line profiles in Figure 15.3 show a large variation with log g. Although equivalent widths are useful in describing the temperature and gravity behavior of hydrogen lines, it is better to use the profiles for analysis. The far wings of the observed and computed profiles can be forced to agree, thus virtually eliminating the serious problem of properly setting the continuum.

Numerous examples of hydrogen-line use can be found in the literature. Kodaira (1964) used several Balmer lines in his analysis of the horizontal-branch star HD 161817. Olson (1975) found agreement between log g determined from hydrogen lines and log g from binary-star mechanics. Schild *et al.* (1971) incorporated energy distributions, Balmer jumps, and H_α, H_β, and H_γ profiles in their study of Vega and Sirius. Two supergiant B stars were parameterized by Dufton (1972) with the help of H_γ and H_δ profiles. Heasley and Wolff (1983) explored H_α and the Balmer jump as diagnostic tools. Bell and Dreiling (1981) were successful in reproducing the independently known surface gravity of Sirius (log g = 4.3). Their estimated error is 0.1 dex (26%). Monier (1992) found log g = 3.75 $\pm$ 0.25 for 21 Com (A3p) using H_α. The influence of extreme chemical composition was investigated by Leone and Manfrè (1997) and Catanzaro *et al.* (2004). High-resolution observations of several normal B and A stars were used by Adelman *et al.* (2002) in concert with energy distributions to pin down surface gravities. A detailed analysis of Vega, using Balmer line profiles, energy distributions, and other criteria led Yoon *et al.* (2010) to a mass of 2.14 $\pm$ 0.07 $\mathfrak{M}_\odot$. Bergeron *et al.* (1992b) and Good *et al.* (2004) used profiles of Balmer lines in white dwarfs to find log $g \sim 7 - 8$ with errors $\approx$ 0.05 dex.

Photometric filter indices can be used to sample a wavelength band centered on a hydrogen line, thus sensing the strength of the line empirically. The most well-known example is the H_β index, alias β index (Crawford 1975, 1978, 1979), which compares the flux in 30 Å and 150 Å half-width filters, both covering the H_β line. The β index is used routinely, along with temperature constraints, as the gravity discriminant in uvby photometry (see Jordi *et al.* 1997, Ribas *et al.* 1997).

The Balmer-Line Confluence

We have seen in Figures 10.7, 10.10, and 14.11 how the top of the Balmer jump is obliterated by the Balmer lines as they reach the series limit. When the Balmer lines are stronger, the effect is stronger, so the apparent position of the Balmer jump moves to higher wavelengths. This is clearly seen in Figure 14.11, where the apparent Balmer continuum reaches well beyond the actual 3646 Å position. This phenomenon was realized early on and has been embodied in the *Inglis–Teller relation* (Pannekoek 1938, Inglis & Teller 1939), $N_1 = 1.8 \times 10^{23}/n^{7.5}$, in which N_1 is the number of ions/cm^3 (or electrons in this case) estimated from n, the principal quantum number of the last visible Balmer line. A more quantitative but empirically calibrated version of the effect was formulated by Barbier, Chalonge, and Divan (Barbier & Chalonge 1941, Chalonge & Divan 1952), with more recent application by Zorec *et al.* (2009) and Aidelman *et al.* (2012). The system is designed to work on faint stars and so uses a resolution of a few angstroms or resolving power ~500. Errors on log g are moderately large, ~0.2–0.3 dex.

Other Strong Lines: Pressure-Broadened Wings

Lines such as the Ca II H and K lines, Ca I $\lambda 4227$, the Na I D lines, and the Mg I b lines show strong pressure-broadened wings in the spectra of cooler stars. These lines persist over a considerable range in temperature, which makes them useful for the task at hand. Consider the Na I D lines, $\lambda 5890$ and $\lambda 5896$, in Figure 15.4. The best model fit has log $g = 1.7$, and the other grid lines go in steps of 0.5 dex from this value.

Notice how the rate of change become smaller toward lower gravity. Still, the resolution allows us to find log g to ~0.1–0.2 dex, and the error would be smaller if log g for the star were larger, and vice versa. We do better with D_2 because it is twice as strong as D_1 (the f value for D_2 is 0.6405, for D_1 0.3199). All of this assumes no temperature and no metallicity error. In fact, the only real difference from one model to the next is the line strength. And the reason the line strength changes with log g is that $P_g(\tau_0)$ and $P_e(\tau_0)$ change with log g. Therefore, any other variable that changes the pressures will mimic the log g changes. For example, a drop in temperature of 1%, which is a typical error for a well-determined case, shifts the solution to log $g = 1.4$, down 0.3 dex. Or if we had used [Fe/H] = 0.0 instead of −0.60, we would have deduced log $g = 2.1$, a shift of 0.4 dex. And if the sodium abundance were 10% larger, we would have found log $g = 1.5$. So we see, again, that the materials of Chapters 14, 15, and 16 are all tied together. Pressure broadening is a good tool, but it has to be used in concert with others to simultaneously determine temperature and metallicity.

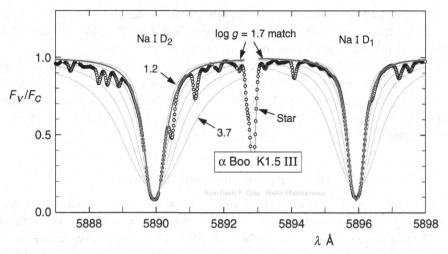

Figure 15.4 The wings of the sodium D lines are pressure sensitive. The D lines of Arcturus (α Boo) from Hinkle *et al.* (2000) are shown with a log *g* grid of model profiles. The best match is for log *g* = 1.7, $S_0 = 0.7164$ or $T_{eff} = 4135$ K, [Fe/H] = −0.60. The other grid lines have log *g* in steps of 0.5 dex from the best match.

Good examples of this process are Bell *et al.* (1985), who employed several pressure-broadened lines, Smith *et al.* (1986) and Smith and Drake (1987), who used the Ca I 6162 Å line, and Fuhrmann *et al.* (1997), who used the Mg b lines. Gravity of pre-main-sequence low-mass stars has been determined by Mohanty *et al.* (2004) using the leverage of strong lines. Figure 15.5 shows the Ca I λ6162 line in τ Cet (G8 V) as measured

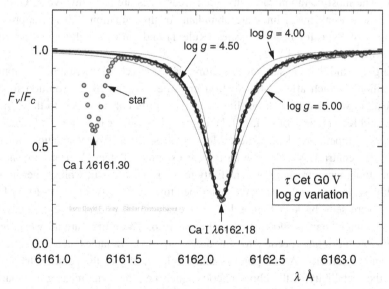

Figure 15.5 The wings of the Ca I λ6162 lines depend upon surface gravity. These models indicate a log *g* value near 4.5 for τ Cet. Observed profile from Smith and Drake (1987).

by Smith and Drake (1987) compared to models. A value of log $g = 4.5$ is indicated. One of the good things about pressure-broadened wings is that the Doppler broadening from rotation and macroturbulence is small for normal slowly rotating stars.

The inverse behavior, i.e., pressure-broadened wings growing weaker with larger gravity, is exhibited by the Ca II H and K lines, as noted earlier in the discussion of Equation 13.14. Lutz *et al.* (1973) found a factor of about two change in equivalent width between dwarfs and supergiants, and the effect to be useful in the F5 to K2 spectral range.

Gravity Determination Using Metal Lines

Since the early days of spectral classification, we knew that lines of ionized metals were enhanced in higher-luminosity, lower g, stars. Certain lines are standards for luminosity classification of cooler stars, such as the Sr II $\lambda4077$ line or Fe II lines, as we saw in Chapter 1. We explored why this happens in Chapter 13, especially Figure 13.12 and Equation 13.13. Recall, it is the ratio of the line absorption to continuous absorption coefficients, ℓ_v/κ_v, that matters (Equation 13.4).

There are significant advantages in using weak lines. By weak lines, we mean those with equivalent widths $\lesssim 0.1$ Å, placing them on the linear portion of the curve of growth. As such, their equivalent widths are independent of the damping constant and only weakly dependent on the microturbulence dispersion, ξ. If we fit line profiles, then the broadening of rotation, macroturbulence, and microturbulence must be included. Here we sidestep these complications and deal only with equivalent widths. So what differentiates one spectral line from the next in this context? Basically the ionization stage and the excitation potential. The most common cases in which weak lines are used involve F, G, and early K stars, where weak metal lines are abundant. In this situation ℓ_v/κ_v is approximately independent of log g for neutral species (Rule 1) and varies roughly as $g^{-1/3}$ for ionic species (Rule 2, Equation 13.12).

Bearing in mind that the star's temperature, which affects the excitation and ionization, and metallicity, which affects the electron pressure, are part of the solution for a log g determination, we need to seek out lines that have different responses to these variables. For F, G, and K stars, lines of Fe I, Ti I, Ni I generally show weak to modest P_e dependence and modest temperature dependence. Most of these lines have excitation potential of ~2–4 eV. By contrast, V I lines, with typical excitation potentials near zero, have high sensitivity to temperature and low sensitivity to P_e. Lines of Fe II, while fewer in number, are highly sensitive to P_e and have a temperature sensitivity opposite to V I. These attributes were seen back in Figure 14.2, for example. The same type of analysis plot, but for a different star, is shown here in Figure 15.6. These lines are all weak, and they exemplify the attributes needed to pin down all three variables: temperature, surface gravity, and metallicity. A key point is that the log $A/A_\odot$ of the solution has to be the same as the metallicity of the photospheric model, i.e., the same relative abundances for the dozen or so elements that contribute electrons.

The parameters in an analysis plot can be permuted. In Figure 15.6, $\log A/A_\odot$ is plotted against temperature for a fixed $\log g$. Alternatively, one could plot $\log g$ versus temperature for constant $A/A_\odot$, or $\log A/A_\odot$ versus $\log g$ for a constant temperature. Depending on the software used, there are computational efficiencies for one of these, but the choice is up to you.

The precision of such solutions depends on the suitability of the spectral lines and the quality of the equivalent widths. In the case of Figure 15.6, the temperature is found to ~1%, $\log g$ to ~0.05 dex, and metallicity to ~0.02 dex. Although only a handful of lines, as in Figure 15.6, is sufficient to establish a solution, some investigations, especially if individual equivalent widths are uncertain, bring together many spectral lines, often in curves of growth. Examples can be seen in Cayrel de Strobel (1968), Van't Veer Menneret (1963), Peterson and Scholz (1971), and Bell *et al.* (1985). Other parameters, such as the Balmer jump and Balmer-line fits, can be added to these analysis diagrams (see Hundt 1972, Lehner *et al.* 2000, Lyubimkov *et al.* 2002).

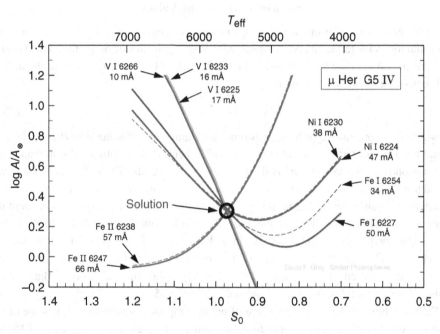

Figure 15.6 The variation of several spectral lines is shown. Abscissa is temperature. Ordinate is the species abundance relative to the solar value that gives the observed equivalent width (as labeled). The V I lines have almost no gravity dependence, while the Fe II lines have strong gravity dependence. These two species have opposite temperature dependence. Based on Gray and Kaur (2019).

The Helium Abundance Ambiguity

The role of helium in shaping stellar spectra is in many respects equivalent to that of surface gravity. Only in the O and B stars can the abundance of helium be measured using spectral lines, since in stars cooler than A0, helium lines are no longer visible. But helium still affects the photosphere, as we saw in Equations 9.23 and 9.24. There is then an ambiguity: the helium abundance, A_{He}, is a free parameter that affects the spectroscopically deduced value of surface gravity. In solar type stars, where $P_g \approx$ constant $g^{1/2}$, Equation 9.23 tells us that $g \approx$ constant $(1 + A_{\text{He}})(1 + 4A_{\text{He}})$. A change in A_{He} from the solar value of 0.085 to double that amount means a change in $\log g$ of ~ 0.13 dex. In hot stars, where $P_g \approx$ constant g, $g \approx (1 + A_{\text{He}})^{1/2}(1 + 4A_{\text{He}})^{1/2}$. The same range in A_{He} corresponds to a $\log g$ change of 0.065 dex. Changes of this size may not be negligible.

Helium-rich stars and the pseudo-gravity effects of helium have been investigated by Zboril (2000). Tanner *et al.* (2013) suggested possible ways to distinguish between helium and gravity effects.

Binaries Give Anchor Values

Just as interferometers, occultations, eclipsing binaries, and oscillations gave us independent calibrations for radii, binary-star mechanics and the physics of oscillations do the same for surface gravity. First the binaries. From Kepler's third law we get the sum of the masses of the two stars in solar mass units,

$$\mathfrak{M}_1 + \mathfrak{M}_2 = a^3/P^2, \tag{15.1}$$

in which a is the semimajor axis in astronomical units and P is the period in years. For visual binaries, a in astronomical units $= a''/p''$, where a'' is the observed semimajor axis in arcseconds on the sky and p'' is the parallax in arcseconds. The individual masses are found using the mass ratio of the two stars, which comes from observing the motion of both components against the background sky, i.e., from the relative sizes of the orbits about the center of mass. For a few systems, the mass ratio can be found from radial-velocity curves (e.g., Eggenberger *et al.* 2008 for 70 Oph). Radii come from techniques described in Chapter 14. The gravity value then follows from the definition, $g = g_\odot \mathfrak{M}/R^2$, where $g_\odot = 2.740 \times 10^4$ cm/s^2 or $\log g_\odot = 4.438$ is the solar surface gravity. Background material can be found in the classic tome of Aitken (1935).

Eclipsing binaries have the advantage of not needing the parallax to get the scale of the system. Absolute dimensions come from the radial velocities, both for radii and for the semimajor axes. The amplitude of the primary's radial velocity curve, say K_1, is related to its orbit semimajor axis, a_1, by

$$K_1 = \frac{2\pi}{P} \frac{a_1 \sin i}{\sqrt{1 - e^2}}$$

in which e is the eccentricity, and $\sin i = 1.0$, or close to it, for an eclipsing system. An analogous equation holds for the secondary's orbit. The two velocity curves are in antiphase and scaled inversely with the masses, so the mass ratio is known from K_2/K_1. Kepler's third law is used with $a = a_1 + a_2$ to get the sum of the masses. Knowledge of the sum and ratio is sufficient to obtain the individual masses. And since eclipsing binaries supply radii from the eclipses, as discussed in Chapter 14, the surface gravity is known. An example of using eclipsing binaries for gravity calibration is Olson (1975). A summary of well-determined binaries and a discussion of techniques is given by Torres *et al.* (2010).

Next we turn to oscillations.

Gravity from Oscillations

The basics were introduced in Chapter 14, largely encapsulated in Figure 14.5. Rearranging Equation 14.8 for our current application gives

$$g/g_\odot = \left(\nu_{max}/\nu_{max}^\odot\right)/\sqrt{S_0}$$

or

$$\log g = \log g_\odot + \log\left(\nu_{max}/\nu_{max}^\odot\right) - 0.5 \log S_0, \tag{15.2}$$

where $\log g_\odot = 4.438$. The square-root dependence on S_0 means that the temperature error is reduced by a factor of two.

Morel and Miglio (2012) found good agreement, but with large scatter, between asteroseismic gravity values and those from ionization balance or evolutionary tracks. A large number of $\log g$ values using oscillations seen in satellite-monitored brightness variations have now been determined (Hekker *et al.* 2009, Huber *et al.* 2014, Pinsonneault *et al.* 2014). An alternative to using the oscillation pattern directly is based on the granulation background (e.g., Bastien *et al.* 2016, Kallinger *et al.* 2016, 2019, Bugnet *et al.* 2018, Ness *et al.* 2018, Pande *et al.* 2018). Remember, these techniques rely on measuring a few parts-per-million photometric variations or a few meters-per-second velocity variations continuously over time spans of days to months, i.e., data from specially designed satellite observations.

These methods exploit the strong gravity dependence of granulation-type motions in stellar photospheres, the main theme of Chapter 17.

Surface Gravity Summary

A summary plot is shown in Figure 15.7. We see that, for the hot stars, surface gravity is nearly constant at $\log g \approx 4.25$ along the main sequence (solid line in the figure) down to $B - V \sim 0.4$, and then it declines slowly, eventually reaching a value of ~5. As mentioned above, the characteristic values for white dwarfs are $\log g \sim 8$. The scatter above the solid line is real and stems from the similar behavior of radii seen in Figure 14.6, i.e., these stars show evolutionary growth in radius with the corresponding decline in gravity. Further evolutionary changes cause the gravity to move from the main-sequence value through the subgiant region and on to the giant and supergiant portion of the plot. Main-sequence stars with $B - V$ greater than about 0.9 have not yet left the main sequence; their lifetimes exceed the age of the universe. So the points populate a web of evolutionary tracks, starting primarily from the left portion of the main sequence, that gradually converge toward the supergiant region at the top.

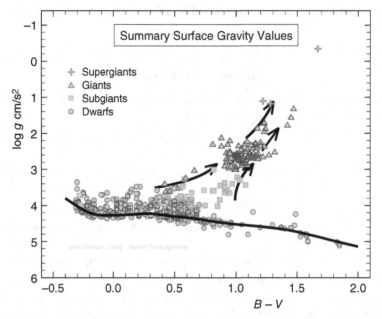

Figure 15.7 Surface gravity versus $B - V$ color index for the different luminosity classes. Arrows show approximate evolutionary trajectories. Data from Popper (1980), Allende Prieto and Lambert (1999), Torres *et al.* (2010), Bennett and Bauer (2015), Jönsson *et al.* (2017), and Gray and Kaur (2019).

Aside from observational error, the stars near the left boundary of points are the more massive ones and vice versa. See also Jönsson *et al.* (2018) for similar diagrams.

And here is a nice part: getting surface gravity values almost for free. If one has the color index and is armed with a luminosity class from a spectral classification or from the absolute magnitude, a value of log g can be estimated from Figure 15.7 with an uncertainty of ~0.3 dex or better. If the luminosity class or the absolute magnitude indicates values higher or lower than the average for the class, a corresponding refinement can be made in estimating log g. For example, if the class is IIIa or IIIb, the corresponding values would be ~2.0 or ~3.0 dex, as read from Figure 15.7.

Empirical Connections to Gravity: the Wilson–Bappu Effect

There are some spectroscopic signatures empirically connected to surface gravity. The first of these concerns the width of the emission seen in the cores of the Ca II H and K lines, the Wilson–Bappu effect. These lines are enormously strong in stars cooler than ~F type and are located in the violet at 3933.68 Å and 3968.49 Å. Emission in these lines was noticed early in the twentieth century by Eberhard and Schwarzschild (1913).

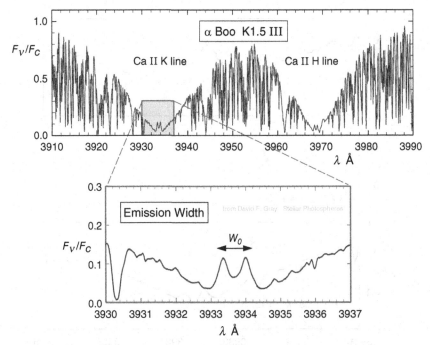

Figure 15.8 The width of the emission in the cores of the Ca II H and K lines, W_0, depends on log g. The exact definition of W_0 varies from one publication to the next. Data from Hinkle *et al.* (2000).

Figure 15.8 illustrates the emission in the K line of Arcturus. The emission width, W_0, varies from star to star; it is wider in lower gravity stars. Three examples are shown in Figure 15.9. Notice how the emission is deep in the cores and often a small fraction of the continuum. Coupled with the low sensitivity of light detectors in this spectral region, getting such data can be difficult. Most studies use only the K line.

The relation between W_0 and absolute magnitude was documented and calibrated by Wilson and Bappu (1957), with subsequent refinements by Wilson (1970), Lutz and Kelker (1975), Wilson (1976), Wallerstein *et al.* (1999), and Pace *et al.* (2003). Parsons (2001) has shown how to carry the calibration to brighter stars. A linear relation of the form $M_V = a \log W_0 + b$ is usually used, but the exact definition of W_0 differs slightly from author to author. Typical values of a and b are ~-16 and $\sim+30$ respectively, when W_0 is expressed in km/s, the traditional units. The Wilson–Bappu relation can be used to estimate M_V with an uncertainty of ~0.5 magnitude. However, with the advent of parallaxes from satellites such as Hipparcos and Gaia, direct determination of the absolute magnitude is often easier than observing the HK emission. The Wilson–Bappu relation remains physically interesting for studies of stellar chromospheres. Some of the points of interest have been summarized by Ayres (1979).

Similar calibrations can be formulated for the strong $\lambda 2804$ and $\lambda 2796$ Mg II h and k lines (e.g., Vladilo *et al.* 1987, Scoville & Mena-Werth 1998, Cassatella *et al.* 2001).

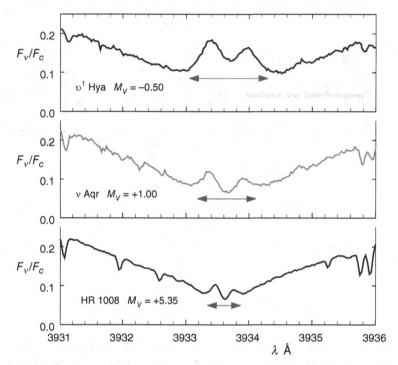

Figure 15.9 The width of the K-line emission is large in more luminous (lower surface gravity) stars. Data taken at the European Southern Observatory with a resolving power of ~60 000; courtesy of G. Pace.

Note in passing: do not confuse the W_0 width of the emission with the amount of emission (area in the emission peaks). The amount of emission is tied to the strength of a star's chromosphere, which in turn depends on the non-thermal power dissipated by acoustic waves and magnetic fields, and which is correlated with age (Wilson 1963, Wilson & Woolley 1970, Pace & Pasquini 2004 for particularly nice spectra). In many stars, perhaps all, the central absorption in the middle of the emission changes position, and the relative strength of the two emission peaks is frequently variable, on timescales of days to decades (Griffin 1963, Chiu *et al.* 1977, White & Livingston 1978, Gray 1980). The change in the relative peak flux is referred to as the V/R variation, V being the left peak, R the right peak. Neither the amount of emission nor the variations affect W_0.

Empirical Connections to Gravity: Macroturbulence and Line Bisectors

Although most of the material concerning macroturbulence and line bisectors is yet to come in Chapter 17, the gravity connection to these spectral characteristics is important.

Macroturbulence is a generic name we give to large-scale mass motions in stellar photospheres. We characterize the size of macroturbulence with a velocity dispersion, ζ_{RT}, that is derived from the Doppler broadening of the spectral lines. This dispersion is larger in lower gravity stars by about 1.8 km/s per decade in log g, but it also depends on the star's temperature. Glance ahead to Figures 17.1, 17.2, and 17.12 to get a flavor of the details. With high-resolution and high signal-to-noise spectra, ζ_{RT} can be determined to ~0.1–0.2 km/s. If there is a unique one-to-one relation between ζ_{RT} and luminosity, and assuming the temperature is well in hand, Figure 17.12 allows us to infer luminosity to about one-fifth of a class or log g to ~0.2 dex.

Macroturbulence in F, G, and K stars has a velocity distribution that is not symmetric. If one looks closely at spectral lines in such stars, the asymmetry can be discerned. This is usually portrayed in the form of a line bisector, which we met briefly in Chapter 12 (e.g., Figure 12.19). The shapes of bisectors depend on surface gravity and temperature. Figure 15.10 compares bisectors for the Fe I $\lambda6253$ line in Arcturus and the Sun. The variation in bisector shape is dramatic. Stars higher in the HR diagram have bisectors with a greater range in velocity, which mean larger asymmetry in the lines, and the left-most, or blue-most, point of the bisector is at a lower flux level.

There is a good correlation between the F_V/F_c height of the "blue bump" and absolute magnitude, as shown in Figure 15.11. In fact, the F_V/F_c value of the blue bump is several times more precise in yielding M_V than the luminosity classification itself. But, as with the Wilson–Bappu relation, this is not a particularly practical way of finding M_V since high-resolution spectroscopy is needed to get the bisectors, meaning that it is applicable only to bright stars for which good parallaxes are already available. Interesting physical behavior for stellar photospheres is implied though, if we can understand what causes the bisectors to behave this way.

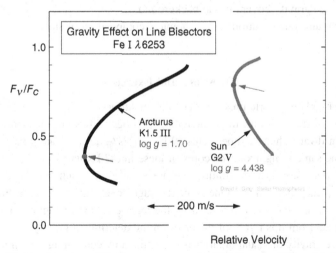

Figure 15.10 Line bisectors differ in shape across the HR diagram. The rough "C" shape shown by F, G, and K stars has a larger velocity span (left–right range) for higher luminosity stars. The blue-most point is also lower (smaller F_V/F_c) in higher luminosity stars, as shown by the dots. Based on Gray (2005).

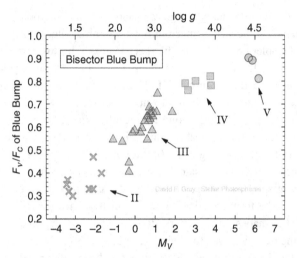

Figure 15.11 The fractional flux level of the blue-most part of the line bisector (blue bump) is shown as a function of absolute magnitude. Approximate log g values are shown on top. The blue bump drops to lower levels with higher luminosity/lower gravity stars.

Gravity and Photospheric Velocity Fields

The connections between photospheric velocities and surface gravity considered in the last section turn out to be just a taste of things to come in Chapter 17, but there we reverse our point of view and think of gravity as a determinant of those velocities: lower log g stronger fields. Although in Chapter 17 we concentrate on methods for measuring these motions, the bridge to surface gravity hovers in the background.

But first we turn our attention to the chemical composition of stars.

Questions and Exercises

1. Explain why photospheric pressure depends on surface gravity.
2. The wings of many strong lines, including hydrogen lines and most metal lines, are stronger in dwarfs than in supergiants. Why is this? Are there any exceptions? Could you say the same thing about the cores of these lines? Explain.
3. There are two main factors limiting the precision with which surface gravity can be spectroscopically determined. One is the leverage or sensitivity of the feature to gravity. The second is the precision in measuring the feature. Consider Figure 15.4. Make your own estimate from this figure of the error on the determined gravity.
4. The surface gravity of visual binaries is determined by combining orbital measurement, Kepler's third law, and the parallax. Write Kepler's third law, Equation 15.1, as $\mathfrak{M} = (a''/p'')^3/P^2$ solar masses, where a'' is the semimajor axis in arcseconds, p'' is the parallax in arcseconds (and we drop the ''), and P is the period in years. Then use

Equation E.7 in Appendix E to show that the fractional error on the mass depends on the fractional errors in the measured quantities according to

$$\frac{\delta \mathfrak{M}}{\mathfrak{M}} = \left[9\left(\frac{\delta a}{a}\right)^2 + 9\left(\frac{\delta p}{p}\right)^2 + 4\left(\frac{\delta P}{P}\right)^2 \right]^{1/2} \approx 3\frac{\delta p}{p}.$$

(In the cases of well-determined orbits, the period error is often negligible and the semimajor axis error is small, leaving parallax as the dominant contributor.) This shows how sensitive the determined mass is to parallax. But we are spared some of this sensitivity when we only need surface gravity. Assume that the radius is determined from combining an angular radius, θ_R in arcseconds, with the parallax, p. Then

$$g = g_\odot \frac{\mathfrak{M}}{R^2} = g_\odot \frac{\mathfrak{M}}{(\theta_R)^2} p^2.$$

In this case, show that

$$\frac{\delta g}{g} = \left[9\left(\frac{\delta a''}{a''}\right)^2 + 4\left(\frac{\delta \theta_R}{\theta_R}\right)^2 + \left(\frac{\delta p}{p}\right)^2 + 4\left(\frac{\delta P}{P}\right)^2 \right]^{1/2},$$

so the parallax error is much less critical.

5. Use the Inglis–Teller expression given in this chapter to estimate the electron pressure in the photosphere of the star in Figure 14.11. How does your answer compare to Figure 9.19?

6. Suppose you have a star classified as K0 III-IV for which you wish to estimate log g. Use Figure 15.7 to reach your goal. You might find Figure 1.5 helpful.

References

Abt, H.A., Meinel, A.B., Morgan, W.W., & Tapscott, J.W. 1968. *An Atlas of Low Dispersion Grating Stellar Spectra* (Tucson: Kitt Peak National Observatory).

Adelman, S.J., Pintado, O.I., Nieva, F., Rayle, K.E., & Sanders, S.E., Jr. 2002. *A&A* **392**, 1031.

Aidelman, Y., Cidale, L.S., Zorec, J., & Arias, M.L. 2012. *A&A* **554**, 64.

Aitken, R.G. 1935. *The Binary Stars* (New York: McGraw-Hill; 1964, New York: Dover).

Allende Prieto, C. & Lambert, D.L. 1999. *A&A* **352**, 555.

Auer, L.H. & Demarque, P. 1977. *ApJ* **216**, 791.

Ayres, T.R. 1979. *ApJ* **228**, 509.

Barbier, D. & Chalonge, D. 1941. *AnAp* **4**, 30.

Bastien, F., Stassun, K.G., Basri, G., & Pepper, J. 2016. *ApJ* **818**, 43.

Bell, R.A. & Dreiling, L.A. 1981. *ApJ* **248**, 1031.

Bell, R.A., Edvardson, B., & Gustafsson, B. 1985. *MNRAS* **212**, 497.

Bennett, P.D. & Bauer, W.H. 2015. *Giants of Eclipse: The ζ Aurigae Stars and Other Binary Systems* (Heidelberg: Springer), T.B. Ake & E. Griffin, eds., p. 85.

Bergeron, P., Wesemael, F., & Fontaine, G. 1992a. *ApJ* **387**, 288.

Bergeron, P., Saffer, R.A., & Liebert, J. 1992b. *ApJ* **394**, 228.

Bessell, M.S. 2007. *PASP* **119**, 605.

352   *15 The Measurement of Surface Gravity*

Bless, R.C. 1960. *ApJ* **132**, 532.
Bond, H.E. 2005. *AJ* **129**, 2914.
Breger, M. 1976. *ApJS* **32**, 7.
Bugnet, L., García, R.A., Davies, G.R., *et al.* 2018. *A&A* **620**, 38.
Cassatella, A., Altamore, A., Badiali, M., & Cardini, D. 2001. *A&A* **374**, 1085.
Catanzaro, G., Leone, F., & Dall, T.H. 2004. *A&A* **425**, 641.
Cayrel de Strobel, G. 1968. *AnAp* **31**, 43.
Chalonge, D. & Divan, L. 1952. *AnAp* **15**, 201.
Chiu, H.Y., Adams, P.J., Linsky, J.L., *et al.* 1977. *ApJ* **211**, 453.
Crawford, D.L. 1975. *AJ* **80**, 955.
Crawford, D.L. 1978. *AJ* **83**, 48.
Crawford, D.L. 1979. *AJ* **84**, 1858.
Dreiling, L.A. & Bell, R.A. 1980. *ApJ* **241**, 736.
Dufton, P.L. 1972. *A&A* **18**, 335.
Eberhard, G. & Schwarzschild, K. 1913. *ApJ* **38**, 292.
Eggenberger, P., Miglio, A., Carrier, F., Fernandes, J., & Santos, N.C. 2008. *A&A* **482**, 631.
Elvey, C.T. & Struve, O. 1930. *ApJ* **72**, 277.
Fuhrmann, K., Pfeiffer, M., Frank, C., Reetz, J., & Gehren, T. 1997. *A&A* **323**, 909.
González Delgado R.M. & Leitherer, C. 1999. *ApJS* **125**, 479.
Good, S.A. Barstow, M.A., Holberg, J.B., *et al.* 2004. *MNRAS* **355**, 1031.
Gray, D.F. 1980. *ApJ* **240**, 125.
Gray, D.F. 2005. *PASP* **117**, 711.
Gray, D.F. & Kaur, T. 2019. *ApJ* **882**, 148.
Griffin, R.F. 1963. *Observatory* **83**, 255.
Hayes, D.S. & Philip, A.G.D. 1988. *PASP* **100**, 801.
Heasley, J.N. & Wolff, S.C. 1983. *ApJ* **269**, 634.
Hekker, S., Kallinger, T., Baudin, F., *et al.* 2009. *A&A* **506**, 465.
Hinkle, K. Wallace, L., Valenti, J., & Harmer, D. 2000. *Visible and Near Infrared Atlas of the Arcturus Spectrum, 3727–9300 Å* (San Francisco: Astronomical Society of the Pacific).
Houtgast, J. 1968. *Solar Phys* **3**, 47.
Huber, D., *et al.* 2014. *ApJS* **212**, 2.
Hulburt, E.O. 1924. *ApJ* **59**, 177.
Hundt, E. 1972. *A&A* **21**, 413.
Inglis, D.R. & Teller, E. 1939. *ApJ* **90**, 439.
Jönsson, H., Ryde, N., Schultheis, M., & Zoccali, M. 2017. *A&A* **598**, 101.
Jönsson, H., *et al.* 2018. *AJ* **156**, 126.
Jordi, C., Ribas, I., Torra, J., & Giménez, A. 1997. *A&A* **326**, 1044.
Kallinger, T., Hekker, S., García, R.A., Huber, D., & Matthews, J.M. 2016. *Sci Adv* **2**, e1500654.
Kallinger, T., Beck, P.G., Hekker, S., *et al.* 2019. *A&A* **624**, 35.
Kodaira, K. 1964. *ZfAp* **59**, 139.
Kunzli, M., North, P., Kurucz, R.L., & Nicolet, B. 1997. *A&AS* **122**, 51.
Lane, M.C. & Lester, J.B. 1984. *ApJ* **281**, 723.
Lehner, N., Dufton, P.L., Lambert, D.L., Ryans, R.S.I., & Keenan, F.P. 2000. *MNRAS* **314**, 199.
Leone, F., & Manfrè, M. 1997. *A&A* **320**, 257.
Lutz, T.E. & Kelker, D.H. 1975. *PASP* **87**, 617.
Lutz, T.E., Furenlid, I., & Lutz, J.H. 1973. *ApJ* **184**, 787.

Lyubimkov, L.S., Rachkovskaya, T.M., Rostopchin, S.I., & Lambert, D.L. 2002. *MNRAS* **333**, 9.

Millward, C.G. & Walker, G.A.H. 1985. *ApJS* **57**, 63.

Mohanty, S., Basri, G., Jayawardhana, R., *et al.* 2004. *ApJ* **609,** 854.

Monier, R. 1992. *A&A* **256**, 166.

Morel, T. & Miglio, A. 2012. *MNRAS* **419**, L34.

Morgan, W.W., Keenan, P.C., & Kellman, E. 1943. *An Atlas of Stellar Spectra* (Chicago: University of Chicago Press).

Ness, M.K., Aguirre, V.S., Lund, M.N., *et al.* 2018. *ApJ* **866**, 15.

Newell, E.B., Rodgers, A.W., & Searle, L. 1969. *ApJ* **156**, 59

Oestreicher, M.O. & Schmidt-Kaler, Th. 1999. *Ast Nach* **320**, 105.

Olson, E.C. 1975. *ApJS* **29**, 43.

Pace, G. & Pasquini, L. 2004, *A&A* **426**, 1034.

Pace, G., Pasquini, L., & Ortolani, S. 2003. *A&A* **401**, 997.

Pande, D., Bedding, T.R., Huber, D., & Kjeldsen, H. 2018. *MNRAS* **480**, 467.

Pannekoek, A. 1938. *MNRAS* **98**, 694.

Parsons, S.B. 2001. *PASP* **113**, 118.

Peterson, D.M. & Scholz, M. 1971. *ApJ* **163**, 51.

Petrie, R.M. 1953. *PDAO Victoria* **9**, 251.

Petrie, R.M. & Lee, E.K. 1965. *PDAO Victoria* **12**, 435.

Pinsonneault, M.H., *et al.* 2014. *ApJS* **215**, 19.

Popper, D.M. 1980. *ARAA* **18**, 115.

Ribas, I., Jordi, C., Torra, J., & Giménez, A. 1997. *A&A* **327**, 207.

Russell, H.N. & Stewart, J.Q. 1924. *ApJ* **59**, 197.

Schild, R., Peterson, D.M., & Oke, J.B. 1971. *ApJ* **166**, 95.

Scoville, F. & Mena-Werth, J. 1998. *PASP* **110**, 794.

Shipman, H.L. & Sass, C.A. 1980. *ApJ* **235**, 177.

Shokry, A., Rivinius, Th., Mehner, A., *et al.* 2018. *A&A* **609**, 108.

Smith, G. & Drake, J.J. 1987. *A&A* **181**, 103.

Smith, G., Edvardsson, B., & Frisk, U. 1986. *A&A* **165**, 126.

Tanner, J.D., Basu, S., & Demarque, P. 2013. *ApJ* **778**, 117.

Torres, G., Andersen, J., & Giménez, A. 2010. *A&AR* **18**, 67.

Underhill, A.B. 1950. *PDAO Victoria* **8**, 385.

Van't Veer Menneret, C. 1963. *AnAp* **26**, 289.

Vladilo, G., Molaro, P., Crivellari, L., *et al.* 1987. *A&A* **185**, 233.

Wallerstein, G., Machado-Pelaez, L., & González, G. 1999. *PASP* **111**, 335.

White, O.R. & Livingston, W.C. 1978. *ApJ* **226**, 679.

Wilson, O.C. 1963. *ApJ* **138**, 832.

Wilson, O.C. 1970. *PASP* **82**, 865.

Wilson, O.C. 1976. *ApJ* **205**, 823.

Wilson, O.C. & Bappu, M.K.V. 1957. *ApJ* **125**, 661.

Wilson, O. & Woolley, R. 1970. *MNRAS* **148**, 463.

Yoon, J., Peterson, D. M., Kurucz, R. L., & Zagarello, R.J. 2010. *ApJ* **708**, 71.

Zabriskie, F.R. 1977. *PASP* **89**, 561.

Zboril, M. 2000. *A&A* **363**, 1051.

Zorec, J., Cidale, L., Arias, M.L., *et al.* 2009. *A&A* **501**, 297.

16

The Measurement of Chemical Composition

When we look at stellar spectra, we soon recognize the presence of many chemicals. The lines of hydrogen dominate the photospheric spectra of hot stars, but with declining temperature, thousands of lines from other elements grow stronger while the lines of hydrogen weaken. Our task in this chapter is to disentangle the effects on the spectra of chemical composition from the effects of temperature and surface gravity, with the goal of learning about the chemical makeup of the stars. In broad strokes, the results of chemical analyses tell us that hydrogen is overwhelmingly the most abundant element, making up ~90% of the atoms in a normal stellar photosphere, and helium comes next with ~10%. The remaining elements that we referred to as "metals," constitute a sprinkling – like salt in a bowl of broth – adding savor and interest. Most of the spectral lines are caused by these low-abundance elements.

And why do we want to know about the chemical composition of stars? There are the philosophical reasons: the natural curiosity and wonder about the nature of stars, the fact that we humans are made of chemicals processed in former generations of stars, what happens when the Sun runs out of energy? There is the basic pragmatic reason: we cannot assign a model to a star without some knowledge of the chemical composition. There are the chemical-tool reasons: we use chemical composition to help understand the nuclear reactions taking place in stars, internal mixing of material, penetration depths of convection zones, diffusion and gravitational settling, the evolutionary stage of a star, or the history and dynamics of our galaxy and other galaxies.

So depending on where our interests lie, there are two broad categories of chemical analysis. The first is simply determining the general metallicity, [Fe/H], which is an average abundance, $\log(A_E/A_E^{\odot})$, for the electron donors. The second is a detailed determination of the abundances of individual chemical species.

Primitive Abundance Indices

The roughest indicators of metallicity are based on the general collective absorption by spectral lines. For a given temperature and surface gravity, stronger lines are produced when there are more metals. Since the number of lines and their total absorption is larger toward shorter wavelengths (see Figure 10.12), photometry across different wavelength bands contains metallicity information. For example, in the UBV system, $U - B$ is more sensitive to metallicity than is $B - V$. The conventional method is to compare a star's position in the $U - B$ versus $B - V$ plane with the "standard curve" based on nearby stars, with $B - V$ being the temperature parameter. A brighter U implies weaker line absorption and lower metallicity. Interstellar reddening also shifts points in this plane, and so corrections must be applied if needed. Examples can be seen in Wallerstein and Carlson (1960), Eggen *et al.* (1962), and Alexander (1967), and a refurbishing is given by Karaali *et al.* (2011). Other magnitude systems have similar indices (e.g., Strömgren 1963, Bond 1980) and direct measurements of line absorption can also be used (Gray *et al.* 2002).

The advantage of photometric indices is that faint stars can be observed, making them particularly useful for studies of galactic halo stars and stars in other galaxies. Photometric indices are basically empirical and require calibration through knowledge of actual chemical composition. Let us move on to see how this is done.

Identification of Spectral Lines

The first step in a chemical analysis is to identify lines of particular chemical species. Much of this work has been done for us by those who compare laboratory spectra to stellar spectra using line positions and strengths. We then use the wavelength positions and strengths given in catalogues and atlases, such as Moore *et al.* (1966), Adelman (1978), or Hinkle *et al.* (1995, 2000) to identify lines of interest in our observations. Once identified, the lines' excitation potentials, χ_n, can be found from Moore *et al.* (1966), the Vienna Atomic Line Database (VALD), and the NIST Atomic Spectra Database internet sites (vald.astro .uu.se and nist.gov/pml/basic-atomic-spectroscopic-data-handbook). In some cases, the excitation potential is specified with the wavenumber, $1/\lambda$, the reciprocal of the wavelength in cm^{-1} units. Convert to electronvolts using $\chi = h\nu = hc/\lambda = 1.2398 \times 10^{-4}/\lambda$ eV, with 1.6022×10^{-12} erg/eV.

Equivalent Widths

The amount of a chemical controls the overall strength of its spectral lines. It does not directly control the shapes of lines, i.e., the line profiles. They are controlled by the shape of the absorption coefficient (Equation 11.48) and by Doppler broadening from gas motions in the photosphere and from rotation. Our measure of the line strength is the equivalent width in angstroms, first given in Chapter 12, Equation 12.21, or

$$W = \int_0^\infty (1 - F_\lambda/F_c) \, d\lambda,$$

where F_λ is the profile flux as a function of wavelength and F_c is the continuum flux in the same units. Recall that $F_\lambda/F_c = F_\nu/F_c$. It behooves us then to focus on the equivalent widths and not the line profiles. This can be advantageous observationally speaking because W can be measured on modest-resolution spectra, meaning shorter exposure times or reaching fainter stars. It also means that we spend little or no effort modeling macro-turbulence and rotation, nor, conversely, does this kind of analysis yield any information about these phenomena. Methods for measuring W were discussed in Chapter 12. Our model value of W comes from the profile using Equation 13.13 or 13.14,

$$\mathfrak{F}_v = 2\pi \int\limits_{-\infty}^{\infty} B_v(T(\tau_0)) \, E_2(\tau_v) \frac{\ell_v + \kappa_v}{\kappa_0} \tau_0 \frac{d\log \tau_0}{\log e},$$

$\mathfrak{F}_c$ from the same equation with ℓ_v set to zero, and integrating as in the previous equation. The abundances of the electron donors can enter in the absorption coefficients, ℓ_v, κ_v, and κ_0, but the abundance of the species of interest enters in ℓ_v.

The abundance dependence of W is called the curve of growth (introduced in Figure 13.4). Every line has its own curve of growth. In chemical abundance studies, the curve of growth can be front and center, as it often was in older work, or it can be in the background, as it often is in more contemporary work. In either case, it is fundamental, so we now turn our attention to the curve of growth and explore some of its characteristics.

Curves of Growth for Analytical Models: an Historical Note

Early curve-of-growth calculations were done by Voigt (1912) and van der Held (1931) with application by Minnaert and Mulders (1931) and Minnaert and Slob (1931). These analyses were made with very simple models. Often the whole line-forming region was characterized by a single temperature and electron pressure. Occasionally these old single-layer models are associated with the phrase "the curve-of-growth method." Curves of growth based on such simple models have been published by Struve and Elvey (1934), Menzel (1936), Strömgren (1940), Pannekoek and van Albada (1946), Greenstein (1948), Wrubel (1949, 1954), and Unsöld (1955). This narrow meaning is not used here however. Our use and understanding is with the full generalization of the curve-of-growth concept to any model photosphere or real star.

Scaling Relations

Although we define a curve of growth as the increase in equivalent width with the increase in the number of absorbers, this is not quite appropriate when we talk about constructing a curve of growth for a star because the star has *one* abundance for a given species and there is no way we can alter it or the number of absorbers. There are, however, other factors that make some lines of a given element stronger than other lines of that same element. Careful handling of these factors allows us to construct a curve of growth for that species using many different lines. Underlying this construction is the assumption that the curves of growth for individual lines have one basic universal shape. This is generally a good approximation except for the damping or strong-line portion of the curves. In essence, the abscissae are scaled, translated on the logarithmic plot, to bring all the lines onto a common curve of growth.

How the variables enter can be seen by looking at the ratio of line to continuous absorption much like we did in Chapter 13. According to Equation 13.4, a weak line profile is given approximately by

$$\frac{\mathfrak{F}_c - \mathfrak{F}_\nu}{\mathfrak{F}_c} \approx \text{constant} \, \frac{\ell_\nu}{\kappa_\nu}.$$

Since the contribution functions for a weak line are localized to a narrow depth range, the relevant physical parameters are essentially single valued. In the current context, we want the equivalent width, so define w as the integral over this profile,

$$w = \frac{\text{constant}}{\kappa_\nu} \int_0^\infty \ell_\nu \, d\nu,$$

where a lower case w explicitly denotes that we are dealing with a weak line. We place κ_ν outside the integral because it is constant over the width of the line. Then use, $\ell_\nu \rho = N\alpha_{\text{nat}}$, in which ρ is the mass density, N is the number of line absorbers per unit volume, and α_{nat} is the atomic absorption coefficient given by Equation 11.9 to replace the integral. Including the f value, we get

$$w \approx \text{constant} \frac{\pi e^2}{m_e c} \frac{\lambda^2}{c} f \frac{N}{\kappa_\nu}, \tag{16.1}$$

where ρ has been rolled into the constant. Introduce the usual number abundance for the element E relative to hydrogen, $A_E = N_E/N_H$, the fraction in the rth stage of ionization, N_r/N_E, and the excitation equation (Equation 1.19) to write N as

$$N = A_E \frac{N_r}{N_E} N_H \frac{g_n}{u(T)} e^{-\chi_n/kT}. \tag{16.2}$$

N_H is the number of hydrogen particles per unit volume. The statistical weight for the level divided by the partition function is $g_n/u(T)$, and χ_n is the excitation potential for the level. Equation 16.1 can then be written as

$$\log \frac{w}{\lambda} \approx \log \left(\text{constant} \frac{\pi e^2}{m_e c^2} \frac{N_r/N_E}{u(T)} N_H \right) + \log \lambda + \log g_n f A_E - \theta \chi_n - \log \kappa_\nu$$

$$\approx \text{constant} + \log \lambda + \log g_n f A_E - \theta \chi_n - \log \kappa_\nu. \tag{16.3}$$

The division of w by λ is customary because it normalizes Doppler-dependent phenomena, in particular the microturbulence and thermal broadening. The temperature from the excitation equation is contained in the usual $\theta = 5040/T$. The "constant" is invariant for a given star and for a given ion because the depth dependence for a weak line is small. Notice that the essential weak-line behavior is retained, namely that w is proportional to A_E or, more generally, the abundance factor $g_n f A_E$.

Equation 16.3 implies that for weak lines, $\log g_n f A_E, \theta \chi_n$, and $\log \kappa_\nu$ are all on the same footing. Different lines can be scaled to the same curve of growth using these factors.

For example, lines having excitation potentials differing by $\Delta\chi$ can be brought onto the same curve by an abscissa shift of $\theta\Delta\chi$. These are the factors that have to be known, controlled, or eliminated in order to find the abundance A_E. They are also the factors that allow us to construct a curve of growth for a star.

Naturally, we ask how good the scaling in Equation 16.3 is compared to a complete and proper modeling of each line. Figure 16.1 shows computations for a range in excitation potential for a model with $S_0 = 0.9$. The curves show small differences in shape as the lines saturate, but by far the strongest change is that higher χ curves are shifted to higher abundances. This occurs because fewer atoms are excited to the absorbing level when χ_n is larger, so a larger abundance is needed to compensate. The amount of shift of each curve from the $\chi_n = 0$ curve, according to Equation 16.3, can be interpreted as $\theta\chi_n$. The inset graph in Figure 16.1 shows the results: θ declines slowly with χ_n, i.e., the spacing between curves diminishes slightly with higher χ_n. This comes about because lines having higher excitation potential are formed deeper in the photosphere, where the temperatures are higher. So then, the scaling with $\theta\chi_n$ is rather good for the weak-line portion of the curves, especially if the small χ_n dependence on θ is included, and it is well within observational error even if this χ_n dependence is neglected.

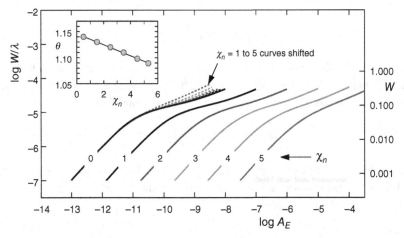

Figure 16.1 The position of the curve of growth shifts to the right with higher excitation potential, χ_n eV. These curves are for a model $\sim$K0 V, having $S_0 = 0.9$, $\log g = 4.5$, [Fe/H] = 0.0, and $\xi = 1.0$ km/s. The shifts are not quite proportional to χ_n, but show $\sim$5% progression as indicated in the inset graph.

The strong-line portions of the curves are a different matter because they depend on the individual line's damping constant and therefore cannot be scaled with $\theta\chi_n$. Furthermore, the van der Waals damping increases with χ_n (Equation 11.30), and we see this in Figure 16.1, where the curves are brought together.

The κ_ν scaling is often small, especially if all the lines are neighbors in λ, and so is often neglected. For example, between 5000 Å and 6000 Å, the change is only 0.1 dex for $T \lesssim 7500$ K. On the other hand, it is easy enough to include this factor using the equations from Chapter 8.

Curve of Growth Temperature Errors

If you have been reading the chapters in order, it will come as no surprise that a star's temperature has to be well established in order to determine chemical abundances. Various methods of determining temperature were discussed in Chapter 14.

Looking at Equation 16.3, we see that temperature enters through the ionization equilibrium, N_r/N_E, through κ_ν, and through the $\theta\chi_n$ excitation term. Altering the temperature causes the curve of growth to be shifted in the $\log A_E$ coordinate through these factors. Depending on the temperature regime and the ionization stage, these factors can combine or cancel, but the change in $\log A_E$ corresponding to a 100 K step is usually $\lesssim 0.2$ dex, as can be seen in Figures 14.2 and 15.6. A simple way to test the sensitivity is to repeat the analysis with a slightly hotter or cooler model. In past times, when the stellar temperature scale was poorly established, serious errors did sometimes make their way into the literature through this portal.

Temperature errors can also enter more deviously through an incorrect model temperature distribution, $T(\tau_0)$. To illustrate the point, consider the two temperature distributions shown in the upper panel of Figure 16.2. (Real stars are not likely to show such large differences.) Since the two distributions are the same in the deep photosphere, the energy distribution and color indices based on the Paschen continuum will hardly differ. The corresponding curves of

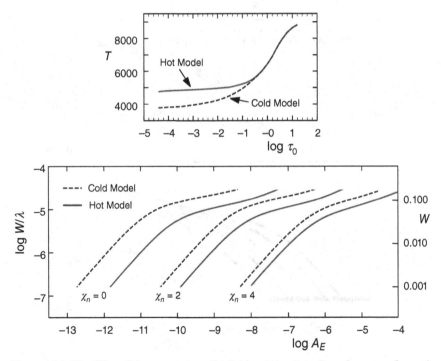

Figure 16.2 The effect of the temperature distribution (upper panel) on the curve of growth (lower panel) for Fe I is shown. Lines having larger excitation potential, χ_n, show a smaller difference because they are formed in deeper layers where the temperature distributions are closer together. $\lambda = 6250$ Å.

growth for Fe I are seen in the lower panel of Figure 16.2. The obvious shift in abscissa means a potential mistake in abundance by the same amount. This is an offshoot of Case 2, Equation 13.8.

It is difficult to reduce abundance uncertainties below about 0.05 dex because of temperature uncertainty.

Curve of Growth Electron Pressure Effects

In F, G, and K stars, the electron pressure depends on both $\log g$ and [Fe/H], and here we take the temperature to be fixed. In these cooler stars, the electron donors are C, Si, Fe, Mg, Ni, Cr, Ca, Na, and K. Their collective abundance is what we normally mean by the metallicity, as in Equation 9.6. In the final analysis, the model photosphere used to determine the abundances of these species must have these same abundances. That implies iteration until the internal agreement is reached.

Examples of the sensitivities are shown in Figure 16.3 for a model having $S_0 = 0.9$. We focus on the horizontal shifts of the weak-line portion. Both N_r/N_E and κ_ν can be pressure sensitive. For neutral lines in solar-type stars, the two effects approximately cancel, as explained in Chapter 13 (Rule 1), while for the ions they do not (Rule 2). This is what we

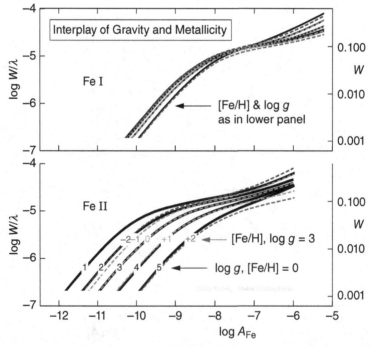

Figure 16.3 The $\log g$ and [Fe/H] dependences of the curve of growth are shown for Fe I (upper) and Fe II (lower). The solid curves have $\log g$ variable while [Fe/H] is held fixed at 0.0. The dashed curves have [Fe/H] variable while $\log g$ is held fixed at 3.0. The wavelength is 6250 Å.

see in Figure 16.3. The curves for Fe I show only small changes. The shift in the Fe II log A_{Fe} from curve to curve is very close to 0.5 dex for a 1.0 dex step in log g, so the abundance shift progresses as the square root of the gravity.

The metallicity effect in Figure 16.3 is slightly more complex. For Fe II, toward higher [Fe/H] values, a 1.0 dex step in [Fe/H] produces close to a 0.5 dex change in log A_{Fe}. But as very low values of [Fe/H] are approached, the electron donors are so few that they no longer dominate the electron pressure and there is a sharp reduction in the log A_{Fe} change per dex in [Fe/H]. Nevertheless, in the log g range of 3 to 5 (giants to dwarfs) and in the [Fe/H] range of -1 to $+2$, these two parameters have comparable leverage in shifting a curve of growth. It is important to register the fact that the normal range of log g is from ~1 to 5, while the range of [Fe/H] is much less, typically -0.5 to $+0.5$ for most Population I stars. We saw previously in Figure 15.6 how the different behavior for Fe I compared to Fe II gives us an important tool for the simultaneous determination of log g and [Fe/H].

Saturation: the Flat Part of the Curve of Growth

Even though weak lines are the best bet for abundance determinations, inevitably stronger lines are pressed into service. The cores of lines cannot go deeper than the flux from the outer boundary of the photosphere, so any growth in equivalent width must come from a widening of the lines (refer back to Figure 13.4). This is saturation. Toward higher abundances, the curve of growth flattens out, the equivalent width no longer grows as fast with abundance increases. In this situation, an error in equivalent width maps into a significantly larger error in abundance compared to the weak-line case. In addition, the saturation portion is complicated by several other factors that we now consider.

Microturbulence (e.g., Equation 11.44) is almost always incorporated into abundance analyses to explain the observed fact that the equivalent widths of saturated lines are greater than predicted by models based on thermal and damping broadening alone. Figure 16.4

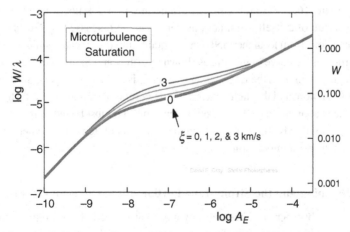

Figure 16.4 The introduction of microturbulence raises the flat part of the curve of growth. Comparison of such curves with observed curves of grow is used to determine ξ.

shows how microturbulence delays the saturation. Microturbulence is assumed to have an isotropic Gaussian velocity distribution with dispersion ξ km/s. In this context, the value of ξ is determined by first fitting the equivalent widths of the weakest lines, where there is little to no change with microturbulence, and then adopting the theoretical curve having the best fit in the saturation portion. In analyses where curves of growth are not explicitly used, one can establish ξ by trial-and-error using the condition that the derived abundance for saturated lines be the same as for weak lines. And, as we shall soon see, the deduced value of ξ is not unique, not trustworthy. The net result from using saturated lines: a lot of work to find ξ and not much gained in finding A_E.

Any other desaturation mechanisms will do much the same as microturbulence. Two of the more common ones are hyperfine and magnetic splitting. In both cases energy-level degeneracy is removed so that the line's absorption is spread over a slightly larger wavelength span. Hyperfine structure, however, is implicit to the atom; odd number elements have an unpaired nucleon and hence a net spin. Examples can be seen in Hartoog *et al.* (1974), Rutten (1978), Booth and Blackwell (1983), Cowley and Frey (1989), Hinkle *et al.* (2000, Figure 7), Mashonkina and Gehren (2000), Vince and Vince (2003), Bergemann and Gehren (2007), and Wood *et al.* (2014). Magnetic or Zeeman splitting is induced by magnetic fields external to the atom. Although the pattern itself is implicit to the atomic structure, the number of components and their separation depend on the orientation of the external field to the line of sight and on its strength (Gray 1988, Stift & Leone 2003). Modeling by Stift and Leone indicated that equivalent widths for complex Zeeman patterns can be increased by as much as a factor of 10 in extreme cases. Magnetic fields of a few gauss or less seem to be common in many cool stars (Aurière *et al.* 2015, Tessore *et al.* 2017).

Not unexpectedly, the saturation portion of the curve of growth also depends on the temperature distribution, especially the temperatures in the higher photospheric layers where the cores of saturated lines are formed. Saturation occurs earlier, i.e., for lines of smaller equivalent widths, in stars having higher temperatures in the outer photosphere. Ambiguity between $T(\tau_0)$ effects and ξ is the result. In the (somewhat exaggerated) example in Figure 16.2, the ambiguity in ξ amounts to 1.3 km/s, a number comparable to the typical value of ξ itself. Similar considerations are discussed by Gustafsson (1983).

Finally, deviations from local thermodynamic equilibrium (LTE) cause similar changes in the saturation portion of the curve of growth, as shown by Athay and Skumanich (1968). Typically, the greater the deviation from the LTE case, the lower S_ν is in the outer layers, the deeper the line grows before saturation, and the higher the saturation portion of the curve of growth. And again, either the deduced abundance or the microturbulence or both must be reduced.

Weak lines obviously have the advantage over saturated lines when it comes to determining chemical composition.

Damping and the Strong-Line Portion of the Curve of Growth

If the number of absorbers is increased beyond what is needed for simple saturation, the line will grow in width approximately as the square root of the abundance (Equation 13.6). How abruptly after saturation this happens depends on the strength of the damping constant. Each line has its own damping constant, so there is no simple scaling relation. Instead, modeling must be done line by line and that requires knowledge of the damping constants. If the line is strong enough in the solar spectrum, it may be possible to find an empirical value of the damping constant. Other sources include the VALD and NIST internet sites given near the beginning of this chapter.

Figure 16.5 illustrates this part of the curve of growth for the sodium D_2 line. Since van der Waals broadening accounts for most of the damping, the position of the curve depends on gas pressure, which in turn depends on surface gravity. Not shown in the figure is the full dependence on metallicity. Increasing [Fe/H] means an increase in electron pressure and that causes an increase in κ_v through the dominant negative hydrogen ion. In response, the photosphere becomes more opaque and produces lower values of P_g at a specified optical depth (e.g., Figure 9.16), reducing the van der Waals broadening. Net effect: increasing metallicity shifts the curve to the right, mimicking smaller surface gravity. The leverage of [Fe/H] is modest compared to gravity. It takes an [Fe/H] of ∼+1.5 to move the log $g = 4$ curve into the log $g = 3$ curve, as shown by the dots in Figure 16.5. Because of these additional complications, using strong lines for abundance work is generally not desirable.

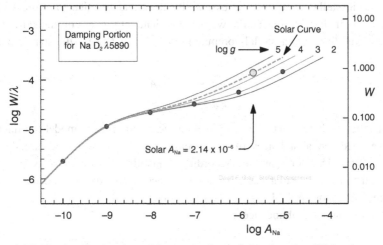

Figure 16.5 The damping portion of the curve of growth involves the additional parameter of the damping constant. For this line, the damping constant is dominated by van der Waals broadening (Figure 11.4). Values of log g are shown for the solar chemical composition, [Fe/H] = 0. Dots show the curve for [Fe/H] = +1.5 and log $g = 4.0$.

With this background on the curve of growth, we now turn more directly to the question of chemical analysis.

Help from the Solar Spectrum

As is so often the case, we use the Sun as a basic reference. The solar composition is the ultimate standard against which other stars are compared. The solar atlases most closely mimicking stellar spectra are for the solar *flux* spectrum, not the center-of-the-disk spectrum. The more comprehensive ones in the visible region are (1) Beckers *et al.* (1976), done at the Sacramento Peak Solar Observatory, (2) Kurucz *et al.* (1984), revised in Kurucz (2005), done with the Brault Fourier-transform spectrometer at Kitt Peak (Brault 1985), (3) Hinkle *et al.* (2000), a useful and convenient reprocessed version of Kurucz *et al.* (1984), (4) Wallace *et al.* (2011), also done with the same equipment, (5) Reiners *et al.* (2016), who used a Fourier-transform spectrometer at the Institut für Astrophysik, Göttingen, and (6) Strassmeier *et al.* (2018), done with a cross-dispersed echelle spectrograph. Some of the characteristics and idiosyncrasies of these data are discussed in Gray and Oostra (2018). The signal-to-noise ratios are ~200–1000 and the resolving powers are ~200 000–1 000 000, so large that the effects of the instrumental profile are almost negligible. See Rutten and van der Zalm (1984) for a list of unblended solar lines.

The majority of atomic spectral lines seen in stellar spectra can, in fact, be seen in the solar spectrum. That is good news because we can use our ability to compute spectral lines using a model solar photosphere to find the line's intrinsic strength, i.e., its $g_n f A_E^\odot$ value. To do this, we measure the equivalent width or use the line profile in the solar flux spectrum and then adjust $g_n f A_E^\odot$ in the calculated profile until it agrees with the observations. Since in stellar analyses it is essentially universal practice to express chemical abundances normalized to the Sun's, we use the identical process on the star, match the stellar observed line with a model-computed one to get $g_n f A_E$ for the star, and we have the answer,

$$\log \frac{g_n f A_E}{g_n f A_E^\odot} = \log \frac{A_E}{A_E^\odot}. \tag{16.4}$$

As nice as this is, there remains a complicating factor. In order to model the lines definitively, we need the abundances of the electron donors, as we have already seen in Figures 14.2 and 15.6 for example. Accordingly, iteration is required to converge to the solution. Further, in the solar case, we want the actual abundances and that means $g_n f$ values are needed for all the lines we want to study.

So what are the solar abundances?

The Solar Chemical Composition

Many studies have been done for the Sun since the pioneering work of Russell (1929), and some years later by Unsöld (1948). Most of the chemical analyses of the Sun have been based on the photospheric spectrum, but some important contributions have also come from chromospheric and coronal measurements, sunspots, and the solar wind. Laboratory measurements of the most careful kind are needed to obtain dependable atomic constants, especially $g_n f$ values. One of the early comprehensive photospheric studies was done by Goldberg *et al.* (1960). Many reviews have been published, including those by Müller (1966), Grevesse and Sauval (1998), and Asplund *et al.* (2009).

Although we have yet to discuss specific techniques, suffice it to say that the equivalent widths or line profiles of solar spectral lines are matched using model photosphere calculations. A summary of the solar composition is given in Table 16.1. These abundances are specified in the usual way as the number of the element relative to hydrogen. Also given in the table is the logarithm of the abundance plus 12, a commonly used notation. Typical uncertainties are ~0.05 dex, but range, as shown in Table 16.1. Systematic errors have been

Table 16.1 *The solar chemical composition*

Element		$\log A$	$\log A + 12$	Error		Element	$\log A$	$\log A + 12$	Error
1	H	0.00	12.00	—	42	Mo	−10.12	1.88	0.09
2	He	−1.07	10.93	0.01	44	Ru	−10.25	1.75	0.08
3	Li	−10.95	1.15	0.10	45	Rh	−11.11	0.89	0.08
4	Be	−10.62	1.38	0.09	46	Pd	−10.45	1.55	0.06
5	B	−9.30	2.70	0.20	47	Ag	−11.04	0.96	0.10
6	C	−3.57	8.43	0.05	48	Cd	−10.23	1.77	0.15
7	N	−4.17	7.83	0.05	49	In	−11.20	0.80	0.20
8	O	−3.31	8.69	0.05	50	Sn	−9.98	2.02	0.10
9	F	−7.44	4.56	0.30	51	Sb	−11.0	1.0	—
10	Ne	−4.07	7.93	0.10	54	Xe	−9.76	2.24	0.06
11	Na	−5.79	6.21	0.04	55	Cs	<−10.2	<1.8	—
12	Mg	−4.41	7.59	0.04	56	Ba	−9.75	2.25	0.07
13	Al	−5.57	6.43	0.04	57	La	−10.89	1.11	0.04
14	Si	−4.49	7.51	0.03	58	Ce	−10.42	1.58	0.04
15	P	−6.59	5.41	0.03	59	Pr	−11.28	0.72	0.04
16	S	−4.88	7.12	0.03	60	Nd	−10.58	1.42	0.04
17	Cl	−6.50	5.50	0.13	62	Sm	−11.05	0.95	0.04
18	Ar	−5.60	6.40	0.30	63	Eu	−11.48	0.52	0.04
19	K	−6.96	5.04	0.05	64	Gd	−10.92	1.08	0.04
20	Ca	−5.68	6.32	0.03	65	Tb	−11.69	0.31	0.10
21	Sc	−8.84	3.16	0.04	66	Dy	−10.90	1.10	0.04
22	Ti	−7.07	4.93	0.04	67	Ho	−11.52	0.48	0.11
23	V	−8.11	3.89	0.08	68	Er	−11.07	0.93	0.05
24	Cr	−6.38	5.62	0.04	69	Tm	−11.89	0.11	0.04
25	Mn	−6.58	5.42	0.04	70	Yb	−11.15	0.85	0.11
26	Fe	−4.53	7.47	0.04	71	Lu	−11.90	0.10	0.09
27	Co	−7.07	4.93	0.05	72	Hf	−11.15	0.85	0.05
28	Ni	−5.80	6.20	0.04	74	W	−11.17	0.83	0.11
29	Cu	−7.82	4.18	0.05	76	Os	−10.60	1.40	0.05
30	Zn	−7.44	4.56	0.05	77	Ir	−10.58	1.42	0.07
31	Ga	−8.98	3.02	0.05	78	Pt	−10.2	1.8	—
32	Ge	−8.37	3.63	0.07	79	Au	−11.09	0.91	0.08
36	Kr	−8.75	3.25	0.06	80	Hg	<−9.0	<3.0	—
37	Rb	−9.53	2.47	0.07	81	Tl	−11.10	0.90	0.20
38	Sr	−9.17	2.83	0.06	82	Pb	−10.08	1.92	0.08
39	Y	−9.79	2.21	0.05	83	Bi	<−11.2	<0.8	—
40	Zr	−9.41	2.59	0.04	90	Th	−11.97	0.03	0.10
41	Nb	−10.53	1.47	0.06	92	U	<−12.47	<−0.47	—

Based on Asplund *et al.* (2009), Scott *et al.* (2015a, 2015b), Grevesse *et al.* (2015).

known to occur, as strikingly illustrated by the case of iron, where a downward revision of the $g_n f$ values led to an order of magnitude increase in the deduced abundance (Garz *et al.* 1969, Baschek *et al.* 1970, Unsöld 1971), but the close agreement with abundances from meteorites is reassuring, as discussed in the reviews cited above. Aside from the very light elements, solar abundances mimic those of the relatively rare carbonaceous chondrite meteorites, thought to be material from the nascent solar system.

We see in Figure 16.6 the general drop in abundance with atomic number. Notice how much more abundant H and He are compared to all others. Deviations from the general trend are sometimes dramatic, such as the approximately nine-order deficiency of lithium, beryllium, and boron, elements destroyed at the bottom of the convection zone. The elements around iron show enhanced abundances, forming a small "iron peak." Also clearly visible is the remarkable alternation caused by even-number elements being more abundant than their adjacent odd-number neighbors. Nucleosynthesis processes are discussed, for example, in Pagel (1997). In the notation of stellar interiors, i.e., the *mass* fraction of hydrogen is X = 0.738, of helium, Y = 0.249, and of all other elements, Z = 0.013.

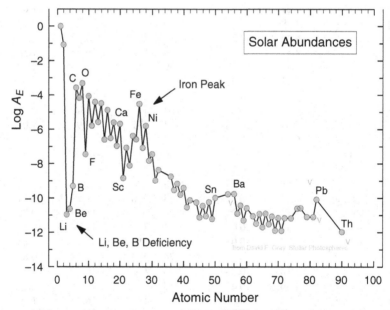

Figure 16.6 Solar chemical composition by numbers relative to hydrogen from Table 16.1. Some elements are missing and a few have only upper limits denoted by V.

With the abundances of the electron donors, C, Si, Fe, Mg, Ni, Cr, Ca, Na, and K, in hand, we can now safely compute the solar model and use observed solar line strengths to find $g_n f A_E^{\odot}$ for any line of interest. On the stellar front, we measure the lines on the highest resolution and signal-to-noise spectra we can get. We can proceed in several different ways, as follows.

Stellar Abundances Using Curves of Growth

Curves of growth can be formed for various elements, but especially for iron because it has so many lines in cool stars. From the point of view of a photospheric model, we choose a spectral line; it has a fixed $g_n f\lambda$ and χ_n, while the model fixes θ and κ_v. We vary A_E to generate the curve of growth for this line. But from the observational point of view, A_E is fixed, and obviously beyond our power to change. So instead we look to a set of lines of the same element that have different strengths because of the other terms in Equation 16.3, that is, the lines span a range in $g_n f\lambda, \chi_n$, and κ_v. We use this fact to construct an empirical curve of growth by plotting $\log W/\lambda$ for the star as a function of the $\chi_n = 0$ abundance, $\log A_E^0 = \log g_n f A_E^\odot - \theta\chi_n + \log \lambda/\lambda_0 - \kappa_v/\kappa_v^0$, where λ_0 is a reference wavelength, typically the mean for the lines being used, and κ_v^0 is taken at this reference wavelength. This amounts to adjusting each line's abscissa for the line-to-line differences in $g_n f$, $\theta\chi_n$, λ, and κ_v, bringing all lines onto the $\chi = 0$ curve for the reference wavelength.

Table 16.2 gives numbers for the example shown in Figure 16.7. The mean wavelength is 6545 Å. Notice that the $\theta\chi_n$ term is much larger than the λ or κ_v terms. We find the value

Table 16.2 *Example of curve of growth for Fe I*

Line	χ (eV)	W (mÅ)	$\log W/\lambda$	$\theta\chi_n$	$\log \lambda/\lambda_0$	$\log \kappa_v/\kappa_v^0$	$W_\odot$	$\log g_n f A_E^\odot$	$\log A_E^0$
6145	3.37	4.3	−6.1551	3.44	−0.027	−0.018	3.6	−8.1766	−11.6229
6148	4.07	43.6	−5.1492	4.15	−0.027	−0.018	49.0	−5.9994	−10.1597
6152	2.18	54.4	−5.0534	2.22	−0.027	−0.018	52.0	−7.6265	−9.8589
6157	3.30	4.0	−6.1873	3.37	−0.027	−0.018	4.0	−8.1986	−11.5734
6158	4.07	62.5	−4.9935	4.15	−0.026	−0.018	67.0	−5.461	−9.6211
6159	4.61	14.5	−5.6282	4.70	−0.026	−0.018	13.0	−6.3574	−11.0683
6165	4.14	44.4	−5.1426	4.22	−0.026	−0.017	47.0	−5.8446	−10.0759
6173	2.22	71.6	−4.9356	2.26	−0.025	−0.017	72.0	−7.0708	−9.3436
6696	4.83	17.0	−5.5954	4.93	0.010	0.006	16.7	−6.0344	−10.9570
6699	4.59	9.4	−5.8529	4.68	0.010	0.006	8.3	−6.6217	−11.2995
6704	2.76	42.2	−5.2010	2.82	0.010	0.006	38.0	−7.4497	−10.2608
6704	4.22	7.1	−5.9751	4.30	0.010	0.006	6.3	−7.1128	−11.4130
6705	4.61	46.9	−5.1552	4.70	0.010	0.006	47.5	−5.4635	−10.1615
6710	1.48	19.0	−5.5480	1.51	0.011	0.006	15.7	−9.3728	−10.8781
6713	4.61	27.1	−5.3939	4.70	0.011	0.007	28.0	−5.8876	−10.5854
6713	4.14	12.1	−5.7441	4.22	0.011	0.007	11.5	−6.8828	−11.1012
6714	4.79	23.1	−5.4633	4.89	0.011	0.007	21.4	−5.8975	−10.7789
6715	4.61	29.9	−5.3514	4.70	0.011	0.007	26.3	−5.9481	−10.6458
6716	4.58	17.3	−5.5891	4.67	0.011	0.007	16.0	−6.2994	−10.9665
6725	4.10	20.0	−5.5267	4.18	0.013	0.007	19.0	−6.6379	−10.8151
6727	4.61	49.5	−5.1332	4.70	0.012	0.007	48.4	−5.4081	−10.1055
6732	4.58	8.0	−5.9251	4.67	0.012	0.007	7.7	−6.6578	−11.3244
6733	4.64	28.6	−5.3719	4.73	0.012	0.007	27.3	−5.8947	−10.6225
6738	4.56	25.0	−5.4306	4.65	0.013	0.008	22.4	−6.1181	−10.7642
6740	1.56	15.8	−5.6300	1.59	0.013	0.008	12.0	−9.4404	−11.0265

$\log A_E^0 = \log g_n f A_E^\odot - \theta\chi_n + \log \lambda/\lambda_0 - \log \kappa_v/\kappa_v^0$

Data from Perrin *et al.* (1988).

of θ to be 1.02 by minimizing the horizontal scatter of the points in the lower panel. This θ value corresponds to an excitation temperature, typically 10–15% cooler than the effective temperature. With $\log A_E^0$ in hand, we now have the observed curve of growth (points in lower panel).

Next we compute the curve of growth for Fe I with $\chi = 0$ at $\lambda_0 = 6545$ Å. In a typical case, the effective temperature is found from an energy distribution, or Balmer line profiles, or color indices, or from the metal lines themselves by exploiting their different sensitivities to temperature, as we saw in Chapter 14. In this example, we adopt $T_{\text{eff}} = 5585$ K or $S_0 = 0.9668$ from Perrin *et al.* (1988) and use the solar $T(\tau_0)$ in Table 9.2. The second parameter, $\log g$, is taken to be 4.45, based on the star's luminosity class of V. The third parameter, [Fe/H], is taken to be 0.0 initially, with the understanding that it will be adjusted to match our derived iron abundance on subsequent iterations.

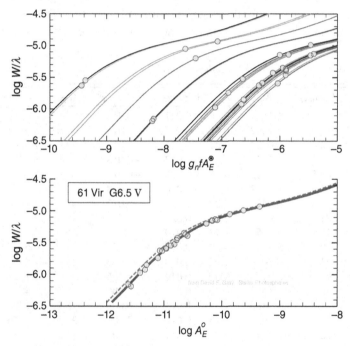

Figure 16.7 Top: each spectral line for 61 Vir gives one point on its own curve of growth. Bottom: the curves of growth are brought together using the scaling in Equation 16.3, namely, $\log A_E^0 = \log g_n f A_E^\odot - \theta \chi_n + \log \lambda/\lambda_0 - \log \kappa_\nu/\kappa_\nu^0$. The dashed curve is for [Fe/H] = 0.0; solid curve is for [Fe/H] = -0.09, the first-iteration solution. The second iteration gives [Fe/H] = -0.10.

Figure 16.7 shows that this model curve (dashed) has to be shifted 0.09 dex to match the observations. The observed equivalent widths are weaker than the original model curve, so the first-iteration iron abundance in 61 Vir is less than the solar value by 0.09 dex. A second iteration with [Fe/H] $= -0.09$ for the model shifts the solution to -0.10 dex. Therefore our analysis gives [Fe/H] $= -0.10$ for the star. The curve with a microturbulence dispersion of 0.0 km/s fits the observations, but as we have seen, this value depends on the temperature distribution.

Examples of this kind of curve-of-growth analysis can be seen in numerous papers, including Cayrel *et al.* (1985), Perrin *et al.* (1988), and Lemasle *et al.* (2007).

Metallicity from Equivalent Widths

A basic metallicity abundance can be found without explicit use of curves of growth, as we have already seen in Figures 14.2 and 15.6. Metallicity is found whenever a suitable set of spectral lines is used to match a model to a star. Uncertainty in [Fe/H] is typically ~0.05 dex. In many cases, this collective abundance of the electron donors is sufficient to reach the scientific goal.

Abundances from Line Profiles and Synthesized Spectra

Another approach is to model the line profiles. Fitting an observed line profile can yield the abundance much like matching the equivalent width does. Essentially all of the same conditions and constraints still apply. But now we need additional information, primarily the size and form of the microturbulence, macroturbulence, and rotation Doppler-shift distributions. In fact, in many applications the shapes of the line profiles are used to determine these broadening parameters from the Doppler imprint they leave on those profiles, as explained in Chapters 17 and 18. But the overall strength of the profile is still dependent on the abundance according to its curve-of-growth behavior. Both theoretical and laboratory $g_n f$ values, as well as empirical solar values, have been used (e.g., Meléndez & Barbuy 2009, Jönsson *et al.* 2017).

Computation of a complete spectral interval in which all the observed lines are included is called a *synthesis*. The method can be applied quite generally, but is especially useful when line blending is severe, such as in cooler stars or stars showing large rotational broadening. Basically one uses Equation 13.15, but now ℓ_v includes all the spectral lines in the synthesized interval. This implies knowledge of the relevant atomic parameters for each of the lines, starting with their wavelengths. The abundances are then varied until an optimum match to the observed spectrum is attained. The literature has many examples: Sneden (1973), Little-Marenin and Little (1984), Youssef and Khalil (1988), Farragiana (1989), Bohlender *et al.* (1998), Thorén and Feltzing (2000), and Burris *et al.* (2000), to cite a few. Synthesis for rotationally broadened lines of ionized oxygen in OB stars is illustrated by Daflon *et al.* (2001) and Villamariz *et al.* (2002). Synthesis is often used to derive beryllium and lithium abundances (e.g., Boesgaard & King 2002, Ford *et al.* 2002).

Figure 16.8 shows a reasonably "clean" case, where an abundance is obtained from an individual profile. Figure 16.9 shows a more complex situation with high-quality solar data and Figure 16.10 shows a synthesis including a number of blended lines in a metal-poor giant.

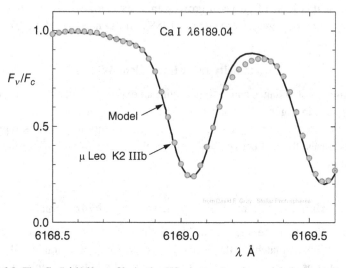

Figure 16.8 The Ca I λ6169 profile in the K2 giant μ Leo is modeled to get the calcium abundance, $\log A_{Ca} = -5.5$. Based on Figure 1 from Jönsson *et al.* (2017). Resolving power is ~67 000, S/N ~ 100.

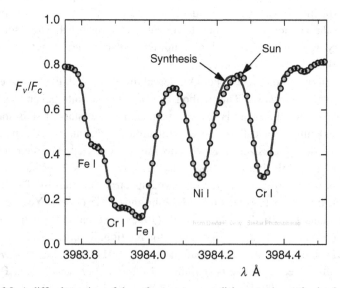

Figure 16.9 A difficult portion of the solar spectrum at disk center is synthesized. Adapted from Ross and Aller (1968) with permission of the *Astrophysical Journal*, copyright 1968 by the University of Chicago Press.

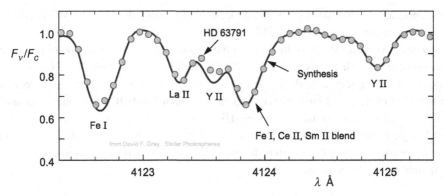

Figure 16.10 Synthesis analysis for HD 63791, an early K giant with a significant metal deficiency, [Fe/H] ~ − 1.7. Resolving power is ~20 000; signal-to-noise ratio ~100. Based on Burris *et al.* (2000).

Stellar Abundances: Summaries and Catalogues

The number of papers citing abundances and metallicities for large numbers of stars has mushroomed in the last decades. Internet searches through ADS (https://ui.adsabs.harvard .edu) or SIMBAD (http://simbad.u-strasbg.fr/simbad) should be part of your research toolbox. Only a few summaries, somewhat arbitrarily chosen, are noted here: Barstow *et al.* (2003), Boeche *et al.* (2011), Hinkel *et al.* (2014), Holtzman *et al.* (2015), Brewer *et al.* (2016), Hawkins *et al.* (2016), Soubiran *et al.* (2016), Petigura *et al.* (2017), and Nissen and Gustafsson (2018). The ultimate precision currently possible was studied by Bedell *et al.* (2014). They found abundance errors ≥ 0.01 dex, with different instruments giving differences of ~ 0.04 dex. Most determinations are not so precise, with errors of ~0.05 dex being more typical of "good" determinations.

Now let us shift our attention to a few selected areas of application.

Galactic Variations

Walter Baade and others developed the concept of two stellar populations nearly a century ago (e.g., Baade 1944, 1958, Baade & Gaposchkin 1963). Population I stars are young, reside in the galactic disk, and have metallicity similar to the Sun's within a factor of two or so. Population II stars are old, occupy the galactic bulge and halo, and are metal poor by a factor of ~100 or more. This basic view has been developed and expanded, and many nuances have been discovered. Population studies help us understand the formation history of the Milky Way, and indeed other galaxies, through the characteristics of the different population types, especially their chemical signatures.

In this context, a distinction is often made between iron-peak species and the alpha-process species, especially C, O, Ne, Mg, Si, S, Ar, and Ca. The alpha process forms these later elements, starting with C and adding He (the alpha particle) at each step. The alpha process does not contribute much to the star's energy generation though, mainly because it proceeds at a slow rate. In this context, the notation [Fe/H] usually means the actual iron abundance, as opposed to the generalized meaning of electron donors. The collective alpha species is sometimes denoted by $[\alpha/\mathrm{H}]$.

Several stars with extremely low metal abundance have been found. One of the lowest reported metal abundances is for CD $-38°245$ with $[\mathrm{Fe/H}] \sim -4$ (Bessell & Norris 1984). A portion of its spectrum is shown in Figure 16.11. This star is a giant. More recently, Christlieb *et al.* (2004) found $[\mathrm{Fe/H}] \sim -5$ for the giant HE 0107-5240, and Nordlander *et al.* (2019) found a halo star to have $[\mathrm{Fe/H}] = -6.2$. Other extreme examples have been found (e.g., Bessell *et al.* 2015, Nordlander *et al.* 2017), but such stars are rare. The abundance patterns (like Figure 16.6) for these stars can be tied to supernovae seeding or other sources of processed material.

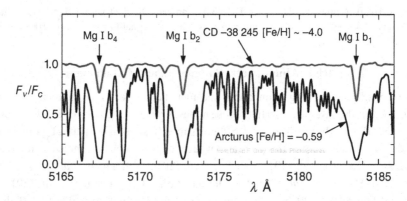

Figure 16.11 The extremely metal-poor giant, CD $-38°245$, compared to the mildly metal-poor giant, Arcturus. CD $-38°245$ data from Cayrel (1996); Arcturus from Hinkle *et al.* (2000).

In the galactic disk, $[\mathrm{Fe/H}]$ declines with distance from the galactic center, R. All Population I tracers show a decline, but the actual value may not be the same for all of them. Population I Cepheids are summarized in Figure 16.12. The individual R positions are uncertain by ~0.7–1.0 kpc, while the $[\mathrm{Fe/H}]$ values are thought to be good to ~0.10–0.15 dex. Early indications of such a gradient were found by Luck and Bond (1980), Harris (1981), and Luck (1982). Subsequent research on Cepheids, other stars, and open star clusters continued to support this general trend, as seen in Figure 16.12 and in Genovali *et al.* (2015), but other variables come into play. For example, Hasselquist *et al.* (2019) found that young stars show larger gradients than old stars. Time variations were measured and discussed by Marsakov *et al.* (2011) and Feuillet *et al.* (2019). A roughly similar pattern was found for open star clusters (Carraro *et al.* 1998, Gozha *et al.* 2012). The implication is that closer to the galactic center more generations of stars run through their evolutionary cycles, presumably owing to the higher density of stars and gas. Minchev *et al.* (2019) emphasize the confusion that can occur when many variables are involved.

The gradient disappears as one considers stars above and below the disk, when that coordinate, z, exceeds about 1 kpc (Cheng *et al.* 2012, Xiang *et al.* 2015), as might well be expected since one is moving from Population I into Population II. Ness and Freeman (2016) found a gradient $d[\text{Fe}/\text{H}]/dz = -0.45$ dex/kpc.

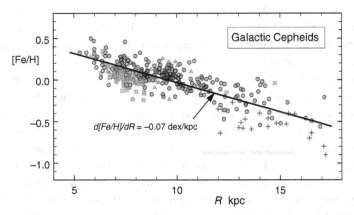

Figure 16.12 Cepheids in the galactic disk show a metallicity gradient. Triangles: Luck *et al.* (2006); crosses: Yong *et al.* (2006); squares: Luck *et al.* (2011); circles: Luck and Lambert (2011). The Sun lies at $R = 8.2$ kpc (Baade 1951, van de Hulst *et al.* 1954, Abuter *et al.* 2019).

Metallicity plays a central role in studies of the central bulge of the Milky Way. Values range from $[\text{Fe}/\text{H}] \sim -3$ to $+1$, but most lie in the -1.5 to $+0.5$ range (Gonzalez & Gadotti 2016, Ness & Freeman 2016, Barbuy *et al.* 2018). The star cluster at the very center of the Milky Way also shows a range in metallicity (Do *et al.* 2015), and metallicity differences between this nuclear star cluster and the galactic bulge were investigated by Schultheis *et al.* (2019).

Globular clusters are classical Population II. They have the spherical or halo distribution around the galactic center, have ages in excess of ~10 billion years, and are generally metal deficient by one to two orders compared to the Sun. But there is a considerable range in metallicity from cluster to cluster, and they can even be classified with abundances playing a central role, as done by Carretta *et al.* (2010). Nor is the non-unique metallicity restricted to the clusters in our galaxy (e.g., Harris *et al.* 2006). Further, careful studies show multiple populations within individual clusters, implying more than one generation of stars (Smith 1987, Ivans *et al.* 2001). The massive cluster ω Cen has as many as five distinct populations that span two orders in metallicity. Johnson *et al.* (2015) give a concise review of clusters having complex star formation processes.

Evidence for two age–metallicity relations was found by Dotter *et al.* (2011). So how does metallicity change with time?

The Fuzzy Time Gauge: [Fe/H]

Elements heavier than lithium are formed inside stars and in novae and supernovae events. Each successive generation of stars is more metal rich, being formed from material that has been enriched by the previous generations. But low-mass stars have very long lifetimes, so low-metallicity stars from early generations are still around. Metal-rich stars, by contrast, could only have been formed out of the many-generation-processed material and so must have formed more recently than the processing time. In this way metallicity is a measure of age. Metal poor implies old. Metal rich implies young. The timescale that we attach to [Fe/H] could, in principle, be established by looking at star clusters, for example, where nuclear-burning ages or white dwarf cooling ages can be found. But, there are complications. In particular, the more dense inner regions of the galaxy cycle material more rapidly, so there the "clock" runs faster, and vice versa. And processed material is not necessarily homogeneously distributed or mixed. That means that pockets of unprocessed material may exist that form low-metallicity stars at recent epochs. Add to this the fact that stars migrate, mixing stars formed in different locations. So [Fe/H] is a complicated time gauge.

Early studies were done by Twarog (1980) and Edvardsson *et al.* (1993). A review of more recent work can be found in Nissen and Gustafsson (2018) and Feuillet *et al.* (2019). Some of the correlations between [Fe/H] and age are shown in Figure 16.13. Ages are determined from evolutionary models and have errors of ~0.5–1.0 Gyr. (Gyr = 10^9 years.) The general trend is for a steep increase in metallicity during the first 2–3 Gyr years followed by, perhaps, a gentle peak ~8 Gyr ago. Such a peak would be consistent with open cluster metallicity values (Boesgaard 1989).

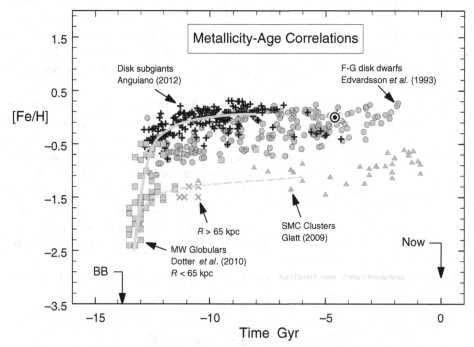

Figure 16.13 Metallicity as a function of time for different groups of objects, as labeled. There is a large scatter at any given time. BB stands for Big Bang, the starting time, while "Now" is our present era. The Sun's point is shown by the circumpunct.

The galactic globular clusters that are in the outskirts of the galaxy ($R > 65$ kpc points) seem to show a tapering off at lower metallicities in continuity with the clusters from the Small Magellanic Cloud. Perhaps these galactic globulars were born elsewhere and captured by the Milky Way.

Aside from the initial rise earlier than about 12 Gyr ago, the relations are too flat and the scatter too large to allow a dependable age estimate from the metallicity. Rather, it seems the best we can do is to say that metal-rich stars have been formed from materials that have been through more processing cycles compared to metal-poor stars that have not.

Lithium in Population II Stars: Part 1

Lithium is an interesting element. Some is formed during the Big Bang, some comes from supernovae, and some is destroyed as stars age. Lithium destruction in main-sequence stars is the most obvious result seen in the observations of stars with convective envelopes (~F5 and cooler). When envelope convection reaches deep enough to encounter temperatures of $\sim 2.5 \times 10^6$ K, lithium is destroyed by conversion back to helium. As the convective circulation continues, lithium is gradually purged from the convective envelope and the photosphere. The observed abundance of lithium is thus a combination of the circulation depth, temperature profile into the star, and age of the star. Figure 16.14 shows lithium abundances for old, halo (Population II) stars selected on the basis of metallicities and galactic orbits.

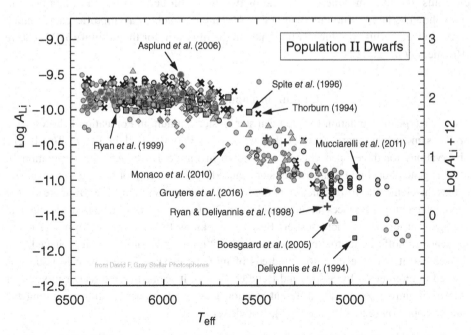

Figure 16.14 Lithium abundance in Population II stars varies with the effective temperature.

Within observation/analysis errors, all observers find the same relation. The decline toward cooler effective temperatures results from the deeper convection zones in cooler stars, which means a monotonic increase in temperature and pressure at the base of the convection zone and the subsequent destruction of more lithium (e.g., Spite & Spite 1982, Deliyannis *et al.* 1990). The nature and meaning of the flat portion on the left has been the subject of considerable discussion, and we take that up a bit later in the chapter.

Most determinations of lithium abundance depend on the single blended feature of the ^{7}Li resonance doublet at 6708 Å, as listed in Table 16.3. The only other useful line is at 6104 Å, but it is very weak (Ford *et al.* 2002).

Table 16.3 *The lithium resonance lines*

	λ	$g_n J$
^{7}Li	6707.7637	0.996
	6707.9147	0.497
^{6}Li	6707.9219	0.996
	6708.0728	0.497

From Yan *et al.* (1998) and Hobbs *et al.* (1999).

These lines are blended together in stellar spectra and also have significant hyperfine structure, so the line profiles are asymmetric and often only a few mÅ in strength (see, for example, Hobbs *et al.* 1999). Some caution should be exercised in dealing with elements like lithium, where only one or two lines can be measured. For example, in stars showing larger amounts of processed elements, such as asymptotic-giant-branch stars, ionized cerium produces a line that can be mistaken for the lithium λ6708 feature (Reyniers *et al.* 2002).

Lithium in Population I Stars

The archetype of Population I is the open star cluster. Lithium in Population I shows some of the same evolutionary changes that we saw in Population II in Figure 16.14. The observations for three open clusters are shown in Figure 16.15. As with Population II, there is general decline toward cooler stars caused by the purging of lithium as convection cycles the element to its destruction at the bottom of the convection zone. But here we see some interesting differences among the clusters. Younger clusters show less depletion, and younger clusters show more scatter. Because we know the cluster ages from the main-sequence turnoff, an approximate timescale for lithium depletion can be established, and we see that it is in the range of hundreds of millions of years. The timescale issue was studied in more detail by Sestito and Randich (2005). It seems that the scatter in younger clusters is introduced by the different rotation rates, with higher rotation inhibiting the lithium depletion (e.g., Gondoin 2014, Bouvier *et al.* 2018).

The maximum abundances in such diagrams, i.e., the values toward the left side of Figure 16.15, may be the abundance of lithium when the stars were formed. No depletion has taken place for these stars because their convective envelopes are too shallow.

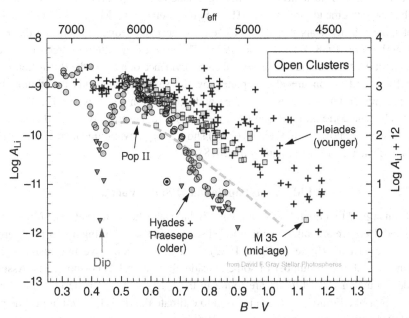

Figure 16.15 Lithium abundances in Population I stars vary with effective temperature. The age of the Hyades is ~650–800 Myr (Perryman *et al.* 1998, Brandt & Huang 2015, Gossage *et al.* 2018, Martín *et al.* 2018); Praesepe is about the same. M35 is ~180 Myr (Kalirai *et al.* 2003); the Pleiades is ~110–160 Myr (Dahm 2015, Gossage *et al.* 2018). Downward-pointing triangles indicate upper limits. Data for this figure come from Boesgaard *et al.* (2016), Cummings *et al.* (2017), Anthony-Twarog *et al.* (2018), and Soderblom *et al.* (1993a). The dashed line is the Population II relation in Figure 16.14. The solar point is shown by the circumpunct; age ~4.6 Gyr.

The Population II relation from Figure 16.14 is shown in Figure 16.15 as the dashed line. It lies about one full dex below the Pleiades values, which is in line with Population II being metal poor compared to Population I. But it has about the same slope as the Pleiades and M35. Puzzling. The Hyades and Praesepe are younger than Population II stars, but their coolest stars have somehow purged more of their lithium than their Population II counterparts. The favored explanation is that the convective envelopes in metal-rich stars extend to deeper/hotter layers, thus making the purging run at a faster rate.

Another striking feature seen in Figure 16.15 is the lithium "dip," where Hyades and Praesepe lithium is depleted by a factor of 100 or more in a narrow temperature range between about 6300 and 6800 K (~F5). The lithium dip was discovered by Boesgaard and Tripicco (1986) and a short review of subsequent investigations is given in Boesgaard *et al.* (2016); see also Soderblom *et al.* (1993b). The dip is barely visible in younger clusters, but from studies of clusters spanning a range in age, ~5 × 10⁷ years is apparently needed for the dip to form.

The physical reason for the dip is presumably related to the internal stellar structure and the convection zones, and may also depend on diffusion and rotation (Stephens *et al.* 1997, Sills & Deliyannis 2000). Although the Population II data in Figure 16.14 just reach to these temperatures, there is no indication of (the right side of) a dip, even though they have had more than enough time to produce one. The very old open cluster, M67, age ~4 Gyr, shows no sign of lithium in the stars that have evolved off the main sequence from the dip interval (Pilachowski *et al.* 1988, Balachandran 1995). The conclusion: lithium purged in the dip is permanent; it does not return. Why then is there no evidence of a dip in the Population II stars?

Beryllium and boron are also depleted by the same mechanisms as lithium, but at somewhat higher temperatures: 2.5, 3.5, and 5.0 million degrees for Li, Be, and B, respectively. The correlation among these elements is reviewed in Boesgaard *et al.* (2016).

For reasons unknown, a few giants (~1%) show a large overabundance of lithium (e.g., Takeda & Tajitsu 2017, Smiljanic *et al.* 2018).

Lithium in Population II Stars: Part 2

It has been argued that the lithium abundance in halo stars hotter than about 5700 K may be undepleted. Here we are talking about the flat section on the left in Figure 16.14, referred to as the Spite Plateau (Spite & Spite 1982). If true, we then have the cosmic-creation abundance of lithium, a number of some considerable interest to cosmologists. Assuming the oldest abundances of lithium are found in the least processed material, as gauged by [Fe/H], a plot like Figure 16.16 can be used to estimate the [Fe/H] = 0.0 limit. The points on the left delineate this pristine limit, ~ − 10.0 dex.

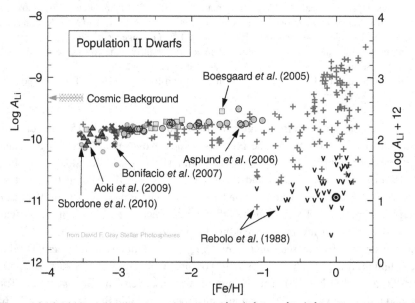

Figure 16.16 Lithium abundance as a function of [Fe/H]. Low [Fe/H] implies more pristine abundances, with the most metal poor perhaps being the creation values. V symbols are upper limits.

It turns out that the cosmic microwave background also leads to a creation value for the abundance of lithium, namely -9.33 ± 0.03 dex (Pizzone *et al.* 2014, Cyburt *et al.* 2015, Takács *et al.* 2015). The obvious disagreement between the cosmic background value and the stellar value has been the cause of considerable consternation. A review is given by Spite and Spite (2010). Recent observations show a downturn and increased scatter at the smallest observed metallicity values (Sbordone *et al.* 2010, Matsuno *et al.* 2017, Bonifacio *et al.* 2018). Ingenious suggestions have been made, such as the separation of ionized lithium by an intergalactic magnetic field in the early stages of the formation of the galaxy (Kusakabe & Kawasaki 2019).

Lithium enrichment and depletion for metal-rich stars, those on the far right side of Figure 16.16, is discussed by Guiglion *et al.* (2016) and Fu *et al.* (2018).

Chemical Evolution in Massive Stars: Nuclear Processes

Since the CNO (carbon, nitrogen, oxygen) cycle dominates the energy generation in hot stars, i.e., those with mass exceeding about 1.3 solar masses, the CNO abundances give a direct handle on stellar evolution. (A short summary of stellar evolution is given by Beccari and Carraro 2015.) Basically, nitrogen is enriched, carbon reduced, and oxygen stays nearly the same, decreasing slowly. That means that the abundances of nitrogen and carbon are anti-correlated. Often plots of N/C versus N/O are used to expresses this, as championed by Maeder *et al.* (2014). Alas, there are other variables that may enter, namely, rotation, binarity, magnetic fields, and mass loss. We are hampered by the fact that nuclear processing in O, B, and A stars is largely hidden from view owing to their radiative envelopes, that is, there is no convective transport from the interior to the surface. The picture is blurred by the fact that a significant range in nitrogen enhancement is found in hot stars. Apparently some core material gets to the surface without convection, likely by rotational mixing (Hunter *et al.* 2009, Maeder *et al.* 2014). A modest fraction of O and B stars, ~7%, have kilogauss fossil magnetic fields. The effect of such fields on rotational mixing was investigated by Keszthelyi *et al.* (2019).

During the course of evolution, hot main-sequence stars eventually become cool giants and supergiants, at which time they develop convective envelopes, cycle deep material to the surface, and reveal what they have been processing in their cores while on the main sequence and in the brief time they have been cool stars. This lifting of deep material to the surface is termed "dredge-up."

Early on Iben (1964, Becker & Iben 1979) predicted changes in carbon, nitrogen, and oxygen abundances in evolved stars from one or more dredge-ups. CNO-processed material dredged to the surface of class III giant stars was confirmed by Lambert and Ries (1981), namely carbon was reduced, nitrogen increased, and oxygen remained the same. Mixing spurred by rotation was shown to account for observed CNO chemical signatures early on by Sweigart and Mengel (1979). Similar results were found for nearby evolved class II to Ib stars by Lyubimkov *et al.* (2019 and references therein). Figure 16.17 shows their result. Although the relation is not particularly tight, the scatter comes mostly from observational error, which is ~0.2 dex in each coordinate.

The direction of more processed material is shown by the arrow. Why should some stars process more of their material? Lyubimkov *et al.* suggest more rapid rotation during the main-sequence stay. We have no simple way to know what the main-sequence rotation rates used to be for evolved stars. But if this hypothesis is true, then rapid rotators stay on the main sequence longer than slow rotators, since they run more of their mass through the CNO cycle before core exhaustion. Most open star clusters do show a spread in their main-sequence turn-offs, and this has been attributed to rotation (e.g., Cordoni *et al.* 2018, Marino *et al.* 2018). Although some open clusters do not support the idea (Tautvaišienė *et al.* 2015), most fall in line with it (e.g., Mikolaitis *et al.* 2012, Tautvaišienė *et al.* 2016). Sometimes the observed anti-correlation between carbon and nitrogen is less than perfect, as in the globular cluster M10 (Gerber *et al.* 2018).

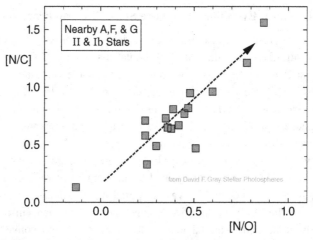

Figure 16.17 The anti-correlation of carbon and nitrogen, expected for the CNO energy cycle, is expressed here using the observed logarithmic abundance ratios of nitrogen to carbon versus nitrogen to oxygen. (Remember, the square brackets mean logarithms normalized to the Sun.) Since these stars all have approximately solar metallicities, the unevolved or starting point is (0.0,0.0). The arrow shows the direction of more CNO-cycle processing. Data from Lyumbimkov *et al.* (2019).

Other anti-correlations are found, such as sodium and oxygen in globular clusters (Carretta *et al.* 2010, Gratton *et al.* 2012).

Additional signatures of the CNO cycle become available once the convective envelope links the core to the surface. One of the most prominent signatures is the ^{12}C to ^{13}C isotope ratio as detected with carbon molecular bands of CH, CN, and CO (e.g., Lambert 1976, Smith & Lambert 1986). The solar-system ^{12}C/^{13}C ratio is 90 (Meibom *et al.* 2007, Woods 2010) and this value is sometimes presumed to be the primordial ratio. Other evidence supports a smaller starting value of ~77 (Woods & Willacy 2009). The equilibrium value for the CNO cycle is 3.4. Convective mixing is predicted to produce ratios of ~20 during first ascent up the red-giant branch, and further mixing at later stages leads to lower values.

There is reasonable agreement with observations. Field giants show a distribution peaking at ~15–25, and the values for supergiants span the same range (Sneden & Pilachowski 1986, Böcek Topcu *et al.* 2016, Szigeti *et al.* 2018).

Halo and globular-cluster giants indicate the position of first dredge-up, with an abrupt transition from high values, $^{12}C/^{13}C$ ~ 20–40, to low values, ~4–6, at log g ~2 or M_V ~ 1 (Keller *et al.* 2001, Shetrone 2003). In older stars more generally, the equilibrium value is approached, as with the globular cluster ω Centauri, which has $^{12}C/^{13}C$ ~ 4 for its bright giants (Smith *et al.* 2002) and M10 with a ratio of ~5 (Maas *et al.* 2019).

Other isotopes also have stories to tell, such as oxygen (e.g., Lebzelter *et al.* 2015) and lithium (e.g., Asplund *et al.* 2006).

Chemical Evolution by Diffusion

Surface chemical composition can be altered by diffusion, specifically a process in which radiation pressure fights against gravity to winnow different elements upward or downward in the photosphere (Vauclair *et al.* 1979, Vauclair & Vauclair 1982, Michaud *et al.* 1983, Michaud *et al.* 2015, but see Kochukhov & Ryabchikova 2018). The vigorous and short-timescale mixing by convective envelopes has generally been assumed to override diffusion, relegating the process to early-type stars. This assumption may be overly restrictive. Recent investigation indicates that [Fe/H] may be depleted by ~0.05 dex in Sun-like stars (Dotter *et al.* 2017). Older stars show more depletion. However, the diffusion process is unique to each element and some will show an increase with time.

Certain early type stars, those classed as chemically "peculiar," seem to show extreme cases of diffusion.

Chemically Peculiar Stars

Peculiar is the word we use for "different" or possibly "not understood." In this section we deal with A and B stars, mostly on or near the main sequence, that show puzzling chemical makeup. The generic term "Ap star" comes from the spectral classification of the star as peculiar. Approximately 15–30% of late B and A stars show strange chemical anomalies in which certain species are mildly to wildly overabundant. There are several sub-classes of these peculiar stars (e.g., Smith 1996), and here we consider only a few of the broad themes. Although elements in the iron group are affected, the most striking peculiarities occur with the rare earths, such as strontium and europium, whose lines can dominate the star's spectrum quite unlike their normal, nearly obscure behavior. One of the most extreme peculiar-abundance stars is Przybylski's star (HD 101065), where some rare earths show overabundances of ~4 dex (Cowley & Mathys 1998). As a group, the Ap stars rotate more slowly than their more normal counterparts, although many still show substantial rotation, up to ~100 km/s. Abt (2000) has advocated the idea that the defining difference between peculiar and normal A stars is their rotation rates. A review of Ap stars is given by Kurtz (1990). Figure 16.18 shows the general run of chromium abundance for Ap stars compared to normal A stars. Not only are the Ap-star abundances typically two orders of magnitude higher, but they also show an order of magnitude spread.

Many (all?) of the Ap stars are spectroscopically variable and periodic. The most successful explanation of this is rotational modulation, where chemically segregated areas are carried across the visible disk of the star as it rotates. The chemical segregation likely arises through diffusion. A quiescent photosphere is needed for diffusion to work, and magnetic fields can play a role in defining diffusion pathways. Strong magnetic fields, in some cases up to ~40 kilogauss (4 tesla), are deduced from the Zeeman splitting of the lines (e.g., Mathys 1990). These fields are thought to be fossil fields embedded in the stars early in their life and invariant during their main-sequence phase. The magnetic fields shows global organization (unlike the magnetic fields of cool stars), typically a dipole field with some distortions. The magnetic axis is usually at an angle to the rotation axis, giving rise to the name "oblique rotator" (Stibbs 1950, Deutsch 1956, Babcock 1960, Mathys 2017). About a quarter of the magnetic Ap stars show photometric variations of a few milli-magnitudes (Hümmerich *et al.* 2018).

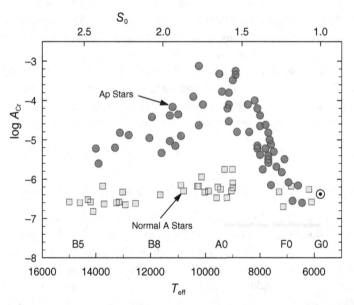

Figure 16.18 Chromium abundance for Ap stars compared to normal A stars. The solar value is shown by the circumpunct. Based on Kochukhov and Ryabchikova (2018).

High-resolution spectroscopy coupled with mathematical inversion techniques yield maps of the chemical distributions across the stellar surface. This is called "rotational mapping" or "Doppler imaging" (Wehlau *et al.* 1982, Khokhlova *et al.* 1986, Landstreet 1988, Rice & Wehlau 1994, Bailey & Landstreet 2013, Silvester *et al.* 2017, and related material in Chapter 18). As anticipated, the location of chemical concentrations is found to be connected to the magnetic geometry. Chemical concentrations of neodymium and europium, for example, tend to form around the magnetic poles. The fact that evolved stars, some of which were Ap stars on the main sequence but now have convective envelopes that mix the surface and subsurface layers, do not show these chemical peculiarities is evidence that the Ap-star phenomenon is localized to the superficial surface layers.

A handful of Ap stars have also been found to be pulsating with periods of a few minutes, amplitudes ranging up to a few km/s (Kurtz 1982, Baldry *et al.* 1998, Kurtz *et al.* 2003, Elkin *et al.* 2015), and photometric variations reaching a few milli-magnitudes (Dorokhova & Dorokhov 1998, Freyhammer *et al.* 2009, Kochukhov *et al.* 2019). These pulsation variations are superimposed on the rotational modulation that differs from one chemical species to the next, making for a complex puzzle to unravel. Three or four of these are in binary systems (Holdsworth *et al.* 2019).

Metallic-line A stars, or Am stars, also show peculiar overabundances, but curiously calcium and scandium are underabundant. In fact, it was through the discordant spectral classification based on the H and K Ca II lines compared to the general metal-line classification that led to their special classification. The Am stars make up ~1/3 of the A-star dwarf and subgiant populations (Abt 1981, Gray *et al.* 2016). The bright star Sirius belongs to this group. These stars are invariably relatively slow rotators (no more than ~100 km/s), perhaps because they are all binaries and affected by tidal coupling (Abt 1961). Further, Am stars usually have only weak magnetic fields (van't Veer-Meneret 1963, Blazère *et al.* 2016).

Several other groups of peculiarities have been identified, but we do not take them up here. Ghazaryan *et al.* (2018) discuss the nuances of classification in their catalogue of chemically peculiar stars.

Chemical Connections to Exoplanets

Early on in the unfolding saga of exoplanets it was suspected (Gonzales 1997) and then shown (Santos *et al.* 2004, Fischer & Valenti 2005, Sousa *et al.* 2011, Osborn & Bayliss 2020) that large planets in small orbits are found more often around metal-rich stars. Numerous subsequent studies have investigated how widespread this connection is for exoplanets more generally. Lu *et al.* (2020) found that even small planets, those with radii down to ~$2R_\oplus$, prefer metal-rich stars, while still smaller planets in the range of $1.0 - 1.7\,R_\oplus$ show no preference according to Petigura *et al.* (2018). Investigators have to watch for systematic uncertainties and sampling biases. As part of the backdrop, observations indicate that there is no detectable metallicity difference between stars with planets and those without (Buchhave & Latham 2015).

Why might metallicity be connected to planet characteristics? Stellar metallicity is a measure of the initial composition of the protoplanetary disk. Large [Fe/H] implies more solids are available to form planets. And planet formation, composition, and size might well be expected to depend on the strength of the near-UV radiation from the host star, and, as we have seen in Figures 8.5 and 8.6, that radiation depends on the metallicity. Oishi and Kanaya (2016) also consider how this UV radiation might affect the generation and destruction of life on a planet.

In the broader picture, there are many variables affecting the situation. Radiation pressure from the star will push away the inner boundary of the debris disk from which planets form. The strength of the radiation pressure goes up with the size, temperature, and, for dwarfs, the mass of the star, and the crucial UV component is smaller when the metallicity is larger. Planets farther from the star suffer less photo-evaporation and larger planets can resist photo-evaporation better than small planets. These variables are slowly being sorted out by grouping the planet systems in bins according to stellar temperature, planet radius or mass, and orbit size. For example, Owen and Murray-Clay (2018) explore the orbital periods (proxy for orbit dimensions through Kepler's third law) for small planets, $R < 6R_\oplus$; Petigura *et al.* (2018) find unique metallicity dependences for planets binned by radius; Wilson *et al.* (2018) find orbital period/planet size/metallicity interactions; Sousa *et al.* (2019) invoke a metallicity/period/mass diagram; Mills *et al.* (2019) find more eccentric orbits around metal-rich stars. Bolmont *et al.* (2017) find that tidal interactions and their affects on planetary orbits are affected by metallicity owing to the change in transparency in the convection envelope. Courcol *et al.* (2016) find a maximum planet mass increasing with host-star metallicity.

Evolved stars mimic dwarfs in their frequency of giant planets as a function of metallicity (Jofré *et al.* 2015, Reffert *et al.* 2015).

Brown dwarfs, which lie between stars and planets, apparently go their own way and do not have the metallicity affinity shown by giant planets (Mata Sánchez *et al.* 2014).

Questions and Exercises

1. Draw a typical curve of growth. Label the axes and label the different parts of the curve. Indicate in a short sentence or two the main characteristics of the different parts of the curve. What is the principal use of the curve of growth? What other uses can it be put to?

2. Do an approximate differential abundance analysis relative to the Sun for
   (a) η Cas A, a G0 V star with $T_{\text{eff}} = 5940$ K, and
   (b) β Vir, an F9 V star with $T_{\text{eff}} = 6150$ K.

   For this exercise, use the λ6226.74 line of Fe I, $\chi_n = 3.884$ eV. In the Sun, this line has an equivalent width of 31.8 mÅ. In η Cas A, the line's strength is 20.0 mÅ, and in β Vir it is 31.5 mÅ. Then subtract off the $\theta\chi_n$ terms and ratio to the solar value (difference in logs) to get [Fe/H]. Published values are [Fe/H] ~ -0.25 for η Cas A and $\sim +0.20$ for β Vir. How do your results compare? Stellar temperatures can have errors of ~100 K. Run your numbers again using temperatures for η Cas A and β Vir that are 100 K hotter to see how the [Fe/H] values change.

3. What effect might you expect granulation to have on a chemical analysis of a star? The stellar surface will be covered with hot granules and cool lanes (see Chapter 17). The spectrum from the granules will obviously be for hotter gas, while the spectrum from the lanes will be for cooler gas. Do you think this will all "average out," so that an average temperature (distribution) can be used for a metallicity analysis? Will the spectroscopic effects be the same for all lines?

4. Suppose you want to do a chemical study of a giant star in the galactic disk. If the random error in your [Fe/H] values is expected to be ~0.05 dex, what is the faintest giant that should be included in your targets if you wish to keep the galactic radial metallicity gradient from biasing your results, say a contribution of less than ~0.01 dex? Call on Figure 16.12 for help.

References

Abt, H.A. 1961. *ApJS* **6**, 37.
Abt, H.A. 1981. *ApJS* **45**, 437.
Abt, H.A. 2000. *ApJ* **544**, 933.
Abuter, R., *et al.* (GRAVITY collaboration). 2019. *A&A* **625**, L10.
Adelman, S.J. 1978. *PASP* **90**, 766.
Alexander, J.B. 1967. *MNRAS* **137**, 41.
Anguiano, B. 2012. *The Age-Metallicity-Velocity Relation in the Nearby Disk*. PhD thesis, Potsdam University.
Anthony-Twarog, B.J., Deliyannis, C.P., Harmer, D., *et al.* 2018. *AJ* **156**, 37.
Aoki, W., Barklem, P.S., Beers, T.C., *et al.* 2009. *ApJ* **698**, 1803.
Asplund, M.N., Lambert, D.L., Nissen, P.E., Primas, F., & Smith, V.V. 2006. *ApJ* **644**, 229.
Asplund, M., Grevesse, N., Sauval, A.J., & Scott, P. 2009. *ARAA* **47**, 481.
Athay, R.G. & Skumanich A. 1968. *ApJ* **152**, 211.
Aurière, M., Konstantinova-Antova, R., Charbonnel, C., *et al.* 2015. *A&A* **574**, 90.
Baade, W. 1944. *ApJ* **100**, 137.
Baade, W. 1951. *Pub Observatory Michigan* **10**, 7.
Baade, W. 1958. *Ricerche Astronomiche, 5, Specola Vaticana, Proceedings of a Conference at Vatican Observatory, Castel Gandolfo, May 20–28, 1957* (Amsterdam: North-Holland and New York: Interscience), D.J.K. O'Connell, ed., p. 3.
Baade, W. & Gaposchkin, C.H.P. 1963. *Evolution of Stars and Galaxies* (Cambridge, Mass.: Harvard University Press).
Babcock, H.W. 1960. *Stars and Stellar Systems, Stellar Atmospheres*, Vol. VI (Chicago: University of Chicago Press), J.L. Greenstein, ed., p. 282.
Bailey, J.D. & Landstreet, J.D. 2013. *MNRAS* **432**, 1687.
Balachandran, S. 1995. *ApJ* **446**, 203.
Baldry, I.K., Kurtz, D.W., & Bedding, T.R. 1998. *MNRAS* **300**, L39.
Barbuy, B., Chiappini, C., & Gerhard, O. 2018. *ARAA* **56**, 223.
Barstow, M.A., Good, S.A., Holberg, J.B., *et al.* 2003. *MNRAS* **341**, 870.
Baschek, B., Garz, T., Holweger, H., & Richter, J. 1970. *A&A* **4**, 229.
Beccari, G. & Carraro, G. 2015. *Ecology of Blue Straggler Stars*, Astrophysics and Space Science Library, Vol. **413** (Berlin, Heidelburg: Springer), H.M.J. Boffin, G. Carroro, & G. Beccari, eds., Chapter 1.
Becker, S.A. & Iben, I., Jr. 1979. *ApJ* **232**, 831.
Beckers, J.M., Bridges, C.A., & Gilliam, L.B. 1976. *High Resolution Spectral Atlas of the Solar Irradiance from 380 to 700 Nanometers*, Environmental Research Papers (Hanscom, Mass.: Air Force Geophysics Laboratory).
Bedell, M., Melendez, J., Bean, J.L., *et al.* 2014. *ApJ* **795**, 23.
Bergemann, M. & Gehren, T. 2007. *A&A* **473**, 291.
Bessell, M.S. & Norris, J. 1984. *ApJ* **285**, 622.

Bessell, M.S., *et al.* 2015. *ApJL* **806**, L16.

Blazère, A., Petit, P., Lignières, F., *et al.* 2016. *A&A* **586**, 97.

Böcek Topcu, G., Afşar, M., & Sneden, C. 2016. *MNRAS* **463**, 580.

Boeche, C., *et al.* 2011. *AJ* **142**, 193.

Boesgaard, A.M. 1989. *ApJ* **336**, 798.

Boesgaard, A.M. & King, J.R. 2002. *ApJ* **565**, 587.

Boesgaard, A.M. & Tripicco, M.J. 1986. *ApJL* **302**, L49.

Boesgaard, A.M., Stephens, A., & Deliyannis, C.P. 2005. *ApJ* **633**, 398.

Boesgaard, A.M., Michael Lum, G., Deliyannis, C.P., *et al.* 2016. *ApJ* **830**, 49.

Bohlender, D.A., Dworetsky, M.M., & Jomaron, C.M. 1998. *ApJ* **504**, 533.

Bolmont, E., Gallet, F., Mathis, S., *et al.* 2017. *A&A* **604**, 113.

Bond, H.E. 1980. *ApJS* **44**, 517.

Bonifacio, P., Molaro, P., Sivarani, T., *et al.* 2007. *A&A* **462**, 851.

Bonifacio, P., Caffau, E., Spite, M., *et al.* 2018. *A&A* **612**, 65.

Booth, A.J. & Blackwell, D.E. 1983. *MNRAS* **204**, 777.

Borrero, J.M., Bellot Rubio, L.R., Barklem, P.S., & del Toro Iniesta, J.C. 2003. *A&A* **404**, 749.

Bouvier, J., Barrado, D., Moraux, E., *et al.* 2018. *A&A* **613**, 63.

Brandt, T.D. & Huang, C.X. 2015. *ApJ* **807**, 24.

Brault, J. W. 1985. *High Resolution Astronomy* (Sauverny: Swiss Society of Astrophysics & Astronomy, Geneva Observatory), A. Benz, M. Huber, & M. Mayer, eds., p. 3.

Brewer, J.M., Fischer, D.A., Valenti, J.A., & Piskunov, N. 2016. *ApJS* **225**, 32.

Buchhave, L.A. & Latham, D.W. 2015. *ApJ* **808**, 187.

Burris, D.L., Pilachowski, C.A., Armandroff, T.E., *et al.* 2000. *ApJ* **544**, 302.

Carraro, G., Ng, Y.K., & Portinari, L. 1998. *MNRAS* **296**, 1045.

Carretta, E., Bragaglia, A., Gratton, R.G., *et al.* 2010. *A&A* **516**, 55.

Cayrel, R. 1996. *A&AR* **7**, 217.

Cayrel, R., Cayrel de Strobel, G., & Campbell, B. 1985. *A&A* **146**, 249.

Cheng, J.Y., Rockosi, C.M., Morrison, H.L., *et al.* 2012. *ApJ* **746**, 149.

Christlieb, N., Gustafsson, B., Korn, A.J., *et al.* 2004. *ApJ* **603**, 708.

Cordoni, G., Milone, A.P., Marino, A.F., *et al.* 2018. *ApJ* **869**, 139.

Courcol, B., Bouchy, F., & Deleuil, M. 2016. *MNRAS* **461**, 1841.

Cowley, C.R. & Frey, M. 1989. *ApJ* **346**, 1030.

Cowley, C.R. & Mathys, G. 1998. *A&A* **339**, 165.

Cummings, J.D., Deliyannis, C.P., Maderak, R.M., & Aaron Steinhauer, A. 2017. *AJ* **153**, 128.

Cyburt, R.H., Fields, B.D., Olive, K.A., & Yeh, T.-H. 2015. *Rev Mod Phys* **88**, 5004.

Daflon, S., Cunha, K., Butler, K., & Smith, V. 2001. *ApJ* **563**, 325.

Dahm, S.E. 2015. *ApJ* **813**, 108.

Deliyannis, C.P., Demarque, P., & Kawaler, S.D. 1990. *ApJS* **73**, 21.

Deliyannis, C.P., Ryan, S.G., Beers, T.C., & Thorburn, J.A. 1994. *ApJL* **425**, L21.

Deutsch, A.J. 1956. *PASP* **68**, 92.

Do, T., Kerzendorf, W., Winsor, N., *et al.* 2015. *ApJ* **809**, 143.

Dorokhova, T.N. & Dorokhov, N.I. 1998. *Contr Ast Obs Skalnate Pleso* **27**, 338.

Dotter, A., Sarajedini, A., Anderson, J., *et al.* 2010. *ApJ* **708**, 698.

Dotter, A., Sarajedini, A., & Anderson, J. 2011. *ApJ* **738**, 74.

Dotter, A., Conroy, C., Cargile1, P., & Asplund, M. 2017. *ApJ* **840**, 99.

Edvardsson, B., Andersen, J., Gustafsson, B., *et al.* 1993. *A&A* **275**, 101.

Eggen, O.J., Lynden-Bell, D., & Sandage, A.R. 1962. *ApJ* **136**, 748.

Elkin, V.G., Kurtz, D.W., & Mathys, G. 2015. *MNRAS* **446**, 4126.

Farragiana, R. 1989. *A&A* **224**, 162.

Feuillet, D.K., Frankel, N., Lind, K., *et al.* 2019. *MNRAS* **489**, 1742.

Fischer, D.A. & Valenti, J. 2005. *ApJ* **662**, 1102.

Ford, A., Jefferies, R.D.,Smalley, B., *et al.* 2002. *A&A* **393**, 617.

Freyhammer, L.M., Kurtz, D.W., Elkin, V.G., *et al.* 2009. *MNRAS* **396**, 325.

Fu, X., *et al.* 2018. *A&A* **610**, 38.

Garz, T., Holweger, H., Kock, M., & Richter, J. 1969. *A&A* **2**, 446.

Genovali, K., Lemasle, B., da Silva, R., *et al.* 2015. *A&A* **580**, 17.

Gerber, J.M., Friel, E.D., & Vesperini, E. 2018. *AJ* **156**, 6.

Ghazaryan, S., Alecian, G., & Hakobyan, A.A. 2018. *MNRAS* **480**, 2953.

Glatt, K. 2009. *Star Clusters as Age Tracers of the Age–Metallicity Relation of the Small Magellanic Cloud*, PhD thesis, University of Basel.

Goldberg, L., Müller E.A., & Aller, L.A. 1960. *ApJS* **5**, 1.

Gondoin, P. 2014. *A&A* **566**, 72.

Gonzalez, G. 1997. *MNRAS* **285**, 403.

Gonzalez, O.A. & Gadotti, D.A. 2016. *ApSS* **418**, 199.

Gossage, S., Conroy, C., Dotter, A., *et al.* 2018. *ApJ* **863**, 67.

Gozha, M.L., Koval',V.V., & Marsakov, V.A. 2012. *Ast Lett* **38**, 519.

Gratton, R.G., Lucatello, S., Carretta, E., *et al.* 2012. *A&A* **539**, 19.

Gray, D.F. 1988. *Lectures on Spectral-Line Analysis: F, G, and K Stars* (Arva, Ontario: The Publisher).

Gray, D.F. & Oostra, B. 2018. *ApJ* **852**, 42.

Gray, D.F., Scott, H.R., & Postma, J.E. 2002. *PASP* **114**, 536.

Gray, R.O., Corbally, C.J., De Cat, P., *et al.* 2016. *AJ* **151**, 13.

Greenstein, J.L. 1948. *ApJ* **107**, 151.

Grevesse, N. & Sauval, A.J. 1998. *Space Sci Rev* **85**, 161.

Grevesse, N., Scott, P., Asplund, M., & Sauval, A.J. 2015. *A&A* **573**, 27.

Gruyters, P., Lind, K., Richard, O., *et al.* 2016. *A&A* **589**, 61.

Guiglion, G., de Laverny, P., Recio-Blanco, A., *et al.* 2016. *A&A* **595**, 18.

Gustafsson, B. 1983. *PASP* **95**, 101.

Harris, H.C. 1981. *AJ* **86**, 707.

Harris, W.E., Whitmore, B.C., Karakla, D., *et al.* 2006. *ApJ* **636**, 90.

Hartoog, M.R., Cowley, C.R., & Adelman, S.J. 1974. *ApJ* **187**, 551.

Hasselquist, S., *et al.* 2019. *ApJ* **871**, 181.

Hawkins, K., Masseron, T., Jofré, P., *et al.* 2016. *A&A* **594**, 43.

Hinkel, N.R., Timmes, F.X., Young, P.A., Pagano, M.D., & Turnbull, M.C 2014. *AJ* **148**, 54.

Hinkle, K., Wallace, Ll., & Livingston, W.C. 1995. *Infrared Atlas of the Arcturus Spectrum, 0.9–5.3 Microns* (San Francisco: Astonomical Society of the Pacific).

Hinkle, K., Wallace, Ll., Valenti, J., & Harmer, D. 2000. *Visible and Near Infrared Atlas of the Arcturus Spectrum, 3727–9300 Å* (San Francisco: Astronomical Society of the Pacific).

Hobbs, L.M., Thorburn, J.A., & Rebull, L.M. 1999. *ApJ* **523**, 797.

Holdsworth, D.L., Saio, H., & Kurtz, D.W. 2019. *MNRAS* **489**, 4063.

Holtzman, J.A., *et al.* 2015. *AJ* **150**, 148.

Hümmerich, S., Mikulášek, Z., Paunzen, E., *et al.* 2018. *A&A* **619**, 98.

Hunter, I., Brott, I., Langer, N., *et al.* 2009. *A&A* **496**, 841.

Iben, I., Jr. 1964. *ApJ* **140**, 1631.

Ivans, I.I., Kraft, R.P., Sneden, C., *et al.* 2001. *AJ* **122**, 1438.

Jofré, E., Petrucci, R., Saffe, C., *et al.* 2015. *A&A* **574**, 50.

Johnson, C.I., Rich, R.M., Pilachowski, C.A., *et al.* 2015. *AJ* **150**, 63.
Jönsson, H., Ryde, N., Nordlander, T., *et al.* 2017. *A&A* **598**, 100.
Kalirai, J.S., Fahlman, G.G., Richer, H.B., & Ventura, P. 2003. *AJ* **126**, 1402.
Karaali, S., Bilir, S., Ak, S., Yaz, E., & Coskunoglu, B. 2011. *PAS Australia* **28**, 95.
Keller, L.D., Pilachowski, C.A., & Sneden, C. 2001. *AJ* **122**, 2554.
Keszthelyi, Z., Meynet, G., Georgy, C., *et al.* 2019. *MNRAS* **485**, 5843.
Khokhlova, V.L., Rice, J.B., & Wehlau, W.H. 1986. *ApJ* **307**, 768.
Kochukhov, O. & Ryabchikova, T.A. 2018. *MNRAS* **474**, 2787.
Kochukhov, O., Shultz, M., & Neiner, C. 2019. *A&A* **621**, 47.
Kurtz, D.W. 1982. *MNRAS* **200**, 807.
Kurtz, D.W. 1990. *ARAA* **28**, 607.
Kurtz, D.W., Elkin, V.G., & Mathys, G. 2003. *MNRAS* **343**, L5.
Kurucz, R.L. 2005. http://kurucz.harvard.edu/Sun.html.
Kurucz, R.L., Furenlid, I., Brault, J., & Testerman, L. 1984. *Solar Flux Atlas from 296 to 1300 nm*, N.S.O. Atlas No. 1.
Kusakabe, M. & Kawasaki, M. 2019. *ApJ* **876**, 30.
Lambert, D.L. 1976. *Mem Roy Ast Soc Liege* **9**, 405.
Lambert, D.L. & Ries, L.M. 1981. *ApJ* **248**, 228.
Landstreet, J.D. 1988. *ApJ* **326**, 967.
Lebzelter, T., Straniero, O., Hinkle, K.H., Nowotny, W., & Aringer, B. 2015. *A&A* **578**, 33.
Lemasle, B., François, P., Bono, G., *et al.* 2007. *A&A* **467**, 283.
Little-Marenin, I.R. & Little, S.J. 1984. *ApJ* **283**, 188.
Lu, C.X., Schlaufman, K.C., & Cheng, S. 2020. *AJ* **160**, 253.
Luck, R.E. 1982. *ApJ* **256**, 177.
Luck, R.E. & Bond, H.E. 1980. *ApJ* **241**, 21
Luck, R.E. & Lambert, D.L. 2011. *AJ* **142**, 136.
Luck, R.E., Kovtyukh, V.V., & Andrievsky, S.M. 2006. *AJ* **132**, 902.
Luck, R.E., Andrievsky, S.M., Kovtyukh, V.V., Gieren, W., & Graczyk, D. 2011. *AJ* **142**, 51.
Lyubimkov, L.S., Korotin, S.A., & Lambert, D.L. 2019. *MNRAS* **489**, 1533.
Maas, Z.G., Gerber,J.M., Deibel, A., & Pilachowski, C.A. 2019. *ApJ* **878**, 43.
Maeder, A., Przybilla, N., Nieva, M.F., *et al.* 2014. *A&A* **565**, 39.
Marino, A.F. Milone, A.P., Casagrande, L., *et al.* 2018. *ApJ* **863**, 33.
Marsakov, V.A., Koval', V.V., Borkova, T.V., & Shapovalov, M.V. 2011. *Ast Rep* **55**, 667.
Martín, E.L., Lodieu, N., Pavlenko, Y., & Béjar, V.J.S. 2018. *ApJ* **856**, 40.
Mashonkina, L. & Gehren, T. 2000. *A&A* **364**, 249.
Mata Sánchez, D., González Hernández, J.I., Israelian, G., *et al.* 2014. *A&A* **566**, 83.
Mathys, G. 1990. *A&A* **232**, 151.
Mathys, G. 2017. *A&A* **601**, 14.
Matsuno, T., Aoki, W., Suda, T., & Li, H. 2017. *PASJ* **69**, 24.
Matteucci, F. & Greggio, L. 1986. *A&A* **154**, 279.
Meibom, A., Krot, A.N., Robert, F., *et al.* 2007. *ApJ* **656**, 33.
Meléndez, J. & Barbuy, B. 2009. *A&A* **497**, 611.
Menzel, D.H. 1936. *ApJ* **84**, 462.
Michaud, G., Tarasick, D., Charland, Y., & Pelletier, C. 1983. *ApJ* **269**, 239.
Michaud, G., Alecian, G., & Richer, J. 2015. *Atomic Diffusion in Stars* (Switzerland: Springer).
Mikolaitis, Š., Tautvaišienė, G., Gratton, R., Bragaglia, A., & Carretta, E. 2012. *A&A* **541**, 137.

Mills, S.M., Howard, A.W., Petigura, E.A., *et al.* 2019. *AJ* **157**, 198.

Minchev, I., Matijevic, G., Hogg, D.W., *et al.* 2019. *MNRAS* **487**, 3946.

Minnaert, M. & Mulders, G.F.W. 1931. *Zeits Ap* **2**, 165.

Minnaert, M. & Slob, C. 1931. *Proc K Acad Amst* **34**, 542.

Monaco, L., Bonifacio, P., Sbordone, L., Villanova, S., & Pancino, E. 2010. *A&A* **519**, 3.

Moore, C.E., Minnaert, M.G.J., & Houtgast, J. 1966. *The Solar Spectrum 2935 Å to 8770 Å*, National Bureau of Standards Monograph 61 (Washington, DC: US Government Printing Office).

Mucciarelli, A., Salaris, M., Lovisi, L., *et al.* 2011. *MNRAS* **412**, 81.

Müller, E.A. 1966. *Abundance Determinations in Stellar Spectra*, IAU Symposium 26 (London: Academic Press), H. Hubenet, ed., p. 171.

Ness, M. & Freeman, K. 2016. *PAS Australia* **33**, 22.

Nissen, P.E. & Gustafsson, B. 2018. *A&AR* **26**, 6.

Nordlander, T., Amarsi, A.M., Lind, K., *et al.* 2017. *A&A* **597**, 6.

Nordlander, T., Bessell, M.S., Da Costa, G.S., *et al.* 2019. *MNRAS* **488**, 109.

Oishi, M. & Kanaya, H. 2016. *ApJ* **833**, 293.

Osborn, A. & Bayliss, D. 2020. *MNRAS* **491**, 4481.

Owen, J.E. & Murray-Clay, R. 2018. *MNRAS* **480**, 2206.

Pagel, B.E.J. 1997. *Nucleosynthesis and Chemical Evolution of Galaxies* (Cambridge: Cambridge University Press).

Pannekoek, A. & van Albada, B.B. 1946. *Pub Ast Inst Univ Amst* **6**, part 2.

Perrin, M.-N., Cayrel de Strobel, G., & Dennefeld, M. 1988. *A&A* **191**, 237.

Perryman, M.A.C., Brown, A.G.A., Lebreton, Y., *et al.* 1998. *A&A* **331**, 81.

Petigura, E.A., Howard, A.W., Marcy, G.W., *et al.* 2017. *AJ* **154**, 107.

Petigura, E.A., Marcy, G.W., Winn, J.N., *et al.* 2018. *AJ* **155**, 89.

Pilachowski, C., Saha, A., & Hobbs, L.M. 1988. *PASP* **100**, 474.

Pizzone, R.G., Sparta, R., Bertulani, C.A., *et al.* 2014. *ApJ* **786**, 112.

Rebolo, R., Beckman, J.E., & Molaro, P. 1988. *A&A* **192**, 192.

Reffert, S., Bergmann, C., Quirrenbach, A., Trifonov, T., & Künstler, A. 2015. *A&A* **574**, 116.

Reiners, A., Mrotzek, N., Lemke, U., Hinrichs, J., & Reinsch, K. 2016. *A&A* **587**, 65.

Reyniers, M., Van Winckel, H., Biémont, E., & Quinet, P. 2002. *A&AL* **395**, L35.

Rice, J.B. & Wehlau, W.H. 1994. *A&A* **291**, 825.

Ross, J. & Aller, L. 1968. *ApJ* **153**, 235.

Russell, H.N. 1929. *ApJ* **70**, 11.

Rutten, R.J. 1978. *Solar Phys* **56**, 237.

Rutten, R.J. & van der Zalm, E.B.J. 1984. *A&AS* **55**, 171.

Ryan, S.G. & Deliyannis, C.P. 1998. *ApJ* **500**, 398.

Ryan, S.G., Norris, J.E., & Beers, T.C. 1999. *ApJ* **523**, 654.

Santos, N.C., Israelian, G., & Mayor, M. 2004. *A&A* **415**, 1153.

Sbordone, L., Bonifacio, P., Caffau, E., *et al.* 2010. *A&A* **522**, 26.

Schultheis, M., Rich, R.M., Origlia, L., *et al.* 2019. *A&A* **627**, 152.

Scott, P., Grevesse, N., Asplund, M., *et al.* 2015a. *A&A* **573**, 25.

Scott, P., Asplund, M., Grevesse, N., Bergemann, M., & Sauval, A.J. 2015b. *A&A* **573**, 26.

Sestito, P. & Randich, S. 2005. *A&A* **442**, 615.

Shetrone, M.D. 2003. *ApJ* **585**, L45.

Sills, A. & Deliyannis, C.P. 2000. *ApJ* **544**, 944.

Silvester, J., Kochukhov, O., Rusomarov, N., & Wade, G.A. 2017. *MNRAS* **471**, 962.

Smiljanic, R., *et al.* 2018. *A&A* **617**, 4.

Smith, G.H. 1987. *PASP* **99**, 67.

Smith, K.C. 1996. *ApSS* **237**, 77.

Smith, V.V. & Lambert, D.L. 1986. *ApJ* **311**, 843.

Smith, V.V., Terndrup, D.M., & Suntzeff, N.B. 2002. *ApJ* **579**, 832.

Sneden, C. 1973. *ApJ* **184**, 839.

Sneden, C. & Pilachowski, C.A. 1986. *ApJ* **301**, 860.

Soderblom, D.R., Jones, B.F., Balachandran, S., *et al.* 1993a. *AJ* **106**, 1059.

Soderblom, D.R., Fedele, S.B., Jones, B.F., Stauffer, J.R., & Prosser, C.F. 1993b. *AJ* **106**, 1080.

Soubiran, C., Le Campion, J.-F., Brouillet, N., & Chemin, L. 2016. *A&A* **591**, 118.

Sousa, S.G., Santos, N.C., Israelian, G., Mayor, M., & Udry, S. 2011. *A&A* **533**, 141.

Sousa, S.G., *et al.* 2019. *MNRAS* **485**, 3981.

Spite, F. & Spite, M. 1982. *A&A* **115**, 357.

Spite, M. & Spite, F. 2010. *Light Elements in the Universe*, IAU Symposium **268** (Cambridge: Cambridge University Press), C. Charbonnel, M. Tosi, F. Primas & C. Chiappini, eds.

Spite, M., Françoism, P., Nissen, P.E., & Spite, F. 1996. *A&A* **307**, 172.

Stephens, A., Boesgaard, A.M., King, J.R., & Deliyannis, C.P. 1997. *ApJ* **491**, 339.

Stibbs, D.W.N. 1950. *MNRAS* **110**, 395.

Stift, M.J. & Leone, F. 2003. *A&A* **398**, 411.

Strassmeier, K.G., Ilyin, I., & Steffen, M. 2018. *A&A* **612**, 44.

Strömgren, B. 1940. *Festschrift für Elis Strömgren* (Copenhagen:Munksgaard), p. 218.

Strömgren, B. 1963. *Basic Astronomical Data* (Chicago: University of Chicago Press), K.A. Strand, ed., p. 123.

Struve, O. & Elvey, C.T. 1934. *ApJ* **79**, 409.

Sweigart, A.V. & Mengel, J.G. 1979. *ApJ* **229**, 624.

Szigeti, L., Mészáros, S., Smith, V.V., *et al.* 2018. *MNRAS* **474**, 4810.

Takács, M.P., Bemmerer, D., Szücs, T., & Zuber, K. 2015. *Phys Rev D* **91**, id. 123526.

Takeda, Y. & Tajitsu, A. 2017. *PASJ* **69**, 74.

Tautvaišienė, G., *et al.* 2015. *A&A* **573**, 55.

Tautvaišienė, G., Drazdauskas, A., Bragaglia, A., Randich, S., & Ženovienė, R. 2016. *A&A* **595**, 16.

Tessore, B., Lèbre, A., Morin, J., *et al.* 2017. *A&A* **603**, 129.

Théado, S. & Vauclair, S. 2001. *A&A* **375**, 70.

Thorburn, J.A. 1994. *ApJ* **421**, 318.

Thorén, P. & Feltzing, S. 2000. *A&A* **363**, 692.

Twarog, B.A. 1980. *ApJ* **242**, 242.

Unsöld, A. 1948. *ZfAp* **24**, 306.

Unsöld, A. 1955. *Physik der Sternatmosphären* (Berlin: Springer-Verlag), 2nd ed.

Unsöld, A. 1971. *Phil Trans R Soc London Ser A* **270**, 23.

van de Hulst, H.C., Muller, C.A., & Oort, J.H. 1954. *BAN* **12**, 117.

van der Held, E.F.M. 1931. *ZfPhys* **70**, 508.

van't Veer Menneret, C. 1963. *AnAp* **26**, 289.

Vauclair, S., Hardorp, J., & Peterson, D.M. 1979. *ApJ* **227**, 526.

Vauclair, S. & Vauclair, G. 1982. *ARAA* **20**, 37.

Villamariz, M.R., Herrero, A., Becker, S.R., & Butler, K. 2002. *A&A* **388**, 940.

Vince, O., & Vince, I. 2003. *Pub Ast Obs Belgrade* **75**, 79.

Voigt, W. 1912. *Münch. Ber.* **603**.

Wallace, L., Hinkle, K.H., Livingston, W.C., & Davis, S.P. 2011. *ApJS* **195**, 6.

Wallerstein, G. & Carlson, M. 1960. *ApJ* **132**, 276.

Wehlau, W., Rice, J., Piskunov, N.E., & Khokhlova, V.L. 1982. *Soviet Ast Lett* **8**, 15.

Wilson, R.F., *et al.* 2018. *AJ* **155**, 68.

Wood, M.P, Lawler, J.E., Den Hartog, E.A., Sneden, C., & Cowan, J.J. 2014. *ApJS* **214**, 18.

Woods, P.M. 2010. arXiv:0901.4513v5.

Woods, P.M. & Willacy, K. 2009. *ApJ* **693**, 1360.

Wrubel, M.H. 1949. *ApJ* **109**, 66.

Wrubel, M.H. 1954. *ApJ* **119**, 51.

Xiang, M.-S., Liu, X.-W., Yuan, H.B., *et al.* 2015. *Res Astron Astrophys* **15**, 1209.

Yan, Z.-C., Tambasco, M., & Drake, G.W.F. 1998. *Phys Rev A* **57**, 1652.

Yong, D., Carney, B.W., Teixera de Almeida, M.L., & Pohl, B.L. 2006. *AJ* **131**, 2256.

Youssef, N.H. & Khalil, N.M. 1988. *A&A* **203**, 378.

17

Velocity Fields in Stellar Photospheres

It is amazing to think that the routine Sun that lights our days has a surface churning and buffeted with rising and falling gases, often with magnetic loops, many pairs of dark spots, and detonations that dwarf anything on Earth. And compared to other stars, our Sun is relatively calm. Stellar photospheres are very dynamic places indeed. The Doppler shifts from these motions and from rotation of the star dominate the shapes and alter the positions of spectral lines. The distributions of the Doppler shifts are imprinted on the spectral lines, and our task is to work backward from the observed line profiles and deduce the nature of these velocity fields. One goal is to separate the Doppler shifts of rotation from those of the internal photospheric velocities. Rotation is the topic of Chapter 18. We explore the attributes of the photospheric velocities such as their characteristic sizes, geometry, temperature heterogeneities, and depth dependence, from the spectral line shapes and positions, with an eye toward understanding more of the physics of stellar photospheres. These are the topics of the current chapter. The emphasis here is on ~F5 stars and cooler, that is, stars with convective envelopes, and where rotation does not overwhelm the line broadening.

Aside from the solar example, the first clue that stars have dynamic photospheres is that the observed spectral lines are broader than expected. The existence of supra-thermal line broadening was realized early in the history of stellar spectroscopy. Rosseland (1928) called these motions "turbulence," and the name has continued to be used even though most of the velocities involved have little or nothing to do with aerodynamic turbulence. In dealing with this turbulence in stellar spectra, it is useful to conceptualize two approximations: (1) the sizes of the turbulent elements are small compared to unit optical depth, called *micro*turbulence, and (2) the sizes of the turbulent elements are large compared to unit optical depth, called *macro*turbulence. The practical significance is that microturbulence can be brought into the radiation transfer by increasing the thermal velocity in the atomic absorption coefficient, while the macroturbulence bypasses the radiation transfer altogether. Both cases incorporate only the kinematic effects, namely the velocities and their associated Doppler shifts, and neglect the dynamics, such as changes in temperature and pressure caused by the motions. These approximations have been used extensively and are still useful even though more rigorous hydrodynamical computations can now be done.

For the Sun, we have a wealth of information to guide our thinking. We see granulation, supergranulation, P-mode oscillations, magnetic fields, sunspots, and various transient events. Remarkably, we can infer the characteristics of many of these same phenomena on other stars even though we deal with disk-integrated light, i.e., flux. The driver of these motions is fundamentally the convection below the surface. Rotation too plays a dynamical role as a mixing agent and in coupling with convection to generate magnetic fields.

First we take a look at the extra line broadening we are dealing with in the stellar spectra.

Examples of Velocity Broadening

Velocity fields broaden all spectral lines. Examples of photospheric velocity broadening are shown in Figures 17.1 and 17.2. Thermal widths are often only a fraction of the full observed widths. The triangular look of weaker lines, like the one shown in Figure 17.1, is characteristic of radial–tangential macroturbulence discussed below. Since we are dealing with velocities in this chapter, the wavelength scales are expressed in velocity units, using the usual Doppler conversion, $c\Delta\lambda/\lambda$, where $\Delta\lambda$ is the wavelength difference from line center or some other chosen reference.

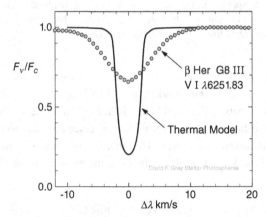

Figure 17.1 The thermal profile computed from a model photosphere with only thermal broadening does not look much like the real stellar profile. The data for β Her were taken at the Elginfield Observatory.

We know that the non-thermal broadening arises from Doppler shifts for a number of reasons: (1) the broadening increases proportional to wavelength, as expected for the Doppler effect; (2) in some stars, large broadening, up to hundreds of km/s, is seen, reasonably interpreted as rotational broadening; (3) spectral lines are shallower when they are broader, as in Figure 17.2, such that the equivalent width is conserved, just as expected from macroturbulence and rotation; (4) small asymmetries are seen in many line profiles, a signature we expect from a surface covered with granulation and therefore its velocities; (5) spectral lines show small blue shifts that vary systematically with core line depth, another offshoot of granulation; and (6) we see first hand the motions of the photospheric gas on the surface of the Sun.

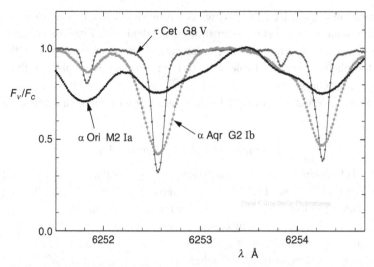

Figure 17.2 Non-thermal broadening increases with luminosity. Notice how the lines of the supergiant α Aqr are wider but less deep than those of the dwarf τ Cet. These two stars are roughly the same temperature. The supergiant α Ori is considerably cooler, resulting in different line strengths, but it illustrates the extreme broadening of spectral lines by macroturbulence. Data from the Elginfield Observatory.

To study the photospheric velocities effectively, we seek out suitable lines, of which there are many. The lines should not be blended, nor should they be intrinsically broad like the hydrogen lines of hotter stars or the Ca II H and K lines of cooler stars. The intrinsic line, i.e., the line before being reshaped by macroturbulence, is essentially a marker of Doppler shifts, and therefore the narrower, the better. To reiterate, rotation should not dominate the broadening, or it will mask the Doppler shifts from the motions in the photosphere.

Solar Velocity Fields

Since we have a direct view of the solar surface, we examine the velocity structures we see there. In struggling to objectively describe the mottled appearance of the solar surface, Dawes (1864) coined the term granulation. Granules, the brighter patches like those shown in Figure 17.3, are the hot rising bubbles. The darker lanes in between are the cooler falling material. Granulation is powered by the convection just below the surface, and the granulation itself is the overshoot region in which the granules are propelled upward but decelerate as they rise, gradually coming to a halt near the top of the photosphere. There the gas cools and flows back down at the edges of the granule cells. Granules have characteristic lifetimes of about 10 minutes and cover a range of sizes. The larger cells are about 2000 km across, while the average size is about 1100 km. There are $\sim 2 \times 10^6$ such cells on the solar disk. The mean temperature difference between granules and dark lanes surrounding the granules is at least 200 K. This corresponds to a brightness difference of $\sim 15\%$ (Abramenko *et al.* 2012). Granules occupy roughly half the area, or perhaps slightly less, and show ~1.4 times larger brightness variation compared to the lanes (Abdussamatov & Zlatopol'skii 1997).

Velocities are difficult to measure because the dark-lane structure is at the limit of spatial resolution, but characteristic velocities are ~2–3 km/s. At the same time, there is a wide range of velocities, and it is their Doppler-shift distribution that is imprinted on spectral lines. For early reviews of granulation, see Beckers (1981), Bray *et al.* (1984), Keil (1984), and Rutten and Severino (1989).

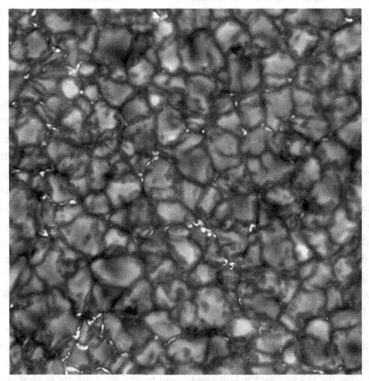

Figure 17.3 Solar granulation shows us the overshoot region above the convective envelope. The larger bright areas are the rising hot cells, ~2000 km across. The darker lanes in between the cells are where the cooled gas falls back down. The tiny bright points are magnetic concentrations. Image courtesy of Luc Rouppe van der Voort (University of Oslo), recorded with the Swedish 1-meter Solar Telescope on La Palma.

When we look at the Sun, granulation is obvious, but velocity fields at the larger scales are not. Traditional names and essential characteristics are summarized in Table 17.1. These are approximate numbers and there is a significant range around each of these values such that their domains merge. Hathaway *et al.* (2000) and Nordlund *et al.* (2009) even argue that there is a smooth continuous variation in geometrical scales. While granulation velocities are largely vertical, larger-scale motions are mainly horizontal and much smaller. Other large-scale flows have been studied by Roudier *et al.* (2018).

Table 17.1 *Solar velocity fields*

	Length (Mm)	Velocity (m/s)	Lifetime (minutes)	Number of cells	Full-disk velocity (m/s)
Oscillations	2.5	<500	5	—	—
Granulation	1	2500	10	2×10^6	1.8
Mesogranulation	8	500	150	45 000	2.4
Supergranulation	30	350	2500	3000	6.4
Giant cells	100	4	150 000	300	0.2

References: Evans & Michard (1962), Howard *et al.* (1968), Nordlund *et al.* (2009), Rieutord & Rincon (2010), Roudier *et al.* (2014),

If we view the disk-integrated Sun as if it were an unresolved distant star, the velocities, being random in nature, average out. From basic statistics, the average disk-integrated variation can be expected to be $\sim \delta v / \sqrt{N}$, where δv is the actual mean cell velocity and N is the number of cells on the stellar disk. For example, granulation velocities with $\delta v \sim 2500$ m/s averaged over $\sim 2 \times 10^6$ cells gives ~ 1.8 m/s, which means that the statistical variation of the disk-averaged granulation velocity is in the range of only two meters per second. Or in stellar terms, the Sun's radial velocity has a statistical variation ~ 2 m/s arising from granulation. The corresponding disk-integrated numbers for each classification are given in the last column of the table. Global brightness fluctuations sometimes accompany these velocity fields, and such fluctuations have been measured in many other stars. On the other hand, solar supergranulation has almost no brightness contrast, with temperature structure of only ~ 1 K (Schrijver *et al.* 1997, Meunier *et al.* 2008), but the situation changes for lower gravity stars, as we shall see.

A fundamental point for stellar-like spectroscopy is that the *distributions* of these velocities produce the broadening of the spectral lines. Macroturbulence is the collective combination of all these distributions, the envelope of the total velocity distribution. For the Sun and other dwarfs, granulation is the most relevant velocity field. Later in the chapter, as we explore velocity fields across the HR diagram, it becomes clear that there is a shift to larger cell sizes for stars of higher luminosity, and by the time we get to cool supergiants, giant cells with granulation embedded in them is the norm.

Then there are the magnetic fields. In most regions where the photospheric magnetic field is relatively weak, granulation buffets and wiggles the field lines. The magnetic field is tied to the gas because the gas is ionized. The magnetic lines are twisted up as the Coriolis forces rotate convective cells (Parker 1982a, 1982b, Spruit *et al.* 1990). Conversely, when the magnetic field is stronger it can alter the convection. Sunspots illustrate the extreme case where convection is blocked completely by fields of 1–2 kilogauss (0.1–0.2 tesla). Blocking the convection reduces the energy flow to the surface, making the spots cooler and darker. Velocity fields are tiny in the central (umbra) region of sunspots and barely seen in the outer portion (penumbra).

In addition, the solar surface is covered in weak waves (oscillations in Table 17.1). Since the restoring force is pressure, they are called P-mode oscillations. These waves have a rich spectrum of frequencies (e.g., Leighton 1960, Howard *et al.* 1968, Broomhall *et al.* 2009), but their characteristic period is about five minutes. Indeed, they are sometimes referred to as the five-minute oscillation. They are resonant sound waves, where the resonant cavity is formed by the top and bottom of the convective envelope, and they are driven by the convection. A great deal of information has been gleaned from their study both for the Sun and for many other stars. Some of these results were discussed in Equations 14.7, 14.8, and 15.2 and more are presented later in this chapter.

With the solar example in mind, we proceed to consider those aspects that are relevant to stellar spectra.

Including Velocities in Models

We start with some simple formulations that allow us to put velocity fields into our spectral line computations. The micro–macroturbulence formulation postulates the forms of the velocity distributions, and the observations are used to fix the free parameters of those distributions. The physics of how the velocities are generated is put to the side and the focus is on matching the observed line profiles.

Since rotation is also a major player in line broadening, there is a great deal to be learned when its contribution can be separated from the turbulence broadening. The micro–macroturbulence model allows us to do this and consequently to lay out the general behavior of both rotation and turbulence across the HR diagram.

From Velocity to Spectrum

Any velocity distribution must first be projected onto the line of sight to obtain the distribution of Doppler shifts. The observed Doppler shifts are spatially averaged over the disk of the star and temporally averaged over the time of the observation. As a basic example, consider the model of a non-isotropic statistical distribution of motions *along stellar radii*. At the limb ($\theta = 90°$), no Doppler shifts are introduced, while at the disk center ($\theta = 0°$), the full velocities enter. Suppose we specify the velocity distribution to be Gaussian of dispersion v_0 given by

$$N(v)\, dv = \frac{1}{\sqrt{\pi} v_0} e^{-(v/v_0)^2}\, dv, \qquad (17.1)$$

where $N(v)\, dv$ is the fraction of material having velocities in the range v to and $v + dv$, and v is directed along stellar radii. We assume that $N(v)$ is statistically uniform over the stellar surface and therefore holds for any small portion of the surface.

Next take the projection of all velocities onto the line of sight by writing

$$\Delta\lambda = \frac{\lambda}{c} v \cos\theta; \quad \beta = \frac{\lambda}{c} v_0 \cos\theta \equiv \beta_0 \cos\theta,$$

where, as usual, θ is the angle between the surface normal and our line of sight. When we express $\Delta\lambda$ in velocity units, the λ/c can be omitted from the above equations. Velocity units are natural because we are dealing with velocity fields, and when combining results from lines of different wavelength, no wavelength scaling is needed. Then it follows from Equation 17.1 that the Doppler-shift distribution is

$$N(\Delta\lambda)\,d\Delta\lambda = \frac{1}{\sqrt{\pi}\,\beta}\,e^{-(\Delta\lambda/\beta)^2}\,d\Delta\lambda$$

$$= \frac{1}{\sqrt{\pi}\,\beta_0\,\cos\theta}\,e^{-(\Delta\lambda/\beta_0\,\cos\theta)^2}\,d\Delta\lambda. \qquad (17.2)$$

Notice that β, the width parameter, is a function of θ, while β_0 is a constant. At the center of the disk, $N(\Delta\lambda)$ reflects $N(v)$ directly, but away from the center, the Doppler distribution becomes narrower. At the limb, $N(\Delta\lambda)$ is a δ function. This example illustrates the general way in which $N(\Delta\lambda)$ and $N(v)$ are related. Specifically, it is $N(\Delta\lambda)$, not the original $N(v)$, that shapes the line profiles.

Microturbulence in Line Computations

When the line of sight penetrates through many cells of motion in the photosphere, the velocity distribution of the cells molds the line profile in the same way the particle thermal velocity distribution does. This is the reason for defining microturbulence as having element sizes that are small compared to the mean free path of a photon. In practice, characteristic microturbulence velocities are ~1–2 km/s, small enough compared to the other components of line broadening to make it impossible to measure the *shape* of the distribution. Consequently, it has become standard practice (since Struve and Elvey 1934) to assume the microturbulence velocity distribution is isotropic and Gaussian. The dispersion is ξ. Then the broadening from microturbulence can be rolled into the equation for thermal broadening as we already did in Equation 11.44,

$$\Delta\lambda_D = \frac{\lambda}{c}\left(\frac{2kT}{m} + \xi^2\right)^{1/2}\text{Å}. \qquad (17.3)$$

The dispersion ξ is frequently referred to as *the* microturbulence.

Theoretical weak lines have a Gaussian shape, as we saw in Chapter 11, reflecting the thermal component α_{therm} or $H(u, a)$ in Equation 11.49. Increasing ξ broadens this Gaussian, and since the line is without saturation, this produces a wider, more shallow absorption line still having a Gaussian shape. The equivalent width is approximately conserved, but only because the line is weak. When the line has saturation, increasing ξ widens the wavelength range covered by the absorption and reduces the saturation, thus increasing the total absorption and raising the flat part of the curve of growth, as discussed in Chapter 16.

Large-scale motion, the *macro*turbulence, behaves very differently.

Macroturbulence: the First Signature of Stellar Granulation

When the turbulence cells in the photosphere are large enough so that photons remain in them from the time they are created until they escape from the star, we have the case of macroturbulence. In order to have line broadening rather than radial velocity variations, the individual spectra from many macro-cells must be viewed simultaneously. This is exactly the case for an unresolved stellar disk. Each macro-cell gives a complete spectrum that is displaced by the Doppler shift corresponding to the line-of-sight velocity of the cell. In the solar spectrum, we can resolve these macro-cells. They are the granulation structure. If an image like the one in Figure 17.3 is placed on a long entrance slit of a spectrograph, the spectral lines have a wiggly appearance, as in Figure 17.4. This spectrum is really a composite of many individual spectra stacked one above the other according to the imaging of the macro-cells along the spectrograph slit. The wiggles are the Doppler shifts of the granulation structure. Measurements of these "wiggly lines" lead to interesting results for the solar photosphere (e.g., McMath *et al.* 1956, Ichimoto *et al.* 1989, Nesis *et al.* 1992). The corresponding situation for other stars is to vertically collapse the resolved spectra in Figure 17.4 into one spectrum, so the wiggles turn into broadening. Macroturbulence broadening does not alter the total absorption of the spectral lines. It makes them wider and shallower (Figure 17.2). This is our first signature of stellar granulation.

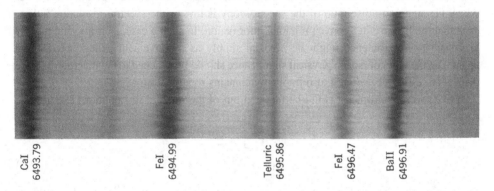

CaI 6493.79   FeI 6494.99   Telluric 6495.86   FeI 6496.47   BaII 6496.91

Figure 17.4 Solar spectra show the individual Doppler shifts of convective cells according to the position of the cells along the entrance slit of the spectrograph (vertically oriented). Notice the telluric line with no wiggles. Taken in the fifth order using the vacuum spectrograph at the McMath-Hulbert Observatory. Courtesy of R.G. Teske and G. Elste.

We can put macroturbulence into our computations by imagining the stellar surface to be subdivided into many smaller areas in which there are numerous macro-cells. We can then specify the macroturbulence velocity distribution for any position on the disk by defining $\Theta(\upsilon)\, d\upsilon$ to be the fraction of the emitted photons in the velocity range υ to $\upsilon + d\upsilon$. In the usual way, the intensity spectrum coming from that position is then given by the convolution

$$I_{\mathrm{v}} = I_{\mathrm{v}}^{0} * \Theta(\lambda)$$

in which I_{v}^{0} is the intensity spectrum without macroturbulence and the velocity distribution $\Theta(\upsilon)$ has been converted to a Doppler-shift distribution $\Theta(\lambda)$ as discussed above. We conserve the equivalent width of the line by normalizing $\Theta(\lambda)$ to unit area.

The flux, being the integral of I_ν over the disk, is then given by

$$\mathfrak{F}_\nu = \oint I_\nu^0 * \Theta(\lambda) \cos \theta \, d\omega, \tag{17.4}$$

which is Equation 5.5 adapted to the present case. The line profiles expressed in I_ν^0 normally include microturbulence broadening. So Equation 17.4 can be used to compute spectral lines that have both small- and large-scale turbulence, and it is our basic starting equation.

(Fictitious) Isotropic Macroturbulence

Isotropic macroturbulence probably does not exist in photospheres of real stars, and certainly does not exist in stars with convective envelopes. Isotropic macroturbulence means that the Doppler-shift distribution at any spot on the stellar surface is independent of the direction from which it is viewed. Obviously granulation, the major component of macroturbulence in cool stars, is not isotropic. Although isotropic macroturbulence is not generally applicable to real stars, we still see it used in the literature. When might this be justified? There are cases where the signal-to-noise ratio of the spectroscopy is low or the spectral resolution is low, such that the line profiles convey minimal information. In such a case, isotropic and usually Gaussian macroturbulence might be invoked simply as a convenient generic broadening agent. But then, of course, no physical significance should be attached to the macroturbulence dispersion, especially when it also implicitly contains rotational broadening.

If $\Theta(\lambda)$ were isotropic, it could be factored out of the integral in Equation 17.4, giving

$$\mathfrak{F}_\nu = \Theta(\lambda) * \oint I_\nu^0 \cos \theta \, d\omega$$
$$= \Theta(\lambda) * \mathfrak{F}_\nu^0. \tag{17.5}$$

In other words, the profile of the star without macroturbulence, $\mathfrak{F}_\nu^0$, would simply be blurred by convolving it with the Doppler-shift distribution of the fictitious macroturbulence. This can be a convenient expedient as long as one does not abuse the situation by attaching physical meaning to the broadening parameter.

In real stars, we need to use *anisotropic* macroturbulence. As such, it cannot be factored out of Equation 17.4 and the simple convolution of Equation 17.5 is then physically wrong. Instead, we must use Equation 17.4 with a full and proper disk integration. One way to approach this is as follows.

Radial–Tangential Anisotropic Macroturbulence

Even a quick glance at the granulation in the solar photosphere tells us that macroturbulence is not isotropic. Convective cells rise and fall, and there are horizontal motions between the rising and falling areas. With this picture in mind, let us postulate motion with velocity vectors restricted to being either along stellar radii or tangent to the surface, but having Gaussian distributions of speeds. This is called the "radial–tangential" model (Gray 1975).

Temperature inhomogeneities are small and for line broadening modeling we ignore them. In which case, the number of photons contributed to the flux is proportional to the emitting area. A certain fraction of the stellar surface area is taken to have radial motion, A_R, and the rest tangential motion, A_T. Any depth variation is ignored. Elaborating on Equation 17.2, the full Doppler-shift distribution for radial–tangential macroturbulence is the weighted sum of the radial distribution, $\Theta_R(\lambda)$, and the tangential distribution, $\Theta_T(\lambda)$,

$$\Theta(\lambda) = A_R \Theta_R(\lambda) + A_T \Theta_T(\lambda)$$

$$= \frac{A_R}{\sqrt{\pi}\, \zeta_R \cos\theta}\, e^{-(\lambda/\zeta_R \cos\theta)^2} + \frac{A_T}{\sqrt{\pi}\, \zeta_T \sin\theta}\, e^{-(\lambda/\zeta_T \sin\theta)^2}. \quad (17.6)$$

Here λ is in velocity units and measured from the center of the distribution. Notice that Θ_R becomes a δ function at the limb, while Θ_T becomes a δ function at disk center. From Equation 17.4 we have the spectrum,

$$\tilde{\mathcal{F}}_\nu = 2\pi\, A_R \int_0^{\pi/2} \Theta_R(\lambda) * I_\nu^0 \, \sin\theta \cos\theta \, d\theta + 2\pi\, A_T \int_0^{\pi/2} \Theta_T(\lambda) * I_\nu^0 \, \sin\theta \cos\theta \, d\theta, \quad (17.7)$$

in which the $*$ stands for convolution. Knowing that the thermal profile is much narrower than the macro-broadening, we keep to the simplest case and take I_ν^0 to be independent of position on the disk. Then the I_ν^0 convolution can be factored out giving

$$\tilde{\mathcal{F}}_\nu = \pi I_\nu^0 * \left[2A_R \int_0^{\pi/2} \Theta_R(\lambda) \sin\theta \cos\theta \, d\theta + 2A_T \int_0^{\pi/2} \Theta_T(\lambda) \sin\theta \cos\theta \, d\theta \right].$$

With $\Theta_R(\lambda)$ and $\Theta_T(\lambda)$ given in Equation 17.6, we are led to evaluate the integrals

$$\frac{1}{\pi^{1/2}\, \zeta_R} \int_0^{\pi/2} e^{-(\lambda/\zeta_R \cos\theta)^2} \sin\theta \, d\theta$$

and

$$\frac{1}{\pi^{1/2}\, \zeta_T} \int_0^{\pi/2} e^{-(\lambda/\zeta_T \sin\theta)^2} \cos\theta \, d\theta.$$

By substituting $u = \zeta_R \cos\theta/\lambda$ in the first integral and $u = \zeta_T \sin\theta/\lambda$ in the second, we obtain

$$\tilde{\mathcal{F}}_\nu = \pi I_\nu^0 * \left[\frac{2A_R \lambda}{\pi^{1/2}\zeta_R^2} \int_0^{\zeta_R/\lambda} e^{-1/u^2} du + \frac{2A_T \lambda}{\pi^{1/2}\, \zeta_T^2} \int_0^{\zeta_T/\lambda} e^{-1/u^2} du \right]$$

$$\equiv \pi I_\nu^0 * M(\lambda) \approx \tilde{\mathcal{F}}_\nu^0 * M(\lambda). \quad (17.8)$$

In other words, the radial and tangential components behave in the same way when integrated across the disk, and the macroturbulence broadening can be expressed as a convolution with the macro-broadening function $M(\lambda)$ given in the square brackets. $M(\lambda)$ is normalized to unit area so the equivalent width of the line is conserved. In most cases, one takes $\zeta_R = \zeta_T \equiv \zeta_{RT}$, and $A_R = A_T$. Figure 17.5 shows the shape of this radial–tangential Doppler-shift distribution. Clearly the disk integration has a very significant effect. The implicit velocity distributions are Gaussian, but the net result after disk integration is a shape with a sharp peak and wide wings. This "cuspy" shape, as we shall see, is the opposite of the shape induced by the rotation of the star, which is flat in the middle and has no wings (see Figures 18.2, 18.5, or 18.6).

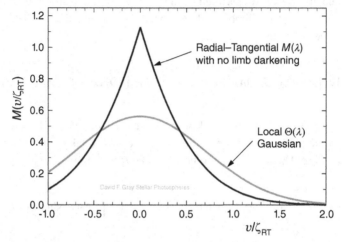

Figure 17.5. Disk integration produces the cuspy shape of the radial–tangential Doppler-shift distribution. The implicit Gaussian distribution is shown for comparison.

Since this $M(\lambda)$ is symmetric, its use is in modeling line broadening and line shapes, but not asymmetries. However, it is easy enough to introduce more elaborate versions of Equation 17.6 with, for example, separate unequal hot and cold radial flows in which the positions of the Gaussians have an offset.

Note that although Equation 17.5 and Equation 17.8 are both convolutions, they differ dramatically in concept. Specifically, we will see that the Doppler shifts of macroturbulence and rotation must be combined point by point across the stellar disk and not with a second convolution.

Including Rotation

In the observed spectra, the Doppler shifts of rotation are combined with those of the photospheric velocity fields. Rotation typically dominates the broadening of lines for hot stars. A few are seen near pole-on, so their rotational broadening is small, and a few others will be truly slow rotators. These can be studied for photospheric velocity fields.

For many cool stars we are in better shape because for them the rotational broadening is comparable to, or even smaller than, the macroturbulence broadening. Even then, we are generally forced to analyze profiles for the combined broadening of rotation and macroturbulence.

To include rotation in our calculations, go back to Equation 17.4 and replace $I_\nu^0(\lambda)$ with its Doppler-shifted form,

$$\mathfrak{F}_\nu = \oint I_\nu^0(\lambda - \Delta\lambda_{\text{rot}}) * \Theta(\lambda) \cos\theta \, d\omega, \tag{17.9}$$

where I_ν^0 is still the specific intensity profile without macroturbulence and $\Delta\lambda_{\text{rot}}$ is the rotational Doppler shift as a function of position on the disk. We shall see in Chapter 18 exactly how to calculate $\Delta\lambda_{\text{rot}}$ and that for rigid rotation $\Delta\lambda_{\text{rot}}$ is constant along lines on the stellar disk that are parallel to the projected rotation axis (Equation 18.1). The convolution places $\Theta(\Delta\lambda)$ at the wavelength position of the $\Delta\lambda_{\text{rot}}$-shifted spectral line, so Equation 17.9 could also be written

$$\mathfrak{F}_\nu = \oint I_\nu^0 * \Theta(\lambda - \Delta\lambda_{\text{rot}}) \cos\theta \, d\omega. \tag{17.10}$$

Numerical disk integrations could be carried out using Equation 17.10 by computing the integrand quantities at numerous points on the disk. But since the thermal profile is significantly narrower than the macroturbulence broadening, its center-to-limb variations can be ignored. Then the I_ν^0 thermal profile can be taken out of the integral and converted to its flux counterpart (Equation 5.7), $\pi I_\nu^0 = \mathfrak{F}_\nu^0$,

$$\mathfrak{F}_\nu = \mathfrak{F}_\nu^0 * \oint \Theta(\lambda - \Delta\lambda_{\text{rot}}) \cos\theta \frac{d\omega}{\pi}$$

$$= \mathfrak{F}_\nu^0 * M(\lambda), \tag{17.11}$$

which is in the form of Equation 17.8, but now oriented toward numerical integration, i.e., $M(\lambda)$ is a full disk integration in which the Doppler shifts of macroturbulence and rotation are combined. Although Equation 17.11 could be used with arbitrary forms for $\Theta(\lambda)$, we are going to use the radial–tangential formulation.

A reminder: the convolution brings $M(\lambda)$ to the position of the line given by $\mathfrak{F}_\nu^0$, the line profile takes on the form of $M(\lambda)$, and since the line is in absorption, $M(\lambda)$ is turned upside down.

One step remains. Limb darkening must still be included.

Limb Darkening

Areas near disk center are brighter than those near the limb (see Figure 9.2), and this needs to be taken into account when calculating the disk-integrated macro-broadening. We only need the relative limb darkening, $L(\theta) = I_c(\theta)/I_c^0$, i.e., the continuum specific intensity at any θ divided by the specific intensity at disk center ($\theta = 0$ or $\cos\theta = 1$), so that we can properly weight the contribution within the integral. Introducing this factor, Equation 17.11 becomes

$$\mathfrak{F}_v = \mathfrak{F}_v^0 * \oint L(\theta)\,\Theta(\lambda - \Delta\lambda_{\rm rot})\cos\theta\,\frac{d\omega}{\pi}$$

$$= \mathfrak{F}_v^0 * M(\lambda), \tag{17.12}$$

where, as before, $\mathfrak{F}_v^0$ is the thermal profile and $\Theta(\lambda - \Delta\lambda_{\rm rot})$ is the macroturbulence Doppler-shift distribution shifted by the rotational Doppler shift, $\Delta\lambda_{\rm rot}$. But now $M(\lambda)$ includes the limb darkening, $L(\theta)$. $M(\lambda)$ is still normalized to unit area. Since the radial distribution contributes most of the broadening near disk center, while the tangential distribution dominates at the limb, limb darkening affects the two components asymmetrically.

We saw early on in Chapter 9 how the temperature distribution is intimately connected to the limb darkening. In the current context, we have a model photosphere that has been chosen to match the star we are analyzing. It is, therefore, straightforward to compute the specific intensity at various limb distances, i.e., $\cos\theta$ values, using Equation 17.10 and make plots like the examples in Figure 17.6. Polynomial fits of fourth order or lower adequately specify $L(\theta)$ and are convenient to use in numerical disk integrations. In some cases a linear approximation is used, $L(\theta) = (1 - \epsilon) + \epsilon\cos\theta$, in which the limb-darkening coefficient, ϵ, has a value of ~0.4–0.6 for visible wavelengths.

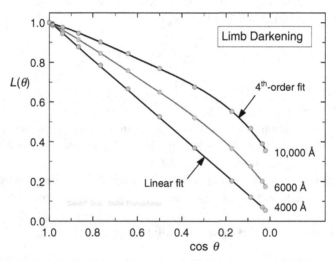

Figure 17.6 Limb darkening curves show the specific intensity normalized by the value at disk center, $L(\theta) = I_c(\theta)/I_c^0$ for a K giant model photosphere. Points are computed values. Lines are least-squares fits. The 4000 Å points are adequately represented by a linear fit. The points for 6000 Å and 10 000 Å are shown with fourth-order polynomial fits.

Preparation for numerical disk integration is now complete.

Disk Integration Mechanics

Numerical disk integration deserves some comment. Assume polar coordinates so that the disk is divided into sectors and annuli, as indicated in Figure 17.7. An increment of disk area is $\Delta A = \rho \, \Delta\phi \, \Delta\rho$, where ρ is the fractional radius from the center of the disk. The numerical step in the two coordinates may have to be quite small in order to properly sample the velocity distributions at each disk location. A variable step in ϕ is advantageous. For example, consider the radial component of macroturbulence, shown Figure 17.7 at three disk locations. Near the limb, the distribution becomes very narrow. Here a fine grid is needed to ensure adequate overlap of the distributions, otherwise $M(\lambda)$ will be "lumpy," leading to numerical errors. Rotation enhances the problem since its Doppler shifts spread out the macroturbulence distributions of the grid areas into a sort of "bed of nails." The "nails" or distributions need to overlap. The cusp of the disk-integrated radial–tangential macroturbulence distribution (Figure 17.5) can serve as a guide here. When it is sharp and well defined, the integration grid is fine enough. Typical disk integrations might have $\sim 10^5$ or more grid points.

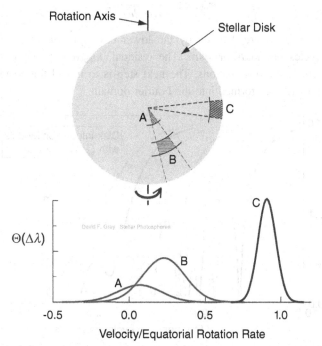

Figure 17.7 The Doppler-shift distributions for radial (only) velocities are shown for three sample positions on the stellar disk. The shift of each individual distribution is caused by the rotation of the star. The dispersion of the distributions decreases toward the limb owing to projection on to the line of sight, as in Equations 17.2 and 17.6.

We now move on to analyzing line profiles for their velocity information.

Fourier Analysis for Turbulence

The *shape* of the line profiles is our central concern at this point. Fourier analysis usually has the advantage over direct profile analysis, although the wavelength domain also has its merits. We assume that the observations are of high signal-to-noise ratio and high resolving power so that the instrumental blurring does not filter away the signals we need. From these good observations, individual line profiles are extracted with the continuum and zero levels firmly established. Implied here is that badly blended lines are disqualified from this kind of analysis and the qualifying profiles have been corrected for minor blends. Expressing the observational data in the usual notation, we write the profile in line-depth form, $D(\lambda) = 1 - F_v/F_c$. The instrumental profile, $I(\lambda)$, enters the observations as a convolution (e.g., Equation 12.2), and so can be divided out (as vectors) in the Fourier domain. Similarly call the line-depth version of the thermal profile $H(\lambda) = 1 - \mathfrak{F}_v^0/\mathfrak{F}_c$, computed from a model photosphere. Thus, in the Fourier domain we get an observed residual transform

$$m_{\text{obs}}(\sigma) = \frac{d(\sigma)}{i(\sigma)\,h(\sigma)}, \tag{17.13}$$

in which the transforms are denoted by the lower case symbols and σ is the Fourier frequency in cycles per km/s or s/km. The residual, $m_{\text{obs}}(\sigma)$, is the transform of the macro-broadeners in the observations. The next step is to model this residual transform using Equation 17.12 transformed into the Fourier domain.

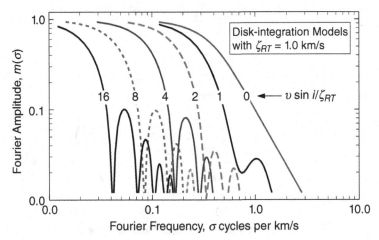

Figure 17.8 Fourier transforms of disk integrations of the combined Doppler shifts of rotation and radial–tangential macroturbulence for several ratios of $v \sin i/\zeta_{RT}$. The 0 curve on the right is for pure radial–tangential macroturbulence, zero rotation. Sidelobe structure develops and strengthens with higher rotation rates. Note the logarithmic coordinates.

So what do these model transforms look like? We had a sample back in Figure 12.16, and Figure 17.8 shows additional curves computed by disk integration to get $M(\lambda)$ and then its transform, $m(\sigma)$. Reminders: the ordinate shows the *absolute values* of the Fourier amplitudes (no logs of negative numbers); since the functions are all symmetric, only the positive frequencies are shown. The two variables are ζ_{RT} and $\upsilon \sin i$. It is their ratio that determines the shape of $m(\sigma)$, and their magnitudes that determine the scale or left–right shift in $\log \sigma$. The transform for radial–tangential macroturbulence (without any rotation) has no sidelobes. Its shape can also be calculated analytically for the case of no limb darkening, as done by Durrant (1979),

$$m(\sigma) = \int_{-\infty}^{\infty} M(\lambda) \cos 2\pi\lambda\sigma \, d\lambda$$

$$= \frac{1 - e^{-(\pi\zeta_{RT}\sigma)^2}}{(\pi\zeta_{RT}\sigma)^2}, \qquad (17.14)$$

in which λ is measured from the center of the line in velocity units.

The curve on the far left of Figure 17.8, labeled 16, is close to the pure rotation case that will be considered in more detail in the next chapter. The main lobe of the curve with $\upsilon \sin i/\zeta_{RT} = 1$ is not far from Gaussian in shape.

The question naturally arises: how do we choose the appropriate model $H(\lambda)$ to go with the observed $D(\lambda)$? First the equivalent width is matched by adjusting the element's abundance in the model. Recall that in the transform domain the equivalent width is the ordinate value at $\sigma = 0$. Second, the value of microturbulence dispersion, ξ, is chosen. For the stronger lines, those that have a sidelobe in their transforms, the position of the first zero can be used for this since it is sensitive to the value of ξ.

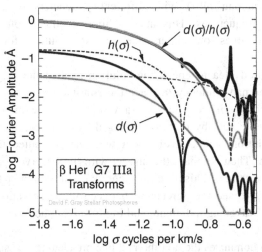

Figure 17.9 Removal of the thermal components, $h(\sigma)$, illustrated for Fe I $\lambda6253$ (stronger line) and V I $\lambda6225$ (weaker line). Lower solid curves: transforms of the observed line profiles; dashed: model thermal profiles, adjusted to match the observed values (off to the left). The first zero of the thermal profile for $\lambda6253$ is matched to that of the observed profile by tuning ξ to 1.6 km/s. The residual transforms agree within errors.

For weaker lines, those without sidelobes, the residual transform, $d(\sigma)/h(\sigma)$, is only weakly dependent on ξ, so any reasonable value can be used. Figure 17.9 illustrates the process for two lines in the G7 giant β Her. The two residual transforms agree, as expected. Near the first zero, $d(\sigma)/h(\sigma)$ goes through zero divided by zero, and consequently the ratio there is noisy.

This same process is repeated for any number of lines, as illustrated in Figures 17.10a and 17.10b. Since the lines span a large range in strength, and therefore depth of formation, agreement of the residual transforms supports the assumption that any depth dependence of macroturbulence can be ignored within the context of line broadening. All recent analyses confirm this result, e.g., Gray (2018a and references therein), including a similar analysis done on the exceptionally good data for the Sun (Gray 2018b).

The final step in getting $m_{\text{obs}}(\sigma)$ is to divide out the transform of the instrumental profile, $i(\sigma)$. Instead of doing this for each line separately, it is convenient to average the $d(\sigma)/h(\sigma)$ ratios for many lines to beat down the noise before dividing by $i(\sigma)$. Figure 17.10b shows this part of the analysis. The model $m(\sigma)$ is established by varying the $\upsilon \sin i$ and ζ_{RT} parameters in disk integrations of $M(\lambda)$ in Equation 17.12 until the shape and position of $m_{\text{obs}}(\sigma)$ are matched. The results for β Her, $\upsilon \sin i = 3.27$ and $\zeta_{RT} = 6.43$, are uncertain by ~0.2 and 0.1 km/s, respectively.

Notice that high-frequency white noise in $d(\sigma)$ is amplified by the division of $i(\sigma)$ and $h(\sigma)$ because the amplitudes of both functions decline with σ, although not necessarily monotonically. Regions containing amplified noise, $\log \sigma \gtrsim -0.9$ in this example, are generally discarded in the analysis. Since the top portions of the transforms are all similar in shape, as can be seen in Figure 10.8, the leverage for distinguishing among $m(\sigma)$ with different parameters requires following the transforms to high frequencies. This implies high spectral resolution and low noise levels. The dashed lines on either side of the solution for β Her emphasize this point.

It is always a good idea to check the solution against the original observations. In fact, there is a human tendency to place the continuum slightly too low (typically $\lesssim 0.5$–1.0%), especially in K-type spectra where there are many faint lines. This problem can be ameliorated by first computing the disk-integrated macro-broadening using an initial solution, and convolving it with the computed thermal profile and instrumental profile. Then compare this model with the observed profile to check and adjust the continuum setting. This iteration technique can tighten up the solution. In any case, it is gratifying to see the agreement between the observations and the Fourier solution back in the wavelength domain, and Figure 17.11 shows such a comparison for this analysis of β Her.

Since granulation dominates the non-thermal line broadening, the size of ζ_{RT} is a direct measure of the range in velocities that exist in the granulation. We have then in ζ_{RT} a quantitative measure of the first signature of stellar granulation.

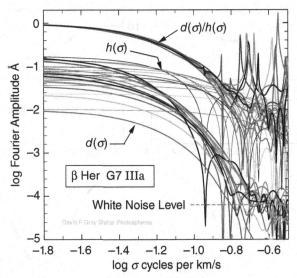

Figure 17.10a The first part of a Fourier analysis for β Her is shown. The transforms of individual lines, $d(\sigma)$, show a wide range in equivalent widths. The residuals, $d(\sigma)/h(\sigma)$, after division by the thermal profiles are tightly grouped. The average white noise level is indicated by the dashed line.

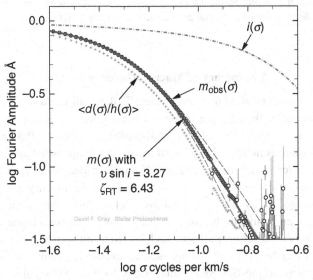

Figure 17.10b The average residual for all the lines in the previous figure, $\langle d(\sigma)/h(\sigma)\rangle$, is shown by the small crosses. The small circles show $m_{\mathrm{obs}}(\sigma)$, the result after division by $i(\sigma)$. A model $m(\sigma)$ with radial–tangential macroturbulence and rotation are matched to these points yielding the broadening parameters $\upsilon \sin i$ and ζ_{RT}. The dashed lines show models having $\upsilon \sin i = 2.0, \zeta_{RT} = 7.0$ and $\upsilon \sin i = 4.0, \zeta_{RT} = 6.0$. Based on Gray (2016).

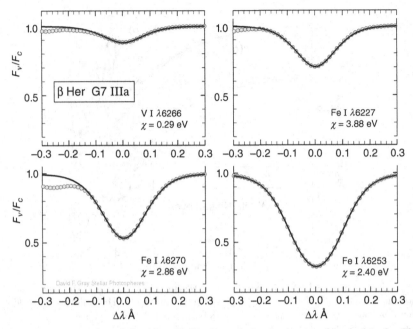

Figure 17.11 Examples of checking the Fourier solution in the wavelength domain. The circles show the observed line profiles of β Her. The solid lines show the model with the disk-integrated macro-broadening ($v \sin i = 3.27, \zeta_{RT} = 6.43$ km/s) convolved with the thermal profiles and the instrumental profile. Based on Gray (2016).

A Summary of Macroturbulence Results

Macroturbulence dispersion varies across the HR diagram as shown in Figure 17.12. Values for stars hotter than F5 are lacking primarily because Doppler broadening of strong rotation overwhelms that of macroturbulence. Errors are ~0.5 km/s. There is a smooth and continuous increase in ζ_{RT} with increasing temperature and with increasing luminosity, i.e., decreasing surface gravity. If one is armed with a star's temperature, macroturbulence dispersion can be used to estimate the luminosity or surface gravity, but the large range in ζ_{RT} for each luminosity class limits the precision. Technical notes: (1) Direct comparisons between ζ_{RT} and the macroturbulence numbers from other formulations of macroturbulence, such as (fictitious) isotropic Gaussian, is not appropriate and can be very misleading. Radial–tangential values are often approximately double the isotropic-Gaussian values (Gray 1978). (2) Convolving the rotational Doppler-shift distribution (to be derived in Chapter 18) with a macroturbulence Doppler-shift distribution, rather than doing proper disk integrations, can lead to spurious results.

Microturbulence ξ-values are generally found to be small for cool dwarfs, 0.0–1.5 km/s, up to as much as 5 km/s in supergiants. These values come either from the saturation portion of the curve of growth or from modeling the shapes of line profiles directly or from

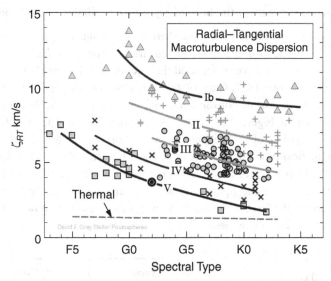

Figure 17.12 On the cool half of the HR diagram, non-thermal velocities become more vigorous toward higher temperatures and higher luminosities. The characteristic thermal dispersion for iron is shown near the bottom for comparison. Different symbols show different luminosity classes; lines show approximate averages for each class as labeled. The solar value is shown by the circumpunct. Data from Gray (1984, 1989b, 2018a, 2018b and references therein), Gray and Nagar (1985), Gray and Toner (1986b, 1987).

the position of the first sidelobe in the Fourier transforms, as discussed above. The main weakness here is with the uncertainty in the temperature distribution (see the discussion with Figure 16.4), which leads to ambiguity with all methods.

A Word about Macroturbulence in Hot Stars

Although rotational broadening dominates the line broadening of most O, B, and A stars, Struve (1952), Huang and Struve (1954), and Slettebak (1956) were quick to realize that another broadening agent was also in play. In particular, as one considers higher luminosity classes, rotation diminishes and a lower bound to the line broadening is revealed, i.e., there appear to be no small $v \sin i$ values. This residual line broadening was attributed to macroturbulence, although it is not likely to be connected to the macroturbulence for cool stars (but see Cantiello and Braithwaite (2019) for indications of shallow convection zones in late B and A stars). Evidence for hot-star macroturbulence has been accumulated by Conti and Ebbets (1977), Howarth *et al.* (1997), Dufton *et al.* (2006), and Simón-Díaz *et al.* (2010), among others, where both direct line-profile analyses and Fourier analyses have been used to separate the macroturbulence component from rotation. These putative macroturbulence dispersions can reach very high values, up to ~50 km/s, well beyond the sound speed. The physical meaning of hot-star macroturbulence is unknown, but oscillations are high on the list of suspects.

Line Asymmetries: the Second Signature of Stellar Granulation

The first signature of granulation, the supra-thermal broadening of spectral lines, is striking, as we have seen. The second signature, asymmetries of spectral lines, is more subtle and not so easy to see, but we have understood their connection to granulation for some time (e.g., Voigt 1956, Schröter 1957, Gray 1980, 1981, 1982, Dravins *et al.* 1981). Bisectors of spectral lines have become the conventional measure of asymmetry. A line profile and its bisector are shown in Figure 17.13 (see also Figure 12.19). The bisector is a locus connecting the midpoints of line segments running horizontally between the sides of the profile. A symmetrical profile has a straight vertical bisector. You have to look closely at the bisector in the left panel of the figure to see that it is not a straight vertical line. This same bisector is shown in the right panel with an expanded wavelength scale. Now the classic "C" shape of the bisector is clearly seen. Spectral resolving power of $\gtrsim 100\,000$ is needed to see asymmetries like these.

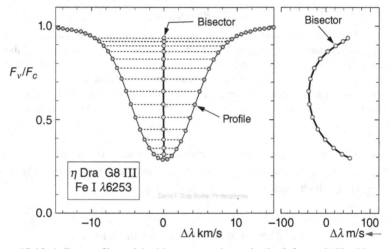

Figure 17.13 A line profile and its bisector are shown in the left panel. The bisector is composed of the midpoints of horizontal line segments running horizontally across the line profile (dashed lines). Interpolation is needed on the right side of the spectral line. The right panel shows the same bisector on a wavelength scale that is expanded about 80 times. Now the shape of the bisector is seen to have the classic C shape associated with granulation. Data from the Elginfield Observatory.

The asymmetry arises because *un*equal numbers of photons are contributed to the flux spectrum by rising and falling material. More light comes from the rising granules, less from the falling lanes in between. Figure 17.14 shows this idea pictorially. The bulk of the profile comes from the granules and is blue shifted; the redshifted portion comes from the inter-granular lanes. Since the redshifted part is weaker, it acts mainly as a perturbation on the profile from the granules by depressing the red wing. Therefore, the bisector of such a profile sweeps toward the red near the continuum, giving the top of the C. This simple picture has no depth dependence, but in real stars, the granulation diminishes with height.

When the core of the line is deep enough, $F_\nu/F_c \lesssim 0.2$, it is formed mainly above the granulation, the blue shift of the granules is gone, and the bottom (line core) portion of the bisector is then near zero velocity, giving the bottom of the C. If the line is weaker than this, the core is formed deeper where the blue shifts are still strong and the bottom of the C is weakened or missing.

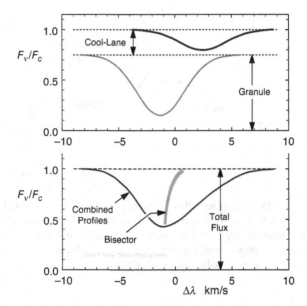

Figure 17.14 A simple two-stream model helps us visualize how granulation produces asymmetries in lines. Averaged over the stellar disk, fall velocities are directed away from us (redshifted), while rise velocities are toward us (blue shifted). The spectrum from the cool falling material alone is shown at the top. Under it is the profile from the hot rising material alone. At the bottom is the sum of the two, a profile that is mostly blue shifted with a depressed red wing.

Asymmetries in solar lines were seen long before those for other stars (e.g., Voigt 1956, 1959, Schröter 1957). With the Sun, we have the luxury of seeing the center-to-limb variation of the bisectors. (Some spatially resolved stellar information is now being obtained using extra-solar planet transits. See Dravins *et al.* 2018.) Although each spectral line is unique, the most common shape of a solar bisector is the somewhat distorted C shape, like those shown in Figure 17.15 for disk center and flux. Reversal in shape near the limb and a steep decline in the blue shift is common (the $\cos\theta = 0.2$ curve). Observations near disk center have the granulation velocities along the line of sight, giving the maximum blue shift. Toward the limb, the projection factor reduces them to near zero. Early investigations, which did not distinguish between granulation blue shift and gravitational redshifts, called this the "limb effect" or the "redshift problem" (e.g., Jewell 1896, Beckers & Nelson 1978). Solar bisector shapes are also influenced by magnetic field (Cavallini *et al.* 1988, Brandt & Solanki 1990). Magnetic fields inhibit the granulation, typically resulting in a smaller bisector velocity span. The effect is also evident in bisector changes during the solar magnetic cycle (Livingston 1982).

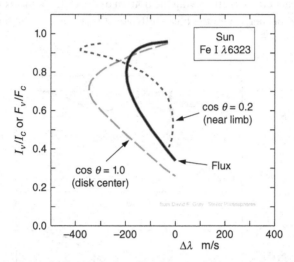

Figure 17.15 Typical solar lines have flux (disk-integrated) bisectors shaped like a distorted C (solid curve) There is a considerable variation with limb distance, arising from a combination of changing depth of formation and geometry of view. The line profiles themselves become shallower and wider toward the limb. Based on Stathopoulou and Alissandrakis (1993) and Löhner-Böttcher *et al.* (2018).

Line Bisector Shapes: Cool Stars

Bisectors for normal cool stars all show some version of the C shape, indicating granulation-type motion. In this section, we discuss only the shapes of line bisectors. It is challenging to determine absolute positions owing to the radial velocity of the whole star (but see Gray 1986), although techniques for doing so are discussed later in the chapter. To first order, the lines in any given star show the same shape bisector, that is, they show that portion of a mean bisector that corresponds to their line depth. This is illustrated in Figure 17.16. So, to improve the signal-to-noise ratio, we sometimes bring together measurements from many lines to construct a mean bisector shape for a star. Other examples are given by Allende Prieto *et al.* (1999), Ramírez *et al.* (2008), and Dravins (2008).

The details of bisectors depend on the wavelength region. For example, limb darkening is greater at shorter wavelengths, proportionally enhancing the contribution from the disk center. Extremes of excitation or ionization also alter the depth of formation of lines and therefore the photospheric depth and the velocities being sampled.

Stellar granulation, while having basic characteristics similar to the Sun's, can display distinct differences. The first granulation signature, macroturbulence, already indicated this, and now we see additional information in the bisectors. In the temperature sequence in Figure 17.17 (lower panel), velocity span drops markedly with declining temperature, mirroring the drop in macroturbulence ζ_{RT} shown in Figure 17.12. In the luminosity

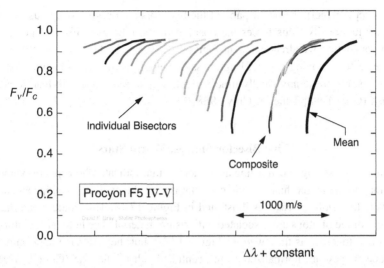

Figure 17.16 The shapes of individual line bisectors are shown on the left. They are arbitrarily positioned in wavelength and ordered approximately by line depth. Although they do not match exactly, the bisectors of weaker lines generally correspond to the upper portion of the bisectors of stronger lines. This is shown in the composite, which is formed by translating each bisector in wavelength. On the far right we see the mean bisector derived from the composite. Based on Gray (1989a).

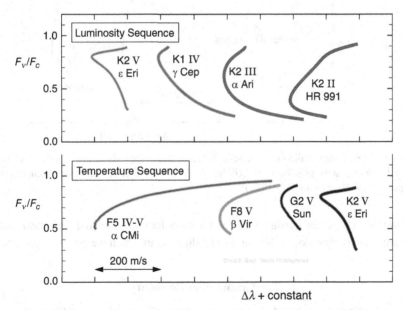

Figure 17.17 This sample of bisectors for $\lambda 6253$ illustrates the systematic changes in shape with luminosity and with temperature on the cool half of the HR diagram. The absolute zero velocities are arbitrary and have no meaning in this diagram. Based on Gray (2005).

sequence (upper panel), the upper parts of the bisectors are progressively pushed red-ward with higher luminosity. This forces the position of the blue-most bisector point to lower F_v/F_c, a connection we already saw in Figures 15.10 and 15.11.

Surface features having distinct brightness or velocity structure can cause distortion of line bisectors, but these are usually time variant, as rotation moves the feature across the stellar disk (Gray 1988, Toner & Gray 1988).

Line Bisector Shapes: Warm Stars

Warm stars show strong spectral-line asymmetries that indicate photospheric velocities an order of magnitude larger than those in cool-star photospheres. But, the slope and curvature are opposite the cool-star case, as illustrated in Figure 17.18. This could mean that only a few percent of the photons are associated with rising material. The hot-star bisectors cannot be telescoped together as they were in Figure 17.16, but they do have approximately the same shape if they are normalized to the central depth of the line (Gray 1989a). Other examples are given by Dravins (1987).

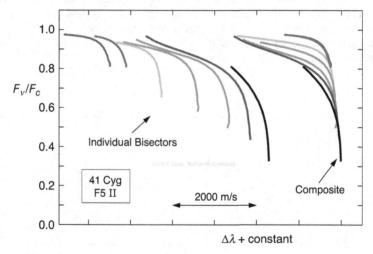

Figure 17.18 Reversed bisectors are seen in warm stars. They do not fit together the way they do for cool stars, as in Figure 17.16. Notice the much larger velocity spans compared to Figure 17.17. Based on Gray (1989a).

The transition from the standard C-shape bisectors for cool stars to the reversed shape for these warm stars occurs along a line in the HR diagram to which we now turn our attention.

The Granulation Boundary

There is a continuous but rapid change in line asymmetry with spectral type through the middle of the HR diagram. The behavior for supergiants is shown in Figure 17.19. Each luminosity class shows a similar change, and the locus of the switch-over from left-leaning (as in Figure 17.18) to right-leaning (as in Figure 17.6) bisectors is dubbed the "granulation boundary," as shown in Figure 17.20. In effect this boundary divides the HR diagram into "warm" stars on the left and "cool" stars on the right. An explanation of this change in bisector behavior, based on tools we have yet to discuss, probably lies mainly with a

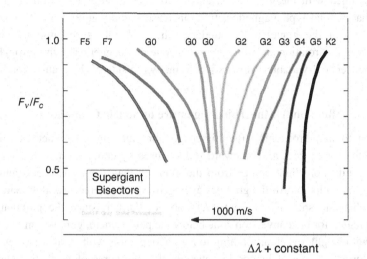

Figure 17.19 Line bisectors show a continuous change in curvature and slope across the central region of the HR diagram. These bisectors are for Ib supergiants. Based on Gray and Toner (1986a).

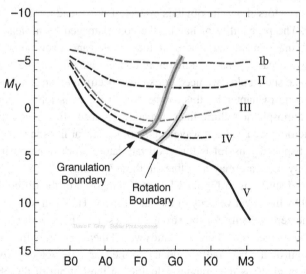

Figure 17.20 The granulation boundary runs through the center of the HR diagram, defined by the transition from normal C-shaped bisectors on the cool side to reversed bisectors on the warm side. The rotation boundary will be discussed in the next chapter, but generally stars on the warm side of it rotate rapidly, while those on the cool side rotate slowly.

steeper photospheric velocity gradient for the warm compared to the cool stars (Gray 2010b). The "rotation boundary" lies to the cool side of the granulation boundary, separating rapid rotators on the left from slow rotators on the right. It is discussed in detail in the next chapter.

Closely allied with the granulation and rotation boundaries is the onset of magnetic activity, that is, solar-type magnetic activity increases rapidly going from the warm side to the cool side of the granulation boundary. This is hardly surprising in view of our understanding that rotation and convection generate the magnetic activity through dynamo action. The Cepheid-variable strip lies just to the warm side of the granulation boundary.

Blue Shifts: the Third Signature of Stellar Granulation

In addition to line broadening (the first signature) and line asymmetries (the second signature) there are systematic blueward shifts caused by granulation. The basic reason is straightforward: more light comes from the granules than from the inter-granular lanes. That is, the rise velocities of the granules averaged over the stellar disk dominate the lines, like the schematic shown in Figure 17.14 above. Further, since the granulation is an overshoot region for the convection zone below the photosphere, granulation rise velocities diminish with height, eventually falling to zero. Couple this with what we know about the depth of formation (e.g., Chapter 13), namely that the cores of weak lines are formed deeper in the photosphere than the cores of strong lines, and we expect to see larger blue shifts for weaker lines. And this is exactly what we see. Figure 17.21 shows line bisectors for lines having a wide range in core depth. In each case, we see the same general pattern: weaker lines, larger blue shift.

Some of the bisectors show the obvious effects of line blending, and this brings out an important point. The probability of having the core damaged by a blend is considerably smaller than having damage anywhere in the profile, and even some badly distorted bisectors still have useable cores. We therefore concentrate on the blue shifts of the cores.

In order to make such a plot, we need our stellar spectrum on an absolute velocity scale. As noted at the end of Chapter 12, this can be done using emission-line lamps, absorption cells, terrestrial atmospheric telluric lines, water-vapor absorption inside the spectrograph, and so on. In addition, we need rest wavelengths of the stellar lines in order to calculate the Doppler shifts. Especially useful is the good laboratory work done on iron (Nave *et al.* 1994), and happily there are may iron lines in these spectra.

So the observed shifts, as in Figure 17.21, are composed of the differential granulation blue shifts offset by the radial velocity of the star in space. How can we separate these? One way is to assume zero velocity for the strongest lines, i.e., those with cores formed at the top of the overshoot layer. Such core velocities would then be largely free of the convective blue shifts and tell us the true radial velocity of the star. The velocity *span* in the third-signature plot then gives us the granule velocities at the bottom of the photosphere. The velocity span for ε Eri, for example, is ~400 m/s, or the granules are rising ~400 m/s deep in the photosphere. By contrast, α Ari has a velocity span about twice as large, implying granulation velocities of ~800 m/s in the deep layers. Naturally, these are disk-averaged

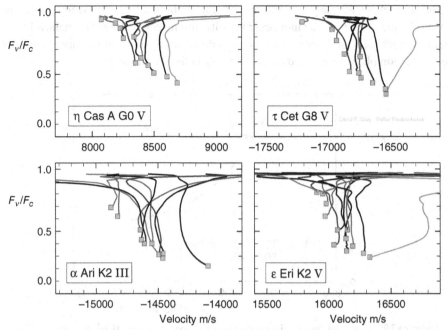

Figure 17.21 Examples of third-signature plots for the λ6250 region. Line bisectors are plotted on an absolute velocity scale. Weaker lines show larger blue shifts. The core points of the bisectors are shown by the squares. The large differences in the abscissae velocity scales reflect the radial velocities of the stars in space. All four plots cover the same velocity span. Adapted from Gray (2009).

values, and the true physical velocities will be larger, but the ratio remains nearly the same. However, there is an even better way to handle things. Nature has given us a handle. It turns out that the shape of the third signature in late F, G, and early K stars from dwarfs through bright giants is the same within observational error (Gray 2009, Gray & Pugh 2012). That is, they differ only by a scale factor (the granulation part) and a shift in the velocity coordinate (the radial velocity part). And, the Sun belongs to this group, so we can use its third-signature plot as a reference, a calibration for other third-signature plots.

The Solar Third Signature

Based on the solar flux observations, especially the atlas references early in Chapter 16, one can study the granulation blue shifts for the Sun (Glebocki & Stawikowski 1971, Dravins *et al.* 1981, Allende Prieto & Garcia Lopez 1998, Gray 2009, Gray & Oostra 2018). The solar plot is given in Figure 17.22. The standard curve shown there is the average solar relation. An important point is that not only do we get the shape of a standard curve from the Sun, we also get the absolute position of the curve (uncertainty $\sim \pm 100$ m/s). Note in passing that the gravitational redshift, $G\mathfrak{M}/(Rc^2) = 636$ m/s, for light leaving the Sun has been added on, and the gravitational blue shift of light coming to the Earth, 3 m/s, has been taken off.

The solar blue shift range is ~500 m/s. Owing to the shallower slope for strong lines, changes in line depth translate into larger velocity differences compared to shallow lines. In granulation terms, it means that the velocities decrease rapidly in the upper photosphere; the granulation phenomenon ends fairly abruptly at the top of the photosphere.

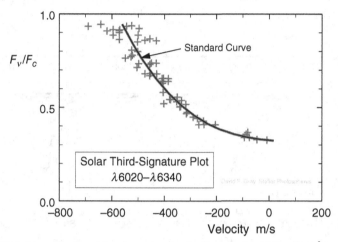

Figure 17.22 The solar flux third-signature plot for lines in the 6020–6340 Å region. The standard curve (line) is the reference to which other stars' plots are compared. Based on Gray and Oostra (2018).

Three other points: all third-signature plots shift slightly if the wavelength region is changed (Dravins *et al.* 1981), owing to the wavelength dependence of the continuous absorption coefficient (see Figure 8.7). In the Paschen continuum, for example, we see deeper at shorter wavelengths into higher velocity depths of the photosphere. Second, granulation blue shifts are diminished in magnetic areas, sometimes down to near zero (Brandt & Steinegger 1990, Meunier *et al.* 2017). Third, the details of these curves depend on the spectral resolution (Gray & Oostra 2018), so there is a risk in mixing data from different sources or with different rotational broadening. The effects are small for resolving power higher than ~10^5 and $v \sin i \lesssim 5\,\mathrm{km/s}$.

Scaling and Shifting: Granulation Velocities and Absolute Radial Velocities

With this background, we can find some interesting results. First, take the solar standard curve, scale it (expand or contract the velocity scale) and shift it in velocity to match a stellar plot, as in Figure 17.23. The scale factor of 1.10 says that the granulation velocities in ε Vir are 10% larger than they are in the Sun. Scaling is more rigorous than simply comparing velocity spans in third-signature plots, as we did earlier, for the obvious reason that lines change strength from one star to the next, or between a star and the Sun. That means they change their position on the third-signature curve. Scaling takes this into account, while simple velocity spans (or the average slope) do not. So the scale factor is an important basic determination of the size of the granulation velocities.

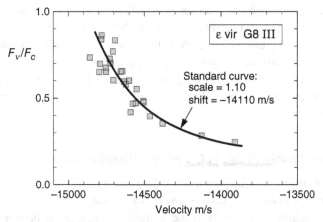

Figure 17.23 The third-signature data for ε Vir (squares) is compared to the scaled and shifted solar standard curve. Based on Gray (2017).

Third-signature scale values for dwarfs show systematic variations with spectral type, as shown in Figure 17.24. In other words, granulation velocities are larger in hotter stars. A similar result was subsequently found by Meunier *et al.* (2017). Giants may show a more complex behavior.

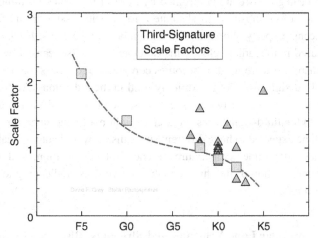

Figure 17.24 Dwarfs (squares) show a clear decline in granulation velocity with spectral type; evolved stars (triangles) less so. Based on Gray (2009 and 2018b and references therein).

Second, the shift in Figure 17.23 tells us the radial velocity of the star on an absolute scale, −14 110 m/s for ε Vir. Absolute scale in this context means that we have the true center-of-mass velocity of the star with full allowance for the granulation blue shifts, but not the gravitational redshift. Ignoring the blue shifts can lead to errors and inconsistencies of hundreds of meters per second, and the error differs from star to star and with the mix of line strengths used in the measurements (see Gray 2009, Figure 13).

By coincidence, the gravitational redshifts for dwarfs are approximately the same size as the convective blue shifts, i.e., they nearly cancel each other. Some empirical calibrations partially compensate for these errors (e.g., Massarotti *et al.* 2008). The gravitational redshift for ε Vir is estimated to be ~120 m/s, using $636\, \mathfrak{M}/R - 3$ m/s, with the mass and radius in solar units.

Amplification of Issues Concerning Precision Radial Velocities

Radial velocities determined by means of the Doppler shift fall into two broad categories: (1) radial-velocity variation and (2) absolute radial velocity.

In the first category, consistency is the byword. If the spectral lines are intrinsically constant, then the onus is on the spectroscopic equipment and technique, and rather small error levels, ~1 m/s, can be reached (e.g., Mayor *et al.* 2003, Fischer *et al.* 2016). But stars can be devious and their spectral lines do vary when the precision is pushed to the limit. For example, the granulation blue shift can vary with the rotational phase of magnetic areas and spots on the stellar surface or during magnetic cycles (e.g., Saar *et al.* 1998, Allende Prieto *et al.* 2013, Bauer *et al.* 2018). And, there is the statistical variation of the granulation, with velocity fluctuations ranging from a few meters per second for dwarfs to a few kilometers per second for supergiants (discussed later in the chapter).

In the second category, we are confronted by most of the issues from the first category plus the challenge of establishing an absolute zero velocity. Aside from the gravitational redshift, this means properly allowing for the granulation blue shifts. If a larger fraction of strong lines is used in the radial-velocity measurement, the offset caused by the blue shifts is less. When many lines are used, it becomes complicated to assess the net effect, unless, of course, the third-signature plot is explicitly used in the reductions. And further, even if the same set of lines is used for two different stars, the net blue shift is not likely to be the same because (1) the third-signature scale factor might not be the same for both stars, and (2) the lines can be expected to have different strengths, as would automatically occur if the stars do not have the same temperature or chemical composition, and then the lines populate a different portion of the third-signature curve. It is challenging to push absolute errors below 100 m/s.

Intergranular Lane Velocities and Strengths: Bisector Mapping

So we have seen that the third signature gives us information on the rise velocities of granulation. What about the fall velocities of the intergranular lanes? The key here is to think about the fact that the depth of formation should be the same for the core point of a weak line and a point in the side of a strong line that has the same F_v/F_c as the core of the weak line. Consequently, if there were no influence from the lane material, the line bisectors would then all have the shape of the third signature. They obviously do not (as in Figure 17.21). The reason is that the lane material is redshifted and depresses the red wing of the profile (Figure 17.14). Therefore, we can work backward and ask what profile alterations are needed to map the observed bisector back onto the third signature. This process is illustrated in Figure 17.25. The right side of the observed profile has been raised by just the right amount to move the bisector onto the standard third-signature curve that has been scaled and shifted appropriately for the star. Although most lines show consistent behavior when their bisectors are mapped (Gray 2010a), here we concentrate on the best-determined line, Fe I λ6253.

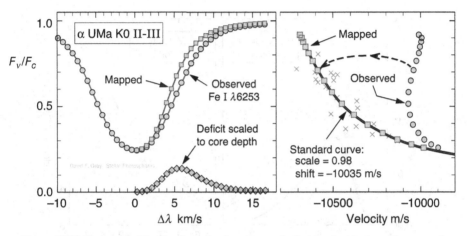

Figure 17.25 Example of the bisector mapping process. Left side: observed profile, mapped profile, and flux deficit. Right side: observed bisector (circles), standard curve (solid line) scaled and shifted to match core points ($\times$), and mapped bisector (squares). Based on Gray (2018b).

The difference between the observed profile and the mapped profile is called the flux deficit. It is a deficit in the sense that if the lanes produced as much light as the granules, the profile would be symmetric, but they produce less. So the flux deficit area is a measure of how much less flux comes from the lanes, and the deficit position is a statement about the velocity difference between lanes and granules (again, integrated over the stellar disk).

Figure 17.26, left panel, compares deficits for dwarfs and giants. Dwarf deficits are smaller and peak at lower velocities. In other words, there is more flux contrast between granules and lanes in giants compared to dwarfs, and the velocity difference between granules and lanes is larger in giants. Direct measurements of solar velocity differences at disk center are ~6 km/s (Oba *et al.* 2017), which implies a disk-averaged value compatible with the dwarf values shown in Figure 17.26. The right panel of the figure shows the rapid increase of the deficit-peak velocity with luminosity; again, stars higher in the HR diagram have a larger velocity difference between lanes and granules.

The flux difference between granules and lanes can be summarized in the deficit area, which is a measure of the disk-integrated brightness contrast between granules and lanes. Deficit area varies systematically with position in the HR diagram, as shown in Figure 17.27. This behavior is similar to what is found in model calculations (e.g., Tremblay *et al.* 2013). Since fundamentally the flux is a combination of area and specific intensity, we can put some constraints on the temperature difference between granules and lanes. For example, studies show that the fraction of the solar surface covered by granules and lanes is comparable (Abdussamatov & Zlatopol'skii 1997, Sánchez Cuberes *et al.* 2000). Assuming equal areas for other stars, granule–lane temperature differences amount to ~100 K for dwarfs and ~200 K for giants.

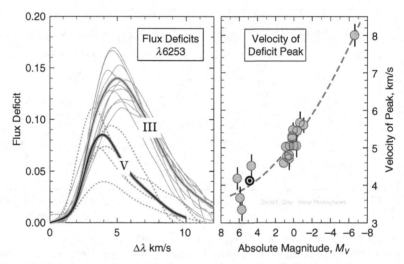

Figure 17.26 Left panel: flux deficits for a few giants (solid thin lines) and dwarfs (dashed thin lines) and characteristic means (thick lines). Right panel: the velocities of the deficits are higher for stars higher in the HR diagram. Based on Gray (2010a, 2010b, and 2017).

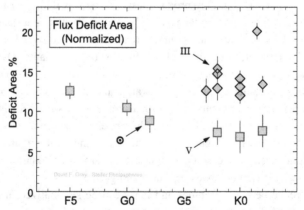

Figure 17.27 Flux deficit area, as a percentage of half the equivalent width of the line, shows systematic differences with temperature and luminosity. The solar point is indicated. Updated from Gray (2010a).

Although bisector mapping is a powerful tool, a modicum of common sense should be used. While the depth-of-formation argument presented at the start of this section is certainly true, it technically applies to the thermal profile, not the macroturbulence-broadened profile that is used in the analysis. Bisector mapping is best viewed as an *empirical* handle on the granulation parameters, but connected a bit indirectly to the real physical parameters.

Explanation of the Granulation Boundary?

Before leaving bisector mapping, it is worth noting that the third-signature plot and bisector mapping suggest an explanation of the granulation boundary. First, stars on the warm side of the granulation boundary have reversed curvature in their third-signature plots (Gray 2010b). This implies that the granulation velocities die off rapidly in the deep photospheric layers and more slowly in higher layers, opposite the cool-star case. It also means that the foundation for warm-star bisectors is this same reversed shape. Second, unlike cooler stars, the flux deficit does not extend into the red wings of the lines, but is localized to the middle portion of the profile. In other words, the redshifted contribution from the lanes occurs midway down the profile instead of near the top, enhancing the reversed shape instead of forming the top of the C. The physical reason for this is that the velocity dispersion of the granulation has increased more than the granule-lane velocity difference compared to cooler stars.

High Luminosity Stars: Rougher Terrain

Some time ago Schwarzschild (1975) predicted larger granule sizes at low surface gravities, and subsequent observations show this to be true. In dwarfs, the numbers of granules is so large that the Doppler-shift distribution they generate is nearly stable in time. In such cases, it takes a concerted effort to see the "granulation noise," ~2 m/s (Table 17.1). With higher luminosity stars, the number of cells is smaller and instabilities in flux (flicker) and in velocity (jitter) become more evident. As we saw in discussing the solar values in Table 17.1, the flux variations go as $\sqrt{N}$, where N is the number of cells on the stellar disk. N is expected to drop more-or-less directly with surface gravity. Going from the main sequence to supergiants drops the surface gravity by about five orders, bringing N to ~10. In such cases, the observed velocity variations are not far from the actual velocities of the individual cells – there is almost no averaging of the velocity distribution with such small statistics. It is not surprising then that supergiants like Betelgeuse (α Ori, M2 Ib) show several km/s velocity variations as large cells rise to the surface (Gray 2008, 2010c, 2013). Figure 17.28 shows an example of the observed profile variations of Betelgeuse. The range in velocity is ~9 km/s, and there are significant differences in line depth (left panel). The giant-cell origin of the shifts is exposed when the individual profiles are all scaled to the same core depth and arbitrarily shifted in wavelength to match (right panel). Then we see that the variation is composed mainly of shifts, with relatively small changes in the profile shape, just what we would expect when individual cells dominate.

Within these individual profiles, there is macroturbulence broadening. Here velocity fields apparently come in two regimes: large and very large, perhaps the supergiant-star version of granulation and giant cells. Less extreme, but similar behavior is shown by α Hya (K3 II-III, Gray 2013). Taking a somewhat different approach, Kiss *et al.* (2006) interpret the observations in terms of long-period oscillations with lifetime of only a few cycles.

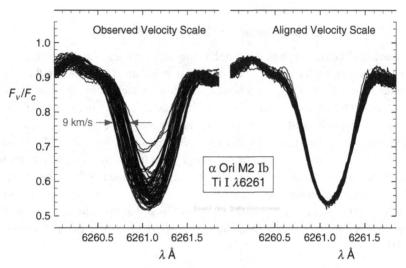

Figure 17.28 An example of velocity "jitter" as seen in Betelgeuse during the 2002–2006 interval. Left panel shows the measured profiles positioned at their observed wavelengths (after correction for barycentric motion). Right panel shows the same profiles scaled to the same core depth and aligned in wavelength. Based on Gray (2008).

Bisectors of supergiants can also show sizeable variations, and third-signature plots can even have variable slope (Gray & Pugh 2012). Photometric and temperature variations can be followed as large cells reach the surface (Percy *et al.* 2003, Lebzelter *et al.* 2005, Gray 2008, Chiavassa *et al.* 2009). Some examples of coordinated temperature and velocity variations are shown for Betelgeuse in Figure 17.29. The emergent cells in such cases are large enough to completely dominate the spectrum. A heat surge accompanies the appearance of the cell. The cell rises, then cools and falls back down. The timescales vary, but are in the range of months. The temperature change indicated by the line-depth ratios is a few hundred degrees, and that is commensurate with the observed brightness variations of ~25% (Krisciunas & Luedeke 1996, Gray 2008). Similar results are inferred from polarimetry (Ariste *et al.* 2018), and interferometer measurements have also shown large hot spots (e.g., Haubois *et al.* 2009, Montargès *et al.* 2016).

In other cases, such as γ Cyg (F8 Ib), ~1–2 km/s excursions in velocity and bisector shape are observed on timescales of ~100 days. These are likely caused by large convection cells, although one cannot rule out pulsation-like motion in which the whole star expands and contracts for a cycle or two but dwindles for lack of resonance (Gray 2010c). Direct

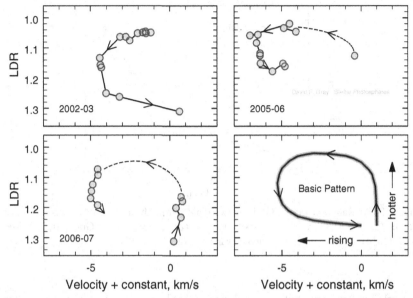

Figure 17.29 Velocity and temperature (LDR: line-depth ratio) are seen to change systematically, indicating the rise and fall of large convection cells. Smaller LDR values indicate higher temperatures. $\Delta T \sim 300 - 400$ K. The velocity zero point is unknown for these data. The arrows show the time sequence. The observing seasons are indicated in each panel. The lower right panel indicates the conceptual pattern. Based on Betelgeuse observations from Gray (2008).

imaging of giant convection cells on the surface of π^1 Gruis, a prototype S star, have been made using an interferometer working in the near infrared (Paladini *et al.* 2018). The star has a radius $R \sim 660\, R_\odot$, $T_{\text{eff}} \sim 3200$ K, and $\log g \sim -0.4$. The cell dimensions are about one-quarter of the stellar diameter and show $\sim 12\%$ brightness contrast.

Large stars with low surface gravity clearly have rougher, more chaotic surfaces. Their large cells, greater velocity differences in the flux deficits, and their fewer cell numbers all conspire to produce velocity and brightness fluctuations, to which we now turn our attention.

Jitter and Flicker

Velocity *jitter*, i.e., random velocity variations, in luminous stars was recognized some years ago (Gunn & Griffin 1979, Pryor *et al.* 1988, Carney *et al.* 2003, Eaton *et al.* 2008). Recent observations have now helped delineate these parameters for lower luminosity stars. Jitter varies mainly with surface gravity, rising from a few meters per second for dwarfs to kilometers per second for supergiants, as shown in Figure 17.30. *Flicker*, or random brightness variations, has a similar gravity dependence, showing a small growth in amplitude from near zero to a few parts per ten thousand in going from $\log g \sim 4.5$ to ~ 3.0 (Bastien *et al.* 2014), and on up to $\sim 10\%$ for supergiants. Flicker may affect Cepheid light curves (Neilson & Ignace 2014).

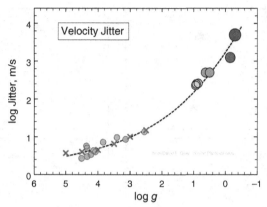

Figure 17.30 Radial-velocity jitter increases rapidly from dwarfs to supergiants. Data from Eaton *et al.* (2008), Gray (2008, 2010c), Ohnaka *et al.* (2013), Pugh & Gray (2013), Bastien *et al.* (2014), Pugh *et al.* (2015), and Tayar *et al.* (2019).

A correlation between photometric flicker and velocity jitter is easy to see in stars like Betelgeuse (Gray 2008), but the situation gets more complicated for dwarfs, where the amplitudes of variation become much smaller. At the dwarf level, additional jitter and flicker from starspots and magnetically active areas come into the picture (Saar *et al.* 1998). All this stellar "noise" is particularly frustrating for extra-solar planet hunters looking for reflex orbital motion or planetary transits. But it is valuable information about stellar photospheres.

The main physical reasons for jitter and flicker lie with the statistical variation of the photospheric velocity fields. As we consider stars with lower gravity, the dispersion of granulation velocities goes up (Figure 17.12), the velocity difference between granules and lanes rises (Figure 17.26), the contrast between granules and lanes increases (Figure 17.27), and the number of velocity cells goes down. In short, the velocity fields become rougher and bigger with decreasing log *g*.

Nature has provided a full spectrum of jitter and flicker variations, so we can learn more by looking at the Fourier spectra of these phenomena.

The Fourier Transform Tool

The composite nature of stellar variations, i.e., different velocity regimes and character-istics, as in Table 17.1, are often explored in the Fourier domain. For example, the Fourier spectrum of velocity variations shown by γ Cyg (F8 Ib) is matched equally well by dispersion and Gaussian shapes with a characteristic timescale of ~20–35 days (Gray 2010c). This is vastly longer than the 10-minute solar granulation timescale or even the 1.7-day timescale of solar supergranulation. An even longer timescale, ~50 days, is found from the Fourier spectrum of the photometric variations of the M2 supergiant Betelgeuse (Gray 2008).

The attainable frequency range and time resolution have been greatly aided by satellite observations that do not suffer the diurnal observing cycle and that can be made continuously for long periods of time. However, interpretation of the Fourier spectra can be somewhat ambiguous, and there has been some debate concerning the form and number of functions to use in fitting the solar and stellar observations (e.g., Mathur *et al.* 2011, Karoff *et al.* 2013, Kallinger *et al.* 2014, Corsaro *et al.* 2015, de Assis Peralta *et al.* 2018). Originally Harvey (1985) used three dispersion profiles and an active-region component that had a steep drop toward high frequencies to describe the solar background. But two components, attributed to granulation and mesogranulation, are often sufficient. For example, Figure 17.31 shows the Fourier transform (amplitude squared or power) of SOHO/Virgo satellite observations of the solar flux (irradiance, disk-integrated sunlight).

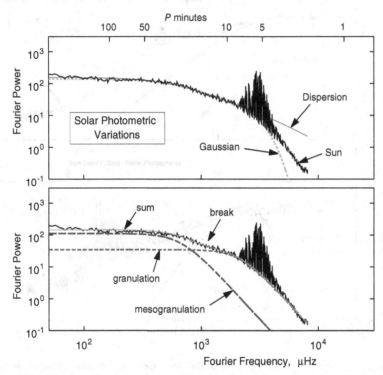

Figure 17.31 The same Fourier spectrum of the solar flux is shown in both panels. Period in minutes is show at top. Upper panel: sum of two Gaussians and of two dispersion profiles, neither of which match the observations. Lower panel: supergranulation and granulation represented by functions that vary inversely as Fourier frequency to the fourth power. Their sum matches the observations quite well, even reproducing the inflection labeled "break." Based on Svensson and Ludwig (2005).

Here we are interested not in the P-mode oscillation peaks centered at ν_{max} (as we were with Figure 14.5), but in the continuous background. The upper panel in Figure 17.31 shows that by themselves dispersion profiles are too high and Gaussian profiles too low at high frequencies compared to the observations. Functions that vary inversely as the fourth power of the frequency match the data, as shown in the lower panel. These functions are centered at zero frequency and have two adjustable parameters, amplitude (in power units) and width. As usual, only the positive-frequency side is shown in these log–log plots.

The slight dip just above 10^3 μHz $\approx 0.5\nu_{max}$ (labeled "break" in the lower panel) is taken to indicate a demarcation between mesogranulation on the left and granulation on the right. The reality of mesogranulation on the Sun has been doubted (e.g., Rieutord & Rincon 2010), but we use it here to give a name to this second granulation-like function. Supergranulation is off to the left in the frequency range ~6 μHz, e.g., see Corsaro *et al.* (2015, Appendix A).

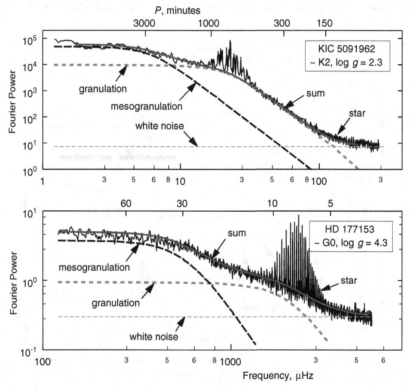

Figure 17.32 Fourier spectra for giant (top) and dwarf (bottom). Periods in minutes are given on the upper axis. The smoothed observations for the stars are shown. Three background components are used: mesogranulation, granulation, and white noise. The sum of the three is shown as the smooth solid line. Note the drastically different range in both axes. The granulation power is ~10 000 times (~100 in Fourier amplitude) stronger in the giant compared to the dwarf and at ~100 times lower frequency. Based on Kallinger *et al.* (2014).

Fourier spectra like these have been determined for a large number of stars using satellite photometry. Amplitudes in the time domain range from ~3 parts per million for stars near the main sequence up to ~1000 parts per million for cool bright giants (Huber *et al.* 2011). Two examples are shown in Figure 17.32. The noisy curves show the smoothed observations; the unsmoothed power values range over about two orders of magnitude. Notice the striking difference in power and frequency range for the two stars that largely results from their difference in surface gravity. From basic physics, we expect larger, less dense stars to show longer periods and lower frequencies, and indeed, both coordinates scale inversely with gravity. In fact, the run of Fourier amplitude (or power) is so steep that it makes a good gravity index, and several papers have addressed this point (e.g., Bugnet *et al.* 2018, Kallinger *et al.* 2019, Chapter 15).

Mesogranulation, granulation, and v_{max} all maintain their relative frequency positions, i.e., they closely track each other in frequency from star to star, as seen in Figure 17.32 (Svensson & Ludwig 2005, Mathur *et al.* 2011, Kallinger *et al.* 2014). But as these frequencies decline, that is, as one moves to stars having lower log *g*, the amplitudes of the mesogranulation and granulation components increase dramatically, corroborating what we already know from the spectroscopy discussed earlier in the chapter and adding many more stars to the picture. But simply scaling things up inversely with gravity does not account for the observed fact that macroturbulence, normally associated with granulation, is contained within the shifted profiles seen in the largest stars (e.g., Figure 17.28). Perhaps the stars are showing us a transition from the dwarf-like case, where granulation dominates and supergranulation has almost no brightness contrast, to the opposite in supergiants, where the supergranulation or giant cells dominate and granulation is nestled within the big cells.

Models of Motions

Models allow us to link together some of the information gleaned from the observations and gain some understanding of the underlying physics. Kinematic models, the first baby steps, are successful in reproducing the gross features of the profiles and bisectors. In these models, the velocity fields are specified in an ad hoc manner and little physics enters. We saw early on in the chapter the success of radial–tangential macroturbulence in reproducing the widths and shapes of line profiles, the first signature of granulation. Somewhat less successful are kinematic models invoked to explain bisector shapes, the second signature of granulation (Gray & Toner 1985, Gray 1986, 1988, Dravins 1990). Far more complex are hydrodynamical models that start with basic physics and numerically calculate the heterogeneous structures that result. Once the physical parameters are in place, the spectral lines are computed and the observable parameters extracted. For examples of hydrodynamical models for the Sun see Gadun *et al.* (1999), Nordlund and Dravins (1990), Asplund *et al.* (2000), as well as the symposium edited by Piskunov *et al.* (2003). You may wish to explore more of the physics of granulation as given by Nordlund *et al.* (2009).

Hydrodynamical models are generally successful in reproducing the widths of line profiles (e.g., Asplund *et al.* 2000 for the Sun, Allende Prieto *et al.* 2013 for Procyon, F5 IV-V), so the first granulation signature is accommodated. Models computed by Allende Prieto *et al.* (2013) also show granulation velocities increasing with effective temperature, in reasonable agreement with the observed behavior of macroturbulence dispersion (Figure 17.12).

Line bisectors for individual lines of sight through model photospheres show a wide range of slopes, curvature, and Doppler shifts (Dravins *et al.* 1981, Asplund *et al.* 2000). An example for the solar disk center is shown in Figure 17.33. It is the averaging of these many individual bisectors that produces the classic C shape. In some respects, these bisector shifts add support to the basic concept of macroturbulence, where complete spectra are formed in individual cells and the distribution of the cell velocities gives the Doppler-shift distribution shaping the observed line profile.

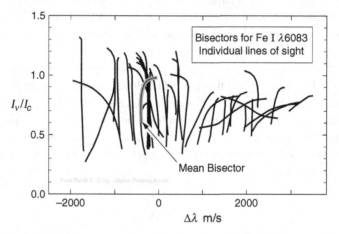

Figure 17.33 Solar line bisectors computed for the center of the disk show a wide range of shapes and shifts, but the average comes back to the standard C shape (or at least the top and middle of the C, as expected for a line of this strength). Granules give the bisectors toward the left. Intergranular lanes give the bisectors toward the right. The excitation potential of the line is 2.22 eV. Adapted from Asplund *et al.* (2000).

In the solar case, we can also compare the absolute bisector shifts as well as the shapes, although even here the velocity zero remains uncertain by ~50–100 m/s (see Molaro & Monai 2012, Reiners *et al.* 2016, Gray & Oostra 2018). Bisectors for three iron lines are shown in Figure 17.34. Owing to a touch of good fortune, the models actually reproduce the observations better than the uncertainties.

For stars other than the Sun, observed bisector shapes are reasonably reproduced by hydrodynamic models. Portrayals similar to Figure 17.33 for a range of model stars are given by Dravins and Nordlund (1990). Figure 17.35 illustrates the successful modeling of bisector shapes by Ramírez *et al.* (2010). They also used the collective shift of 58 iron lines relative to their models to obtain the radial velocity of the star. The bisectors, shapes and relative shifts, for Procyon (F5 IV-V) were modeled by Allende Prieto *et al.* (2013). In this same investigation, granulation blue shifts were found to be in qualitative agreement with observations.

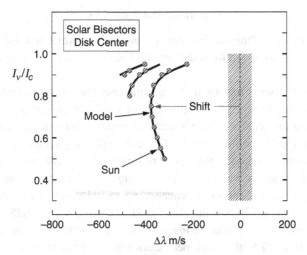

Figure 17.34 Hydrodynamical models reproduce the shapes and absolute shifts of the solar iron line bisectors (λ6804.30, λ6271.28, and λ6240.65 left to right). The shaded section indicates the uncertainty in the zero velocity position. Based on Asplund *et al.* (2000). See also Gadun *et al.* (1999).

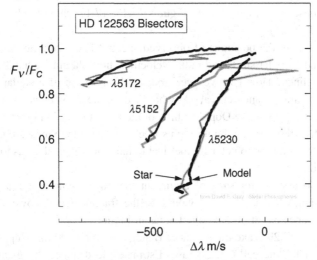

Figure 17.35 Observed iron line-bisector *shapes* are matched by hydrodynamical models, but here the observed bisectors have been shifted to match the positions of the models. Based on Ramírez *et al.* (2010).

The models of Trampedach *et al.* (2013) and Tremblay *et al.* (2013) agree with the observed patterns of larger granulation being rougher and bigger in higher luminosity stars. Models by Samadi *et al.* (2013) are successful in reproducing the observed scaling relations from Fourier spectra.

In the next chapter we delve into the second major shaper of lines: rotational broadening.

Questions and Exercises

1. What do we mean by "thermal broadening" and "non-thermal broadening"?
2. Why are the equivalent widths of the spectral lines conserved when they are broadened by macroturbulence?
3. Sketch a typical Fourier transform of a spectral line. Use log–log coordinates. On your diagram, illustrate how (a) noise, (b) spectral resolving power, and (c) the Nyquist frequency limit the acquisition of physical information from the spectral line.
4. Extract the Fe I $\lambda6227$ profile from the Hinkle *et al.* (2000) solar atlas (or any line from a convenient source). Take the Fourier transform of the profile and plot it on log–log coordinates. Now go back and raise the continuum by 1%, renormalize the profile, and recompute its transform. What differences do you see? How accurately do you think you can set the continuum? Repeat the exercise for the stronger Fe I $\lambda6253$ line.
5. The velocity span of the bisector in Figure 17.13 is about 80 m/s. If the spectrograph has a dispersion at 6250 Å of 13 mÅ per detector pixel, what fraction of a pixel is this velocity span? What linear dimension does the velocity span correspond to if the pixel is 15 μm across? Would you agree that the spectrograph must be well built, have small optical aberrations, and be extremely stable in order to do these kind of measurements? Support your conclusion.
6. Why does the conceptual bisector in Figure 17.14 show only the top of the C shape, while most observed bisectors, e.g., Figures 17.13, 17.17, show the bottom portion of the C as well?
7. When we compare the positions of observed spectral lines with laboratory rest wavelengths to find Doppler shifts and radial velocities, how should we handle asymmetries in the stellar lines? How can we even specify the position of a stellar line when its bisector is not straight and vertical? At what level of precision does this problem arise? How can we disentangle the Doppler shifts of the bisectors, as in Figure 17.21, from the Doppler shift of the star moving through space? Is it feasible to measure the change in a star's distance with time to get a geometrical radial velocity, analogous to measurement of proper motions?
8. If we were to assume for sake of argument that there is no temperature difference between granules and lanes, what would be the fractional area coverage of granules compared to lanes based on the deficit areas in Figure 17.27?
9. Refer to Figure 17.29. Take the radius of Betelgeuse to be 800 $R_\odot$. Suppose a duration of one month for the cell-rising phase. Estimate the distance the giant cell rises in kilometers, in solar radii, and as a fraction of the star's radius. What do you make of this rise distance?

References

Abdussamatov, H.I. & Zlatopol'skii, A.G. 1997. *Ast Lett* **23**, 752.
Abramenko, V.I., Yurchyshyn, V.B., Goode, P.R., Kitiashvili, I.N., & Kosovichev, A.G. 2012. *ApJL* **756**, L27.

Allende Prieto, C. & Garcia Lopez, R.J. 1998. *A&AS* **129**, 49.

Allende Prieto, C., Garcia Lopez, R.J., Lambert, D.L., & Gustafsson, B. 1999. *ApJ* **526**, 991.

Allende Prieto, C., Koesterke, L., Ludwig, H.-G., Freytag, B., & Caffau, E. 2013. *A&A* **550**, 103.

Ariste, A.L., Mathias, P., Tessore, B., *et al.* 2018. *A&A* **620**, 199.

Asplund, M., Nordlund, Å., Trampedach, R., Allende Prieto, C., & Stein, R.F. 2000. *A&A* **359**, 729.

Bastien, F.A., Stassun, K.G., Pepper, J., *et al.* 2014. *AJ* **147**, 29.

Bauer, F.F., Reiners, A., Beeck, B., & Jeffers, S.V. 2018. *A&A* **610**, 52.

Beckers, J.M. 1981. *The Sun as a Star* (Washington, DC: CNRS/NASA), NASA SP-450, S. Jordan, ed., p. 11.

Beckers, J.M. & Nelson, G.D. 1978. *Solar Phys* **58**, 243.

Brandt, P.N. & Solanki, S.K. 1990. *A&A* **231**, 221.

Brandt, P.N. & Steinegger, M. 1990. *Solar Photosphere: Structure, Convection, and Magnetic Fields* (Dordrecht: Springer), IAUS, Vol. 138, J.O. Stenflo, ed., p. 41.

Bray, R.J., Loughhead, R.E., & Durrant, C.J. 1984. *The Solar Granulation* (Cambridge: Cambridge University Press), 2nd ed.

Broomhall, A.-M., Chaplin, W.J., Davies, G.R., *et al.* 2009. *MNRAS* **396**, 100.

Bugnet, L., García, R.A., Davies, G.R., *et al.* 2018. *A&A* **620**, 38.

Cantiello, M. & Braithwaite, J. 2019. *ApJ* **883**, 106.

Carney, B.W., Latham, D.W., Stefanik, R.P., Laird, J.B., & Morse, J.A. 2003. *AJ* **125**, 293.

Cavallini, F., Ceppatelli, G., & Righini, A. 1988. *A&A* **205**, 278.

Chiavassa, A., Plez, B., Josselin, E., & Freytag, B. 2009. *A&A* **506**, 1351.

Conti, P.S. & Ebbets, D. 1977. *ApJ* **213**, 438.

Corsaro, E., De Ridder, J., & García, R.A. 2015. *A&A* **579**, 83.

Dawes, W.R. 1864. *MNRAS* **24**, 161.

de Assis Peralta, R., Samadi, R., & E. Michel, E. 2018. *Ast Nach* **339**, 134.

Dravins, D. 1987. *A&A* **172**, 211.

Dravins, D. 1990. *A&A* **228**, 218.

Dravins, D. 2008. *A&A* **492**, 199.

Dravins, D. & Å. Nordlund 1990. *A&A* **228**, 184.

Dravins, D., Lindegren, L., & Nordlund, Å. 1981. *A&A* **96**, 345.

Dravins, D., Gustavsson, M., & Ludwig, H.-G. 2018. *A&A* **616**, 144.

Dufton, P.L., Ryans, R.S.I., Simón-Díaz, S., Trundle, C., & Lennon, D.J. 2006. *A&A* **451**, 603.

Durrant, C.J. 1979. *A&A* **76**, 208.

Eaton, J.A., Henry, G.W., & Odell, A.P. 2008. *ApJ* **679**, 1490.

Evans, J.W. & Michard, R. 1962. *ApJ* **136**, 493.

Fischer *et al.* 2016. *PASP* **128**, 066001.

Gadun, A.S., Solanki, S.K., & Johannesson, A. 1999. *A&A* **350**, 1018.

Glebocki, R. & Stawikowski, A. 1971. *Acta Ast* **21**, 185.

Gray, D.F. 1975. *ApJ* **202**, 148.

Gray, D.F. 1978. *Solar Phys* **59**, 193.

Gray, D.F. 1980. *ApJ* **235**, 508.

Gray, D.F. 1981. *ApJ* **251**, 583.

Gray, D.F. 1982. *ApJ* **255**, 200.

Gray, D.F. 1984. *ApJ* **281**, 719.

Gray, D.F. 1986. *PASP* **98**, 319.

Gray, D.F. 1988. *Lectures on Spectral-Line Analysis: F, G, and K Stars* (Arva, Ontario: The Publisher).
Gray, D.F. 1989a. *PASP* **101**, 832.
Gray, D.F. 1989b. *ApJ* **347**, 1021.
Gray, D.F. 2005. *PASP* **117**, 711.
Gray, D.F. 2008. *AJ* **135**, 1450.
Gray, D.F. 2009. *ApJ* **697**, 1032.
Gray, D.F. 2010a. *ApJ* **710**, 1003.
Gray, D.F. 2010b. *ApJ* **721**, 670.
Gray, D.F. 2010c. *AJ* **140**, 1329.
Gray, D.F. 2013. *ApJ* **764**, 100.
Gray, D.F. 2016. *ApJ* **832**, 68.
Gray, D.F. 2017. *ApJ* **845**, 62.
Gray, D.F. 2018a. *ApJ* **857**, 139.
Gray, D.F. 2018b. *ApJ* **869**, 81.
Gray, D.F. & Nagar, P. 1985. *ApJ* **298**, 756.
Gray, D.F. & Oostra, B. 2018. *ApJ* **852**, 42.
Gray, D.F. & Pugh, T. 2012. *ApJ* **143**, 92.
Gray, D.F. & Toner, C.G. 1985. *PASP* **97**, 543.
Gray, D.F. & Toner, C.G. 1986a. *PASP* **98**, 499.
Gray, D.F. & Toner, C.G. 1986b. *ApJ* **310**, 277.
Gray, D.F. & Toner, C.G. 1987. *ApJ* **322**, 360.
Gunn, J.E. & Griffin, R.F. 1979. *AJ* **84**, 752.
Harvey, J. 1985. *ESASP* **235**, 199.
Hathaway, D.H., Beck, J.G., Bogart, R.S., *et al.* 2000. *Solar Phys* **193**, 299.
Haubois, X., Perrin, G., Lacour, S., *et al.* 2009. *A&A* **508**, 923.
Hinkle, K., Wallace, Ll., Valenti, J., & Harmer, D. 2000. *Visible and Near Infrared Atlas of the Arcturus Spectrum, 3727–9300 Å* (San Francisco: Astronomical Society of the Pacific).
Howard, R., Tanenbaum, A.S., & Wilcox, J.M. 1968. *Solar Phys* **4**, 286.
Howarth, I.D., Siebert, K.W., Hussain, G.A.J., & Prinja, R.K. 1997. *MNRAS* **284**, 265.
Huang, S.-S. & Struve, O. 1954. *AnAp* **17**, 85.
Huber, D., *et al.* 2011. *ApJ* **743**, 143.
Ichimoto, K., Hiei, E., & Nakagomi, Y. 1989. *PAS Japan* **41**, 333.
Jewell, L.E. 1896. *ApJ* **3**, 89.
Kallinger, T., De Ridder, J., Hekker, S., *et al.* 2014. *A&A* **570**, 41.
Kallinger, T., Beck, P.G., Hekker, S., *et al.* 2019. *A&A* **634**, 25.
Karoff, C., Campante, T.L., Ballot, J., *et al.* 2013. *ApJ* **767**, 34.
Keil, S.L., ed. 1984. *Small Scale Dynamical Processes in Quiet Stellar Atmospheres* (Sacramento: Sacramento Peak Observatory).
Kiss, L.L., Szabó, Gy.M., & Bedding, T.R. 2006. *MNRAS* **372**, 1721.
Krisciunas, K. & Luedeke, K. 1996. *IBVS* 4355.
Lebzelter, T., Griffin, R.F., & Hinkle, K.H. 2005. *A&A* **440**, 295.
Leighton, R.B. 1960. *IAUS* **12**, 321.
Livingston, W.C. 1982. *Nature* **297**, 208.
Löhner-Böttcher, J., Schmidt, W., Stief, F., Steinmetz, T., & Holzwarth, R. 2018. *A&A* **611**, 4.
Massarotti, A., Latham, D.W., Stefanik, R.P., & Fogel, J. 2008. *AJ* **135**, 209.
Mathur, S., Hekker, S., Trampedach, R., *et al.* 2011. *ApJ* **741**, 119.

Mayor, M., *et al.* 2003. *The Messenger*, **114**, 20.

McMath, R.R., Mohler, O.C., Pierce, A.K., & Goldberg, L. 1956. *ApJ* **124**, 1.

Meunier, N., Roudier, T., & Rieutord, M. 2008. *A&A* **488**, 1109.

Meunier, N., Mignon, L., & Lagrange, A.-M. 2017. *A&A* **607**, 124.

Molaro, P. & Monai, S. 2012. *A&A* **544**, 125.

Montargès, M., Kervella, P., Perrin, G., *et al.* 2016. *A&A* **588**, 130.

Nave, G., Johansson, S., Learner, R.C.M., Thorne, A.P., & Brault, J.W. 1994. *ApJS* **94**, 221.

Neilson, H.R. & Ignace, R. 2014. *A&A* **563**, 4.

Nesis, A., Hanslmeier, A., Hammer, R., *et al.* 1992. *A&A* **253**, 561.

Nordlund, Å. & Dravins, D. 1990. *A&A* **228**, 155.

Nordlund, Å., Stein, R.F., & Asplund, M. 2009. *Living Rev Solar Phys* **6**, 2.

Oba, T., Riethmüller, T.L., Solanki, S.K., *et al.* 2017. *ApJ* **849**, 7.

Ohnaka, K., Hofmann, K.-H., Schertl, D., *et al.* 2013. *A&A* **555**, 24.

Paladini, C., Baron, F., Jorissen, A., *et al.* 2018. *Nature* **553**, 310.

Parker, E.N. 1982a. *ApJ* **256**, 292.

Parker, E.N. 1982b. *ApJ* **256**, 302.

Percy, J.R., Besla, G., Velocci, V., & Henry, G.W. 2003. *PASP* **115**, 479.

Piskunov, N., Weiss, W.W., & Gray, D.F. (eds.) 2003. *Modelling of Stellar Atmospheres*, IAU Symposium 210 (San Francisco: Astronomical Society of the Pacific).

Pryor, C.P., Latham, D.W., & Hazen, M.L. 1988. *AJ* **96**, 123.

Pugh, T. & Gray, D.F. 2013. *ApJ* **777**, 10.

Pugh, T., Gray, D.F., & Griffin, R.F. 2015. *MNRAS* **454**, 2344.

Ramírez, I., Allende Prieto, C., & Lambert, D.L. 2008. *A&A* **492**, 841.

Ramírez, I., Collet, R., Lambert, D.L., Allende Prieto, C., & Asplund, M. 2010. *ApJL* **725**, L223.

Reiners, A., Mrotzek, N., Lemke, U., Hinrichs, J., & Reinsch, K. 2016. *A&A* **587**, 65.

Rieutord, M. & Rincon, F. 2010. *Living Rev Solar Phys* **7**, 2.

Rosseland, S. 1928. *MNRAS* **89**, 49.

Roudier, Th., Švanda, M., Rieutord, M., *et al.* 2014. *A&A* **567**, 138.

Roudier, Th., Švanda, M., Ballot, J., Malherbe, J.M., & Rieutord, M. 2018. *A&A* **611**, 92.

Rutten, R.J. & Severino, G. (eds.) 1989. *Solar and Stellar Granulation* (Dordrecht: Kluwer), NATO ASI Series, Vol. 263.

Saar, S.H., Butler, R.P., & Marcy, G.W. 1998. *ApJ* **498**, L153.

Samadi, R., Belkace, K., Ludwig, H.-G., *et al.* 2013. *A&A* **559**, 40.

Sánchez Cuberes, M., Bonet, J.A., & Vázquez, M. 2000. *ApJ* **538**, 940.

Schrijver, C.J. & Hagenaar, H.J. 1997. *ApJ* **475**, 328.

Schröter, E.H. 1957. *ZfAp* **41**, 141.

Schwarzschild, M. 1975. *ApJ* **195**, 137.

Simón-Díaz, S., Herrero, A., Uytterhoeven, K., *et al.* 2010. *ApJL* **720**, L174.

Slettebak, A. 1956. *ApJ* **124**, 173.

Spruit, H.C., Nordlund, Å., & Title, A.M. 1990. *ARAA* **28**, 263.

Stathopoulou, M. & Alissandrakis, C.E. 1993. *A&A* **274**, 555.

Struve, O. 1952. *PASP* **64**, 117.

Struve, O. & Elvey, C.T. 1934. *ApJ* **79**, 409.

Svensson, F. & Ludwig, H.G. 2005. *Proceedings of the 13th Cool Stars Workshop* (ESA SP-560), F. Favata, G. Hussain & B. Battrick, eds.

Tayar, J., Stassun, K.G., & Corsaro, E. 2019. *ApJ* **883**, 195.

Toner, C.G. & Gray, D.F. 1988. *ApJ* **334**, 1008.

Trampedach, R., Asplund, M., Collet, R., Nordlund, Å., & Robert F. Stein, R.F. 2013. *ApJ* **769**, 18.

Tremblay, P.E., Ludwig, H.-G., Freytag, B., Steffen, M., & Caffau, E. 2013. *A&A* **557**, 7.

Voigt, H.H. 1956. *ZfAp* **40**, 157.

Voigt, H.H. 1959. *ZfAp* **47**, 144.

18

Stellar Rotation

Stars rotate. Even the slightest angular momentum in interstellar clouds is magnified by many orders of magnitude with contraction down to stellar dimensions. Stellar rotation is a driving force for diverse phenomena in stellar photospheres including circulation currents, mass loss, and magnetic field generation along with its offshoots of starspots, flares, chromospheres and coronae, and activity cycles. The Sun's rotation, first measured in the early 1600s by Galileo, is very slow, amounting to only 2 km/s at the equator. Such slow rotation rates are typical of cool stars. The Sun also shows differential rotation, with the equator having an angular velocity about 25% larger than at higher latitudes of ~75°.

The effects of rotation on the continuous spectrum are small except when rotation is very near the break-up rate, where centrifugal force balances the surface gravity. The spectral lines, on the other hand, are strongly changed by the Doppler shifts of the light coming from different parts of the stellar disk. The approaching limb corresponds to the short-wavelength portion of a rotation-broadened line profile, the receding limb to the long-wavelength portion, a concept first discussed in 1877 by Abney. In fact, the line profile becomes a one-dimensional map of the stellar surface through the one-to-one coupling of position on the stellar disk and Doppler shift within the profile. Actually, this was the basis for an early proof that stars rotate – eclipsing binaries were expected to have radial velocities go through zero during eclipse when the orbital motion is across the line of sight, but some showed a steep rise as the eclipse started, and the inverse behavior as the eclipse ended (McLaughlin 1924, Rossiter 1924, Shajn and Struve 1929, Albrecht *et al.* 2011). Figure 18.1 illustrates the effect. As the eclipse progresses, the observable portion of the star being eclipsed has a net redshift. The remainder of the line is missing because it comes from the hidden, blue-shifted portion of the partially eclipsed star. Upon egress, the mirror image of the process plays out and the redshifted portion of the profile is masked by the eclipse. The resolution of the early spectrograms was too low for changes in line shape to be discerned, so the lines appeared to show a shift in position that the observers translated into radial velocity.

Naturally the rotational line broadening depends on the orientation of the star's rotation axis to the line of sight. There are no Doppler shifts from rotation for a star viewed pole-on. It is the rotation velocity projected onto our line of sight that broadens the spectral lines, which means that our spectroscopic analyses yield the actual rotation rate times the sine of the angle between the rotation axis and our line of sight, i.e., $v \sin i$. In most cases $\sin i$ remains unknown.

Rotation can also be detected photometrically if the star has a long-lived spot on its surface. Each time rotation brings the spot across the disk, the total brightness varies so the period can be determined. Inclination of the rotational axis is also important here since stars seen near pole-on will show no modulation.

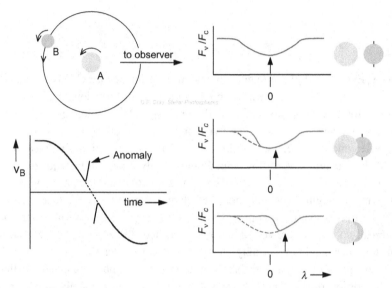

Figure 18.1 The radial-velocity anomaly occurs when low spectral resolution is used to measure the radial velocity of rotating eclipsing stars. The spectral lines of the B component disappear starting on the short-wavelength side as the disk of the star is covered by the A component during ingress into the eclipse. This gives the impression of a line shift. The A component generally has a different temperature and does not have this line in its spectrum. The effect has also proved useful with exoplanet transits (e.g., Bayliss *et al.* 2010).

In this chapter, we concern ourselves primarily with the spectroscopic effects of stellar rotation, including the line broadening caused by rotation, techniques for extracting the rotation rate from the broadening, and some of the results. Many of the analysis tools are the same as those discussed in Chapter 17. As we did in Chapter 17, let us first take a quick look at what we are dealing with.

Examples of Rotational Broadening

Rotationally broadened profiles are shown in Figure 18.2. Rotational broadening can be larger than, comparable to, or smaller than macroturbulent broadening. Rotational broadening can also be much larger than in these examples, reaching up to hundreds of kilometers per second. When the rotation is larger, the lines are wider and shallower, i.e., the equivalent widths are conserved, just as they are with macroturbulence, and for the same reason: in any local area of the stellar surface, the complete spectrum is formed and then it is Doppler shifted by the rotation. The light from the receding limb of the star is the most redshifted and it is found at the extreme right of these broadened profiles. And vice versa for the approaching limb. So in principle it is straightforward to get the projected rotation rate from the profile. We just measure the width from the center to the edge, although as you can see from Figure 18.2, this might not be so easy owing to the smooth transition from the profile to the continuum. You likely also noticed the basin-like shape of the profiles, especially the flat core sections. This is the characteristic shape of the Doppler-shift distribution for rotation, and we now investigate this shape in detail.

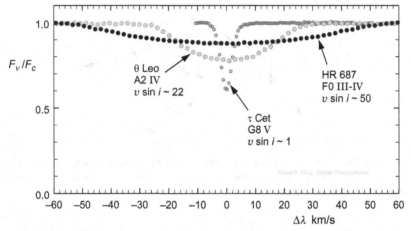

Figure 18.2 Rotationally broadened profiles for θ Leo ($\lambda4559$ Cr II) and HR 687 ($\lambda6142$ Ba II) compared to the much narrower profile of τ Cet ($\lambda6241$ Fe I).

The Doppler-Shift Distribution for Rotation

Just as the Doppler-shift distribution of macroturbulence shapes spectral lines, as we saw in Chapter 17, so does the Doppler-shift distribution of rotation. We start out by ignoring macroturbulence and investigate the Doppler-shift distribution of rotation on its own. Eventually we will combine the two distributions using disk integrations.

The first step is to find the Doppler shift as a function of position on a stellar disk. Assume that the star is spherical and rotates as a rigid body with angular velocity Ω. In Figures 18.3 and 18.4, the $x-y$ plane is the plane of the sky and the z axis is toward us. The y axis has been chosen to align with the rotation axis. If you prefer the straightforward geometry approach, look at Figure 18.3. If instead you like the more comprehensive vector approach, skip to the next paragraph. In Figure 18.3, we take the star to be viewed equator-on, with the rotation axis perpendicular to our line of sight. But it is clear that if the rotation axis is inclined to the line of sight (in the $y-z$ plane) by the inclination angle i, then the Doppler shifts are reduced by the projection factor $\sin i$. The projection of the stellar radius, R, onto the $x-z$ plane is $\rho = R \cos \alpha$, and it makes an angle ϕ with the x axis. The x component of R

is therefore $R \cos \alpha \cos \phi$. The velocity of the surface at the end of R is $v = \Omega \rho$. The component of v along the line of sight is then $v_z = v \cos \phi = \Omega \rho \cos \phi = \Omega R \cos \alpha \cos \phi = \Omega x$, or if the inclination is included, $v_z = x \Omega \sin i$. This is the result we need.

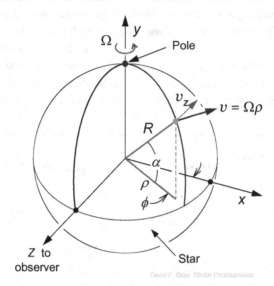

Figure 18.3 The surface point at the end of the radius R has a linear velocity $v = \Omega \rho$, where Ω is the angular velocity. The projection of R onto the x–z plane is $\rho = R \cos \alpha$ and onto the x axis is $x = \rho \cos \phi = R \cos \alpha \cos \phi$.

With the vector approach, we use the arrangement shown in Figure 18.4, where now the inclination, i, is shown explicitly, and Ω and R are vectors. The linear velocity of any point on the surface of the star is the vector product $v = \Omega \times R$. The Doppler shift arises from the z component of the vector product, namely, $v_z = y\Omega_x - x\Omega_y$, but there is no x component of Ω by choice of the coordinate arrangement, so $\Omega_x = 0$. Since $\Omega_y = \Omega \sin i$, the z component of velocity becomes $v_z = -x\Omega \sin i$. We ignore the negative sign and adopt the standard astronomical convention of receding velocity being positive, thus the Doppler shift is

$$\Delta \lambda_{\text{rot}} = -v_z = x\,\Omega \sin i = \frac{x}{R}\, v_{\text{eq}}\, \sin i, \qquad (18.1)$$

where $\Delta \lambda_{\text{rot}}$ is in velocity units and v_{eq} is the equatorial velocity. This is an important result. It says that all parts of the stellar disk having the same x value have the same v_z, and the light has the same Doppler shift. This $\Delta \lambda_{\text{rot}}$ is the same Doppler shift introduced back in Equation 17.9.

Now that we know that the Doppler shifts vary only with x, imagine the stellar disk divided into vertical strips, i.e., strips of constant x, like the one shown in Figure 18.5. The largest shifts occur at the limbs where $x = \pm R$ with a value

$$\Delta \lambda_L = R\Omega \sin i = v_{\text{eq}} \sin i, \qquad (18.2)$$

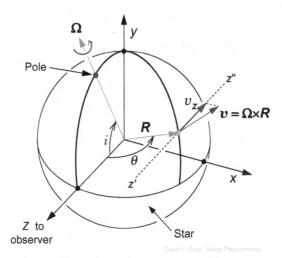

Figure 18.4. The rotation axis of the star has an inclination i from the line of sight. The surface point at the end of the radius vector R has a velocity $v = \Omega \times R$. The dashed line labeled $z' - z''$ runs through this point and is parallel to the z axis. The component of v in the z direction is v_z.

where $\Delta\lambda_L$ is in velocity units. We usually omit the subscript on v_{eq} when there is no ambiguity, writing simply $v \sin i$ as the projected rotation rate. If for now we assume there is no limb darkening, the amount of light from each strip is proportional to the strip area, $2y\,\delta x = 2\,\delta x\sqrt{R^2 - x^2}$, or normalized to the area of the disk, the distribution is

$$G(x)\,\delta x = \frac{2y\,\delta x}{\pi R^2} = \frac{2}{\pi R}\left[1 - \left(\frac{x}{R}\right)^2\right]^{\frac{1}{2}}\delta x.$$

In the limit, $\delta x \to dx$ and the Doppler-shift distribution becomes

$$G(\Delta\lambda)\,d\Delta\lambda = \frac{2}{\pi\Delta\lambda_L}\left[1 - \left(\frac{\Delta\lambda}{\Delta\lambda_L}\right)^2\right]^{\frac{1}{2}}d\Delta\lambda, \tag{18.3}$$

where we invoke Equations 18.1 and 18.2 to convert from x to $\Delta\lambda$. You will recognize this as the equation of an ellipse with a semimajor axis $\Delta\lambda_L$.

At this point we simplify the notation slightly. With the understanding that when we are dealing with velocity distributions, we measure wavelength from the center of the distribution, and therefore in most cases we can drop the Δ from $\Delta\lambda$ and just write λ. Then $G(\Delta\lambda)$ becomes $G(\lambda)$.

As we did with Equation 17.11, let us assume the local thermal profile shows negligible variation over the disk. Then the line profile broadened (only) by rotation is

$$\mathfrak{F}_v = \mathfrak{F}_v^0 * G(\lambda), \tag{18.4}$$

that is, we have a convolution between the thermal profile, $\mathfrak{F}_v^0$, and the rotational Doppler-shift distribution, $G(\lambda)$. The convolution moves $G(\lambda)$ to the position of the line and turns it

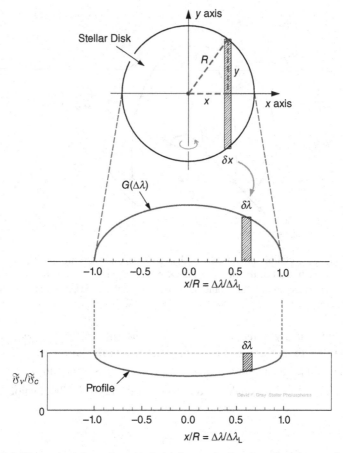

Figure 18.5 The projected rotation axis is in the *y* axis direction. Rotation produces blue shifts on the left half of the disk and redshifts on the right. Specifically, the Doppler shift of rotation is constant along each vertical strip on the stellar disk, and that Doppler shift is proportional to the *x* value, so the light from each strip corresponds to a specific position in the Doppler-shift distribution, $G(\lambda)$. The convolution in Equation 18.4 turns $G(\lambda)$ upside down and scales its area to the equivalent width of the thermal profile to give the rotationally broadened line profile.

over (as in Figure 18.5), so we have a broadened absorption line, and since $G(\lambda)$ is normalized to unit area, the broadened line has the same equivalent width as $\mathfrak{F}_\nu^0$. As can be seen from Equation 18.3, $G(\lambda)$ gets wider and shallower linearly with $\Delta\lambda_L = v \sin i$. Every line (almost) in a star's spectrum is another replication of $G(\lambda)$.

The form of Equation 18.3 is an ellipse and so has the basin shape noted in Figure 18.2. But to be more realistic, limb darkening and macroturbulence need to be included.

Rotational Broadening with Macroturbulence and Limb Darkening

The rotation Doppler-shift distribution in Equation 18.3 dominates stellar line profiles only when $v \sin i \gg \zeta_{RT}$, i.e., macroturbulent broadening is negligible. More generally we must revert to disk integrations, previously expressed in Equation 17.12, and properly combine the Doppler-shift distributions of rotation and macroturbulence with weighting by the limb darkening,

$$\mathfrak{F}_\nu = \mathfrak{F}_\nu^0 * \oint L(\theta)\Theta(\lambda - \Delta\lambda_{rot})\cos\theta\frac{d\omega}{\pi}$$

$$= \mathfrak{F}_\nu^0 * M(\lambda), \tag{18.5}$$

where $\Theta(\lambda)$ is the macroturbulence Doppler-shift distribution, $\Delta\lambda_{rot}$ is the rotational Doppler shift (Equation 18.1), and $L(\theta)$ is the continuum limb-darkening factor, as in Figure 17.6. Here $M(\lambda)$ is the combined *disk-integrated* Doppler-shift distribution of rotation and macroturbulence including limb darkening, normalized to unit area.

Figure 18.6 gives some examples using Equation 18.5. When $\upsilon\sin i$ reaches large values, like the 160 km/s curve in the figure, it becomes a challenge to distinguish the profile from the continuum, especially if the signal-to-noise ratio is not high.

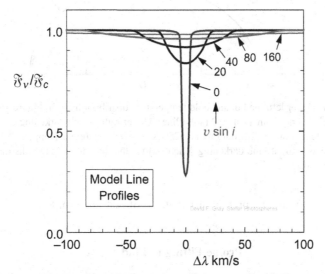

Figure 18.6 Model-computed rotational broadening, including a thermal profile, based on Equations 18.3 and 18.5 with $\zeta_{RT} = 1.0$ km/s and $\upsilon\sin i$ as shown.

When we choose a model photosphere for a star, we have established $T(\tau_0)$. The continuum limb darkening is computed, as in Figure 17.6, giving $L(\theta)$ to use in Equation 18.5. It is useful to understand what limb darkening does to $G(\lambda)$. Figure 18.7 shows examples. The continuum limb-darkening curves (upper right) are shown for three temperature distributions (upper left). The corresponding $G(\lambda)$ distributions are shown in the lower panel. The small differences in limb darkening among the three cases hardly affect $G(\lambda)$, but there is a very significant difference between the limb-darkened cases and no limb darkening. And it goes as expected: limb darkening reduces the light from the equatorial-limb areas that have the largest rotational Doppler shift, so $G(\lambda)$ near the extremes,

$\pm\Delta\lambda/\Delta\lambda_L$, has smaller amplitude. Since the area is normalized to unity, the central portion balances the weaker edges by having higher amplitude. Line profiles with limb darkening have deeper cores and weaker wings.

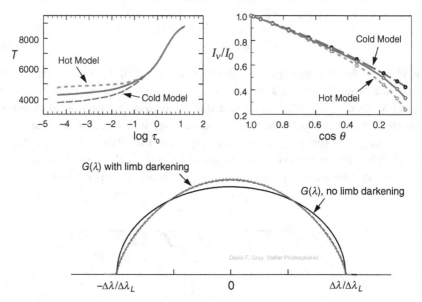

Figure 18.7 Upper left: the hot and cold temperature distributions from Figure 16.2 and the standard solar $T(\tau_0)$ from Table 9.2 (solid line). Upper right: Limb darkening curves for the same three temperature distributions at 6250 Å. Lower: rotational Doppler-shift distributions with and without limb darkening. The $G(\lambda)$ for the three temperature distributions are nearly identical.

Now we move on to interpreting observations of real line profiles.

Profile Fitting to Find $\upsilon \sin i$

The obvious way to measure $\upsilon \sin i$ is to compare computed line profiles directly with the observed ones. The best lines for the job are free of strong pressure broadening, and yet strong enough so that the broadened profile can still be measured. In spectra of O and early B stars, relatively few lines are suitable. The He I $\lambda4026$, $\lambda4388$, and $\lambda4471$ lines have commonly been used even though this third line has a substantial blend. For slightly cooler stars, the late B through early A types, the Mg II $\lambda4481$ line is nearly ideal as long as the rotational broadening is large compared to the 0.197 Å (13 km/s) splitting of this doublet. In A and F spectral types, Fe I $\lambda4405$, $\lambda4473$, $\lambda4476$, $\lambda4549$, and Fe II $\lambda4508$ were often chosen in days past. Lines at longer wavelengths can now be measured routinely thanks to the advent of red-sensitive light detectors.

An example is shown in Figure 18.8. Here the simple geometry of Equation 18.3 and of a full disk integration that includes limb darkening and macroturbulence are compared to the observations. Again we see limb darkening producing the general change in shape by lessening the contribution from the equatorial limb. Macroturbulence rounds the corners near the continuum. Other examples of the fitting of line profiles are found in numerous papers, e.g., Abt (1957), Hill (1995), Johns-Krull (1996), Chauville *et al.* (2001), Gray (2014).

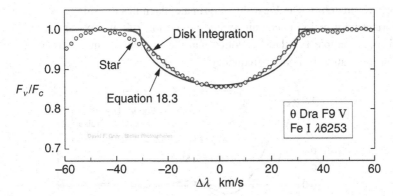

Figure 18.8. The computed profiles using Equation 18.3 ($v \sin i = 30.8$) and Equation 18.5 (Disk Integration, $v \sin i = 30.5$) are compared to the observed profile of Fe I $\lambda 6253$. The star is sometimes classified as F8 IV. Elginfield Observatory data.

Historically, many of the determinations of $v \sin i$ were based on line-width measurements done with modest- to low-resolution spectroscopy, instead of full profile fitting. This has some justification from the fact that Equation 18.3 scales linearly with $v \sin i$, and it is a straightforward approach to the observations. A modern extension of this technique is to measure the width of the peak in a cross correlation between the stellar spectrum and a narrow-lined standard star (e.g., Simkin 1977, Benz & Mayor 1981, Penny 1996, De Medeiros & Mayor 1999). Typically, profile modeling was done for a few standard stars that were then used to calibrate the simple width measurements. Examples of such work are Slettebak and Howard (1955), Slettebak (1956), Slettebak *et al.* (1975), Abt and Morrell (1995), Abt *et al.* (2002), Mochnacki *et al.* (2002), and Fekel (2003).

Width-type measurements fail at low rotation rates ($\lesssim 10$ km/s) because macroturbulent broadening takes over, and they fail at high rotation rates ($\gtrsim 200$ km/s) owing to gravity darkening (a word on that later), significant variations in the thermal profile across the stellar disk, and non-spherical stellar shapes that come with high rotation (Slettebak 1949). In other words, the rotational broadening no longer scales as $v \sin i$. In fact, we saw back in Figure 12.15 that we can do better than using just line width. Figure 18.8 illustrated the technique of direct profile fitting. However, analysis in the Fourier domain has certain advantages.

The Basics of Rotation Transforms

Sample transforms of the rotation profile, $G(\lambda)$, are displayed in Figure 18.9. These are computed using disk integrations and include limb darkening. The first sidelobe has an amplitude of about one-tenth the main lobe, and is negative in sign. The sidelobes alternate in sign and slowly diminish in amplitude toward larger Fourier frequency, reminiscent of a sinc function. The horizontal scale of the rotation profile is proportional to $\upsilon \sin i$, so of course the scale of $g(\sigma)$ is inversely proportional to $\upsilon \sin i$. Normally, log–log scales are used. We plot the absolute value of the amplitudes in order to use a logarithmic ordinate with negative sidelobes. The logarithmic ordinate magnifies the sidelobes, where much of the important information is contained, and with a logarithmic abscissa, $\upsilon \sin i$ scaling is changed into a simple translation. Notice that the two curves in the figure are identical except for the scaling in Fourier frequency.

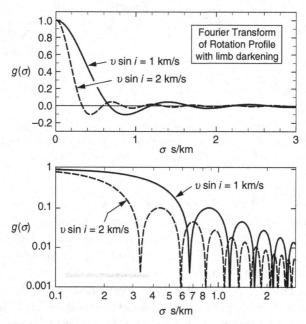

Figure 18.9. Transforms of the rotation profile including limb darkening (6250 Å) for projected rotation rates of 1 and 2 km/s. Upper panel: linear plot. Lower panel: log–log plot.

Recall, it is the line depth of observed profiles that we transform, $D(\lambda) = 1 - F_\nu/F_c$, and likewise for the thermal profile, $H(\lambda) = 1 - \mathfrak{F}_\nu^0/\mathfrak{F}_c$. Following Equations 18.5 and 12.2, the observed profile is

$$D(\lambda) = H(\lambda) * M(\lambda) * I(\lambda), \tag{18.6}$$

in which we conceptualize the true stellar profile as $H(\lambda) * M(\lambda)$ and $I(\lambda)$ is the instrumental profile. As usual, the convolutions become products in the Fourier domain (one of the advantages), giving

$$d(\sigma) = h(\sigma)\, m(\sigma)\, i(\sigma). \tag{18.7}$$

In the special case when the rotational broadening far exceeds all others, we can replace $m(\sigma)$ with $g(\sigma)$ giving

$$d(\sigma) = h(\sigma) \, g(\sigma) \, i(\sigma). \tag{18.8}$$

This is the simplest case. And here is the important point: the characteristic shape of $g(\sigma)$, especially its sidelobes and their zeros, is imprinted on the observations. Meanwhile $h(\sigma)$ and $i(\sigma)$ have much smaller variation with σ and they certainly do not introduce zeros anywhere near the low frequencies involved with rapid rotators.

Fourier Analysis When $\upsilon \sin i$ Is Large

Large $\upsilon \sin i$ implies that macroturbulence broadening is small or negligible compared to the broadening by rotation. In other words, we can use Equation 18.8 instead of 18.7. Figure 18.10 shows the determination of $\upsilon \sin i$ for a rapid rotator. The log $g(\sigma)$ shape is translated vertically to match the observed equivalent width (top of main lobe) and horizontally to match the sidelobe and zero positions. The horizontal positioning gives $\upsilon \sin i$. The sharpness of the sidelobe structure enhances the precision of the match, another of the advantages of the Fourier domain.

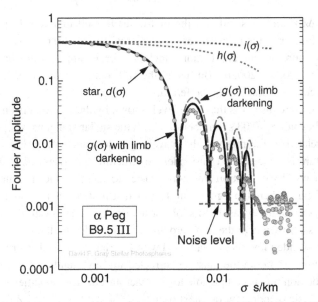

Figure 18.10. The Fourier transform of the mean of Mg II $\lambda4481$ and Fe II $\lambda4352$ profiles for α Peg is shown by the circles, labeled $d(\sigma)$. The transform of the rotation profile with limb darkening, $g(\sigma)$, scaled in both coordinates, is shown by the solid line. A $\upsilon \sin i$ of 140 km/s is found. The faint dashed line just above it shows the transform without limb darkening, and in this case $\upsilon \sin i$ is 135 km/s. The transforms of the thermal and instrumental profiles are shown at the top. Based on Gray (1977).

So when we are dealing with high rotation rates, $\upsilon \sin i$ tumbles out almost for free. We don't have to compute the thermal profile. We don't have to measure the instrumental profile. And as long as the macroturbulence velocities are only a few kilometers per second, we don't have to worry about them either. Naturally, the filtering of these other factors in Equation 18.8 reduces the amplitude by increasing amounts as we move to larger σ, as we can see in Figure 18.10 by the divergence in amplitude between $g(\sigma)$ and the observations, but that does not affect our ability to find $\upsilon \sin i$.

The transform of Equation 18.3, the case with no limb darkening, is also shown in Figure 18.10, but the rotation rate is 3.7% too small. Notice also the increase in the mismatch of the zero positions as σ increases. This gives some indication of the importance of limb darkening. It also shows that the derived rotation rate is not overly sensitive to the limb darkening. On the other hand, interpreting ratios of just zero positions on their own can clearly be risky.

The concept of using only zeros stems back to the early 1900s, well before the observations had the necessary signal-to-noise ratios to make it work (Carroll & Ingram 1933). For example, Carroll (1933a, 1933b) derived an analytical form for $g(\sigma)$ in which a linear approximation for limb darkening was included,

$$g(\sigma) = \frac{2J_1(u\beta)}{u\beta} - \frac{3 \cos u\beta}{2u^2\beta^2} + \frac{3 \sin u\beta}{2u^3\beta^3}. \tag{18.9}$$

Here $u\beta = 2\pi \Delta\lambda_L \sigma$ and J_1 stands for the first-order Bessell function. In particular, the σ positions of the first three zeros are given by $\Delta\lambda_L\sigma_1 = 0.66$, $\Delta\lambda_L\sigma_2 = 1.16$, and $\Delta\lambda_L\sigma_3 = 1.66$. More recent applications include, for example, Ramella et al. (1989) and Royer et al. (2002a, 2002b). But using the full transform match, as in Figure 18.10 and figures to come, is much to be preferred.

Another factor to consider is the noise level, shown by the data points on the far right in Figure 18.10, beyond $\sigma \sim 0.03$ s/km, below which the stellar signal has dropped. The noise level here is related to the noise in the wavelength domain through Equation 2.42. The rotation signature is concentrated into sidelobe structure that rises above the noise level, making the shape of the rotation transform easier to identify and measure compared to direct profile fitting, where the signal is uniformly distributed. This is another advantage of a Fourier analysis. Naturally, as the sidelobe structure descends toward the noise level, it is to be trusted less. So we give the first zero and the first sidelobe the most credence.

A reminder: blended lines are not allowed with Fourier analyses. Either clean unblended lines must be used or the profile must be corrected for any blend.

When the rotation rate is comparable to the macroturbulence velocities, then we have to take a more complete approach, our next topic.

Fourier Analysis for Moderate to Small Rotation

When the rotational broadening does not exceed the other broadeners by a factor of two or more, it becomes necessary to model the observed line profile more completely. Every broadening factor in Equation 18.6 has to be included: the instrumental profile, $I(\lambda)$, the thermal profile, $H(\lambda)$, and especially the combined Doppler shifts of macroturbulence and rotation, $M(\lambda)$. Although some of the following material has already been covered in Chapter 17, it is expanded here with the emphasis on rotation. The residual transform of the observed line profile is found by dividing out the transforms of the instrumental profile and thermal profile that has been computed from a model photosphere (Equation 17.13),

$$m_{\text{obs}}(\sigma) = \frac{d(\sigma)}{i(\sigma)h(\sigma)}.$$ (18.10)

Then we construct the model $M(\lambda)$, compute its transform, $m(\sigma)$, and compare it to $m_{\text{obs}}(\sigma)$. Constructing $M(\lambda)$ typically means doing a disk integration using Equation 18.5 with the two adjustable parameters $v \sin i$ and ζ_{RT}.

Understanding how to adjust $v \sin i$ and ζ_{RT} in the computed $M(\lambda)$ is helped by looking at plots like Figure 18.11. As increasing amounts of macroturbulence are added to the rotation, the main lobe narrows and changes shape, the sidelobes decline in amplitude, and the frequencies of the zeros diminish. When ζ_{RT} is ~0.5 $v \sin i$ and larger, the sidelobe structure alters dramatically. The first sidelobe is squeezed out and the second sidelobe grows slightly in amplitude while continuing to move toward lower frequencies. For still larger macroturbulence, the rotation sidelobe is replaced by a positive-amplitude sidelobe from the macroturbulence distribution.

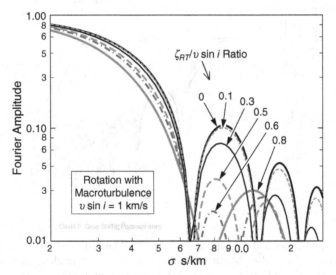

Figure 18.11 These examples show how the rotation transform is reshaped by increasing amounts of macroturbulence. Pure rotation is the 0 curve. Small amounts of macroturbulence mostly reduce the rotation sidelobe amplitudes. Larger amounts depress the main lobe and change the sidelobe structure completely. This graph is similar to Figure 17.8.

For the majority of cool single stars, macroturbulence broadening is larger than rotational broadening and any sidelobe information is buried in the noise. The remaining information is in the main lobe, both its shape and position.

Although individual spectral lines can be analyzed (as in Figure 17.9) and their results averaged, it is advantageous to combine residual transforms for many lines (as in Figure 17.10). Another example is given in Figure 18.12, in which transforms of the individual lines are shown. The transform of each observed profile, $d(\sigma)$, is divided by the transform of the matching thermal profile, $h(\sigma)$, to give the residual transforms, $d(\sigma)/h(\sigma)$. The agreement or otherwise of residual transforms gives us information about how good our observed profiles are and how well we have modeled the thermal profiles. Their scatter gives us the error bars on the mean residual transform. The rest of the analysis is shown in Figure 18.13. Here the residual, $m_{\mathrm{obs}}(\sigma)$, which is the mean of the residuals in the previous figure divided by the transform of the instrumental profile, $i(\sigma)$, is matched with a model $m(\sigma)$. The result is $\upsilon \sin i = 22.5$ and $\zeta_{RT} = 4.5$ km/s.

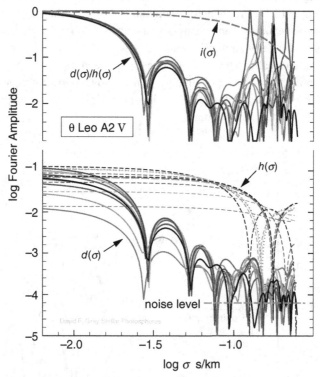

Figure 18.12 First step in the Fourier analysis for θ Leo. Lowest set of curves: transforms of eight observed profiles, $d(\sigma)$. Middle set (dashed): transforms of matching thermal profiles, $h(\sigma)$. Top set: residual transforms, $d(\sigma)/h(\sigma)$. Based on Gray (2014).

    In this example, the rotation is still large enough compared to ζ_{RT} to show the sidelobes of $g(\sigma)$. The match is not perfect, but is typical of real observations. The amplified noise rises rapidly toward higher frequencies, restricting the analysis to $\log \sigma < -1$ dex.
    Another example, one with slightly lower rotation and higher macroturbulence dispersion, is shown in Figure 18.14. Here we are in the intermediate range in which $\zeta_{RT}/\upsilon \sin i$ is ~0.5–0.6 and the first sidelobe is being extinguished, as we saw in Figure 18.11. The sidelobes are barely above the noise, which rises rapidly for $\log \sigma$ exceeding -1 (right panel).

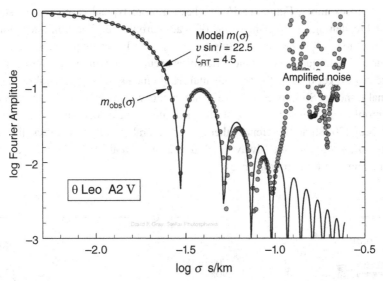

Figure 18.13 Second step in the analysis for θ Leo is to divide the mean residual transform from Figure 18.12 by $i(\sigma)$. This gives $m_{obs}(\sigma)$, the filled circles. The model $m(\sigma)$ is then matched to the observations by adjusting $v \sin i$ and ζ_{RT}. The white noise in Figure 18.12 is amplified here at high frequencies by the division of $h(\sigma)$ and $i(\sigma)$. Based on Gray (2014).

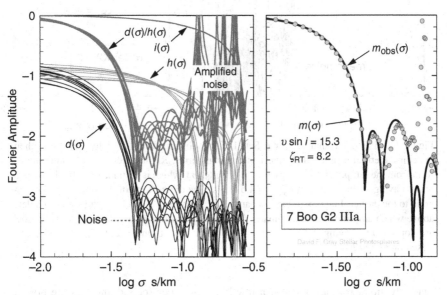

Figure 18.14 Left panel: transforms of the observed profiles, $d(\sigma)$, thermal profiles, $h(\sigma)$, and residual transforms, $d(\sigma)/h(\sigma)$. The sidelobes in the individual transforms are at the noise level. Right panel: the solution, $m_{obs}(\sigma) = d(\sigma)/[h(\sigma)\,i(\sigma)]$ shown as points, is fit with a model $m(\sigma)$ having $v \sin i = 15.3$ and $\zeta_{RT} = 8.2$ km/s shown as the line. Notice that the first sidelobe is smaller than the second. Observations from the Elginfield Observatory.

In the third example, Figure 18.15, we have a more typical case for cool stars, with rotational broadening smaller than that of macroturbulence. The amplified noise rises abruptly near $\log \sigma = -0.8$ dex, limiting the analysis to lower frequencies. But, it is only toward the higher frequencies that the differences occur, as shown by the outrider models in the figure. So we get back to the fundamental requirements of high spectral resolution and high signal-to-noise. Still, meeting these requirements will carry us only so far because the thermal-profile transforms have zeros starting near this same cutoff. Similar analyses can be found in Gray (2018b and references therein). In favorable Fourier analyses, the errors on $v \sin i$ are $\sim 0.1 - 0.2$ km/s. It is always hard to measure a weaker broadener (rotation) in the face of a larger one (macroturbulence).

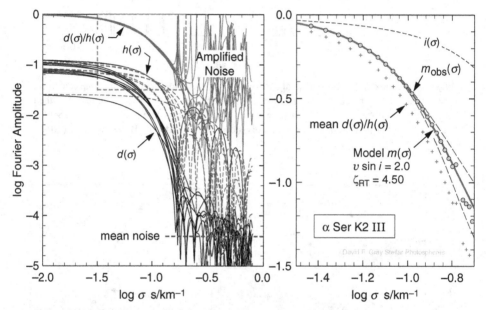

Figure 18.15 Fourier analysis for α Ser. Left panel: transforms of individual lines, their matching thermal profiles, and their residual transforms. Right panel: mean residual transform from the left panel divided by $i(\sigma)$ to get $m_{\text{obs}}(\sigma)$, which is then matched by the model, $m(\sigma)$, having $v \sin i = 2.0$ and $\zeta_{RT} = 4.5$ km/s. The dimensions of the right panel correspond to the dashed-line box in the left panel. The analysis is restricted to $\log \sigma \lesssim -0.8$ dex, above which amplified noise dominates. The dashed outrider-model lines in the right panel are for models having $v \sin i = 0.0$, $\zeta_{RT} = 4.97$ and $v \sin i = 2.9$, $\zeta_{RT} = 3.95$ km/s. Based on Gray (2016).

Relatively small rotation rates of a few kilometers per second and less are normal for most single cool stars, a topic we explore later in this chapter. The Sun fits right into the cool-star pattern, but as usual, it gives us some additional information that is hard to extract for other stars.

The Solar Rotation

The rotation of the solar photosphere is measured in several ways. The rate at which sunspots are carried across the solar disk is the earliest technique historically (see Livingston 1969 for an informal historical perspective). In the mid-1800s, Carrington set up a rotation count system that is still used today (see Ulrich & Boyden 2006). It is based on a fixed rotation period of 27.2753 days with a start date of November 9, 1853. Carrington also initiated the use of the standard value for the inclination of the rotation axis to the ecliptic of 7.25 degrees (Wöhl 1978, LaBonte 1981). With the appearance of spectrographs in astronomy, Doppler shifts offered an alternative approach for measuring the rotation, the simplest of which is to look at the Doppler shift of the equatorial limbs. But when observations were extended to the full solar disk, progressively slower angular rotation was found away from the equator. This *differential rotation*, as this latitude effect is called, has been measured by many observers, and a sampling is given in Figure 18.16. Early observations are summarized by De Lury (1939). In the upper panel, we see the average change in *angular* velocity with latitude according to several observers. Notice that zero is at the left side of the figure; the variation is modest, $\lesssim 30\%$. To calculate line broadening with Equation 18.5, we need the linear velocity for the Doppler shift.

The linear velocity even for a rigid rotator declines toward the poles in proportion to the distance of a surface point from the rotation axis, i.e., with the cosine of the latitude. The lower left panel shows the linear velocities for the angular velocities of the upper panel. The solid curve above the others shows the cosine drop, which accounts for most of the decline. The small additional reduction is from the differential rotation. All these determinations are nearly the same, especially when normalized to their equatorial rates, as shown in the lower right panel. Simple expressions are useful when working numerically with Equation 18.5, such as

$$\Omega(\ell) = 2.88 - 0.48 \sin^2 \ell - 0.40 \sin^4 \ell \; \mu\text{rad/s},$$
$$v(\ell)/v_{\text{eq}} = 1.00 - 0.71 \sin^2 \ell - 0.15 \sin^6 \ell,$$

(18.11)

in which v_{eq} is the equatorial rate and ℓ is the latitude. Points for this equation are shown as circles in the figure.

The equatorial rate is hard to pin down to better than ~1%, mainly owing to the other motions on the surface, such as wave motion and supergranulation. A comparison of selected determinations is shown in Figure 18.17.

These are sidereal rates, but synodic rates are measured. By sidereal we mean the "real" rate that would be seen by an observer outside the solar system. It is the value we use when comparing the Sun to other stars. But we observe the Sun from our planet in orbit, and that orbital motion is in the same direction as the solar rotation (as one might expect). So from our moving vantage point, the Sun appears to be rotating more slowly. The difference in these rates can be found by taking the projection of Earth's orbital velocity along the line of sight to the solar limb, as indicated in Figure 18.18. With the average orbital velocity of 29.873 km/s, and the semidiameter of the solar disk of 0.2666 degrees, this gives 0.139 km/s. A sidereal equatorial rate of 1.98 km/s corresponds to a synodic rate of 1.84 km/s.

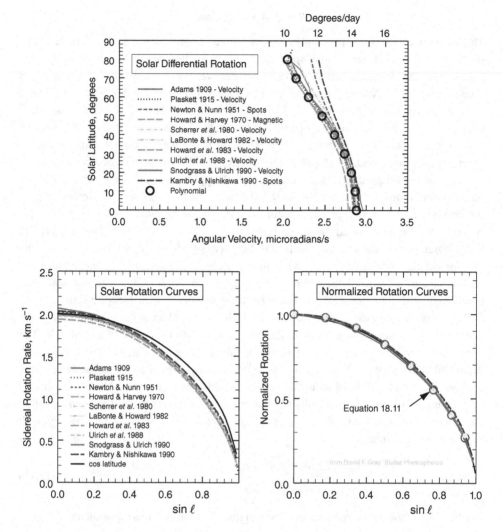

Figure 18.16 Upper panel: angular velocity declines by ≲30% from equator to pole. Different tracers show slightly different behavior. Lower left panel: linear velocity drops from equator to pole. Lower right panel: curves from the left panel normalized to their equatorial velocity. Circles show the values from Equations 18.11.

Negligible variations, ~0.005 km/s, arise from the 0.0167 eccentricity of the orbit. The 7.25 degree inclination of the rotation axis introduces no more than 0.8% variation, 0.016 km/s.

If we treat the Sun like other stars and look at its disk-integrated spectrum, differential rotation is undetectable. The features impressed on the Doppler-shift distribution occur at Fourier frequencies too high to be reached when rotation is small like the Sun's. But knowing the differential rotation for the Sun, as in Equation 18.11, we can calculate an $m(\sigma)$ to compare with $m_{obs}(\sigma)$ from the solar spectral lines. The result is a sidereal

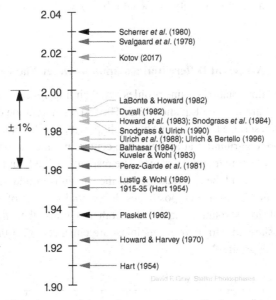

Figure 18.17 Published equatorial rotation rates (sidereal) show a considerable range. Based on Gray (2018a).

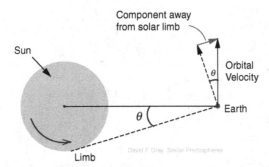

Figure 18.18 The lower limb is moving toward the Earth. The component of orbital motion in the limb direction is shown by the short arrow. Based on Gray (2018a).

equatorial velocity of 2.06 ± ~0.1 km/s ($\zeta_{RT} = 3.69$ km/s) (Gray 2018a). Although this $v \sin i$ is ~4% too high, it does fit into Figure 18.17 within the estimated errors. The solar observations can be fit equally well with a rigid-rotation model having $v \sin i = 1.96 \pm$ ~0.1 km/s ($\zeta_{RT} = 3.72$ km/s). Since most stellar $v \sin i$ values assume rigid rotation, 1.96 is the solar value to use when comparisons are made with other stars. This corresponds to a period of 25.8 days.

Because differential rotation is an important part of the solar behavior and an integral part of generating magnetic activity, it would be nice to know this information for other stars.

A Peek at Differential Rotation in Other Stars

The effect of differential rotation on line profiles and their transforms can be explored using models, i.e., disk integrations via Equation 18.5. The faster that rotation declines away from the equator, the more surface there is contributing light with small rotational Doppler shifts. Then $G(\lambda)$ has more contribution at and near the center, and the flat bowl-shaped characteristic is replaced by a more triangular shape. The response in the Fourier domain is seen in the broadening of the main lobe and reduction of the strengths of odd numbered sidelobes, as illustrated in Figure 18.19. If one postulates differential rotation of the opposite sense, with the equatorial zones rotating slower than other regions, then the response of the transforms is opposite and odd-number sidelobes are enhanced. Can these characteristics be seen in stellar line profiles?

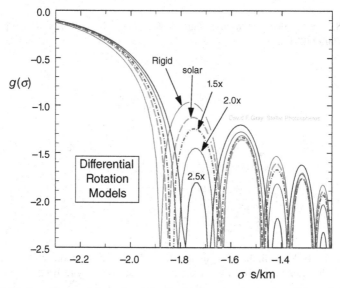

Figure 18.19 Transforms of Doppler-shift distributions incorporating differential rotation using Equation 18.11 with enhancements as labeled. A $\upsilon \sin i$ of 50 km/s was used in the disk integrations. These are equator-on models; larger changes are seen in non-equator-on models. (See also Bruning 1981, Garcia-Alegre *et al.* 1982, and Gray 1999.)

If we are to see such sidelobe structure, the rotation rate has to be high enough to move the sidelobes well out of the amplified noise region seen in Figure 18.15, for example. This requires $\upsilon \sin i \gtrsim 25$ km/s. A similar restriction on $\upsilon \sin i$ is needed to avoid the weakening of the first sidelobe by macroturbulence, as illustrated in Figure 18.11. Because most cool stars are slow rotators, with $v \sin i \lesssim 10$, with them we are out of luck. We cannot use their line profile shapes to measure differential rotation. Some years ago a sample of A and F stars, where rotation rates are higher, was investigated, but with negative results (Gray 1977, 1982a, Dravins *et al.* 1990). For the more rapid rotators, the ratio of zero positions has also been tried (Reiners & Schmitt 2003). More recently, Ammler-von Eiff and Reiners (2012) found that a small percentage of mostly F stars on the cool side of the granulation boundary show differential rotation.

Variation in rotational modulation can also be interpreted as differential rotation (e.g., Gray & Baliunas 1997), as can multiple periods (Karoff *et al.* 2018). In some cases, non-radial oscillations (as in Chapter 17) can be used to extract differential rotation. The particularly well-studied cases of 61 Cyg A and B, stars similar to the Sun, seem to have solar-like differential rotation (Bazot *et al.* 2019), while some others seem to have about five times more differential rotation, at least for lower latitudes (Benomar *et al.* 2018). Indirect evidence from magnetic field monitoring suggests that the Am star γ Gem A has a small differential rotation (Blazère *et al.* 2020).

Still, in most cases, differential rotation is kept in the background and we focus on $v \sin i$ and rotation periods as if all stars were rigid rotators.

Statistical Corrections for Axial Projection

In order to see a star pole-on, we have to be near the direction in which the rotation axis points. But to see the star equator on, we can be anywhere near the equatorial band. In other words, the least probable orientation is pole-on and the most probable orientation is equator-on. Evidence for random orientation of rotation axes is found in the lack of correlation between $v \sin i$ and galactic latitude or longitude and in the frequency distribution of $v \sin i$ for homogeneous groups of stars (Struve 1945, Slettebak 1949, Huang and Struve 1954, Abt 2001). Under the assumption of random orientation, we can convert the average $v \sin i$ characterizing a group of stars to an average v. Consider Figure 18.20, a celestial sphere of unit radius. All axes oriented within an increment di around the angle i to the line of sight are represented by the shaded ring. The fraction of randomly oriented axes between i and $i + di$ is proportional to the solid angle subtended by the ring,

$$P(i)\, di = 2\pi \sin i\, di.$$

The average $\sin i$ is then given by

$$\langle \sin i \rangle = \frac{\int_0^{\pi/2} P(i) \sin i\, di}{\int_0^{\pi/2} \sin i\, di} = \frac{\pi}{4}. \tag{18.12}$$

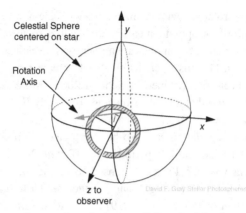

Figure 18.20 The rotating star is at the center of the celestial sphere with its rotation axis pointing in the direction of the arrow and making an angle *i* to the line of sight (*z* axis). For a fixed *i*, the axis can be anywhere in the shaded ring and still produce the same Doppler broadening.

The integration need only be carried out over the hemisphere facing us, so *i* ranges from 0 to $\pi/2$. Thus, for a homogeneous group of stars, the average rotational velocity is $\langle v \rangle = \langle v \sin i \rangle \, 4/\pi$.

More complete discussions of statistics and velocity distributions are given in the literature (e.g., Chandrasekhar & Münch 1950, Bernacca & Perinotto 1974, Gray 1988). The axial orientation of individual stars is also of some interest for extra-solar planetary studies (Doyle *et al.* 1984) and stellar magnetic activity (Gray 1989b). For a small number of stars having large apparent angular sizes, there may be just enough resolution of their surfaces to detect the orientation of their rotation axes from the slant of their spectral lines. Essentially one is measuring the differential Doppler shifts along the equator, much like early observations of the planets (Keeler 1895a, 1895b, Moore & Menzel 1928, Uitenbroek *et al.* 1998, Lesage & Wiedemann 2014).

Rotational Modulation

Just as Galileo could watch sunspots move across the disk of the Sun and determine the solar rotation period, so can we monitor modulation arising from rotation of stars more generally. The star need only have a non-uniform surface, invariant on a timescale that is long compared to the period of rotation. The physical source of the non-uniformity need not be understood physically. The signal itself can be photometric, such as a diminution in photometric flux as a dark spot traverses the disk, or it can be spectroscopic, for example, a change in the emission strength of the Ca II H and K lines (Horne & Baliunas 1986, Choi *et al.* 1995, Gray & Baliunas 1997), or changes in the shape of line profiles (e.g., Toner & Gray 1988). Magnetic modulation can also be followed (e.g., Cotton *et al.* 2019). In some cases, the timescale for intrinsic variation of the surface structure is only a few rotations, and this limits the precision. Stars seen close to pole-on show no rotational modulation, and stars without surface features are not going to show anything either. But when the period measurements can be made, they have the advantage of not being tied to a sin *i*-projection factor the way the line broadening results are. Conversion to velocity requires a value for the radius. If a star's $v \sin i$, the period, and the radius are all known, the inclination can be found from

$$\sin i = \frac{v \sin i}{2\pi R/P}. \tag{18.13}$$

There are several techniques for period determination. One of the best is to minimize the scatter in a phase diagram. With this technique, the periodic variation can have any shape and that shape will appear as a matter of course in the final phase diagram. See Jurkevich (1971), Stellingwerf (1978), Marraco and Muzzio (1980), and Cuypers (1986). Fourier techniques, such as the periodograms discussed in Chapter 2, are often useful (Gray & Desikachary 1973, Scargle 1982, Kurtz 1985), but should not be used without careful consideration of the data-sampling window that often produces confusing peaks in the transform. Furthermore, the signal itself is rarely a pure sinusoidal, i.e., composed of a single frequency, so harmonics can be expected in the periodogram.

Simple photometric measurements are particularly effective for faint stars, although they require long sequences of telescope time. Starspots as small as sunspots might largely escape detection, but many stars have much larger spots, making photometric modulation a fairly common occurrence. Examples in the literature are numerous, for example, Lockwood *et al.* (1984), Stauffer (1984), Frasca *et al.* (1998), Hartman *et al.* (2010), or Douglas *et al.* (2016). One must also be on guard for multiple surface features that could fake a period of half or a third the true value. Another delicate issue is differential rotation. Spots modulating the light curve are probably at latitudes away from the equator and likely have a period somewhat different from the equatorial value (e.g., Lurie *et al.* 2017).

An extension of rotational modulation is rotational mapping of the stellar surface, a topic briefly taken up at the end of the chapter.

Rotation of Hot Stars

Most early determinations of rotation for hot stars used half widths of lines, and more recently profile fitting and Fourier transforms have been invoked (e.g., Slettebak 1966, Abt & Morrell 1995, Howarth *et al.* 1997, Royer *et al.* 2002a, Simón-Díaz & Herrero 2007, Simón-Díaz *et al.* 2010). The hot stars are the speedy rotators, with dwarfs reaching speeds up to $v \sim 450$ km/s. And that is close to the break-up velocity, estimated to be ~500 km/s. The fastest rotators are Be or emission-line B stars. The emission is mainly in the Balmer lines (e.g., Evans *et al.* 2006) and comes from a disk surrounding the star. Numerous models have been proposed, e.g., McLaughlin (1961), Poeckert and Marlborough (1978), Kjurkchieva *et al.* (2016), Arcos *et al.* (2017). A review of Be stars is given by Rivinius *et al.* (2013).

Published rotation rates are shown as a function of spectral type in Figure 18.21. Notice how the Be stars (Slettebak points) populate the upper portion. The $v \sin i$ distributions for

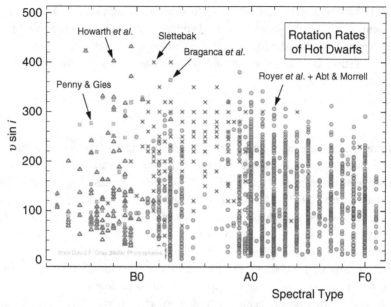

Figure 18.21 The results of six surveys are shown. Howarth *et al.* (1997), Penny and Gies (2009), Slettebak (1982, Be stars), Bragança *et al.* (2012), Royer *et al.* (2002a, 2002b) including the values from Abt and Morrell (1995).

B and A dwarfs are broad, often with Maxwell–Boltzmann-like shapes (e.g., Royer *et al.* 2007). But less than 1% are above 80% of the break-up velocity (Abt *et al.* 2002). In some cases, a second population of slow rotators makes the distributions bimodal. In addition, peculiar A stars and their relatives also have relatively slow rotation with a mean of ~50 km/s. Figure 18.22 shows typical distributions for hot stars.

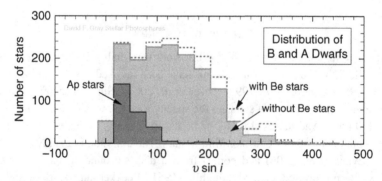

Figure 18.22 The main distribution is for B and A dwarfs with Be stars excluded. The dashed distribution includes the Be stars. The small dark distribution is for Ap and related stars. Although the main distribution has Ap stars excluded, the anomalous height of the second interval from the left suggests that some Ap stars may not have been recognized as such.

As one might expect, rotation rates decline as stars evolve off the main sequence and grow in size. Dwarfs in open clusters, which are generally younger than field stars, have slightly higher rotation because some of the field stars have already started evolving off the zero-age main sequence (Huang & Gies 2008). Somewhat further along their evolutionary trek, O supergiants have mean $v \sin i$ of ~100 km/s, early B supergiants ~60 km/s, and late B supergiants range down to ~25 km/s (Howarth *et al.* 1997, Dufton *et al.* 2006, Simón-Díaz & Herrero 2007, Brott *et al.* 2011). As noted in Chapter 17, the O supergiant rates are probably systematically high owing to a "macroturbulence" floor on the line broadening amounting to ~30 km/s. The physical nature of this macroturbulence is unknown, but unlikely to be connected to the velocity fields producing macroturbulence in cool stars.

Stellar parameters are altered when stars spin rapidly, especially so near the break-up rate. When rotating rapidly, the star has the shape of an oblate spheroid causing the surface gravity to be larger at the poles and weaker at the equator. This results in the so-called *gravity darkening*. In its simplest theoretical formulation, gravity darkening says that the bolometric flux is proportional to the local surface gravity (von Zeipel 1924, Tassoul 1978, Espinosa Lara & Rieutord 2011), which means that the star is brightest at the poles and dimmest at the equator. It follows that the poles are also the hottest and the equator the coolest. Needless to say, the shape change and the darkening play havoc with the shape of the Doppler-shift distributions and the rotational line broadening (Figure 18.5). Even more irksome, all observed parameters are now a function of the aspect of view. This includes color indices and energy distributions, M_V, T_{eff} (e.g., Collins 1966, Faulkner *et al.* 1968, Mathew & Rajamohan 1992), and altered line strengths and shapes (e.g., Stoeckley 1968, Collins & Truax 1995, Zorec *et al.* 2017, Lipatov & Brandt 2020). Divining the rotation rate in such situations is a challenge.

Occasionally one sees unusual line profiles caused by a combination of rapid rotation and aspect of view. One such case is Vega (α Lyr, HR 7001, A0 V), our photometric standard of Chapter 10. The star has relatively narrow lines with $v \sin i \sim 22$ km/s, but actually it is a rapid rotator with $v \sim 245$ km/s seen nearly pole-on (Gulliver *et al.* 1994). With this aspect of view, the limb and equator are nearly the same. At the equator, the cooler temperature and lower gravity alter the ionization balance and strengthen the local thermal profile of certain lines. The Doppler-shift distribution for such lines is dominated by the two equatorial regions near the limb, i.e., it is bimodal. Figure 18.23 illustrates one of these curious profiles.

As noted earlier, even though hot stars do not have convective envelopes, their rotation induces circulation currents that mix envelope material into the core. This results in more available fuel and a longer main-sequence lifetime, so stars with faster rotation stay longer on the main sequence. And some surface chemical signatures are influenced by rotation, such as the lithium and CNO abundances (e.g., Brott *et al.* 2011, and as noted in Chapter 16).

The strong radiation pressure produced by O stars results in mass loss strong enough, $\sim 10^{-5}$ $\mathfrak{M}_\odot$/yr, to carry away significant angular momentum (de Jager *et al.* 1988). Since this process hinges on line opacity, spin-down rates depend on metallicity. Galactic O stars typically spin down to ~1/2 their original value before evolving off the main sequence (e.g., Meynet & Maeder 2000, Penny & Gies 2009). Mass leaving from the equatorial regions is most effective in slowing the star since it has the most angular momentum per unit mass.

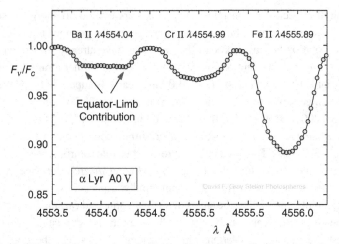

Figure 18.23 The Ba II line on the left shows enhanced contributions from the equatorial limb, where owing to the lower temperatures there, the line is stronger. Data from the Elginfield Observatory.

Magnetic fields must be small or without open field lines (see below) because such large mass loss combined with magnetic fields would produce much more severe spin-down. B and A stars are cool enough to avoid this whole process and consequently conserve their angular momentum for their entire main-sequence lifetimes. Rotation behaves very differently for the cooler stars. Let us take a look.

Rotation of Cool Dwarfs

The behavior of rotation along the main sequence is quite remarkable, as shown in Figure 18.24. This transition from high to low rotation rates was discovered and reflected on long ago (Westgate 1934, Struve 1945, Herbig & Spalding 1955). Early-type stars show a wide distribution of rotation rates, ranging up to hundreds of km/s. Through the F stars, we see a dramatic drop. Precisely in this same spectral interval the convective envelopes develop and deepen, reaching ~1/3 the radius in solar-like stars, and marching on to fully convective stars by cool M types. In addition, magnetic activity, especially the enhancement of x-ray emission, UV emission lines, and the emission cores of the Ca II H and K lines, become increasingly strong over this very same interval (Pallavicini *et al.* 1981, Güdel 2004, Marsden *et al.* 2009). This is all tied together with the hypothesis of magnetic fields and a magnetic brake that is generated by the interaction of rotation with the convective envelope in a process called a dynamo, as we shall see. The rotation rates of typical mid-G and cooler stars are only a few km/s, not unlike the solar value.

Open clusters are not only younger than the field stars shown in Figure 18.24, their different ages give us time snapshots of rotation that can be used to map out the time sequence of events similar to what we do with evolutionary changes in the HR diagram. A wealth of rotational-modulation data has been obtained in recent years, especially from the Kepler satellite (e.g., Nielsen *et al.* 2013, Douglas *et al.* 2016, 2017). Figure 18.25

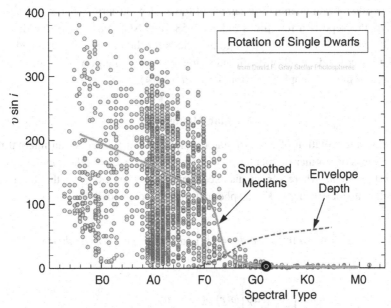

Figure 18.24 Circles: projected rotation rates for individual field stars. The solid line shows the approximate smoothed medians. The dashed line shows the growth in depth of the convective envelope. The solar position is shown by the circumpunct. Data from Slettebak *et al.* (1975), Conti and Ebbets (1977), Soderblom (1982), Gray (1984b), Halbedel (1996), Penny (1996), Fekel (1997, 2003), and Royer *et al.* (2002a, 2002b).

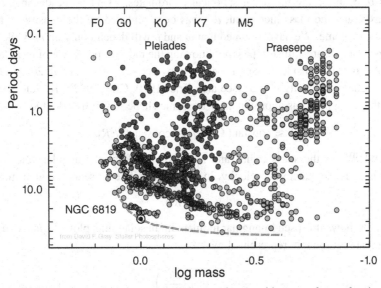

Figure 18.25 Rotation periods in open star clusters show a wide range, from a fraction of a day at the top to tens of days at the bottom. Data from Hartman *et al.* (2010), Douglas *et al.* (2017), and Meibom *et al.* (2015). The solar position is shown by the circumpunct. The dashed line is for field stars, age ~4.5 Gyr based on McQuillan *et al.* (2014).

displays period measurements for a sample of three open clusters. Pleiades is the youngest, ~0.12 Gyr. Praesepe is a few times older, ~0.58 Gyr. NGC 6819 is very old, ~2.5 Gyr. (Gyr = 10^9 years.) We see a large range in periods for the first two clusters and a densely populated lower band for all of them. Stars slow down moving downward in Figure 18.25 with time. For more examples see Bouvier *et al.* (2014).

Angular Momentum

When the surface rotation of a dwarf star slows, there must be either an internal redistribution or a loss of angular momentum. For a single point mass, m, the angular momentum is $L = xmv = \omega x^2 m$, where x is the distance from the rotation axis, v is the linear velocity, and ω is the angular velocity. For a spherical star, this expands into

$$L = \oint \omega(r) r^2 \, \sin^2\phi \, dm = \int_0^{2\pi} \int_0^{\pi} \int_0^{R} \omega(r) \, \rho(r) \, \sin^3\phi \, r^4 \, dr \, d\theta \, d\phi$$

$$= \int_0^{2\pi} d\theta \int_0^{\pi} \sin^3\phi \, d\phi \int_0^{R} \omega(r) \, \rho(r) \, r^4 \, dr$$

$$= \frac{8\pi}{3} \int_0^{R} \omega(r) \, \rho(r) \, r^4 \, dr, \qquad (18.14)$$

in which θ, ϕ, and r are the usual spherical coordinates, $r \sin i$ is the distance from the rotation axis, and the mass increment, dm, has been replaced by the density, ρ, times an increment of volume. We have assumed that ω and ρ will depend only on r. While this is a nice expression, we are still stumped, not knowing the radial dependence of $\omega(r)$, although the complex behavior inside the Sun is being uncovered (e.g., Howe *et al.* 2005). But if we follow the lead of McNally (1965) (see also Kraft 1970, Gray 1982b, Kawaler 1987) and suppose that $\omega(r) \, \rho(r)$ has approximately the same shape for all dwarfs, then

$$L \approx \text{constant } \Omega R^2 \mathfrak{M} \approx \text{constant } \mathfrak{M} R v.$$

Here R and $\mathfrak{M}$ are the star's radius and mass, $\Omega = \omega(R)$ is the angular velocity of the surface, and v is the equatorial velocity. In effect, we define a pseudo-angular momentum

$$\mathcal{L} = \mathfrak{M} R v. \qquad (18.15)$$

Figure 18.26 shows this pseudo-angular momentum in solar units plotted against mass. The hotter stars are reasonably represented by

$$\frac{\mathcal{L}}{\mathcal{L}_\odot} = 100 \left(\frac{\mathfrak{M}}{\mathfrak{M}_\odot} \right)^{5/3}. \qquad (18.16)$$

Likely all stars lie on or near this line when they approach the main sequence, an expression of the break-up limit, where centripetal force balances gravity (Dicke 1970).

The B and A stars keep their angular momentum, while cooler stars experience magnetic braking, first a rapid phase, then a slow phase as shown by the distribution of points in both Figures 18.25 and 18.26. When stars make the transition to slow braking, they pile up along the top edge of the slow-braking locus. As we shall see below, the transition depends on the depth of the convection zone and the evolution of the dynamo and its magnetic field. From the observations we find $\mathcal{L}/\mathcal{L}_\odot \sim (\mathcal{M}/\mathcal{M}_\odot)^4$ along the transition locus for G and K dwarfs.

Loss of angular momentum moves stars vertically downward in the diagram, i.e., their masses are sensibly constant. The fast and slow braking is seen most clearly with open clusters, where most stars are found in the slow-braking phase, but some, especially the lower mass stars in younger clusters, are caught in the rapid-braking phase. The evolutionary ages of these clusters point to a timescale of ~30 million years for the rapid-braking stage of G dwarfs, with increasingly longer times needed for lower masses (Stauffer 1987, Stauffer *et al.* 1987, Soderblom *et al.* 2001). In Figure 18,26 we see that the observation of Praesepe stars has been pushed to very small masses, where only modest losses in angular momentum have occurred (Douglas *et al.* 2017).

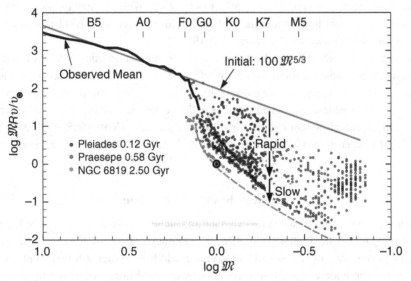

Figure 18.26 Pseudo-angular momentum (Equation 18.15) normalized to the Sun is shown as a function of mass. The mean relation for hot stars follows Equation 18.16. Pleiades (youngest), Praesepe (older), and NGC 6819 (oldest) cluster stars are shown as points. The dashed line near the bottom is for ~4.5 Gyr field stars, as in the previous figure. The majority of cool stars are in the slow braking phase. Some cluster stars are seen in the rapid-braking stage if the cluster is young enough. Modified from Gray (1982b). Cluster data based on Hartman *et al.* (2010), Douglas *et al.* (2017), and Meibom *et al.* (2015).

The range in $\mathcal{L}/\mathcal{L}_\odot$ during the rapid-braking phase can stem from a dispersion in initial rotation rates, an age spread (Gossage *et al.* 2019), or different rates of braking. The transition from rapid to slow braking may be caused by (1) changes in the brake behavior such as reduction in the mass loss or the interaction of the escaping mass with the magnetic field (e.g., Belcher & MacGregor 1976) or (2) intrinsic changes in the dynamo as it transitions from high to low rotation rates (e.g., Gray 1982b). One possibility in the second category is a change in the topology of the magnetic field, complex fields giving hard braking, a more dipolar field weak braking (Holzwarth & Jardine 2005, Cohen *et al.* 2009, Garraffo *et al.* 2018). Semi-empirical modeling of rotation changes has been done by Gallet and Bouvier (2013) and van Saders and Pinsonneault (2013). Detailed models that include the effects of rotation on the stellar structure and magnetic braking are given by Amard *et al.* (2019). Many of these models show a fast braking phase followed by a slow braking phase, but the details vary with mass, chemical composition, and assumed initial rotation rate.

Turning back to the observations, weak braking has a timescale of gigayears, dropping roughly as

$$v = 4.2 \, \mathfrak{M}^2 / \text{age}^{1/2} \tag{18.17}$$

for mass, $\mathfrak{M}$, in solar masses and age in Gyr (Skumanich 1972, Soderblom 1983, Barry *et al.* 1987, Pognan *et al.* 2018). The Sun is well along in the weak-braking stage, rotating near the lower limit observed for G dwarfs. But there are likely more twists and turns in the story. For reasons not understood, the K dwarfs in the cluster NGC 6811 are not slowing down as rapidly as expected (Curtis *et al.* 2019).

As we have seen, most B and A dwarfs seem to be content spinning at their birth rate for their entire main-sequence lifetime (e.g., Wolff *et al.* 2007). This makes sense if magnetic braking requires a convective envelope, which they lack, and the dynamo process. The exceptions to this general pattern are some peculiar A stars and the metallic-line A stars that show anomalously slow rotation and fossil magnetic fields (Conti 1965, Abt 1979, Wolff 1981, Kholtygin *et al.* 2010, Sikora *et al.* 2018). So how does magnetic braking work?

Magnetic Braking of Rotation

Angular momentum is lost when an extended magnetic field is coupled to escaping mass, exerting drag on the star (Schatzman 1959,1962; Durney & Latour 1978). Escaping mass is ionized and so spirals in helical paths around the field lines (Lorentz force). With the Sun, for example, there are coronal holes from which the field lines extend out to many solar radii, the so-called open field lines. Ionized particles are free to move in the direction of the field lines, so mass can escape from coronal-hole areas. Figure 18.27 shows an example.

The solar wind, i.e., the mass escaping from the Sun, is well known (e.g., Verscharen *et al.* 2019). Meanwhile, these field lines are anchored in the convection zone and co-rotate with the star. Therefore, the escaping material is spun up as it moves outward. Angular momentum is transferred from the star to the escaping particles. Well away from the star, the spun-up particles decouple from the magnetic field at a point where the kinetic energy of the particle exceeds the magnetic energy. This distance is called the Alfvén radius, r_A. Weber and Davis (1967) gave a useful approximation for the rate at which angular momentum is drained from the star,

Figure 18.27 The dark areas on the solar surface are coronal holes from which mass escapes along magnetic field lines that extend out several solar radii. This was taken 20-21OC2016 at extreme UV wavelengths (typically a few hundred Å). Lines are models of magnetic field lines. Courtesy NASA/JPL-Caltech.

$$\frac{dL}{dt} = -\frac{2}{3}\,\Omega r_A^2\,\frac{d\mathfrak{M}}{dt},$$

(18.18)

where $d\mathfrak{M}/dt$ is the mass loss rate. Variations on this theme are explored by Kawaler (1988), and more complex geometries are considered by Garraffo *et al.* (2015). Equation 18.18 says that angular momentum is extracted in proportion to the rotation rate, so rapid rotators slow faster, leading to a sort of convergence effect. The mass loss also slows over time and contributes to the time variation of L. For the Sun, $d\mathfrak{M}/dt$ is currently ~2×10^{-14} $\mathfrak{M}_\odot$/yr (compared to ~7×10^{-14} $\mathfrak{M}_\odot$/yr lost in the solar luminosity). This is a low mass-loss rate. Values for other dwarfs are roughly similar to the Sun's, while stars ~100 times younger have ~10 times higher rates (Wood *et al.* 2002, Pognan *et al.* 2018). Evolved stars show much stronger mass loss, ranging up to ~10^{-4} $\mathfrak{M}_\odot$/yr for M supergiants (van Loon *et al.* 2005).

In order for all this to happen, a star must have, or generate, a magnetic field. In cool stars the generation mechanism is called a dynamo. Although stars know how to do this, mortals still struggle to understand its intricacies. Basic to the process is (1) convection that twists magnetic field lines by means of the Coriolis force and (2) differential rotation that wraps field lines roughly along latitude lines. Differential rotation is assumed to scale with rotation, so rotation is normally taken as the second factor. Durney and Latour (1978) employed the Rossby number, N_R, as a simple conceptual statement that rotation can influence convection only if the convective motion is slow compared the rotation. Using the rotation period for the rotation timescale and t_{con} for the time for convection to cycle or "turn over," the Rossby-number condition is

$$N_R = P_{\text{rot}}/t_{\text{con}} \lesssim 1 \qquad (18.19)$$

for a dynamo to function. If the characteristic velocity of convection is v_{con} and the characteristic length scale is ℓ, then the condition becomes

$$N_R = \frac{2\pi R}{v_{\text{rot}}} \frac{v_{\text{con}}}{\ell} \lesssim 1,$$

where R is the stellar radius and v_{rot} is the rotation velocity, or more simply,

$$v_{\text{rot}} \gtrsim 2\pi v_{\text{con}}/(\ell/R). \qquad (18.20)$$

Since this is a conceptual argument, the factor of 2π is sometimes dropped. So this tells us that when rotation is too slow the dynamo will not run.

When the rotation is very large compared to the Expression 18.20 criterion, there is evidence from x-rays (one proxy for magnetic fields) of magnetic saturation (e.g., Wright *et al.* 2011). When v_{rot} is below the saturation level, x-rays seem to be proportional to the square of the rotation rate (Pallavicini *et al.* 1981). On the opposite end, does v_{rot} ever get small enough to terminate magnetic activity, including braking? At 4.57 Gyr of age, the Sun is still perking along with a meagre rotation just below 2 km/s. But eventual termination of the dynamo is the logical implication of Expression 18.20. The magnetic braking will ultimately bring v_{rot} low enough to no longer produce dynamo action. Evidence of this happening for evolved stars is discussed below (Figures 18.28 and 18.29, Gray 1982c, 1983, 1986, 1988). Recent observations of old field stars now also points in this direction (Metcalfe *et al.* 2016, van Saders *et al.* 2016) and at a rotation rate similar to the Sun's, implying that the solar dynamo may be close to turning off.

The numerical applications of Expression 18.20 or its variants leave some ambiguities. Convective velocities vary with depth in the convection zone, so how is v_{con} to be calculated? Some investigators use the full depth of the convective envelope for ℓ, while others use a pressure scale height, which has its own ambiguity since it too varies with depth. And the dynamo is not just a surface phenomenon, so how much error is there in using the surface rotation rate for v_{rot}? Consequently, it is not the absolute numbers from Expression 18.20 that are of primary interest, but rather the *variation* in the parameters with spectral type, luminosity, chromospheric activity, x-ray emission, and so on. Most particularly v_{rot} in Expression 18.20 is a function of mass and evolutionary state.

Comments on Magnetic Fields, Measured and Inferred

Direct measurement of magnetic fields is useful, but much of the field in cool stars is in the form of loops, the foot points of which are dark starspots. How much of the field is in the form of open field lines, those involved in braking, is not clear. Zeeman broadening of spectral lines and Zeeman polarization signals give us direct measurements of fields (e.g., Robinson *et al.* 1980, Gray 1984a, Marcy 1984, Saar 1988, Rosén *et al.* 2016, Moutou *et al.* 2017, Afram & Berdyugina 2019, Kochukhov *et al.* 2019). Since magnetic fields go hand in glove with chromospheric and coronal emission, proxy indicators are often used in lieu of magnetic fields (Catalano & Marilli 1983, Mangeney & Praderie 1984, Noyes *et al.* 1984, Vilhu 1984, Soderblom 1985, Rucinski & Vandenberg 1986, Mamajek & Hillenbrand 2008, Zhang *et al.* 2020).

Another important aspect of dynamo action is the magnetic cycle. Many dwarfs show cyclic changes corresponding to the ~11 year solar magnetic cycle (Wilson 1968, 1978, Baliunas & Vaughan 1985) and many do not. Rotation and cycles are coupled through the dynamo with interesting dependences (Böhm-Vitense 2007, Metcalfe *et al.* 2016, Nielsen *et al.* 2019). A shift from spot-dominated to faculae-dominated cycles has been suggested by Reinhold *et al.* (2019).

The Rotation Clock

From figures such as 18.25 and 18.26, we infer a time sequence. Rotation is slowing down. If the mechanisms for this process are universal, then rotation can serve as a clock. In practice, we do not know enough about the physics of the slowdown to calculate a time for a given rotation rate. Instead, the Sun and standard evolutionary times from HR diagrams are used for calibration (early papers: Sandage 1962, Demarque & Larson 1964). In effect, rotation becomes a sophisticated interpolation parameter (Soderblom 1985, Kawaler 1989, Lachaume *et al.* 1999). An important application lies in age-dating field dwarfs because evolutionary tracks in the HR diagram lose their leverage for these stars.

Although implicit in earlier works (e.g., Gray 1982b, Soderblom 1985, Kawaler 1989), Barnes (2007) recasts the mass dependence of Equation 18.16 in terms of the $B - V$ color index and brings supporting evidence that the shape of the slow-rotation locus is universal. Estimated uncertainty in rotation ages is ~15%. In this way rotation joins other age indicators such as chromospheric and coronal emission (generated by rotation), lithium abundance, metallicity, and galactic orbit parameters (see also Mamajek & Hillenbrand 2008). But, if the variant shape of the rotation plot for the cluster NGC 6811 (Curtis *et al.* 2019) proves true, precise age estimates from rotation may have a stumbling block. Undetected close binaries could also compromise this clock (Fleming *et al.* 2019). And finally, we expect dynamo braking to turn off and the rotation rate to stabilize when the angular momentum loss drives v_{rot} below the limit given by Expression 18.20.

Rotation of Evolved Stars: Subgiants (IV) and Giants (III)

Figures 18.28 and 18.29 show the observed rotation rates of luminosity class IV and III stars as a function of spectral type. Some considerable similarity with Figure 18.24 for dwarfs is apparent. But the meaning is quite different. In the case of dwarfs, spectral type is a mass sequence, whereas in the case of subgiants and giants, it is a time sequence. Dwarfs stay at nearly the same spectral type until they lift off the main sequence and start evolving toward cooler temperatures and higher luminosities. The IVs and IIIs are already in these evolving stages, and with the exception of various relatively small loops and excursions, generally march systematically toward cooler temperatures, higher luminosities, and larger

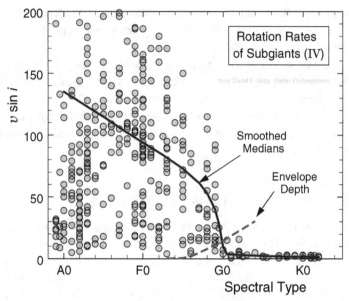

Figure 18.28 Projected equatorial rotation rates of subgiants show a sharp decline near G0 IV, mimicking the behavior of dwarfs. The line shows smoothed median values. Data from Uesugi and Fukuda (1982), Gray and Nagar (1985), Fekel (1997, 2003), and Royer *et al.* (2002a, 2002b).

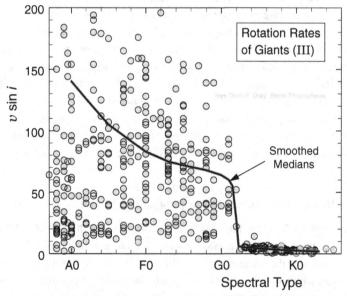

Figure 18.29 Projected equatorial rotation rates of class III giants show a sharp decline near G2 III, mimicking the behavior of dwarfs. The line shows smoothed median values. Data from Alschuler (1975), Gray (1982c, 1989a, 2018b and references therein), Royer *et al.* (2002a, 2002b), and Fekel (2003).

radii with time. The important common factor is the convective envelope. Cool dwarfs have their convective envelopes continuously, in contrast to evolved stars that came from the warm side (~ late B to mid A types) of the main sequence and developed a convective envelope only after cooling to the ~G0 temperatures.

Putting together the declines seen in Figures 18.24, 18.28, and 18.29, we have a "rotation boundary" in the HR diagram. This boundary was shown in the previous chapter in Figure 17.20, on the cool side of the granulation boundary. With few exceptions, stars on the hot side of the rotation boundary rotate rapidly, those on the cool side slowly. Before crossing the rotation boundary, stars have a wide range of rotation rates, but after crossing, rotation is a single-valued function of spectral type. For subgiants and giants on the cool side of the boundary, the rotation rates are

$$v_{IV} = \frac{4}{\pi} \langle v \sin i \rangle \approx 25.9 - 15.4\,Sp + 2.47\,Sp^2,$$

$$v_{III} = \frac{4}{\pi} \langle v \sin i \rangle \approx 40.3 - 21.6\,Sp + 3.07\,Sp^2,$$

$$(18.21)$$

where the numerical spectral type, $Sp = 2.0$ for G0, 2.1 for G1, ..., 2.9 for G9, 3.0 for K0, etc. This is an observational result (Gray 1989a). Just past G0 IV, $v_{IV} \sim 5.0$ km/s; just past G2 III, $v_{III} \sim 7.6$ km/s.

Can we understand the rotation behavior of these evolved stars? Since the majority of evolved stars come from the upper main sequence, they start with high rotation on average. Then, obviously, as the star evolves and grows in size, the moment of inertia increases and the rotation declines accordingly. Historically, two extreme cases were considered: the rigid or solid body case, and angular momentum conserved in shells (e.g., Huang & Struve 1953, Oke & Greenstein 1954, Herbig & Spalding 1955, Sandage 1955, Slettebak 1955, Abt 1957, 1958, Kraft 1968). In recent decades, models have included more physics, such as angular momentum pumping, the dynamics of convection, internal magnetic fields, and so on (e.g., Endal & Sofia 1979, Brun & Palacios 2009, Ekström *et al.* 2012, Fuller *et al.* 2014, Kissin & Thompson 2015, Privitera *et al.* 2016). But magnetic braking remains the crucial factor in the scenario because (1) the drop in rotation occurs when the convective envelope develops, allowing dynamos to function; (2) changes in moment of inertia, internal redistribution of angular momentum, and so on do not reduce the model rotation rates enough to match the observed levels; (3) magnetic activity in the form of chromospheric and coronal emission, and starspots starts at the rotation boundary, so all the hallmarks of magnetic braking seen for dwarfs are found for subgiants and giants; (4) the rapid braking that stars experience as they evolve across the rotation boundary parallels the initial rapid braking seen for dwarfs (as in Figure 18.26).

And there is more. Given the large rotation rates to the left of the break in Figures 18.28 and 18.29, all but the very slowest rotators will fulfill the dynamo criterion in Expression 18.20. So we can expect the dynamo and the magnetic braking to start functioning for virtually all of the stars as soon as the convective envelope becomes sufficiently developed. Now comes the interesting point. Braking reduces the rotation for all these stars until $v_{rot} \lesssim 2\pi v_{con}/(\ell/R)$, at which time the dynamo criterion is no longer satisfied, the dynamo and the brake turn off, and the star is left spinning at this υ_{rot}. These values, as we have seen, are ~5.0 km/s at ~G0 for class IV and ~7.6 km/s at ~G2 for class III, according to the observations (Equations 18.21). The rotation boundary is the great leveler, reducing all rotation rates to the same dynamo turn-off value. Any distribution of rotation velocities that existed before stars crossed the rotation boundary is destroyed, reduced to a single value.

Rotation rates continue to decline toward later spectral types, possibly owing to further increases in moment of inertia or possibly under the control of convection through the dynamo criterion as ℓ/R continues to increase, or both (Gray 1982b, 1982c, 1986, Ekström *et al.* 2012).

A small fraction of cool giants are mavericks and rotate rapidly (Carlberg *et al.* 2011, 2012). Suggested explanations for their enhanced rates include undetected binaries, coalescence of binaries or engulfment of large planets, and dredge-up of angular moment from a still rapidly-rotating core.

Rotation of Cool Bright Giants (II) and Supergiants (I)

These high-luminosity stars show no rotational break. In the G- and K-star region, their rotation rates are ≲5–8 km/s, which is consistent with modeled conservation of angular momentum and evolutionary changes in moment of inertia (Gray & Toner 1986, 1987; Ekström 2012). Such low rotation rates likely fall below the dynamo criterion in Expression 18.20. The evidence from proxy indicators of dynamo activity is mixed. Many supergiants emit x-rays, but at much lower levels than seen in giants, suggesting weak or absent dynamos, but the high-excitation lines like those of C IV emission are present, indicating some residual magnetic activity (Linsky & Haisch 1979, Ayres *et al.* 1981, Maggio *et al.* 1990, Güdel 2004, Ayres 2005). In the case of the M supergiant Betelgeuse, its giant convection cells are suggested by Mathias *et al.* (2018) to be the generator of the ~1 G magnetic field they found.

Rotation of stars in the extreme upper right of the HR diagram have not been extensively studied spectroscopically because of the severe blending of lines and because the macro-turbulence of these stars is rather large, ~15 km/s (e.g., Gray 2000, 2008, Chapter 17). Rotation of Betelgeuse was investigated by Kervella *et al.* (2018) at $\lambda = 0.88$ mm and found to have $\upsilon \sin i = 5.5 \pm 0.3$ km/s, in reasonable agreement with the model calculations (Ekström *et al.* 2012) and with the earlier rotation rate based on chromospheric observations (Uitenbroek *et al.* 1998). The period of Betelgeuse's rotation is uncertain owing to a small and uncertain parallax and uncertainty in the axial inclination, but is likely tens of years (~10^4 days, ~500 times longer than the solar period).

Rotation of White Dwarfs and Neutron Stars

Some of the latest evolutionary stages for which we can measure rotation are the white-dwarf and neutron-star phases. These are vast areas of study and are just mentioned here. Even though stars have shrunk to very small dimensions upon becoming white dwarfs, they rotate rather slowly. Rotation periods have been determined for a handful of white dwarfs, and these range widely from a few minutes to ~18 days, but the most common values are ~1 day. This corresponds to $v \sim 0.7$ km/s or $v \sin i \sim 0.6$ km/s (Kawaler 2015). Not only are these values of interest in their own right, but they give us a window on the internal angular momentum distribution of their forerunners, the highly evolved red giants (e.g., Cantiello *et al.* 2014).

Neutron stars, seen as pulsars, have observed rotation periods between ~0.001 and 30 seconds. With estimated radii of ~10 km, equatorial velocities range from $\gtrsim 10^4$ (!) to ~2 km/s. Their angular momentum is dissipated rapidly according to the timing of pulsar pulses (Ho & Andersson 2012, Antonopoulou *et al.* 2018). The shortest *orbital* period for a binary pulsar is 38 minutes (Strohmayer *et al.* 2018).

Rotation of Binary Stars

The binary stars we consider in this section have their components close enough together to have tidal interaction, but not so close as to be in contact. Tides are the conduit through which angular momentum is exchanged between the orbit and the stars. Wide binaries, those too far apart to have significant tidal interaction, simply rotate as single stars. Rotation for a close binary is at the mercy of its orbit since the orbital angular momentum is ~100–1000 times larger than the stellar angular momentum. In short, the stars are coerced into having the same angular velocity as the orbital motion. Rapid rotators are slowed; slow rotators are sped up. The simplest case occurs for orbits with eccentricity approaching zero. Then tidal coupling makes the two periods equal and there is exact synchronization. But a binary's life becomes more complicated when the orbit is eccentric. The tidal forces, being differential gravitational forces, vary as the inverse cube of the separation. Consequently, coupling forces at periastron dominate. So the star's rotation tries to match the orbital angular velocity near periastron, which is higher than at other orbital positions according to Kepler's second law (conservation of orbital angular momentum). The resulting rotational period is therefore shorter than the orbital period (e.g., Hall 1986).

The degree of tidal interaction depends on the size of the orbit compared to the size of the stars, expressed as semimajor axis over stellar radius, a/R, and it takes time for the angular-momentum transfer to occur (Zahn 1989, Meibom *et al.* 2006). An example given by Tassoul (1987) indicates a synchronization time for a 9 solar mass star with $a/R \sim 20$ to be ~2 × 10^7 years, which is comparable to the main-sequence lifetime. So one expects altered rotation for binaries having orbits smaller than this, unless they are very new arrivals on the main sequence, and this is borne out by the observations (e.g., Giuricin *et al.* 1984b, 1984c). The tidal interaction depends only modestly on mass, whereas the main-sequence lifetime depends strongly on mass. Synchronization effects are therefore more complete for lower mass stars. In binaries with evolved components, a/R is much smaller so tidal interactions work faster.

Figure 18.30 shows rotation rates of close binaries as a function of spectral type. On average close binaries of early spectral type rotate slower than single stars, while those of later type often rotate many times faster than their single counterparts. The distribution of equilibrium rotation rates hinges on the distribution of orbital angular velocities or orbital periods, which in turn is constrained by these stars being close binaries. Comparisons of rotation characteristics with coupling theory are given by Hall (1990) and Meibom *et al.* (2006). Although there is no steep drop in the rotation near F5 like the one seen for single stars, magnetic braking of cool stars does not go away. It continues to drain angular momentum from the star(s), which is replenished from the orbital reservoir. It is clear from Figure 18.30 that magnetic braking has not been successful in overpowering tidal spin-up, but there may be evidence that it is not far behind (Fleming *et al.* 2019).

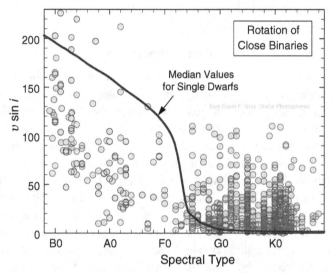

Figure 18.30 Rotation rates of interacting binaries are controlled by exchange of angular momentum through tides. Their rotation rates do not show the abrupt drop near F5. Based on data from Koch *et al.* (1965), Levato (1974), Huisong and Xuefu (1987), De Medeiros *et al.* (2004), and Eker *et al.* (2008).

The results of the "extra" rotation are strikingly seen in the G and K binaries, where their magnetic activity is often greatly enhanced compared to single stars. Most notably, their chromospheric emission lines are very strong, flares are seen, and they can have giant dark starspots. Stars in this situation are typically α CVn stars, BY Dra stars, and similar types (Bopp & Fekel 1977, Strassmeier *et al.* 1990, Cutispoto & Rodono 1992, Rodono 1992, Kunt & Dal 2017, Cao *et al.* 2019, Pi *et al.* 2019). A few super-active single stars of the FK Com type likely have orbital angular momentum turned into stellar rotation when the binary components coalesce (Ayres *et al.* 2016).

Tidal forces also circularize binary orbits. More angular momentum is dissipated at periastron because the tidal interaction is strongest there. This amounts to a reduction in orbital velocity near periastron and a corresponding decrease in eccentricity. Since the orbital angular momentum is larger than the stellar angular momentum, the timescales for circularization are longer than for synchronization (Giuricin *et al.* 1984a, Tassoul 1988).

The observations show that binaries having dwarf components with periods less than ~10 days have circular orbits, while longer period orbits show a wide range of eccentricity (Duquennoy & Mayor 1991, Latham *et al.* 2002, Raghavan *et al.* 2010). Binaries with giant components have smaller a/R, and orbital periods up to ~100 days have small eccentricity (Price-Whelan & Goodman 2018). And tides eventually force the axis of rotation to align itself with the orbital angular momentum vector, i.e., perpendicular to the orbital plane (e.g., Albrecht *et al.* 2013).

Contact binaries are considerably more complicated (Lucy 1968a, 1968b, Mochnacki 1981, Plavec 1986, Rucinski 2015, Eaton 2016), and because of the mass exchange across the Roche lobe, sporadic variations in the rotation may even occur (e.g., Olson 1984). At least for the simpler cases, the rotation profiles essentially show the shape of the two stars, one more example of the rotational Doppler-shift distribution in Figure 18.5 (Anderson & Shu 1979, Anderson *et al.* 1983). This Doppler-shift distribution also allows us to map stellar surfaces through *rotational mapping*.

Rotational Mapping

Rotational mapping is also called Doppler imaging. In Figure 18.31 we revisit the one-to-one correspondence between position within the Doppler-shift distribution, $G(\lambda)$, and x position on the stellar disk, as in Figure 18.5. In Figure 18.31, we assume an equator-on aspect of view. Any surface feature that alters the amount of light coming from only certain x bands will introduce structure into the rotation profile. In the simple case in Figure 18.31, a single dark spot reduces the amount of light in the strip containing it. This produces a dip in $G(\lambda)$ at the Doppler shift corresponding to that strip. The strength of the dip starts at zero when the spot is coming out around the approaching limb (left side), since its projected area is zero. It then grows to a maximum as the spot crosses the meridian and returns to zero as it slips from view on the receding limb (right side). Of course, when this process is seen in absorption lines, $G(\lambda)$ is upside down and the dip looks like an emission bump in the profile. In real spectra, we are going to be able to see these details only if $G(\lambda)$ dominates the line broadening. In practice that means $v \sin i$ has to be larger than ~15–20 km/s.

More generally, the Doppler shift (in km/s) for a spot of latitude ℓ and longitude L (measured from the central meridian) is given by

$$\Delta\lambda = \frac{x}{R} v \sin i = \cos \ell \, \sin L \, v \sin i. \tag{18.22}$$

The higher the spot latitude, the more restricted is the $\Delta\lambda$ range around line center. Or turning this around, following the range of $\Delta\lambda$ for the spot gives us the latitude of the spot. When the rotation axis is not perpendicular to the line of sight, latitude information is still contained in the fraction of the rotation cycle over which the spot or bump is visible. If the latitude is high and the inclination of the rotation axis is small (pole toward us), the spot may be visible continuously. In such a case, the bump moves back and forth through the profile, its size changing with the spot's projected area.

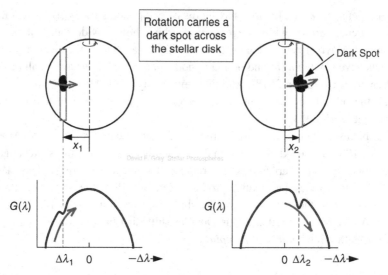

Figure 18.31 Upper panel: a dark spot is shown on the stellar disk at two rotation phases, approaching at position x_1 and receding at position x_2. Lower panel: Doppler-shift distributions for the two cases, where $\Delta\lambda_1$ and $\Delta\lambda_2$ correspond to x_1 and x_2 through Equation 18.1. Reduced light in the strip containing the spot causes a dip in $G(\lambda)$. The dip migrates across $G(\lambda)$ as the star's rotation moves the spot across the disk. The size of the dip is proportional to the projected area of the spot.

The case we have been considering assumes a single time-stable position-stable spot. On the Sun, spots often do not last a full rotation period, there can be more than one spot, and sunspots are small, $\lesssim 1\%$ area. Sun-like spots on other stars are not easy to see. Interestingly, some starspots are often larger and last longer than their solar counterparts. Magnetically active stars in particular have larger spots that last for many rotation cycles. Spectra of the RS CVn type star σ Gem are shown at two rotation phases in Figure 18.32. Spot bumps in line profiles were first recognized in the RS CVn star HR 1099 by Fekel (1983), and photometric variations were first interpreted as starspots by Eaton and Hall (1979).

The decoding of profile structure has been expanded into mapping of the stellar surface. Full mapping, while limited in resolution, reveals spot patterns, their latitude and longitude positions, size, and temperature, and in some cases, the magnetic field (Goncharsky *et al.* 1982, Vogt & Penrod 1983, Vogt *et al.* 1987, Rice *et al.* 1989, Strassmeier & Rice 1998, Morgenthaler *et al.* 2012, Kochukhov & Shulyak 2019). A comprehensive review is given by Strassmeier (2009). Differential rotation has even been mapped (e.g., Barnes *et al.* 2000, Collier Cameron 2002, Kovári *et al.* 2015, Pi *et al.* 2019, Xiang *et al.* 2020). Photometric variations arising from extra-solar planetary transits have been used by Netto and Valio (2020) to map starspots and their changes over time.

In addition to starspots, rotational mapping of chemical areas on Ap stars and their kin has become common, for example, Khokhlova *et al.* (1986), Rice and Wehlau (1990, 1994), and Rusomarov *et al.* (2016).

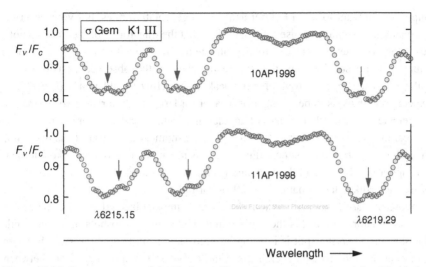

Figure 18.32 Two exposures of the K giant σ Gem taken on consecutive nights. The dark-spot bumps (arrows) move through the profiles as the star rotates. Period = 19.6 days. Elginfield Observatory data.

When $v \sin i$ is less than ~10–15 km/s, most of the spatial resolution is lost, but variations in line broadening and line asymmetries can still yield interesting information about areas having enhanced velocity fields (Toner & Gray 1988) or magnetic fields (Kochukhov *et al.* 2004, 2019).

Questions and Exercises

1. Draw the profiles of a spectral line without rotational broadening and with 30 km/s rotational broadening. Have you conserved the equivalent width? Why is the equivalent width not altered by rotation? Now attempt to draw a rotationally broadened profile when the rotation rate is 300 km/s. What special challenges arise when measuring line profiles with very large rotational broadening?
2. How would the rotation profile, $G(\lambda)$, be altered from the classical case if the star were a rapid rotator showing an elliptical shape (equator-on aspect of view)? If the rotation is rapid enough to significantly distort the shape, then so-called gravity darkening will also come into play, so the equator will be less bright than the poles. Stars can be tricky. How does this brightness variation alter $G(\lambda)$?
3. Notice how in Figure 18.5 you can essentially read off the rotation rate by simply looking at the velocity position of the edges of the profile. Give it a try on the profiles in Figures 18.2 and 18.8. What $v \sin i$ values do you get?
4. In the case illustrated in Figure 18.5, lines of constant Doppler shift are straight and parallel to the projected rotation axis. What would lines of constant Doppler shift look like if the star had differential rotation, i.e., angular rotation depended on stellar latitude?

5. Suppose you want to use a Fourier analysis to get rotation rates, but your observations have a modest signal-to-noise ratio. You find that the transforms of the lines do not even reach the first rotation zero before they plunge below the white noise level. To make the situation more specific, say that the main lobes of the observed transforms can be followed down to only 20% of their peak value. How would you proceed? Do you expect your analysis to be stronger or weaker for large rotation rates versus small ones?

6. To get a feel for the relative sizes of angular momenta, calculate L for (a) the Sun, where you can use $L = I\Omega = \mathfrak{M}k^2R^2\Omega$, where I is the moment of inertia, $\mathfrak{M}$ is the mass, k is the radius of gyration, R is the radius, and Ω is the angular velocity – take k to be 0.25, (b) Jupiter in orbit around the Sun, and (c) the orbital motion of a binary star consisting of two solar-like stars separated by 20 stellar radii.

7. Suppose a rotating star with axis in the plane of the sky has a dark spot ~0.05R in size near its equator, where R is the star's radius. Draw three snapshots of a line profile for this star: (a) when the spot is just coming into view from behind the star, (b) when the spot is at the center of the stellar disk, and (c) when the spot is halfway between the center of the disk and the limb where it will disappear behind the star. How would your diagrams change with inclination of the rotation axis?

   With that mastered, consider now a similar situation but with *two spots* separated by 90° in longitude, one on the equator and the other at 60° latitude. Consider the line profile changes that result from altering the relative sizes of the spots. How would your results be affected if the star rotated differentially in latitude?

8. Use the profiles in Figure 18.32 to make your own estimate of the rotation period of σ Gem. What additional observations or information would you like in order to improve the accuracy of your result?

References

Abney, W.deW. 1877. *MNRAS* **37**, 278.
Abt, H.A. 1957. *ApJ* **126**, 503.
Abt, H.A. 1958. *ApJ* **127**, 658.
Abt, H.A. 1979. *ApJ* **230**, 485.
Abt, H.A. 2001. *AJ* **122**, 2008.
Abt, H.A. & Morrell, N.I. 1995. *ApJS* **99**, 135.
Abt, H.A., Levato, H., & Grosso, M. 2002. *ApJ* **573**, 359.
Adams, W.S. 1909. *ApJ* **29**, 110.
Afram, N. & Berdyugina, S.V. 2019. *A&A* **629**, 83.
Albrecht, S., Winn, J.N., Carter, J.A., Snellen, I.A.G., & de Mooij, E.J.W. 2011. *ApJ* **726**, 68.
Albrecht, S., Setiawan, J., Torres, G., Fabrycky, D.C., & Winn, J.N. 2013. *ApJ* **767**, 32.
Alschuler, W.R. 1975. *ApJ* **195**, 649.
Amard, L., Palacios, A., Charbonnel, C., *et al.* 2019. *A&A* **631**, 77.
Ammler-von Eiff, M. & Reiners, A. 2012. *A&A* **542**, 116.
Anderson, L. & Shu, F.H. 1979. *ApJS* **40**, 667.
Anderson, L., Stanford, D., & Leininger, D. 1983. *ApJ* **270**, 200.
Antonopoulou, D., Espinoza, C.M., Kuiper, L., & Andersson, N. 2018. *MNRAS* **473**, 1644.

Arcos, C., Jones, C.E., Sigut, T.A.A., Kanaan, S., & Curé, M. 2017. *ApJ* **842**, 48.
Ayres, R.R. 2005. *ApJ* **618**, 493.
Ayres, T.R., Linsky, J.L., Vaiana, G.S., Golub, L., & Rosner. R. 1981. *ApJ* **250**, 293.
Ayres, T.R., Kashyap, V., Saar, S., *et al.* 2016. *ApJS* **223**, 5.
Baliunas, S.L. & Vaughan, A.H. 1985. *ARAA* **23**, 379.
Balthasar, H. 1984. *Solar Phys* **93**, 219.
Barnes, J.R., Collier Cameron, A., James, D.J., & Donati, J.-F. 2000. *MNRAS* **314**, 162.
Barnes, S.A. 2007. *ApJ* **669**, 1167.
Barry, D.C., Cromwell, R.H., & Hege, E.K. 1987. *ApJ* **315**, 264.
Bayliss, D.D.R., Winn, J.N., Mardling, R.A., & Sackett, P.D. 2010. *ApJ* **722**, 224.
Bazot, M., Benomar, O., Christensen-Dalsgaard, J., *et al.* 2019. *A&A* **623**, 125.
Belcher, J.W. & MacGregor, K.B. 1976. *ApJ* **210**, 498.
Benomar, O., Bazot, M., Nielsen, M.B., *et al.* 2018. *Science* **361**, 1231.
Benz, W. & Mayor, M. 1981. *A&A* **93**, 235.
Bernacca, P.L. & Perinotto, M. 1974. *A&A* **33**, 443.
Blazère, A., Petit, P., Neiner, C., *et al.* 2020. *MNRAS* **492**, 5794.
Böhm-Vitense, E. 2007. *ApJ* **657**, 486.
Bopp, B.W. & Fekel, F., Jr. 1977. *AJ* **82**, 490.
Bouvier, J., Matt, S.P., Mohanty, S., *et al.* 2014. *Protostars and Planets VI* (Tucson: University of Arizona Press), H. Beuther, R. Klessen, K. Dullemond, & Th. Henning, eds., p. 433.
Bragança, G.A., Daflon, S., Cunha, K., *et al.* 2012. *AJ* **144**, 130.
Brott, I., deMink, S.E., Cantiello, M., *et al.* 2011. *A&A* **530**, 115.
Brun, A.S. & Palacios, A. 2009. *ApJ* **702**,1078.
Bruning, D.H. 1981. *ApJ* **248**, 274.
Cantiello, M., Mankovich, C., Bildsten, L., Christensen-Dalsgaard, J., & Paxton, B. 2014. *ApJ* **788**, 93.
Cao, D., Gu, S., Ge, J., *et al.* 2019. *MNRAS* **482**, 988.
Carlberg, J.K., Majewski, S.R., Patterson, R.J., *et al.* 2011. *ApJ* **732**, 39.
Carlberg, J.K., Cunha, K., Smith, V.V., & Majewski, S.R. 2012. *ApJ* **757**, 109.
Carroll, J.A. 1933a. *MNRAS* **93**, 478.
Carroll, J.A. 1933b. *MNRAS* **93**, 680.
Carroll, J.A. & Ingram, L.J. 1933. *MNRAS* **93**, 508.
Catalano, S. & Marilli, E. 1983. *A&A* **121**, 190.
Chandrasekhar, S. & Münch, G. 1950. *ApJ* **111**, 142.
Chauville, J., Zorec, J., Ballereau, D., *et al.* 2001. *A&A* **378**, 861.
Choi, H.-J., Soon, W., Donahue, R.A., Baliunas, S.L., & Henry, G.W. 1995. *PASP* **107**, 744.
Cohen, O., Drake, J.J., Kashyap, V.L., & Gombosi, T.I. 2009. *ApJ* **699**, 1501.
Collier Cameron, A. 2002. *Ast Nach* **323**, 336.
Collins, G.W., II 1966. *ApJ* **146**, 914.
Collins, G.W., II & Truax, R.J. 1995. *ApJ* **439**, 860.
Conti, P.S. 1965. *ApJ* **142**, 1594.
Conti, P.S. & Ebbets, D. 1977. *ApJ* **213**, 438.
Cotton, D.V., Evensberget, D., Marsden, S.C., *et al.* 2019. *MNRAS* **483**, 1574.
Curtis, J.L., Agüeros, M.A., Douglas, S.T., & Meibom, S. 2019. *ApJ* **879**, 49.
Cutispoto, G. & Rodono, M. 1992.*The Solar Cycle*, ASP Conference Series **27**, p. 465.
Cuypers, J. 1986. *A&A* **167**, 282.
de Jager, C., Nieuwenhuijzen, H., & van der Hucht, K.A. 1988. *A&AS* **72**, 259.
De Lury, R.E. 1939. *JRASC* **33**, 345.

Demarque, P.R. & Larson, R.B. 1964. *ApJ* **140**, 544.

De Medeiros, J.R. & Mayor, M. 1999. *A&AS* **139**, 433.

De Medeiros, J.R., Udry, S., & Mayor, M. 2004. *A&A* **427**, 313.

Dicke, R.H. 1970. *Stellar Rotation* (Dordrecht: Reidel), A. Slettebak, ed., p. 289.

Douglas, S.T., Agüeros, M.A., Covey, K.R., *et al.* 2016. *ApJ* **822**, 47.

Douglas, S.T., Agüeros, M.A., Covey, K.R., & Kraus, A. 2017. *ApJ* **842**, 83.

Doyle, L.R., Wilcox, T.J., & Lorre, J.J. 1984. *ApJ* **287**, 307.

Dravins, D., Lindegren, L., & Torkelsson, U. 1990. *A&A* **237**, 137.

Dufton, P.L., Ryans, R.S.I., Simón-Díaz, S., Trundle, C., & Lennon, D.J. 2006. *A&A* **451**, 603.

Duquennoy, A. & Mayor, M. 1991. *A&A* **248**, 485.

Durney, B.R. & Latour, J. 1978. *Geophys Ap Fluid Dyn* **9**, 241.

Duvall, T.L., Jr. 1982. *Solar Phys* **76**, 137.

Eaton, J.A. 2016. *MNRAS* **457**, 836.

Eaton, J.A. & Hall, D.S. 1979. *ApJ* **227**, 907.

Eker, Z., Filiz Ak, N., Bilir, S., *et al.* 2008. *MNRAS* **389**, 1722.

Ekström, S., Georgy, C., Eggenberger, P., *et al.* 2012. *A&A* **537**, 146.

Endal, A.S. & Sofia, S. 1979. *ApJ* **232**, 531.

Espinosa Lara, F. & Rieutord, M. 2011. *A&A* **533**, 43.

Evans, C.J., Lennon, D.J., Smartt, S.J., & Trundle, C. 2006. *A&A* **456**, 623.

Faulkner, J., Roxburgh, I.W., & Strittmatter. P.A. 1968. *ApJ* **151**, 203.

Fekel, F.C., Jr. 1983. *ApJ* **268**, 274.

Fekel, F.C. 1997. *PASP* **109**, 514.

Fekel, F.C. 2003. *PASP* **115**, 807.

Fleming, D.P., Barnes, R., Davenport, J.R.A., & Luger, R. 2019. *ApJ* **881**, 88.

Frasca, A., Marilli, E., & Catalano, S. 1998. *A&A* **333**, 205.

Fuller, J., Lecoanet, D., Cantiello, M., & Brown, B. 2014. *ApJ* **796**, 17.

Gallet, F. & Bouvier, J. 2013. *A&A* **556**, 36.

Garcia-Alegre, M.C., Vazquez, M., Wöhl, H. 1982. *A&A* **106**, 261.

Garraffo, C., Drake, J.J., & Cohen, O. 2015. *ApJ* **813**, 40.

Garraffo, C., Drake, J.J., Dotter, A., *et al.* 2018. *ApJ* **862**, 90.

Giuricin, G., Mardirossian, F., & Mezzetti, M. 1984a. *A&A* **134**, 365.

Giuricin, G., Mardirossian, F., & Mezzetti, M. 1984b. *A&A* **135**, 393.

Giuricin, G., Mardirossian, F., & Mezzetti, M. 1984c. *A&A* **141**, 227.

Goncharsky, A.V., Stepanov, V.V., Khokhlova, V.L., & Yagola, A.G. 1982. *Soviet Ast* **26**, 690.

Gossage, S., Conroy, C., Dotter, A., *et al.* 2019. *ApJ* **887**, 199.

Gray, D.F. 1977. *ApJ* **211**, 198.

Gray, D.F. 1982a. *ApJ* **258**, 201.

Gray, D.F. 1982b. *ApJ* **261**, 259.

Gray, D.F. 1982c. *ApJ* **262**, 682.

Gray, D.F. 1983. *IAUS* **102**, 461.

Gray, D.F. 1984a. *ApJ* **277**, 640.

Gray, D.F. 1984b. *ApJ* **281**, 719.

Gray, D.F. 1986. *Highlights in Astronomy* (Dordrecht: IAU) **7**, p. 411.

Gray, D.F. 1988. *Lectures on Spectral-Line Analysis: F, G, and K Stars* (Arva, Ontario: The Publisher), Chapter 5.

Gray, D.F. 1989a. *ApJ* **347**, 1021.

Gray, D.F. 1989b. *PASP* **101**, 1126.

Gray, D.F. 1999. *Precise Stellar Radial Velocities*, J.B. Hearnshaw & C.D. Scarfe, eds., Astronomical Society of the Pacific, Conference Series **185**, p. 243.

Gray, D.F. 2000. *ApJ* **532**, 487.

Gray, D.F. 2008. *AJ* **135**, 1450.

Gray, D.F. 2014. *AJ* **147**, 81.

Gray, D.F. 2016. *ApJ* **826**, 92.

Gray, D.F. 2018a. *ApJ* **857**, 139.

Gray, D.F. 2018b. *ApJ* **869**, 81.

Gray, D.F. & Baliunas S.L. 1997. *ApJ* **475**, 303.

Gray, D.F. & Desikachary, K. 1973. *ApJ* **181**, 523.

Gray, D.F. & Nagar, P. 1985. *ApJ* **298**, 756.

Gray, D.F. & Toner, C.G. 1986. *ApJ* **310**, 277.

Gray, D.F. & Toner C.G. 1987. *ApJ* **322**, 360.

Güdel, M. 2004. *A&AR* **12**, 71.

Gulliver, A.F., Hill. G., & Adelman, S.J. 1994. *ApJL* **429**, L81.

Halbedel, E.M. 1996. *PASP* **108**, 833.

Hall, D.S. 1986. *ApJL* **309**, L83.

Hall, D.S. 1990. *Active Close Binaries* (Dordrecht: Kluwer), C. Ibanoglu, ed., p. 95.

Hart, A.B. 1954. *MNRAS* **114**, 17.

Hartman, J.D., Bakos, G.Á., Kovács, G., & Noyes, R.W. 2010. *MNRAS* **408**, 475.

Herbig, G.H. & Spalding, J.F., Jr. 1955. *ApJ* **121**, 118.

Hill, G.M. 1995. *A&A* **294**, 536.

Ho, W.C.G. & Andersson, N. 2012. *Nature Phys* **8**, 787.

Holzwarth, V. & Jardine, M. 2005. *A&A* **444**, 661.

Horne, J.H. & Baliunas, S.L. 1986. *ApJ* **302**, 757.

Howe, R., Christensen-Dalsgaard, J., Hill, F., *et al.* 2005. *ApJ* **634**, 1405.

Howard, R. & Harvey, J. 1970. *Solar Phys* **12**, 23.

Howard, R., Adkins, J.M., Boyden, J.E., *et al.* 1983. *Solar Phys* **83**, 321.

Howarth, I.D., Siebert, K.W., Hussain, G.A.J., & Prinja, R.K. 1997. *MNRAS* **284**, 265.

Huang, W. & Gies, D.R. 2008. *ApJ* **683**, 1054.

Huang, S.-S. & Struve, O. 1953. *ApJ* **118**, 463.

Huang, S.-S. & Struve, O. 1954. *AnAp* **17**, 85.

Huisong, T. & Xuefu, L. 1987. *A&A* **172**, 74.

Johns-Krull, C.M. 1996. *A&A* **306**, 803.

Jurkevich, I. 1971. *ApSS* **13**, 154.

Kambry, M.A. & Nishikawa, J. 1990. *Solar Phys* **126**, 89.

Karoff, C., *et al.* 2018. *ApJ* **852**, 46.

Kawaler, S.D. 1987. *PASP* **99**, 1322.

Kawaler, S.D. 1988. *ApJ* **333**, 236.

Kawaler, S.D. 1989. *ApJ* **343**, L65.

Kawaler, S.D. 2015. *ASP Conference Series* **493**, 65.

Keeler, J.E. 1895a. *PASP* **7**, 154.

Keeler, J.E. 1895b. *MNRAS* **55**, 474.

Kervella, P., Decin, L., Richards, A.M.S., *et al.* 2018. *A&A* **609**, 67.

Khokhlova, V.L., Rice, J.B., & Wehlau, W.H. 1986. *ApJ* **307**, 768.

Kholtygin, A.F., Fabrika, S.N., Drake, N.A., *et al.* 2010. *Ast Lett* **36**, 370.

Kissin, Y. & Thompson, C. 2015. *ApJ* **808**, 35.

Kjurkchieva, D., Marchev, D., Sigut, T.A.A., & Dimitrov, D. 2016. *AJ* **152**, 56.

Koch, R.H., Olson, E.C., & Yoss, K.M. 1965. *ApJ* **141**, 955.

Kochukhov, O. & Shulyak, D. 2019. *ApJ* **873**, 69.

Kochukhov, O., Bagnulo, S., Wade, G.A., *et al.* 2004. *A&A* **414**, 613.

Kochukhov, O., Shultz, M., & Neiner, C. 2019. *A&A* **621**, 47.

Kotov, V.A. 2017. *Solar Phys* **292**, 76.

Kovári, Z., Kriskovics, L., Künstler, A., *et al.* 2015. *A&A* **573**, 98.

Kraft, R.P. 1968. *Stellar Astronomy* (New York: Gordon & Breach), H.-Y. Chiu, R.L. Warasila, & J.L. Remo, eds., Vol. 1, p. 317.

Kraft, R.P. 1970. *Spectroscopic Astrophysics* (Berkeley: University of California Press), ed. G.H. Herbig, p. 385.

Kunt, M. & Dal, H.A. 2017. *Acta Ast* **67**, 345.

Kurtz, D.W. 1985. *MNRAS* **213**, 773.

Küveler, G. & Wöhl, H. 1983. *A&A* **123**, 29.

LaBonte, B.J. 1981. *Solar Phys* **69**, 177.

Labonte, B.J. & Howard, R. 1982. *Solar Phys* **80**, 361.

Lachaume, R., Dominik, C., Lanz, T., & Habing, J.J. 1999. *A&A* **348**, 897.

Latham, D.W., Stefanik, R.P., Torres, G., *et al.* 2002. *AJ* **124**, 1144.

Lesage, A.-L. & Wiedemann, G. 2014. *A&A* **563**, 86.

Levato, H. 1974. *A&A* **35**, 259.

Linsky, J.L. & Haisch, B.M. 1979. *ApJ* **229**, L27.

Lipatov, M. & Brandt, T.D. 2020. *ApJ* **901**, 100.

Livingston, W.C. 1969. *ASP Leaflet 484*, **10**, 265.

Lockwood, G.W., Thompson, D.T., Radick, R.R., *et al.* 1984. *PASP* **96**, 714.

Lucy, L.B. 1968a. *ApJ* **151**, 1123.

Lucy, L.B. 1968b. *ApJ* **153**, 877.

Lurie, J.C., Vyhmeister, K., Hawley, S.L., *et al.* 2017. *AJ* **154**, 250.

Lustig, G. & Wöhl, H. 1989. *A&A* **218**, 299.

Maggio, A., Vaiana, G.S., Haisch, B.M., *et al.* 1990. *ApJ* **348**, 253.

Mamajek, E.E. & Hillenbrand, L.A. 2008. *ApJ* **687**, 1264.

Mangeney, A. & Praderie, F. 1984. *A&A* **130**, 143.

Marcy, G.W. 1984. *ApJ* **276**, 286.

Marraco, H.G. & Muzzio, J.C. 1980. *PASP* **92**, 700.

Marsden, S.C., Carter, B.D., & Donati, J.-F. 2009. *MNRAS* **399**, 888.

Mathew, A. & Rajamohan, R. 1992. *JA&A* **13**, 61.

Mathias, P., Aurière, M., López Ariste, A., *et al.* 2018. *A&A* **615**, 116.

McLaughlin, D.B. 1924. *ApJ* **60**, 22.

McLaughlin, D.B. 1961. *JRASC* **55**, 13 and 73.

McNally, D. 1965. *Observatory* **85**, 166.

McQuillan, A., Mazeh, T., & Aigrain, S. 2014. *ApJS* **211**, 24.

Meibom, S., Mathieu, R.D., & Stassun, K.G. 2006. *ApJ* **653**,621.

Meibom, S., Barnes, S.A., Platais, I., *et al.* 2015. *Nature* **517**, 589.

Metcalfe, T.S., Egeland, R., & van Saders, J. 2016. *ApJL* **826**, L2.

Meynet, G. & Maeder, A. 2000. *A&A* **361**, 101.

Mochnacki, S.W. 1981. *ApJ* **245**, 650.

Mochnacki, S.W., Gladders, M.D., Thomson, J.R., *et al.* 2002. *AJ* **124**, 2868.

Moore, J.H. & Menzel, D.H. 1928. *PASP* **40**, 234.

Morgenthaler, A., Petit, P., Saar, S., *et al.* 2012. *A&A* **540**, 138.

Moutou, C., Hébrard, E.M., Morin, J., *et al.* 2017. *MNRAS* **472**, 4563.

Netto, Y. & Valio, A. 2020. *A&A* **635**, 78.

Newton, H.W. & Nunn, M.L. 1951. *MNRAS* **111**, 413.

Nielsen, M.B., Gizon, L., Schunker, H., & Karoff, C. 2013. *A&A* **557**, L10.

Nielsen, M.B., Gizon, L., Cameron, R.H., & Miesch, M. 2019. *A&A* **622**, 85.

Noyes, R.W., Hartmann, L.W., Baliunas, S.L., Duncan, D.K., & Vaughan, A.H. 1984. *ApJ* **279**, 763.

Oke, J.B. & Greenstein, J.L. 1954. *ApJ* **120**, 384.

Olson, E.C. 1984. *PASP* **96**, 376.

Pallavicini, R., Golub, L., Rosner, R., *et al.* 1981. *ApJ* **248**, 279.

Penny, L.R. 1996. *ApJ* **463**, 737.

Penny, L.R. & Gies, D.R. 2009. *ApJ* **700**, 844.

Perez Garde, M., Vazquez, M., Schwan, H., & Wöhl, H. 1981. *A&A* **93**, 67.

Pi, Q.-F., Zhang, L.-Y., Bi, S-.l., *et al.* 2019. *ApJ* **877**, 75.

Plaskett, J.S. 1915. *ApJ* **42**, 373.

Plaskett, H.H. 1962. *MNRAS* **123**, 541.

Plavec, M.J. 1986. *Instrumentation and Research Programmes for Small Telescopes* (Dordrecht: sReidel), J.B. Hearnshaw & P.L. Cottrell, eds., p. 173.

Pognan, Q., Garraffo, C., Cohen, O., & Drake, J.J. 2018. *ApJ* **856**, 53.

Poeckert, R. & Marlborough, J.M. 1978. *ApJS* **38**, 229.

Price-Whelan, A.M. & Goodman, J. 2018. *ApJ* **867**, 5.

Privitera, G., Meynet, G., Eggenberger P., *et al.* 2016. *A&A* **591**, 45.

Raghavan, D., McAlister, H.A., Henry, T.J., *et al.* 2010. *ApJS* **190**, 1.

Ramella, M., Boehm, C., Gerbaldi, M., & Faraggiana, R. 1989. *A&A* **209**, 233.

Reiners, A. & Schmitt, J.H.M.M. 2003. *A&A* **412**, 813.

Reinhold, T., Bell, K.J., Kuszlewicz, J., Hekker, S., & Shapiro, A.I. 2019. *A&A* **621**, 21.

Rice, J.B. & Wehlau, W.H. 1990. *A&A* **233**, 503.

Rice, J.B. & Wehlau, W.H. 1994. *PASP* **106**, 134.

Rice, J.B., Wehlau, W.H., & Khokhlova, V.L. 1989. *A&A* **208**, 179.

Rivinius, T., Carciofi, A.C., & Martayan, C. 2013. *A&AR* **21**, 69.

Robinson, R.D., Worden, S.P., & Harvey, J.W. 1980. *ApJL* **236**, L155.

Rodono, M. 1992. *IAU Symposium* **151**, 71.

Rosén, L., Kochukhov, O., Hackman, T., & Lehtinen, J. 2016. *A&A* **593**, 35.

Rossiter, R.A. 1924. *ApJ* **60**, 15.

Royer, F., Gerbaldi, M., Faraggiana, R., & Gómez, A.E. 2002a. *A&A* **381**, 105.

Royer, F., Grenier S., Baylac M.-O., Gómez, A.E., & Zorec, J. 2002b. *A&A* **393**, 897.

Royer, F., Zorec, J., & Gómez, A.E. 2007. *A&A* **463**, 671.

Rucinski, S.M. 2015. *AJ* **149**, 49.

Rucinski, S.M. & Vandenberg, D.A. 1986. *PASP* **98**, 669.

Rusomarov, N., Kochukhov, O., Ryabchikova, T., & Ilyin, I. 2016. *A&A* **588**, 138.

Saar, S.H. 1988. *ApJ* **324**, 441.

Sandage, A.R. 1955. *ApJ* **122**, 263.

Sandage, A. 1962. *ApJ* **135**, 349.

Scargle, J.D. 1982. *ApJ* **263**, 835.

Schatzman, E. 1959. *IAU Symposium* **10**, 129.

Schatzman, E. 1962. *AnAp* **25**, 18.

Scherrer, P.H., Wilcox, J.M., & Svalgaard, L. 1980. *ApJ* **241**, 811.

Shajn, G. & Struve, O. 1929. *MNRAS* **89**, 222.

Sikora, J., Wade, G.A., & Power, J. 2018. *Cont Ast Obs Skalnaté Pleso* **48**, 87.

Simkin, S.M. 1977. *A&A* **31**, 129.

Simón-Díaz, S. & Herrero, A. 2007. *A&A* **468**, 1073.

Simón-Díaz, S., Herrero, A., Uytterhoeven, K., *et al.* 2010. *ApJL* **720**, L174.

Skumanich, A. 1972. *ApJ* **171**, 565.

Slettebak, A. 1949. *ApJ* **110**, 498.

Slettebak, A. 1955. *ApJ* **121**, 653.

Slettebak, A. 1956. *ApJ* **124**, 173.

Slettebak, A. 1966. *ApJ* **145**, 126.

Slettebak, A. 1982. *ApJS* **50**, 55.

Slettebak, A. 1985. *Calibration of Fundamental Stellar Quantities*, IAU Symposium 111 (Dordrecht: Reidel), D.S. Hayes, L.E. Pasinetti, & A.G.D. Philip, eds., p. 163.

Slettebak, A., Collins, G.W. II, Boyce, P.B., White, N.M., & Parkinson, T.D. 1975. *ApJS* **29**, 137.

Slettebak, A. & Howard, R.F. 1955. *ApJ* **121**, 102.

Snodgrass, H.B. & Ulrich, R.K. 1990. *ApJ* **351**, 309.

Snodgrass, H.B., Howard, R., & Webster, L. 1984. *Solar Phys* **90**, 199.

Soderblom, D.R. 1982. *ApJ* **263**, 239.

Soderblom, D.R. 1983. *ApJS* **53**, 1.

Soderblom, D.R. 1985. *AJ* **90**, 2103.

Soderblom, D.R., Jones, B.F., & Fischer, D. 2001. *ApJ* **563**, 334.

Stauffer, J.R. 1984. *ApJ* **280**, 189.

Stauffer, J.R. 1987. *LNP* **291**, 182.

Stauffer, J.R., Schild, R.A., Baliunas, S.L, & Africano, J.L. 1987. *PASP* **99**, 471.

Stellingwerf, R.F. 1978. *ApJ* **224**, 953.

Stoeckley, T.R. 1968. *MNRAS* **140**, 149.

Strassmeier, K.G. 2009. *A&AR* **17**, 251.

Strassmeier, K.G. & Rice, J.B. 1998. *A&A* **330**, 685.

Strassmeier, K.G., Fekel, F.C., Bopp, B.W., Dempsey, R.C., & Henry, G.W. 1990. *ApJS* **72**, 191.

Strohmayer, T.E., Arzoumanian, Z., Bogdanov, S., *et al.* 2018. *ApJL* **852**, L13.

Struve, O. 1945. *Popular Ast* **53**, 201 and 259.

Svalgaard, L., Scherrer, P.H., & Wilcox, J.M. 1978. *Proceedings of the Workshop on Solar Rotation*, Catania, G. Belvedere & L. Paternò, eds., p. 151.

Tassoul, J.-L. 1978. *Theory of Rotating Stars* (Princeton: Princeton University Press).

Tassoul, J.-L. 1987. *ApJ* **322**, 856.

Tassoul, J.-L. 1988. *ApJL* **324**, L71.

Toner, C.G. & Gray, D.F. 1988. *ApJ* **334**, 1008.

Uesugi, A. & Fukuda, I. 1982. *Revised Catalogue of Stellar Rotation Velocities* (Kyoto: Kyoto University).

Uitenbroek, H., Dupree, A.K., & Gilliland, R.L. 1998. *AJ* **116**, 2501.

Ulrich, R.K. & Bertello, L. 1996. *ApJ* **465**, 65.

Ulrich, R.K. & Boyden, J.E. 2006. *Solar Phys* **235**, 17.

Ulrich, R.K., Boyden, J.E., Webster, L., Padilla, S.P., & Snodgrass, H.B. 1988. *Solar Phys* **117**, 291.

van Loon, J.Th., Cioni, M.-R.L., Zijlstra, A.A., & Loup, C. 2005. *A&A* **438**, 273.

van Saders, J.L. & Pinsonneault, M.H. 2013. *ApJ* **776**,67.

van Saders, J.L., Ceillier, T., Metcalfe, T.S., *et al.* 2016. *Nature* **529**, 181.

Verscharen, D., Klein, K.G., & Maruca, B.A. 2019. *Living Rev Sol Phys* **16**, 5.

Vilhu, O. 1984. *A&A* **133**, 117.

Vogt, S.S. 1983. *Activity in Red Dwarf Stars* (Dordrecht: Reidel), P.B. Byrne & M. Rodono, eds., p. 137.

Vogt, S.S. & Penrod, G.D. 1983. *PASP* **95**, 565.

Vogt, S.S., Penrod, G.D., & Hatzes, A.P. 1987. *ApJ* **321**, 496.

von Zeipel, H. 1924. *MNRAS* **84**, 665.

Weber, E.J. & L. Davis, Jr. 1967. *ApJ* **148**, 217.

Westgate, C. 1934. *ApJ* **79**, 357.

Wilson, O.C. 1968. *ApJ* **153**, 221.

Wilson, O.C. 1978. *ApJ* **226**, 379.

Wöhl, H. 1978. *A&A* **62**, 165.

Wolff, S.C. 1981. *ApJ* **244**, 221.

Wolff, S.C., Strom, S.E., Dror, D., & Venn, K. 2007. *AJ* **133**, 1092.

Wood, B.E., Müller, H.-R., Zank, G.P., & Linsky, J.L. 2002. *ApJ* **574**, 412.

Wright, N.J., Drake, J.J., Mamajek, E.E., & Henry, G.W. 2011. *ApJ* **743**, 48.

Xiang, Y., Shenghong, G., Wolter, U., *et al.* 2020. *MNRAS* **492**, 364.

Zahn, J.-P. 1989. *A&A* **223**, 112.

Zhang, J., Bi, S., Li, Y., *et al.* 2020. *ApJS* **247**, 9.

Zorec, J., Rieutord, M., Espinosa Lara, F., *et al.* 2017. *A&A* **606**, 32.

Appendix A
Useful Constants

Constant	Symbol	Value	Units 1	Units 2
Speed of light	c	2.997 924 58	10^{10} cm/s	10^8 m/s
Planck's constant	h	6.626 070 15	10^{-27} erg s	10^{-34} J s
	h	4.135 667 7	10^{-15} eV s	
Electron charge	e	1.602 176 634	10^{-19} C	
	e	4.803 204 673	10^{-10} esu	
Electron mass	m_e	9.109 383 701 5	10^{-28} g	10^{-31} kg
Proton mass	m_p	1.672 621 923 7	10^{-24} g	10^{-27} kg
Atomic mass unit	amu	1.660 539 066 6	10^{-24} g	10^{-27} kg
Rydberg constant	R_∞	1.097 373 156 8	10^5/cm	10^7/m
	R_H	1.096 775 834	10^5/cm	10^7/m
Stefan–Boltzmann con	σ	5.670 374 419	10^{-5} erg/$(\mathrm{s\ cm^2\ K^4})$	10^{-8} W/$(\mathrm{m^2\ K^4})$
Wien's law (max)	for I_ν	5.099 442 83	cm K	10^7 Å K
	for I_λ	2.897 771 96	cm K	10^7 Å K
Boltzmann's constant	k	1.380 649	10^{-16} erg/K	10^{-23} J/K
	k	8.617 333	10^{-5} eV/K	
Gravitational constant	G	6.674 2	10^{-8} cm^3/(s^2 g)	10^{-11} m^3/(s^2 kg)
Solar mass	$\mathcal{M}_\odot$	1.988 5	10^{33} g	10^{30} kg
Solar radius	$R_\odot$	6.961	10^{10} cm	10^8 m
Solar surface gravity	$g_\odot$	2.7399	10^4 cm/s^2	10^2 m/s^2
	$\log g_\odot$	4.437 7	cm/s^2	
Solar T_eff	$T_\mathrm{eff}^\odot$	5772	K	
Solar luminosity	$L_\odot$	3.830	10^{33} erg/s	10^{26} W
App. bolometric flux	$F_\mathrm{bol}^\odot$	1.362	10^6 erg/(s cm^2)	10^3 W/m^2
Abs. V magnitude Sun	$M_V^\odot$	4.82		
App. V magnitude Sun	$V_\odot$	−26.75		
Astronomical unit	a.u.	1.495 978 707	10^{11} m	10^8 km
Parsec	pc	2.062 648 06	10^5 a.u.	
	pc	3.085 677 58	10^{16} m	10^{13} km

Appendix B

Approximate Physical Parameters of Stars

The following tables give characteristic physical parameters for normal stars. They are intended to act as a guide, not definitive values. There is real cosmic dispersion around these values. If precise values are needed, stars should be treated individually.

The second column is a numerical spectral type. Color indices and magnitudes are based on Figures 1.7 and 1.8. Effective temperatures are from Figure 14.17. Temperatures normalized to the solar value are $S_0 = T_{eff}/5772$, as in Equation 9.5. Surface gravity values are from Figure 15.7. Radii are from Figure 14.6. Dwarf masses are based on the observed values from Popper (1980). These are independent of tabulated $\log g$ and R values, which means that small inconsistencies exist between the $\log g$ value in the table and $\log\left(\mathfrak{M}/R^2\right) + 4.438$.

Radial–tangential macroturbulence dispersions, ζ_{RT}, are from Figure 17.12. As one can see from that figure, there is ~20–30% range in values for each spectral type. The projected rotation rates, $v\sin i$, are the median values (lines) from Figures 18.24 (dwarfs) and 18.29 (giants).

Table B.1 *Approximate parameters for dwarfs*

Sp	Sp#	B − V	M_V	T_{eff}	S_0	log g	R	$\mathfrak{M}$	ζ_{RT}	$v\sin i$
O6	−1.4	−0.32	−13::	42025	7.28	3.98	14.30	—	—	210
O8	−1.2	−0.31	−5.80	38663	6.70	4.00	10.70	—	—	203
B0	−1.0	−0.29	−4.20	33015	5.72	4.04	6.50	17.00	—	196
B1	−0.9	−0.26	−3.48	26610	4.61	4.10	4.75	10.70	—	192
B2	−0.8	−0.24	−2.76	23359	4.05	4.13	4.20	8.30	—	189
B3	−0.7	−0.21	−2.10	19581	3.39	4.19	3.50	6.30	—	185
B4	−0.6	−0.18	−1.56	16770	2.91	4.22	2.94	5.00	—	182
B5	−0.5	−0.16	−1.08	15293	2.65	4.24	2.69	4.30	—	178
B6	−0.4	−0.15	−0.66	14410	2.50	4.25	2.56	3.70	—	174
B7	−0.3	−0.13	−0.25	13528	2.34	4.26	2.44	3.20	—	170
B8	−0.2	−0.10	0.10	12176	2.11	4.27	2.28	2.90	—	167
B9	−0.1	−0.05	0.45	10572	1.83	4.27	2.07	2.65	—	163
A0	0.0	0.00	0.80	9521	1.65	4.27	1.92	2.46	5:	160
A1	0.1	0.04	1.06	9158	1.59	4.27	1.85	2.31	—	155
A2	0.2	0.06	1.32	8946	1.55	4.26	1.80	2.21	5:	151
A3	0.3	0.09	1.58	8709	1.51	4.26	1.76	2.15	—	147
A4	0.4	0.12	1.80	8475	1.47	4.26	1.73	2.10	—	143

Table B.1 (*cont.*)

Sp	Sp#	$B - V$	M_V	T_{eff}	S_0	log g	R	$\mathfrak{M}$	ζ_{RT}	$v \sin i$
A5	0.5	0.14	2.00	8278	1.43	4.25	1.71	2.04	—	139
A6	0.6	0.16	2.17	8097	1.40	4.25	1.70	1.98	—	134
A7	0.7	0.19	2.35	7910	1.37	4.24	1.68	1.93	—	130
A8	0.8	0.23	2.50	7639	1.32	4.24	1.68	1.85	—	125
A9	0.9	0.27	2.64	7382	1.28	4.22	1.67	1.75	—	119
F0	1.0	0.31	2.78	7155	1.24	4.22	1.63	1.68	—	112
F1	1.1	0.34	2.93	7012	1.21	4.23	1.57	1.61	—	103
F2	1.2	0.36	3.08	6886	1.19	4.24	1.51	1.56	—	88
F3	1.3	0.39	3.23	6760	1.17	4.25	1.44	1.50	8.0	58
F4	1.4	0.41	3.40	6631	1.15	4.26	1.37	1.44	7.0	27
F5	1.5	0.44	3.60	6502	1.13	4.28	1.31	1.39	6.4	18
F6	1.6	0.47	3.74	6376	1.10	4.29	1.25	1.32	5.9	13
F7	1.7	0.50	3.95	6243	1.08	4.30	1.20	1.25	5.4	9.8
F8	1.8	0.53	4.15	6129	1.06	4.31	1.15	1.20	5.0	7.7
F9	1.9	0.56	4.35	6017	1.04	4.32	1.10	1.15	4.6	5.5
G0	2.0	0.59	4.50	5909	1.02	4.33	1.06	1.10	4.3	4.0
G1	2.1	0.61	4.65	5834	1.01	4.34	1.03	1.06	4.0	3.2
G2	2.2	0.63	4.78	5774	1.00	4.34	1.02	1.03	3.7	3.0
G3	2.3	0.65	4.91	5723	0.99	4.35	1.00	1.00	3.5	2.8
G4	2.4	0.66	5.00	5666	0.98	4.36	0.98	0.99	3.2	2.6
G5	2.5	0.68	5.15	5617	0.97	4.37	0.97	0.98	3.0	2.4
G6	2.6	0.70	5.30	5569	0.96	4.37	0.96	0.97	2.8	2.3
G7	2.7	0.71	5.44	5514	0.96	4.38	0.94	0.95	2.6	2.2
G8	2.8	0.74	5.60	5443	0.94	4.40	0.91	0.93	2.4	2.1
G9	2.9	0.78	5.74	5346	0.93	4.42	0.87	0.91	2.2	2.0
K0	3.0	0.82	5.90	5234	0.91	4.44	0.83	0.89	2.0	1.9
K1	3.1	0.87	6.04	5117	0.89	4.47	0.80	0.86	1.9	1.9
K2	3.2	0.92	6.20	5010	0.87	4.49	0.77	0.82	1.8	1.8
K3	3.3	0.96	6.45	4909	0.85	4.51	0.74	0.77	1.6	1.7
K4	3.4	1.03	6.75	4771	0.83	4.53	0.72	0.71	—	1.7
K5	3.5	1.15	7.10	4566	0.79	4.56	0.68	0.65	—	1.6
K7	3.7	1.30	7.80	4328	0.75	4.62	0.60	0.55	—	1.6
M0	4.0	1.41	8.90	4166	0.72	4.66	0.54	0.45	—	1.5
M2	4.2	1.50	9.77	4038	0.70	4.73	0.47	0.40	—	—
M5	4.5	1.60	12.00	3899	0.68	4.81	0.41	0.34	—	—

Values of log g, R, and $\mathfrak{M}$ (mass) for evolved stars in the next two tables have a wide range around the listed values. Mass in Table B2 is computed from the surface gravity and radius estimates listed there, i.e., from Figures 15.7 and 14.17.

Table B.2 *Approximate parameters for Class III giants*

Sp	Sp#	$B-V$	M_V	T_{eff}	S_0	log g	R	$\mathfrak{M}$	ζ_{RT}	$v \sin i$
A7	0.7	0.19	0.8	7910	1.37	3.9	3.0	2.6	—	96
A8	0.8	0.23	0.9	7639	1.32	3.9	3.2	2.6	—	92
A9	0.9	0.27	1.0	7388	1.28	3.8	3.3	2.6	—	87
F0	1.0	0.34	1.1	7000	1.21	3.8	3.5	2.5	—	83
F1	1.1	0.34	1.2	6980	1.21	3.7	3.5	2.4	—	80
F2	1.2	0.36	1.3	6886	1.19	3.7	3.5	2.3	—	76
F3	1.3	0.39	1.3	6736	1.17	3.7	3.7	2.4	—	75
F4	1.4	0.41	1.4	6640	1.15	3.7	3.7	2.3	—	73
F5	1.5	0.44	1.4	6502	1.13	3.6	3.7	2.0	—	72
F6	1.6	0.47	1.4	6372	1.10	3.6	3.8	2.0	—	70
F7	1.7	0.50	1.4	6247	1.08	3.5	3.9	1.9	—	69
F8	1.8	0.53	1.3	6129	1.06	3.5	4.0	1.8	—	68
F9	1.9	0.59	1.3	5909	1.02	3.4	4.2	1.6	—	66
G0	2.0	0.64	1.2	5742	0.99	3.3	4.4	1.4	—	64
G1	2.1	0.70	1.1	5557	0.96	3.2	4.7	1.3	—	60
G2	2.2	0.76	1.1	5389	0.93	3.1	5.0	1.1	6.7	28
G3	2.3	0.81	1.0	5259	0.91	3.0	5.4	1.1	6.4	5.3
G4	2.4	0.86	0.9	5138	0.89	3.0	5.9	1.1	6.1	4.8
G5	2.5	0.90	0.8	5047	0.87	2.9	6.7	1.2	5.8	4.3
G6	2.6	0.92	0.8	5003	0.87	2.9	7.0	1.2	5.6	3.9
G7	2.7	0.94	0.7	4960	0.86	2.8	7.3	1.2	5.4	3.5
G8	2.8	0.96	0.7	4919	0.85	2.8	7.8	1.2	5.1	3.1
G9	2.9	0.99	0.6	4858	0.84	2.7	8.5	1.2	5.0	2.8
K0	3.0	1.02	0.6	4800	0.83	2.6	10.5	1.2	4.8	2.5
K1	3.1	1.09	0.5	4670	0.81	2.5	10.5	1.2	4.6	2.3
K2	3.2	1.17	0.3	4533	0.79	2.2	14.0	1.1	4.5	2.1
K3	3.3	1.28	0.2	4358	0.76	1.8	21.0	1.0	4.4	1.9
K4	3.4	1.37	0.1	4224	0.73	1.5	29.0	0.9	—	—
K5	3.5	1.44	0.0	4123	0.71	—	35.0	—	—	—
K7	3.7	1.53	−0.1	3996	0.69	—	48.0	—	—	—
M0	4.0	1.57	−0.2	3941	0.68	—	58.0	—	—	—
M2	4.2	1.60	−0.2	3899	0.68	—	70.0	—	—	—
M5	4.5	1.58	—	3927	0.68	—	70.0	—	—	—

The parameters for supergiants have the most uncertainty. Mass is too uncertain to list, but based on evolutionary tracks, a reasonable guess is ~10 solar masses. Because the nature of the photospheric velocity fields in supergiants has a different character compared to lower luminosity stars, the listed ζ_{RT} values are suspect. For the same reason, rotation rates are not well established.

Table B.3 *Approximate parameters for Class Ib supergiants*

Sp	Sp#	$B-V$	M_V	T_{eff}	S_0	$\log g$	R	ζ_{RT}	$v \sin i$
A7	0.7	0.13	−4.8	8356	1.45	—	24	—	—
A8	0.8	0.14	−4.8	8278	1.43	—	25	—	—
A9	0.9	0.15	−4.7	8201	1.42	—	25	—	—
F0	1.0	0.16	−4.7	8126	1.41	—	25	—	—
F1	1.1	0.18	−4.7	7980	1.38	—	25	—	—
F2	1.2	0.20	−4.7	7840	1.36	—	25	—	—
F3	1.3	0.24	−4.7	7575	1.31	—	25	—	—
F4	1.4	0.28	−4.7	7328	1.27	—	25	—	—
F5	1.5	0.32	−4.6	7099	1.23	—	25	—	~10
F6	1.6	0.38	−4.6	6785	1.18	—	26	—	—
F7	1.7	0.45	−4.6	6458	1.12	—	26	—	—
F8	1.8	0.55	−4.6	6053	1.05	—	27	—	—
F9	1.9	0.66	−4.6	5678	0.98	—	28	12.6	—
G0	2.0	0.76	−4.6	5389	0.98	—	30	11.7	~8
G1	2.1	0.82	−4.6	5234	0.93	—	35	11.0	—
G2	2.2	0.87	−4.6	5115	0.91	—	40	10.5	—
G3	2.3	0.91	−4.5	5025	0.89	—	47	10.1	—
G4	2.4	0.95	−4.5	4940	0.87	—	55	9.8	—
G5	2.5	1.00	−4.5	4838	0.86	—	60	9.6	~6
G6	2.6	1.04	−4.5	4762	0.84	—	63	9.4	—
G7	2.7	1.08	−4.5	4688	0.82	1.7	68	9.3	—
G8	2.8	1.12	−4.5	4617	0.81	1.6	71	9.2	—
G9	2.9	1.16	−4.5	4549	0.80	1.5	74	9.1	—
K0	3.0	1.20	−4.5	4484	0.79	1.4	80	9.0	~5
K1	3.1	1.23	−4.5	4436	0.78	1.3	85	8.9	—
K2	3.2	1.28	−4.5	4358	0.77	1.0	90	8.9	—
K3	3.3	1.37	−4.5	4224	0.76	0.7	160	8.8	—
K4	3.4	1.49	−4.5	4052	0.73	0.4	250	8.8	—
K5	3.5	1.59	−4.5	3913	0.70	0.1	300	8.7	~4
K7	3.7	1.62	−4.5	3871	0.68	0.0	350	—	—
M0	4.0	1.65	—	3829	0.67	—	400	—	—
M2	4.2	1.65	—	3829	0.66	—	—	—	—

Reference

Popper, D.M. 1980. *ARAA* **18**, 115.

Appendix C
Atomic Data

Ionization Potentials

Table C.1 gives basic atomic parameters including the first three ionization potentials in electronvolts for the 92 natural elements. The "Weight" is the atomic mass in atomic mass units (one amu $= 1.660\,539\,066\,6 \times 10^{-24}$ g based on the ^{12}C $= 12.000$ scale. Values are from International Union of Pure and Applied Chemistry (www.ciaaw.org/atomic-weights .htm). Ionization potentials are from the NIST website (physics.nist.gov/PhysRefData/ ASD/ionEnergy.html).

Table C.1 *Atomic weights and ionization potentials*

No.	Element	Symbol	Weight	I_1	I_2	I_3
1	Hydrogen	H	1.008	13.598	—	—
2	Helium	He	4.003	24.587	54.418	—
3	Lithium	Li	6.968	5.392	75.640	122.454
4	Beryllium	Be	9.012	9.323	18.211	153.896
5	Boron	B	10.813	8.298	25.155	37.931
6	Carbon	C	12.054	11.260	24.383	47.888
7	Nitrogen	N	14.007	14.534	29.601	47.445
8	Oxygen	O	15.999	13.618	35.121	54.936
9	Fluorine	F	18.998	17.423	34.971	62.708
10	Neon	Ne	20.180	21.565	40.963	63.423
11	Sodium	Na	22.990	5.139	47.286	71.620
12	Magnesium	Mg	24.305	7.646	15.035	80.144
13	Aluminum	Al	26.982	5.986	18.829	28.448
14	Silicon	Si	28.085	8.152	16.346	33.493
15	Phosphorus	P	30.974	10.487	19.769	30.203
16	Sulfur	S	32.068	10.360	23.338	34.86
17	Chlorine	Cl	35.452	12.968	23.814	39.80
18	Argon	Ar	39.878	15.760	27.630	40.735
19	Potassium	K	39.098	4.341	31.625	45.803
20	Calcium	Ca	40.078	6.113	11.872	50.913
21	Scandium	Sc	44.956	6.561	12.800	24.757
22	Titanium	Ti	47.867	6.828	13.576	27.492
23	Vanadium	V	50.942	6.746	14.634	29.311
24	Chromium	Cr	51.996	6.767	16.486	30.959
25	Magnesium	Mn	54.938	7.434	15.640	33.668

Table C.1 (*cont.*)

No.	Element	Symbol	Weight	I_1	I_2	I_3
26	Iron	Fe	55.845	7.902	16.199	30.651
27	Cobalt	Co	58.933	7.881	17.084	33.50
28	Nickel	Ni	58.693	7.640	18.169	35.187
29	Copper	Cu	63.546	7.726	20.292	36.841
30	Zinc	Zn	65.38	9.394	17.964	39.723
31	Gallium	Ga	69.723	5.999	20.515	30.726
32	Germanium	Ge	72.630	7.899	15.935	34.058
33	Arsenic	As	74.922	9.789	18.589	28.349
34	Selenium	Se	78.971	9.752	21.196	31.697
35	Bromine	Br	79.904	11.814	21.591	34.871
36	Krypton	Kr	83.798	14.000	24.360	35.838
37	Rubidium	Rb	85.468	4.177	27.290	39.247
38	Strontium	Sr	87.62	5.695	11.030	42.884
39	Yttrium	Y	88.906	6.217	12.224	20.524
40	Zirconium	Zr	91.224	6.634	13.13	23.170
41	Niobium	Nb	92.906	6.759	14.32	25.04
42	Molybdenum	Mo	95.95	7.092	16.16	27.13
43	Technetium	Tc	98.906	7.119	15.26	29.55
44	Ruthium	Ru	101.07	7.360	16.76	28.47
45	Rhodium	Rh	102.905	7.459	18.08	31.06
46	Palladium	Pd	106.42	8.337	19.43	32.93
47	Silver	Ag	107.868	7.576	21.484	34.8
48	Cadmium	Cd	112.414	8.994	16.908	37.468
49	Indium	In	114.818	5.786	18.870	28.044
50	Tin	Sn	118.710	7.344	14.633	30.506
51	Antimony	Sb	121.760	8.608	16.626	25.324
52	Tellurium	Te	127.60	9.010	18.6	27.84
53	Iodine	I	126.904	10.451	19.131	29.570
54	Xenon	Xe	131.293	12.130	20.975	31.05
55	Cesium	Cs	132.905	3.894	23.157	33.195
56	Barium	Ba	137.327	5.539	10.956	36.906
57	Lanthanum	La	138.905	5.577	11.185	19.177
58	Cesium	Ce	140.116	5.387	10.85	20.198
59	Praseodymium	Pr	140.908	5.470	10.631	21.624
60	Neodymium	Nd	144.242	5.525	10.783	22.09
61	Promethium	Pm	146.	5.582	10.938	22.44
62	Samarium	Sm	150.32	5.644	11.078	23.55
63	Europium	Eu	151.964	5.670	11.240	24.84
64	Gadolinium	Gd	157.25	6.150	12.076	20.54
65	Terbium	Tb	158.925	5.864	11.513	21.82
66	Dysprosium	Dy	162.500	5.939	11.647	22.89
67	Holmium	Ho	164.930	6.022	11.781	22.79
68	Erbium	Er	167.259	6.108	11.916	22.70
69	Thulium	Tm	168.934	6.184	12.065	23.66
70	Ytterbium	Yb	170.045	6.254	12.179	25.053
71	Lutetium	Lu	174.967	5.426	14.13	20.959
72	Hafnium	Hf	178.486	6.825	14.61	22.55

Table C.1 (*cont.*)

No.	Element	Symbol	Weight	I_1	I_2	I_3
73	Tantalum	Ta	180.948	7.550	16.2	23.1
74	Tungsten	W	183.84	7.864	16.37	26.0
75	Rhenium	Re	186.207	7.834	16.6	27.
76	Osmium	Os	190.23	8.438	17.0	25.0
77	Iridium	Ir	192.217	8.967	17.0	28.0
78	Platinum	Pt	195.084	8.959	18.56	29.0
79	Gold	Au	196.967	9.226	20.203	30.0
80	Mercury	Hg	200.592	10.438	18.757	34.49
81	Thallium	Tl	204.383	6.108	20.428	29.852
82	Lead	Pb	207.2	7.417	15.032	31.937
83	Bismuth	Bi	208.980	7.286	16.703	25.571
84	Polonium	Po	210.	8.418	19.3	27.3
85	Astatine	At	210.	9.318	17.880	26.58
86	Radon	Rn	222.	10.748	21.4	29.4
87	Francium	Fr	223.	4.073	22.4	33.5
88	Radium	Ra	226.025	5.278	10.147	31.0
89	Actinium	Ac	227.	5.380	11.75	17.431
90	Thorium	Th	232.038	6.307	12.10	18.32
91	Protactinium	Pa	231.036	5.89	11.9	18.6
92	Uranium	U	238.029	6.194	11.6	19.8

Partition Functions

Partition functions, $u(T)$, are defined in Equation 1.18, namely,

$$u(T) = \sum_{i=1}^{m} g_i \, e^{-\frac{\chi_i}{kT}} = \sum_{i=1}^{m} g_i \, 10^{-\theta \chi_i},$$

and are used to calculate excitation and ionization. Here g_i is the statistical weight of level i, equal to $2J + 1$, where J is the inner quantum number or angular-momentum quantum number of the level, χ_i is the energy of the level above the ground state, k is the Boltzmann constant, and T the absolute temperature. Generally the smaller T is, the less contribution there is from terms higher than the ground state, and then a rough approximation for $u(T)$ is simply g_0. This simple statement can be slightly misleading since the ground state itself is often multileveled. For example, consider iron. The ground state is ^5D with five levels having $J = 0, 1, 2, 3$, and 4. One normally says that g_0 is the sum of the $2J + 1$ values, i.e., 25. But actually these levels have small excitation potentials: 0.000, 0.051, 0.087, 0.110, and 0.121 eV. Then if $\theta = 2$, to be specific, the sum of the $2J + 1$ values time their Boltzmann factors amounts to only 20.244, a difference in the logarithm of 0.092 dex. So only in the limit of zero temperature, hardly relevant for stellar photospheres, does g_0 actually equal $u(T)$.

The summation is carried out to some upper limit, m, that is not well defined. In principle $m = \infty$, in which case the sum would diverge, but in practice two things happen to ensure that $u(T)$ is finite. First, the "theoretical" upper levels are not stable because of collisions. Electrons in such levels are easily ionized to the continuum by collisions, and this has been called a lowering of the ionization potential or pressure ionization. Numerical estimates depend on temperature and density, but for a typical stellar photosphere, one finds this $\Delta I \approx 0.005$ to 0.5 eV. The details here only have relevance in high-temperature cases, where the exponent in the above equation begins to grow toward zero. It is precisely here where the second factor comes into play: the species at these temperatures will be largely ionized to the next level. As a result, only in exceptional circumstances does one need to worry about the actual value of m and the complication of $u(T)$ changing with depth in the photosphere.

Partition functions for ions with no excitable electrons are taken to be 1.0, $\log u(T) = 0$ in the notation of Table C.2.

The computation of partition functions can be somewhat laborious and requires a detailed knowledge of the energy levels, e.g., Moore *et al.* (1949, 1952, 1958), Moore (1970). For model photosphere computation, interpolation within Table C.2 is easy and convenient. In many publications, polynomials are given, and while these are also convenient, they can be misleading when used outside the temperature range for which they were intended. The temperature parameter in Table C.2 is the usual $\theta = 5040/T$. The entries are accurate to ~1% and come mainly from Irwin (1981). Other sources include Aller (1963), Bolton (1970), Allen (1973), Irwin (1981), Cowley and Adelman (1983), Sauval and Tatum (1984), Milone and Merlo (1998), Halenka *et al.* (2001), and the NIST atomic database. Sauval and Tatum (1984) also give polynomials for partition functions for 300 diatomic molecules.

Table C.2 *Partition functions,* $\log u(T)$

$\theta =$	0.2	0.4	0.6	0.8	1.0	1.2	1.4	1.6	1.8	2.0	$\log g_0$
H	0.368	0.303	0.301	0.301	0.301	0.301	0.301	0.301	0.301	0.301	0.301
He	0.000	0.000	0.000	0.000	0.000	0.000	0.000	0.000	0.000	0.000	0.000
He$^+$	0.301	0.301	0.301	0.301	0.301	0.301	0.301	0.301	0.301	0.301	0.301
Li	3.858	0.987	0.488	0.359	0.320	0.308	0.304	0.302	0.302	0.302	0.301
Li$^+$	0.000	0.000	0.000	0.000	0.000	0.000	0.000	0.000	0.000	0.000	0.000
Be	1.741	0.328	0.087	0.025	0.007	0.002	0.001	0.000	0.000	0.000	0.000
Be$^+$	0.541	0.334	0.307	0.302	0.301	0.301	0.301	0.301	0.301	0.301	0.301
B	1.191	0.831	0.786	0.778	0.777	0.777	0.777	0.777	0.777	0.776	0.778
B$^+$	0.435	0.051	0.006	0.000	0.000	0.000	0.000	0.000	0.000	0.000	0.000
C	1.163	1.037	0.994	0.975	0.964	0.958	0.954	0.951	0.950	0.948	0.954
C$^+$	0.853	0.782	0.775	0.774	0.773	0.772	0.771	0.770	0.769	0.767	0.778
C^{++}	0.143	0.010	0.000	0.000	0.000	0.000	0.000	0.000	0.000	0.000	0.000
N	1.060	0.729	0.645	0.616	0.606	0.603	0.602	0.602	0.602	0.602	0.602
N$^+$	1.073	0.993	0.965	0.953	0.946	0.942	0.939	0.937	0.934	0.932	0.954
O	1.095	0.991	0.964	0.953	0.947	0.944	0.941	0.939	0.937	0.935	0.954
O$^+$	0.895	0.655	0.614	0.604	0.602	0.602	0.602	0.602	0.602	0.602	0.602
F	0.788	0.772	0.768	0.765	0.762	0.759	0.756	0.753	0.750	0.747	0.778
F$^+$	1.034	0.968	0.949	0.940	0.935	0.930	0.926	0.923	0.919	0.915	0.954

Table C.2 (*cont.*)

$\theta =$	0.2	0.4	0.6	0.8	1.0	1.2	1.4	1.6	1.8	2.0	log g_0
Ne	0.002	0.000	0.000	0.000	0.000	0.000	0.000	0.000	0.000	0.000	0.000
Ne$^+$	0.771	0.766	0.760	0.754	0.748	0.743	0.737	0.732	0.727	0.723	0.778
Na	4.316	1.043	0.493	0.357	0.320	0.309	0.307	0.306	0.306	0.306	0.301
Na$^+$	0.000	0.000	0.000	0.000	0.000	0.000	0.000	0.000	0.000	0.000	0.000
Mg	2.839	0.478	0.110	0.027	0.007	0.002	0.001	0.001	0.001	0.000	0.000
Mg$^+$	0.537	0.326	0.304	0.301	0.301	0.301	0.301	0.301	0.301	0.301	0.301
Mg++	0.000	0.000	0.000	0.000	0.000	0.000	0.000	0.000	0.000	0.000	0.000
Al	2.208	0.973	0.806	0.774	0.768	0.767	0.766	0.764	0.762	0.760	0.778
Al$^+$	0.488	0.053	0.006	0.000	0.000	0.000	0.000	0.000	0.000	0.000	0.000
Al^{++}	0.346	0.304	0.301	0.301	0.301	0.301	0.301	0.301	0.301	0.301	0.301
Si	1.521	1.111	1.030	0.996	0.976	0.961	0.949	0.940	0.932	0.925	0.954
Si$^+$	0.900	0.778	0.764	0.759	0.755	0.750	0.746	0.741	0.736	0.731	0.778
Si^{++}	0.144	0.010	0.000	0.000	0.000	0.000	0.000	0.000	0.000	0.000	0.000
P	1.010	0.886	0.755	0.684	0.645	0.624	0.613	0.607	0.604	0.603	0.602
P$^+$	1.091	1.027	0.984	0.956	0.936	0.921	0.909	0.898	0.889	0.881	0.954
P^{++}	0.794	0.760	0.749	0.740	0.731	0.721	0.712	0.703	0.694	0.685	0.778
S	1.153	1.036	0.990	0.965	0.949	0.937	0.929	0.921	0.915	0.910	0.954
S$^+$	0.967	0.792	0.688	0.640	0.618	0.608	0.604	0.603	0.602	0.602	0.602
Cl	0.845	0.769	0.757	0.751	0.745	0.739	0.733	0.727	0.722	0.716	0.778
Cl$^+$	1.053	1.001	0.963	0.940	0.924	0.912	0.903	0.894	0.887	0.880	0.954
Ar	0.048	0.002	0.000	0.000	0.000	0.000	0.000	0.000	0.000	0.000	0.000
Ar$^+$	0.767	0.756	0.746	0.736	0.727	0.718	0.710	0.702	0.695	0.689	0.778
K	4.647	1.329	0.642	0.429	0.351	0.320	0.308	0.303	0.302	0.302	0.301
K+	0.000	0.000	0.000	0.000	0.000	0.000	0.000	0.000	0.000	0.000	0.000
Ca	5.238	1.332	0.465	0.181	0.073	0.028	0.010	0.003	0.001	0.000	0.000
Ca$^+$	0.825	0.658	0.483	0.391	0.344	0.320	0.309	0.304	0.302	0.301	0.301
Ca^{++}	0.000	0.000	0.000	0.000	0.000	0.000	0.000	0.000	0.000	0.000	0.000
Sc	4.674	2.078	1.422	1.181	1.077	1.027	1.002	0.989	0.982	0.977	1.000
Sc$^+$	1.911	1.595	1.483	1.413	1.362	1.322	1.290	1.265	1.244	1.227	1.176
Sc^{++}	1.027	0.999	0.992	0.988	0.985	0.982	0.980	0.977	0.974	0.971	1.000
Ti	4.159	2.333	1.818	1.596	1.480	1.411	1.367	1.337	1.316	1.300	1.322
Ti$^+$	2.237	1.995	1.872	1.797	1.746	1.708	1.678	1.653	1.632	1.613	1.447
Ti^{++}	1.524	1.435	1.374	1.338	1.315	1.300	1.289	1.280	1.272	1.265	1.322
V	4.497	2.408	1.945	1.770	1.681	1.625	1.584	1.552	1.525	1.502	1.447
V$^+$	2.355	2.023	1.832	1.720	1.648	1.597	1.560	1.531	1.507	1.487	1.398
V^{++}	1.722	1.618	1.518	1.462	1.430	1.410	1.396	1.385	1.377	1.369	1.447
Cr	4.284	1.977	1.380	1.141	1.022	0.956	0.917	0.892	0.875	0.865	0.845
Cr$^+$	1.981	1.489	1.125	0.944	0.856	0.813	0.793	0.784	0.781	0.780	0.778
Cr^{++}	1.845	1.555	1.440	1.390	1.365	1.351	1.341	1.332	1.324	1.316	1.398
Mn	4.804	1.640	1.037	0.865	0.808	0.788	0.781	0.779	0.778	0.778	0.778
Mn$^+$	2.056	1.319	1.052	0.940	0.889	0.866	0.855	0.850	0.848	0.848	0.845
Fe	3.760	2.049	1.664	1.519	1.446	1.402	1.372	1.350	1.332	1.317	1.398
Fe$^+$	2.307	1.881	1.749	1.682	1.638	1.604	1.575	1.549	1.525	1.504	1.477
Fe^{++}	1.887	1.518	1.411	1.370	1.350	1.337	1.328	1.320	1.312	1.304	1.398
Co	3.431	2.012	1.716	1.598	1.529	1.478	1.437	1.402	1.371	1.344	1.447
Co$^+$	2.017	1.714	1.606	1.531	1.470	1.418	1.373	1.334	1.300	1.269	1.322
Co^{++}	1.717	1.503	1.428	1.388	1.361	1.339	1.321	1.305	1.290	1.276	1.447

Table C.2 (*cont.*)

$\theta =$	0.2	0.4	0.6	0.8	1.0	1.2	1.4	1.6	1.8	2.0	log g_0
Ni	2.779	1.753	1.577	1.521	1.490	1.467	1.447	1.428	1.410	1.394	1.322
Ni$^+$	1.659	1.386	1.215	1.108	1.037	0.988	0.953	0.927	0.908	0.893	1.000
Ni^{++}	1.403	1.307	1.226	1.238	1.216	1.196	1.178	1.162	1.147	1.133	1.322
Cu	1.253	0.763	0.536	0.425	0.368	0.337	0.320	0.310	0.305	0.303	0.301
Cu$^+$	0.913	0.379	0.140	0.045	0.009	0.000	0.000	0.000	0.000	0.000	0.000
Cu^{++}	1.016	0.953	0.930	0.912	0.895	0.880	0.867	0.855	0.845	0.836	1.000
Zn	0.904	0.116	0.017	0.001	0.000	0.000	0.000	0.000	0.000	0.000	0.000
Zn$^+$	0.426	0.309	0.301	0.301	0.301	0.301	0.301	0.301	0.301	0.301	0.301
Zn^{++}	0.028	0.001	0.000	0.000	0.000	0.000	0.000	0.000	0.000	0.000	0.000
Ga	2.030	0.944	0.777	0.733	0.715	0.702	0.689	0.677	0.665	0.653	0.778
Ga$^+$	0.226	0.017	0.000	0.000	0.000	0.000	0.000	0.000	0.000	0.000	0.000
Ga^{++}	0.316	0.302	0.301	0.301	0.301	0.301	0.301	0.301	0.301	0.301	0.301
Ge	1.446	1.049	0.965	0.917	0.879	0.844	0.812	0.784	0.758	0.734	0.954
Ge$^+$	0.877	0.725	0.694	0.666	0.641	0.622	0.608	0.599	0.593	0.590	0.778
Ge^{++}	0.062	0.006	0.000	0.000	0.000	0.000	0.000	0.000	0.000	0.000	0.000
As	1.383	0.902	0.765	0.696	0.655	0.632	0.619	0.613	0.610	0.610	0.602
As$^+$	1.124	0.957	0.880	0.818	0.771	0.737	0.714	0.699	0.689	0.683	0.954
As^{++}	1.053	0.841	0.711	0.627	0.569	0.525	0.491	0.463	0.439	0.419	0.778
Se	1.253	1.003	0.943	0.905	0.875	0.851	0.832	0.817	0.805	0.794	0.954
Se$^+$	0.992	0.815	0.710	0.655	0.619	0.593	0.572	0.553	0.537	0.522	0.602
Br	1.705	0.748	0.703	0.688	0.670	0.655	0.647	0.644	0.647	0.653	0.602
Br$^+$	1.034	0.958	0.897	0.856	0.829	0.810	0.797	0.788	0.782	0.777	0.954
Kr	0.092	0.006	0.000	0.000	0.000	0.000	0.000	0.000	0.000	0.000	0.000
Kr$^+$	0.740	0.706	0.681	0.661	0.646	0.635	0.626	0.620	0.615	0.612	0.778
Rb	6.046	1.963	0.890	0.518	0.378	0.325	0.307	0.303	0.304	0.305	0.301
Rb$^+$	0.003	0.000	0.000	0.000	0.000	0.000	0.000	0.000	0.000	0.000	0.000
Sr	5.005	1.441	0.548	0.229	0.098	0.041	0.016	0.005	0.001	0.000	0.000
Sr$^+$	1.115	0.658	0.462	0.375	0.334	0.315	0.306	0.303	0.301	0.301	0.301
Sr^{++}	0.000	0.000	0.000	0.000	0.000	0.000	0.000	0.000	0.000	0.000	0.000
Y	1.716	1.704	1.379	1.183	1.073	1.011	0.975	0.953	0.939	0.929	1.000
Y$^+$	1.784	1.521	1.370	1.272	1.201	1.144	1.097	1.055	1.017	0.982	1.000
Y^{++}	1.061	1.018	0.994	0.975	0.960	0.946	0.934	0.923	0.913	0.903	1.000
Zr	3.289	2.170	1.814	1.630	1.513	1.430	1.366	1.315	1.273	1.237	1.322
Zr$^+$	2.367	2.002	1.849	1.744	1.662	1.593	1.534	1.483	1.438	1.398	1.447
Zr^{++}	1.594	1.477	1.384	1.321	1.276	1.240	1.211	1.185	1.163	1.142	1.322
Nb	3.238	2.330	2.001	1.832	1.727	1.655	1.601	1.558	1.523	1.492	1.813
Nb$^+$	2.459	2.090	1.887	1.747	1.643	1.562	1.496	1.441	1.395	1.355	1.398
Nb^{++}	1.540	1.416	1.371	1.341	1.315	1.291	1.267	1.244	1.222	1.201	—
Mo	—	1.783	1.361	1.098	0.966	0.904	0.876	0.864	0.858	0.855	0.845
Mo$^+$	—	1.636	1.276	1.081	0.956	0.857	0.766	0.678	0.590	0.504	0.778
Mo^{++}	—	1.669	1.485	1.354	1.251	1.163	1.089	1.023	0.964	0.911	1.398
Tc	2.079	1.754	1.536	1.398	1.302	1.228	1.169	1.120	1.077	1.039	0.778
Tc$^+$	1.717	1.433	1.315	1.233	1.170	1.120	1.079	1.046	1.019	0.997	0.845
Tc^{++}	2.459	2.331	2.312	2.300	2.290	2.283	2.279	2.277	2.277	2.278	—
Ru	2.535	2.020	1.782	1.636	1.535	1.463	1.408	1.366	1.332	1.303	1.544
Ru$^+$	2.143	1.790	1.586	1.465	1.382	1.318	1.266	1.220	1.180	1.143	1.447
Ru^{++}	1.391	1.35	1.328	1.314	1.307	1.302	1.298	1.295	1.292	1.290	—

Table C.2 (*cont.*)

θ =	0.2	0.4	0.6	0.8	1.0	1.2	1.4	1.6	1.8	2.0	log g_0
Rh	2.278	1.794	1.609	1.501	1.426	1.369	1.323	1.285	1.252	1.224	1.447
Rh$^+$	1.852	1.503	1.339	1.250	1.192	1.149	1.113	1.082	1.055	1.030	1.322
Rh^{++}	1.795	1.532	1.403	1.320	1.261	1.217	1.182	1.155	1.133	1.114	—
Pd	1.953	1.050	0.791	0.614	0.474	0.364	0.277	0.208	0.153	0.111	0.000
Pd$^+$	1.429	1.035	0.935	0.896	0.874	0.858	0.843	0.829	0.815	0.803	1.000
Pd^{++}	1.519	1.329	1.243	1.192	1.159	1.135	1.118	1.104	1.093	1.084	—
Ag	3.248	0.606	0.334	0.302	0.302	0.302	0.301	0.301	0.300	0.300	0.301
Ag$^+$	0.768	0.079	0.017	0.006	0.001	0.000	0.000	0.000	0.000	0.000	—
Cd	2.825	0.289	0.027	0.000	0.000	0.000	0.000	0.000	0.000	0.000	0.000
Cd$^+$	0.658	0.314	0.302	0.301	0.301	0.301	0.301	0.301	0.301	0.301	0.301
Cd^{++}	0.019	0.005	0.000	0.000	0.000	0.000	0.000	0.000	0.000	0.000	0.000
In	4.313	1.154	0.742	0.651	0.615	0.588	0.562	0.538	0.514	0.493	0.778
In$^+$	0.391	0.028	0.004	0.001	0.000	0.000	0.000	0.000	0.000	0.000	—
In^{++}	0.351	0.304	0.302	0.300	0.299	0.298	0.297	0.297	0.297	0.298	—
Sn	1.428	0.986	0.862	0.781	0.712	0.651	0.595	0.545	0.499	0.457	0.954
Sn$^+$	0.887	0.656	0.594	0.543	0.500	0.465	0.438	0.418	0.403	0.392	0.778
Sn^{++}	0.255	0.008	0.000	0.000	0.000	0.000	0.000	0.000	0.000	0.000	0.000
Sb	1.357	0.936	0.802	0.726	0.679	0.649	0.632	0.621	0.615	0.611	0.602
Sb$^+$	1.085	0.843	0.718	0.613	0.525	0.454	0.395	0.347	0.308	0.274	—
Sb^{++}	0.701	0.590	0.523	0.483	0.457	0.440	0.428	0.419	0.412	0.406	—
Te	1.234	0.959	0.881	0.833	0.798	0.773	0.755	0.742	0.732	0.724	0.954
Te$^+$	1.017	0.854	0.744	0.683	0.644	0.616	0.594	0.575	0.559	0.545	—
Te^{++}	0.815	0.732	0.578	0.474	0.399	0.341	0.294	0.252	0.215	0.182	—
I	1.762	0.720	0.656	0.641	0.628	0.620	0.614	0.610	0.607	0.605	0.778
I$^+$	1.015	0.894	0.823	0.780	0.751	0.731	0.716	0.705	0.697	0.690	—
I^{++}	0.990	0.803	0.703	0.643	0.604	0.575	0.553	0.535	0.521	0.508	—
Xe	2.857	0.079	0.005	0.002	0.000	0.000	0.000	0.000	0.000	0.000	0.000
Xe$^+$	0.745	0.663	0.636	0.622	0.612	0.607	0.604	0.602	0.602	0.602	—
Cs	—	1.459	0.916	0.595	0.433	0.354	0.320	0.306	0.303	0.303	0.301
Cs$^+$	0.008	0.000	0.000	0.000	0.000	0.000	0.000	0.000	0.000	0.000	0.000
Cs^{++}	0.700	0.643	0.662	0.611	0.604	0.600	0.597	0.594	0.592	0.591	—
Ba	3.036	1.842	1.096	0.672	0.419	0.264	0.166	0.103	0.062	0.037	0.000
Ba$^+$	1.585	0.932	0.791	0.699	0.623	0.558	0.504	0.460	0.424	0.396	0.301
Ba^{++}	0.000	0.000	0.000	0.000	0.000	0.000	0.000	0.000	0.000	0.000	0.000
La	2.267	2.079	1.782	1.576	1.433	1.328	1.246	1.180	1.125	1.077	1.000
La$^+$	2.135	1.824	1.651	1.546	1.471	1.413	1.364	1.323	1.286	1.254	1.322
La^{++}	1.324	1.177	1.088	1.018	0.967	0.930	0.903	0.884	0.870	0.860	1.000
Ce	3.705	2.792	2.573	2.406	2.258	2.128	2.017	1.922	1.840	1.771	0.954
Ce$^+$	3.154	2.764	2.567	2.408	2.274	2.158	2.052	1.952	1.852	1.750	1.643
Ce^{++}	1.870	1.673	1.581	1.508	1.454	1.414	1.387	1.369	1.356	1.348	1.519
Pr	3.674	3.010	2.790	2.616	2.467	2.340	2.232	2.140	2.062	1.994	1.716
Pr$^+$	2.608	2.326	2.154	2.028	1.927	1.842	1.768	1.703	1.644	1.591	1.301
Pr^{++}	—	2.158	1.873	1.667	1.433	1.153	0.842	0.521	0.209	0.000	1.716
Nd	—	2.816	2.449	2.135	1.899	1.725	1.597	1.501	1.428	1.372	1.813
Nd$^+$	3.032	2.622	2.347	2.148	2.000	1.886	1.793	1.713	1.639	1.566	1.892
Nd^{++}	2.350	2.052	1.867	1.729	1.619	1.526	1.446	1.374	1.308	1.247	1.813
Pm	—	2.275	1.908	1.716	1.600	1.518	1.455	1.401	1.355	1.313	1.820

Table C.2 (*cont.*)

$\theta =$	0.2	0.4	0.6	0.8	1.0	1.2	1.4	1.6	1.8	2.0	log g_0
Pm$^+$	—	2.145	1.730	1.507	1.377	1.291	1.225	1.170	1.121	1.074	1.886
Pm^{++}	—	2.197	1.929	1.757	1.604	1.446	1.280	1.109	0.939	0.774	—
Sm	—	—	2.051	1.754	1.564	1.440	1.353	1.288	1.235	1.189	1.690
Sm$^+$	2.799	2.407	2.112	1.919	1.783	1.680	1.596	1.525	1.460	1.399	1.748
Sm^{++}	2.124	1.714	1.554	1.469	1.411	1.367	1.331	1.298	1.270	1.245	1.690
Eu	—	—	1.472	1.188	1.040	0.967	0.933	0.917	0.911	0.908	0.903
Eu$^+$	2.245	1.657	1.404	1.277	1.205	1.160	1.128	1.104	1.085	1.069	0.954
Eu^{++}	1.337	0.933	0.906	0.904	0.902	0.900	0.899	0.898	0.898	0.898	0.903
Gd	2.541	2.346	2.053	1.865	1.744	1.661	1.602	1.556	1.519	1.488	1.653
Gd$^+$	2.781	2.418	2.192	2.041	1.930	1.842	1.770	1.708	1.652	1.602	1.699
Gd^{++}	1.834	1.709	1.639	1.585	1.547	1.520	1.503	1.491	1.483	1.478	1.653
Tb	3.148	2.740	2.402	2.205	2.083	1.999	1.937	1.888	1.847	1.813	1.820
Tb$^+$	2.847	2.213	2.006	1.893	1.817	1.761	1.718	1.683	1.655	1.631	1.505
Tb^{++}	2.270	1.947	1.804	1.720	1.663	1.619	1.584	1.555	1.529	1.508	1.820
Dy	2.434	2.518	2.288	2.101	1.965	1.866	1.792	1.736	1.692	1.658	1.813
Dy$^+$	2.483	2.112	1.873	1.743	1.663	1.608	1.567	1.533	1.504	1.479	1.531
Dy^{++}	2.236	1.809	1.654	1.572	1.518	1.476	1.441	1.411	1.385	1.362	—
Ho	3.211	2.364	2.174	2.027	1.899	1.790	1.699	1.624	1.563	1.514	1.716
Ho$^+$	2.299	2.039	1.802	1.636	1.523	1.444	1.389	1.347	1.315	1.289	1.505
Ho^{++}	2.014	1.730	1.515	1.401	1.325	1.264	1.207	1.153	1.102	1.053	1.716
Er	3.128	2.279	1.910	1.663	1.490	1.371	1.289	1.234	1.198	1.174	1.519
Er$^+$	—	2.139	1.776	1.607	1.524	1.478	1.446	1.419	1.393	1.367	1.415
Er^{++}	2.436	1.700	1.523	1.454	1.413	1.380	1.350	1.321	1.295	1.270	1.519
Tm	—	—	—	1.192	1.038	0.964	0.930	0.916	0.911	0.909	1.146
Tm$^+$	—	1.877	1.521	1.344	1.263	1.226	1.209	1.200	1.193	1.186	1.204
Tm^{++}	1.952	1.393	1.101	0.972	0.873	0.769	0.652	0.528	0.402	0.279	1.146
Yb	—	0.522	0.220	0.081	0.027	0.008	0.003	0.002	0.002	0.001	0.000
Yb$^+$	1.446	0.968	0.508	0.355	0.316	0.309	0.305	0.293	0.272	0.246	0.301
Yb^{++}	1.069	0.406	0.089	0.021	0.016	0.010	0.000	0.000	0.000	0.000	0.000
Lu	1.969	1.389	1.131	1.009	0.941	0.897	0.863	0.836	0.811	0.789	1.000
Lu$^+$	1.620	0.918	0.606	0.406	0.265	0.162	0.086	0.030	0.000	0.000	0.000
Lu^{++}	1.723	0.851	0.677	0.576	0.500	0.438	0.390	0.351	0.320	0.296	0.301
Hf	2.337	1.804	1.475	1.286	1.166	1.083	1.021	0.973	0.933	0.899	1.322
Hf$^+$	2.06	1.612	1.372	1.218	1.106	1.015	0.936	0.866	0.801	0.740	—
Hf^{++}	1.663	1.491	1.364	1.268	1.196	1.143	1.104	1.074	1.051	1.033	—
Ta	2.542	1.996	1.630	1.398	1.241	1.130	1.046	0.981	0.928	0.883	1.447
Ta$^+$	2.173	1.900	1.658	1.493	1.365	1.253	1.149	1.050	0.953	0.859	1.544
Ta^{++}	1.701	1.333	1.123	0.978	0.856	0.745	0.642	0.544	0.454	0.371	—
W	2.493	1.857	1.482	1.260	1.111	1.000	0.910	0.831	0.761	0.696	1.398
W$^+$	2.008	1.805	1.508	1.313	1.154	1.006	0.861	0.721	0.587	0.460	1.477
W^{++}	1.994	1.642	1.410	1.235	1.071	0.911	0.755	0.606	0.468	0.341	—
Re	—	1.655	1.238	1.010	0.892	0.833	0.804	0.792	0.787	0.785	0.778
Re$^+$	—	1.436	1.117	0.970	0.900	0.857	0.823	0.790	0.756	0.720	0.845
Re^{++}	2.200	1.521	1.246	1.025	0.824	0.640	0.475	0.328	0.201	0.093	—
Os	—	1.895	1.595	1.419	1.305	1.227	1.170	1.128	1.094	1.067	1.398
Os$^+$	1.858	1.625	1.450	1.337	1.256	1.193	1.141	1.097	1.059	1.024	1.477
Os^{++}	1.958	0.981	0.829	0.753	0.691	0.637	0.591	0.552	0.520	0.495	—

Table C.2 (*cont.*)

$\theta =$	0.2	0.4	0.6	0.8	1.0	1.2	1.4	1.6	1.8	2.0	log g_0
Ir	2.473	1.787	1.550	1.415	1.325	1.260	1.211	1.173	1.142	1.116	1.447
Ir$^+$	1.736	1.334	1.167	1.083	1.028	0.985	0.948	0.916	0.886	0.859	1.544
Ir^{++}	1.778	1.412	1.260	1.158	1.082	1.023	0.975	0.936	0.904	0.877	—
Pt	2.016	1.511	1.390	1.333	1.297	1.270	1.249	1.231	1.216	1.204	1.176
Pt$^+$	1.735	1.388	1.191	1.074	0.991	0.926	0.869	0.819	0.774	0.733	1.000
Pt^{++}	1.683	1.177	1.061	0.983	0.923	0.878	0.846	0.823	0.808	0.798	—
Au	1.397	0.698	0.529	0.444	0.391	0.358	0.337	0.324	0.316	0.311	0.301
Au$^+$	1.015	0.532	0.281	0.139	0.050	0.000	0.000	0.000	0.000	0.000	0.000
Hg	1.862	0.960	0.003	0.003	0.002	0.000	0.000	0.000	0.000	0.000	0.000
Hg$^+$	0.606	0.330	0.306	0.303	0.302	0.300	0.300	0.299	0.299	0.299	—
Hg^{++}	0.293	0.068	0.003	0.000	0.000	0.000	0.000	0.000	0.000	0.000	—
Tl	2.960	0.780	0.479	0.393	0.350	0.323	0.304	0.291	0.283	0.278	0.778
Tl$^+$	0.169	0.008	0.001	0.000	0.000	0.000	0.000	0.000	0.000	0.000	—
Pb	3.467	0.943	0.479	0.298	0.198	0.134	0.091	0.062	0.044	0.032	0.954
Pb$^+$	0.674	0.450	0.375	0.340	0.322	0.312	0.306	0.302	0.300	0.298	0.778
Pb^{++}	0.096	0.004	0.000	0.000	0.000	0.000	0.000	0.000	0.000	0.000	0.000
Bi	1.360	0.840	0.707	0.653	0.628	0.615	0.609	0.606	0.604	0.604	0.602
Bi$^+$	0.667	0.397	0.206	0.107	0.053	0.021	0.000	0.000	0.000	0.000	—
Bi^{++}	0.573	0.376	0.328	0.307	0.296	0.290	0.287	0.287	0.287	0.287	—
Po	1.313	0.824	0.747	0.723	0.713	0.707	0.704	0.702	0.701	0.701	0.954
Po$^+$	1.020	0.854	0.745	0.683	0.643	0.615	0.592	0.573	0.557	0.542	—
Po^{++}	0.843	0.733	0.578	0.468	0.388	0.325	0.274	0.231	0.194	0.162	—
At	1.730	0.720	0.655	0.641	0.628	0.618	0.611	0.609	0.608	0.608	0.778
At$^+$	1.016	0.894	0.824	0.780	0.751	0.730	0.716	0.704	0.696	0.689	—
At^{++}	0.995	0.803	0.704	0.644	0.604	0.576	0.554	0.537	0.523	0.511	—
Rn	1.363	0.049	0.000	0.000	0.000	0.000	0.000	0.000	0.000	0.000	0.000
Rn$^+$	0.692	0.634	0.617	0.608	0.604	0.601	0.600	0.599	0.598	0.598	—
Fr	—	—	—	0.595	0.434	0.355	0.320	0.306	0.303	0.304	0.301
Fr$^+$	0.007	0.000	0.000	0.000	0.000	0.000	0.000	0.000	0.000	0.000	—
Ra	—	0.919	0.585	0.317	0.163	0.080	0.038	0.017	0.008	0.004	0.000
Ra$^+$	1.149	0.689	0.502	0.408	0.357	0.330	0.317	0.311	0.308	0.308	—
Ra^{++}	0.000	0.000	0.000	0.000	0.000	0.000	0.000	0.000	0.000	0.000	—
Ac	1.982	1.585	1.286	1.101	0.982	0.904	0.851	0.812	0.784	0.762	1.000
Ac$^+$	1.713	1.294	1.008	0.812	0.661	0.535	0.428	0.335	0.253	0.181	—
Ac^{++}	1.201	1.013	0.934	0.883	0.847	0.822	0.804	0.792	0.783	0.776	—
Th	3.124	2.359	1.933	1.644	1.439	1.288	1.175	1.089	1.021	0.968	1.322
Th$^+$	2.334	1.644	1.289	1.089	0.967	0.887	0.830	0.787	0.752	0.721	1.447
Th^{++}	1.732	1.211	0.967	0.848	0.769	0.704	0.643	0.583	0.526	0.472	—
Pa	4.030	2.929	2.600	2.414	2.279	2.169	2.074	1.990	1.915	1.846	1.778
Pa$^+$	2.985	2.413	2.169	1.990	1.846	1.723	1.615	1.515	1.419	1.323	1.519
Pa^{++}	—	2.106	1.845	1.652	1.469	1.281	1.088	0.895	0.708	0.532	—
U	3.221	2.696	2.298	2.032	1.849	1.719	1.623	1.552	1.498	1.455	1.929
U$^+$	2.694	2.031	1.719	1.553	1.455	1.392	1.349	1.317	1.292	1.270	1.716
U^{++}	2.137	1.653	1.454	1.362	1.305	1.259	1.217	1.178	1.140	1.105	—

References

Allen, C.W. 1973. *Astrophysical Quantities* (London: Athlone), 3rd ed., pp. 30, 36.

Aller, L.H. 1963. *The Atmospheres of the Sun and Stars* (New York: Ronald), 2nd ed., p. 115.

Bolton, C.T. 1970. *ApJ* **161**, 1187.

Cowley, C.R. & Adelman, S.J. 1983. *QJRAS* **24**, 393.

Halenka, J., Madej, J., Lander, K., & Mamok, A. 2001. *Acta Ast* **51**, 347.

Irwin, A.W. 1981. *ApJS* **45**, 621.

Milone, L.A. & Merlo, D.C. 1998. *ApSS* **359**, 173.

Moore, C.E. 1949, 1952, 1958. *Atomic Energy Levels, US National Bureau of Standards Circular* 467, Vols. **I**, **II**, & **III**.

Moore, C.E. 1970. *Ionization Potentials* (Washington: National Bureau of Standards), NSRDS-NBS34.

NIST, National Institute of Standards and Technology, http://physics.nist.gov.

Sauval, A.J. and Tatum, J.B. 1984. *ApJS* **56**, 193.

Appendix D

The Strongest Lines in the Solar Spectrum

The following summary of strong lines is taken from *The Solar Spectrum 2935 Å to 8770 Å* by C.E. Moore, M.G.J. Minnaert, & J. Houtgast (US National Bureau of Standards Monograph 61, 1966). Only the strongest lines in each spectral region are listed. The equivalent widths should only be used as an approximate indication of the line strength. Under the Name column is the letter designation given originally by Fraunhofer (1787–1826) to the most prominent absorption features.

Table D.1 *Strong lines*

λ	Element	W (Å)	Name	λ	Element	W (Å)	Name
3581.21	Fe I	2.14	N	4920.51	Fe I	0.43	
3719.95	Fe I	1.66		4957.61	Fe I	0.45	c
3734.87	Fe I	3.03	M	5167.33	Mg I	0.65	b_4
3749.50	Fe I	1.91		5172.70	Mg I	1.26	b_2
3758.24	Fe I	1.65		5183.62	Mg I	1.58	b_1
3770.63	H_{11}	1.86		5232.95	Fe I	0.35	
3797.90	H_{10}	3.46		5269.55	Fe I	0.41	E
3820.44	Fe I	1.71	L	5324.19	Fe I	0.32	
3825.89	Fe I	1.52		5328.05	Fe I	0.38	
3832.31	Mg I	1.68		5528.42	Mg I	0.29	
3835.39	H9	2.36		5889.97	Na I	0.63	D_2
3838.30	Mg I	1.92		5895.94	Na I	0.56	D_1
3859.92	Fe I	1.55		6122.23	Ca I	0.22	
3889.05	H_8	2.35		6162.18	Ca I	0.22	
3933.68	Ca II	20.25	K	6562.81	H_α	4.02	C
3968.49	Ca II	15.47	H	6867 band	O_2	telluric	B
4045.82	Fe I	1.17		7594 band	O_2	telluric	A
4101.75	H_δ	3.13	h	8194.84	Na I	0.30	
4226.74	Ca I	1.48	g	8498.06	Ca II	1.46	
4310±10	blend	7.2	G	8542.14	Ca II	3.67	
4340.48	H_γ	2.86	f	8662.17	Ca II	2.60	
4383.56	Fe I	1.01	e	8688.64	Fe I	0.27	
4861.34	H_β	3.68	F	8736.04	Mg I	0.29	
4891.50	Fe I	0.31					

Appendix E

Computation of Random Errors

A knowledge of errors gives us an indication of how well an observation, measurement, or calculation is determined. One should distinguish between accuracy and precision. Accuracy is a statement of how close the result is to the true value. Precision is a statement of how well defined the result is. Accuracy is limited by the precision combined with systematic errors. Systematic errors can only be assessed by comparing completely independent determinations of the same result. The following discussion is limited to statements of precision.

The best estimate of the true value of a set of N measurements, x_j, is the mean given by

$$\bar{x} = \frac{\sum x_j}{N}, \tag{E.1}$$

where the summation is carried over each of the jth measurements from 1 to N. The uncertainty in this value we call the error. Error is a statistical quantity; repeat measurements produce a distribution of x_j values, and that distribution is usually Gaussian in shape. The Gaussian is centered on $\bar{x}$ and the dispersion of the Gaussian is an inverse measure of the goodness of the observations. A narrow distribution means a small dispersion or consistent results. Such observations are generally more trustworthy than those showing wide scatter. Here we use the term dispersion with its statistical meaning, so that it is $\beta/\sqrt{2}$, where β is our usual notation in Equation 2.6. The best estimate of the dispersion is the standard deviation, s, calculated from the set of N measurements using

$$s = \sqrt{\frac{\sum (x_j - \bar{x})^2}{N - 1}}. \tag{E.2}$$

An additional measurement added to the set can be expected to deviate from the mean by s one eth of the time.

We expect the mean to be better determined than any individual measurement, and the standard deviation associated with the mean is

$$\bar{s} = s/\sqrt{N}. \tag{E.3}$$

The relatively slow decline of $\bar{s}$ with N means that it is very laborious to reduce the error with repeated measurements. On the other hand, it is considered only good technique to repeat measurements at least three or four times whenever possible. Not only does this reduce the uncertainty in $\bar{s}$, i.e., the error on the error, but it helps eliminate mistakes and show up erratic behavior sometimes shown by equipment.

Thirty two percent of the measurements fall farther away from $\bar{x}$ than $\pm s$, 16% in each wing of the Gaussian. For $\pm 2s$ it is 2.3% per wing or nearly 5% outside of $\pm 2s$, and for three standard deviations we expect only one measurement in 385 to be in the wings. When we choose to have one-half of the measurements farther away from $\bar{x}$ than an interval p, we call p the probable error. The probable error is related to the standard deviation by

$$p = 0.6745\,s.$$

Sometimes, especially when doing an error calculation by hand, it is convenient to use the average deviation defined by

$$A = \frac{\sum |x_j - \bar{x}|}{N}.$$

This average deviation is related to the other width parameters by

$$p = 0.8453\,A,$$
$$s = 1.2532\,A.$$

The notation s.d. after an error denotes that the standard deviation, s, is being given. The notation m.e., which stands for mean error, denotes the same thing. In a similar manner, p.e. indicates probable error, p, and a.d. indicates average deviation, A.

Should some determinations of x_j be more dependable than others, weights can be used in the computation of $\bar{x}$, s, and $\bar{s}$. Call the weights w_j. Then the weighted mean is

$$\bar{x} = \frac{\sum w_j x_j}{\sum w_j}. \tag{E.4}$$

The uncertainty of one particular measurement now depends on the weight of that particular value. Call this weight w_k. Then

$$s_k = \sqrt{\frac{\sum w_j (x_j - \bar{x})^2}{(N-1)\,w_k}} \tag{E.5}$$

is the standard deviation for this particular measurement. Sometimes one specifies the standard deviation for a unit-weight observation, in which case, $w_k = 1$. Usually one needs the standard deviation on the mean from Equation E.4 for the whole ensemble,

$$\bar{s} = \sqrt{\frac{\sum w_j (x_j - \bar{x})^2}{(N-1)\sum w_j}}. \tag{E.6}$$

In some cases, it is possible to estimate the error on a computed result from known errors on the numbers going into the computation. First express the result, $f(x_1, x_2, \ldots)$ in terms of the independent variables $x_1, x_2, \ldots$. Each of the independent variables is uncertain by the corresponding standard deviation $s_1, s_2, \ldots$. The derivatives, $\partial f/x_i$, specify the leverage each independent variable has on the result, and the error on f is then

$$s_f = \left[\sum \left(\frac{\partial f}{\partial x_i} \right)^2 s_i^2 \right]^{1/2}, \tag{E.7}$$

where the sum is carried over all the independent variables. This can be thought of as the expansion of Equation 2.20 in which the Gaussian error distributions stemming from each independent variable have been convolved together.

Index

Printed in the United States
by Baker & Taylor Publisher Services